# MATHEMATICS STUDY GUIDE Year 8

**Get the Results You Want!**

John Compton, Allyn Jones & Peter Nicolas

First edition 1995
Reprinted 1996, 1998, 2000, 2001
**Revised edition 2006**
Reprinted 2008, 2009, 2011, 2013

**Updated in 2014 for the Australian Curriculum**

Reprinted 2016, 2017, 2019, 2020, 2022, 2023, 2024, 2025

ISBN 978 1 74125 007 7

Pascal Press
PO Box 250
Glebe NSW 2037
www.pascalpress.com.au

Publisher: Vivienne Joannou
Project Editors: Lisa Burrell and Leanne Poll
Edited by Bruce Howarth, Linden Hyatt and Karen Enkelaar
Answers checked by Peter Little
Indexed by Puddingburn Publishing Services
Typeset by Nikki M Group Pty Ltd
Cover by DiZign Pty Ltd
Printed by Vivar Printing/Green Giant Press



**Students**
All care has been taken in compiling this study guide, but please check with your teacher about the exact requirements of the course as these can change from year to year.

**Acknowledgements**
Allyn, Peter and I thank our wives Fiona, Maria and Robyn for their patience over the time that we have spent writing our series of study guides. Without their encouragement and understanding these books would still be but a good idea. Thanks also to our children for their help and understanding.

# CONTENTS

# HOW TO USE THIS BOOK

- This book covers **the topics in the Year 8 Australian Curriculum (Mathematics)**. It is best to work through it bit by bit over the whole year. However, you can also use it for revision before tests or examinations. You will also find it useful for finding out about particular topics.
- Try to work through the book in **chapter order** (Chapter 1 first, then Chapter 2, Chapter 3, and so on). Your teacher may cover topics in a different order. If you are not sure which chapter of the book relates to your classroom work, ask your teacher for assistance.
- If you want to practise a particular topic, then look up that topic in the Table of contents (the pages before this one) or the Index (at the back of the book).

## Chapter by Chapter ...

### PATTERNS AND LINEAR RELATIONSHIPS 5

- Graphing Ordered Pairs on the Number Plane
- Number Patterns
- Graphing Lines on the Number Plane
- Points on Lines
- Horizontal and Vertical Lines
- Points of Intersection
- Parallel and Perpendicular Lines
- Using the Linear Relationships to Solve Problems

- If you want to work through a whole chapter start at the first page. Here you will find an overview of the topics covered in that chapter.

### KEYWORDS

| | |
|---|---|
| **Abscissa** | **Horizontal** |
| **Axes** | **Intersection** |
| **Axis** | **Number plane** |
| **Coefficient** | **Ordered pair** |
| **Collinear** | **Ordinate** |
| **Coordinates** | **Origin** |
| **Cubic** | **Parabola** |
| **Domain** | **Parallel** |
| **Function** | **Perpendicular** |
| **Gradient** | **Range** |
| **Graph** | **Vertical** |

- You should also study the **keywords** at the start of the chapter. Look out for these important terms as you work through the chapter.

## Testing Solutions

A possible solution for an equation can be tested by substituting that value for the pronumeral. ← Explanation

**1** Show that $x = 4$ is the correct solution of $5x - 12 = 8$. ← Example

**1** Consider $5x - 12 = 8$ [$5x$ means $5 \times x$]

$$\begin{aligned} \text{LHS} &= 5x - 12 \\ &= 5 \times 4 - 12 \\ &= 20 - 12 \\ &= 8 \end{aligned}$$

[LHS = left-hand side]

← Worked solution

RHS = 8

As LHS = RHS, then $x = 4$ is a correct solution.

- Each concept is explained using both an **explanation** and an **example question**. Read the explanation and then work through the **example question**. The worked solution to the example question appears straight after the question in blue type. Make sure you understand how the answer was found. If you are not sure, read through the topic again or ask your teacher for help.
- Use an exercise book to make notes as you go along. This will help you remember what you have learned.

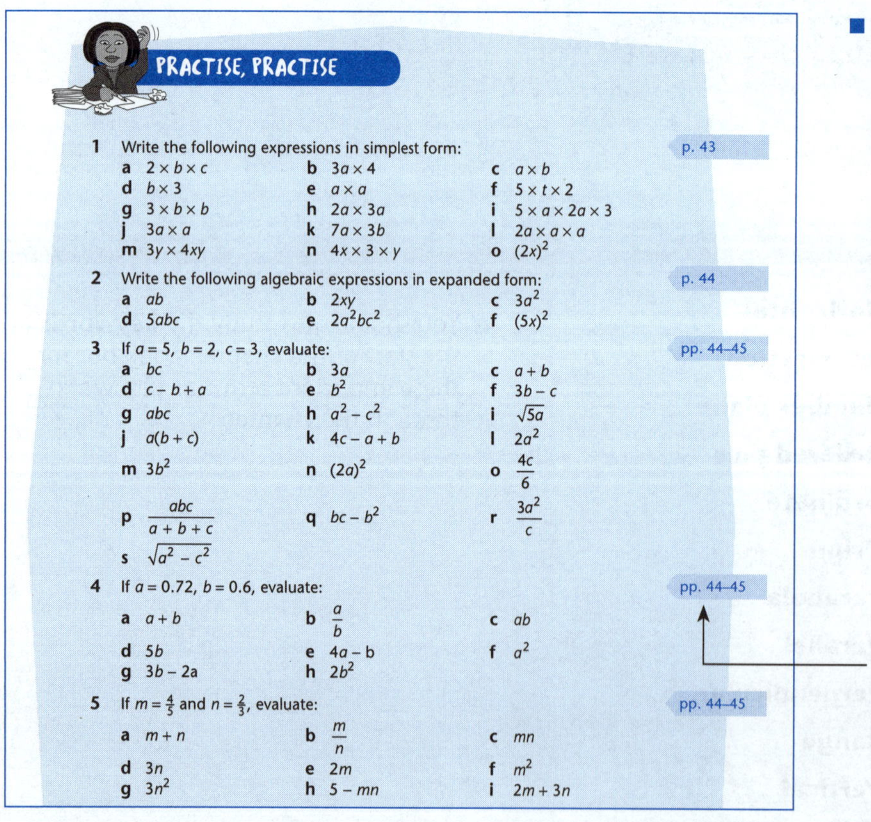

PRACTISE, PRACTISE

**1** Write the following expressions in simplest form: p. 43

| | | |
|---|---|---|
| **a** $2 \times b \times c$ | **b** $3a \times 4$ | **c** $a \times b$ |
| **d** $b \times 3$ | **e** $a \times a$ | **f** $5 \times t \times 2$ |
| **g** $3 \times b \times b$ | **h** $2a \times 3a$ | **i** $5 \times a \times 2a \times 3$ |
| **j** $3a \times a$ | **k** $7a \times 3b$ | **l** $2a \times a \times a$ |
| **m** $3y \times 4xy$ | **n** $4x \times 3 \times y$ | **o** $(2x)^2$ |

**2** Write the following algebraic expressions in expanded form: p. 44

| | | |
|---|---|---|
| **a** $ab$ | **b** $2xy$ | **c** $3a^2$ |
| **d** $5abc$ | **e** $2a^2bc^2$ | **f** $(5x)^2$ |

**3** If $a = 5$, $b = 2$, $c = 3$, evaluate: pp. 44–45

| | | |
|---|---|---|
| **a** $bc$ | **b** $3a$ | **c** $a + b$ |
| **d** $c - b + a$ | **e** $b^2$ | **f** $3b - c$ |
| **g** $abc$ | **h** $a^2 - c^2$ | **i** $\sqrt{5a}$ |
| **j** $a(b + c)$ | **k** $4c - a + b$ | **l** $2a^2$ |
| **m** $3b^2$ | **n** $(2a)^2$ | **o** $\frac{4c}{6}$ |
| **p** $\frac{abc}{a + b + c}$ | **q** $bc - b^2$ | **r** $\frac{3a^2}{c}$ |
| **s** $\sqrt{a^2 - c^2}$ | | |

**4** If $a = 0.72$, $b = 0.6$, evaluate: pp. 44–45

| | | |
|---|---|---|
| **a** $a + b$ | **b** $\frac{a}{b}$ | **c** $ab$ |
| **d** $5b$ | **e** $4a - b$ | **f** $a^2$ |
| **g** $3b - 2a$ | **h** $2b^2$ | |

**5** If $m = \frac{4}{5}$ and $n = \frac{2}{3}$, evaluate: pp. 44–45

| | | |
|---|---|---|
| **a** $m + n$ | **b** $\frac{m}{n}$ | **c** $mn$ |
| **d** $3n$ | **e** $2m$ | **f** $m^2$ |
| **g** $3n^2$ | **h** $5 - mn$ | **i** $2m + 3n$ |

↑ Page references to explanations

- When you have worked through all topics in the chapter, try the **Practise, Practise** section. Quick answers and worked solutions to this section are found at the back of the book. The page reference on the right of each question tells you where each concept is explained in the chapter.

**YOUR CHECKLIST**

**For a complete understanding of this topic you must be able to:**

| | | |
|---|---|---|
| ✓ | Simplify algebraic expressions that involve multiplication and division | pp. 43–44 |
| ✓ | Substitute in algebraic expressions and formulae | pp. 44–45 |
| ✓ | Simplify algebraic expressions by collecting like terms | pp. 46–47 |
| ✓ | Use the laws of indices to simplify algebraic expressions | pp. 47–48 |
| ✓ | Expand algebraic expressions by removing grouping symbols | pp. 48–49 |
| ✓ | Factorise algebraic expressions | p. 49 |
| ✓ | Translate from everyday language to algebraic language and vice versa | pp. 49–50 |
| ✓ | Generate number patterns from an algebraic expression. | p. 50 |

**Now you are ready to do the tests!**

Page references

- Go through the **Checklist** near the end of the chapter to ensure that you understand the key concepts of the chapter. Put a tick next to each checklist item that you have covered. You need to be able to tick every item to be able to do the tests. If you have missed any items, go through the chapter to ensure that you cover the relevant topics. The page reference next to each item tells you where the concept is explained.

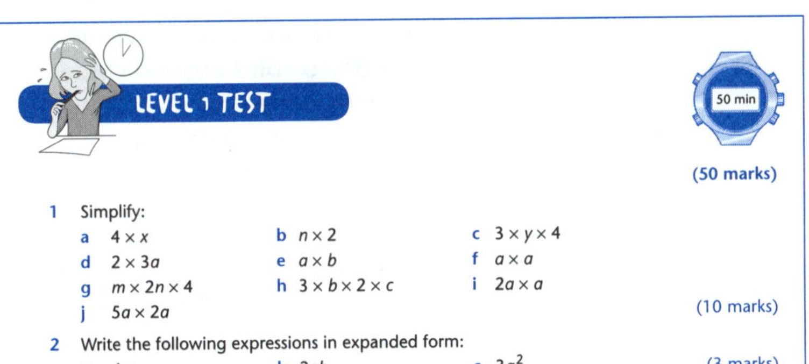

**LEVEL 1 TEST**

50 min

(50 marks)

1 Simplify:
a $4 \times x$ b $n \times 2$ c $3 \times y \times 4$
d $2 \times 3a$ e $a \times b$ f $a \times a$
g $m \times 2n \times 4$ h $3 \times b \times 2 \times c$ i $2a \times a$
j $5a \times 2a$ (10 marks)

2 Write the following expressions in expanded form:
a $4a$ b $2ab$ c $3a^2$ (3 marks)

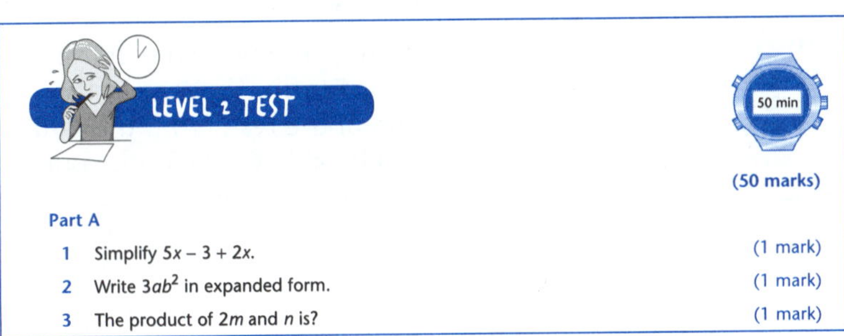

**LEVEL 2 TEST**

50 min

(50 marks)

Part A

1 Simplify $5x - 3 + 2x$. (1 mark)

2 Write $3ab^2$ in expanded form. (1 mark)

3 The product of $2m$ and $n$ is? (1 mark)

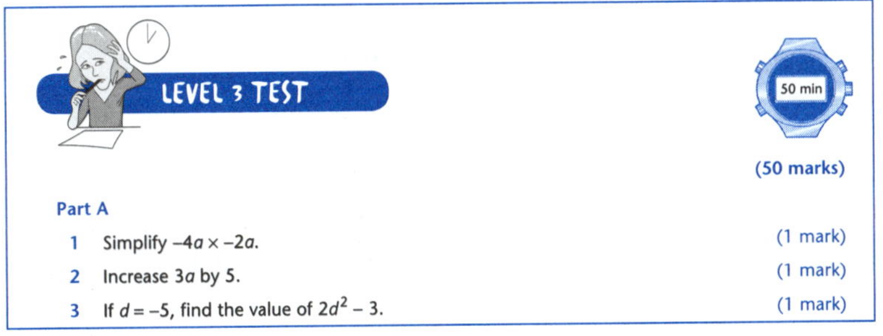

**LEVEL 3 TEST**

50 min

(50 marks)

Part A

1 Simplify $-4a \times -2a$. (1 mark)

2 Increase $3a$ by 5. (1 mark)

3 If $d = -5$, find the value of $2d^2 - 3$. (1 mark)

- Next try to do all the **Tests**. The tests are of three levels of difficulty: the Level 1 tests are straightforward, the Level 2 tests are of average difficulty and the Level 3 tests are challenging. Start with the Level 1 test. If you find the first test easy, it will still be good practice before you try the other tests. If you find Test 1 or Test 2 close to your ability, then try the third test and complete all the questions you can—even if you find them difficult. Because there are worked solutions to each question, you will always be able to find out how to obtain each answer.

- The tests are designed to **prepare you for your school tests or examinations** so try to keep to the maximum time allowed for each one. The time appears on the stopwatch at the start of the test. Avoid looking at your textbook or notes while doing the test—this will help you develop your examination technique.

1 Simplify $5x - 3 + 2x$. (1 mark)

2 Write $3ab^2$ in expanded form. (1 mark)

3 The product of $2m$ and $n$ is? (1 mark)



- **Marks** are allocated for each question. These are similar to the marks you will be trying for in your school tests and exams. Spend more time on the questions that are worth the most marks. For example, if there are 20 marks in total, and 20 minutes have been allocated for completion of the test, then spend about one minute on each mark.

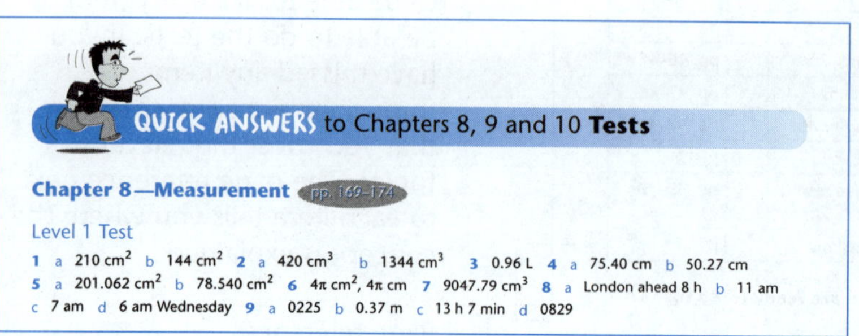

**QUICK ANSWERS to Chapters 8, 9 and 10 Tests**

**Chapter 8—Measurement** pp. 169–174

Level 1 Test

**1** a 210 cm$^2$ b 144 cm$^2$ **2** a 420 cm$^3$ b 1344 cm$^3$ **3** 0.96 L **4** a 75.40 cm b 50.27 cm **5** a 201.062 cm$^2$ b 78.540 cm$^2$ **6** $4\pi$ cm$^2$, $4\pi$ cm **7** 9047.79 cm$^3$ **8** a London ahead 8 h b 11 am c 7 am d 6 am Wednesday **9** a 0225 b 0.37 m c 13 h 7 min d 0829

- **Mark your work** by referring to the answers at the back of the book. The **Quick Answers** section allows you to see straight away if you got the answer right—only the answers are shown. The **Worked Solutions** section sets out the solutions to the question in full, so look at this section if your answer was incorrect. The **ticks** in worked solutions indicate those parts of the working which receive marks. Therefore, even if your answer is wrong, you may be entitled to some marks for the question. Compare the worked solution to your own working to find out if you are entitled to any marks for the question.

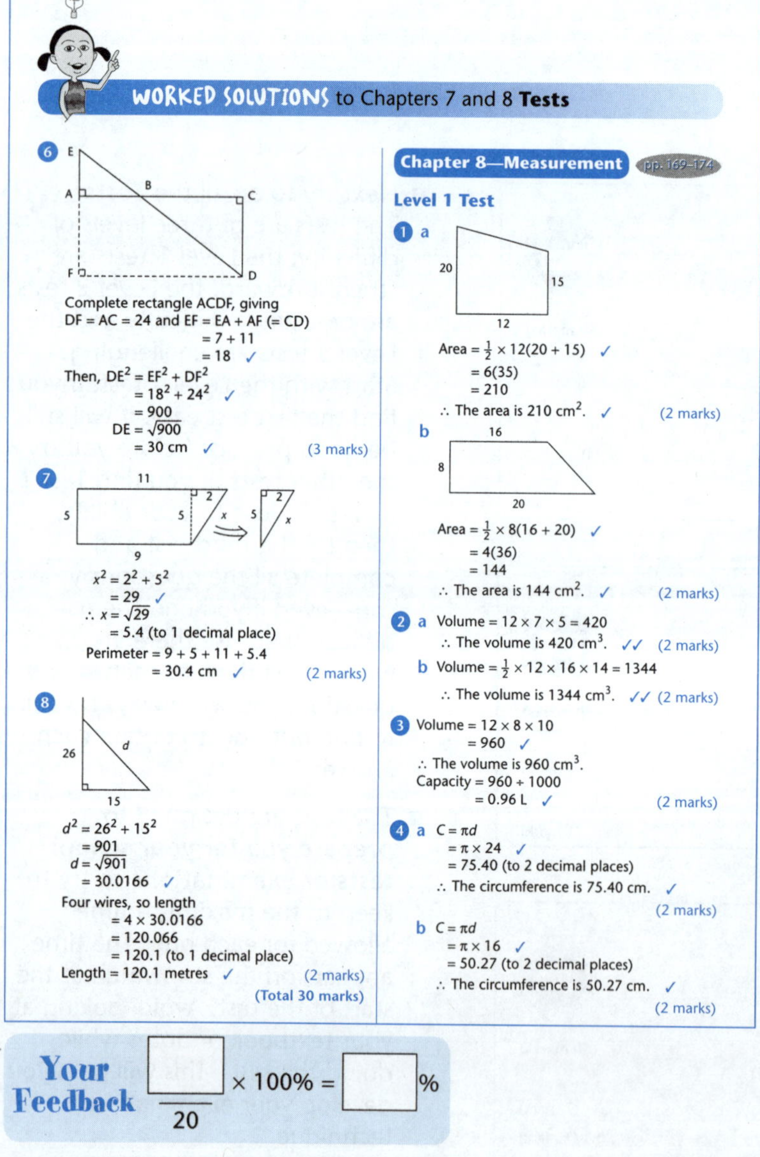

**WORKED SOLUTIONS to Chapters 7 and 8 Tests**

**6** Complete rectangle ACDF, giving DF = AC = 24 and EF = EA + AF (= CD) = 7 + 11 = 18 ✓

Then, DE$^2$ = EF$^2$ + DF$^2$ = $18^2 + 24^2$ ✓ = 900

DE = $\sqrt{900}$ = 30 cm ✓ (3 marks)

**7** $x^2 = 2^2 + 5^2 = 29$ ✓

$\therefore x = \sqrt{29}$ = 5.4 (to 1 decimal place)

Perimeter = 9 + 5 + 11 + 5.4 = 30.4 cm ✓ (2 marks)

**8** $d^2 = 26^2 + 15^2 = 901$

$d = \sqrt{901}$ = 30.0166 ✓

Four wires, so length = 4 × 30.0166 = 120.066 = 120.1 (to 1 decimal place)

Length = 120.1 metres ✓ (2 marks)

(Total 30 marks)

**Chapter 8—Measurement** pp. 169–174

**Level 1 Test**

**1** a Area = $\frac{1}{2} \times 12(20 + 15)$ ✓ = 6(35) = 210

$\therefore$ The area is 210 cm$^2$. ✓ (2 marks)

b Area = $\frac{1}{2} \times 8(16 + 20)$ ✓ = 4(36) = 144

$\therefore$ The area is 144 cm$^2$. ✓ (2 marks)

**2** a Volume = 12 × 7 × 5 = 420

$\therefore$ The volume is 420 cm$^3$. ✓✓ (2 marks)

b Volume = $\frac{1}{3} \times 12 \times 16 \times 14$ = 1344

$\therefore$ The volume is 1344 cm$^3$. ✓✓ (2 marks)

**3** Volume = 12 × 8 × 10 = 960 ✓

$\therefore$ The volume is 960 cm$^3$.

Capacity = 960 ÷ 1000 = 0.96 L ✓ (2 marks)

**4** a $C = \pi d$ = $\pi \times 24$ ✓ = 75.40 (to 2 decimal places)

$\therefore$ The circumference is 75.40 cm. ✓ (2 marks)

b $C = \pi d$ = $\pi \times 16$ ✓ = 50.27 (to 2 decimal places)

$\therefore$ The circumference is 50.27 cm. ✓ (2 marks)

**Your Feedback** $\frac{\square}{20} \times 100\% = \square\%$

- Write your own score in the **Your Feedback** space at the end of the test. From this you will be able to work out your percentage score. If your score for the test was less than 50%, you will need to work through the chapter again. If you are still having trouble, ask your teacher to explain those topics that you can't understand.
- Worked solutions to the tests are found at the back of the book. Check the worked solutions to any test questions that you answered incorrectly.

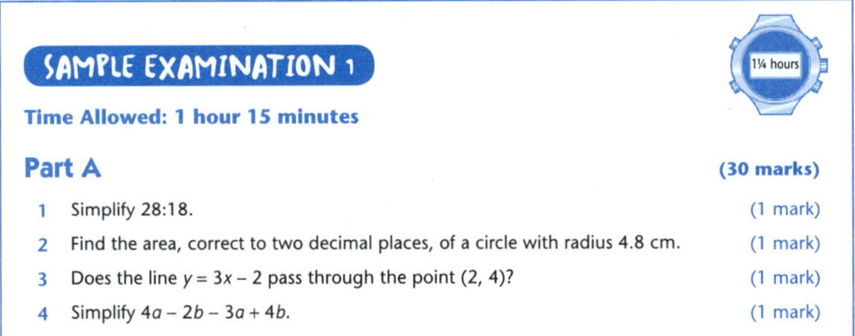
SAMPLE EXAMINATION 1

1¼ hours

Time Allowed: 1 hour 15 minutes

Part A (30 marks)

1 Simplify 28:18. (1 mark)

2 Find the area, correct to two decimal places, of a circle with radius 4.8 cm. (1 mark)

3 Does the line $y = 3x - 2$ pass through the point (2, 4)? (1 mark)

4 Simplify $4a - 2b - 3a + 4b$. (1 mark)

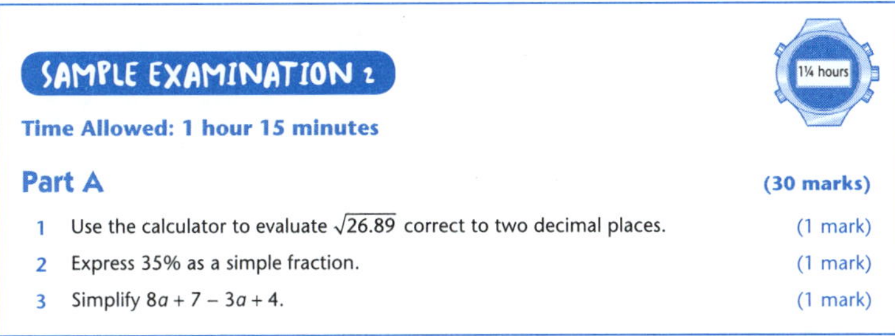
SAMPLE EXAMINATION 2

1¼ hours

Time Allowed: 1 hour 15 minutes

Part A (30 marks)

1 Use the calculator to evaluate $\sqrt{26.89}$ correct to two decimal places. (1 mark)

2 Express 35% as a simple fraction. (1 mark)

3 Simplify $8a + 7 - 3a + 4$. (1 mark)

- When you have completed all of the chapters you can try the **Sample Examination Papers.** The Sample Examination Papers cover the entire year's work, so you should attempt the papers if you have covered all the topics in this book. (Ask your teacher if you are not sure.) The Sample Examination Papers will provide you with good practise for your final Year 8 examination. Like the tests, they are of three levels of difficulty. Set aside the time allowed for each paper and complete them under exam conditions.

## Some general advice on succeeding in Mathematics

- Mathematics is a subject that **builds on knowledge**. Often you need to master one topic in order to be able to understand a later topic. This is why you should work through the book from beginning to end.
- Many students fail to **read questions carefully**. This accounts for a large number of the mistakes in exams. Although your time is limited, you must still take enough time to read each question carefully. If necessary, read the question a second (or even a third) time. Don't start your answer until you are sure you understand the question. For the longer questions, take a moment or two to **plan** your answer.
- Take care with your working. Write all answers down the page and avoid skipping any steps, as this often leads to errors. Your teachers like to see all your working, so it is wise to get into good habits. For questions that require a lot of working (such as equations), **write down each step of the solution on a separate line, lining up any equals signs** (=). This helps your teachers follow your working and will also help you check your work.
- Set out all of your working, because in Mathematics you may get some marks for your working, even if your answer is wrong.
- Always work through the solutions to any questions you answer incorrectly. This helps you learn. Remember: if you have attempted the questions, you are entitled to look at the answers—this is not cheating. If you don't get feedback, you won't be able to improve!
- You cannot expect to score 100% all the time. Remember that this book has been designed to help you identify your strengths as well as your weaknesses, so it is OK to make mistakes. The key to success is to learn from those mistakes.

# 1 SELECTED REVISION OF YEAR 7

- Number
- Fractions
- Decimals
- Percentages
- Basic Algebra
- Basic Geometry
- Statistics
- Probability

Before you start Year 8 Chapters, revise the following topics from Year 7.

## Number

1 Evaluate:
- a $17 + 3 - 6 + 14$
- b $19 - 8 \times 2$
- c $20 - 11 + 17 - 14$
- d $4 + 6 \times 9$
- e $5 \times 7 + 7 \times 7$
- f $5 + 3 \times 6 + 9$
- g $15 - 18 \div 6$
- h $(5 + 4) \times 7$
- i $(7 + 2) \times (9 + 3)$
- j $(19 - 12) \times (16 - 14)$
- k $15 + [17 - (3 \times 4)]$
- l $\{[(19 - 17) + 5] \times 8\} \div 4$

2 Michelle purchased eight packets of squash balls. If each packet contained six balls, how many balls did Michelle purchase?

3
- a Find the sum of 19 and 17.
- b Calculate the product of 9 and 6.
- c By how much does 84 exceed 29?
- d Find the average of 19, 11, 17, 13 and 15.
- e Find the difference between 31 and 18.
- f What is the quotient of 63 and 7?
- g Calculate the remainder when 75 is divided by 8.
- h Total the numbers 16, 19, 41, 34 and 24.
- i Find the product of the sum of 9 and 7 and the difference between 9 and 7.
- j Calculate the sum of the product of 6 and 9 and the product of 4 and 9.

4 If 8 Mattara T-shirts are valued at $48, how much would it cost to buy five T-shirts?

5 Nicholas earns $61.60 for working 8 hours. How much would Nicholas earn if he worked only 6 hours at this wage rate?

6 Greg collects basketball cards. He has 144 cards in one album, 132 in another, 24 in his top drawer, 19 under his pillow and 15 super cards hidden in a secret location. How many cards does Greg have altogether?

7 Big Al scored 19, 26, 42, 17, 35 and 5 in six innings. Calculate Big Al's average score.

8 It is 26 km from Maitland to Maryland. If Dennis drives from Maitland to Maryland and back six days a week, how far does Dennis travel each week?

9 Jack has 247 trees to plant in 18 equal rows. How many trees will be left over?

## Fractions

1 Write as fractions in simplest form:
- a $\frac{7}{21}$
- b $\frac{18}{36}$
- c $\frac{14}{35}$
- d $\frac{75}{100}$
- e $\frac{6x}{8x}$
- f $\frac{40}{60}$
- g $\frac{\frac{1}{2}}{\frac{3}{4}}$
- h $\frac{18}{6}$

i $\frac{15}{7}$

j $\frac{12}{8}$

k $\frac{15}{6}$

l $\frac{20}{15}$

**2** Write as improper fractions:

a $3\frac{1}{3}$

b $6\frac{2}{5}$

c $4\frac{3}{4}$

d $9\frac{1}{11}$

e $8\frac{1}{3}$

**3** Perform these calculations:

a $\frac{7}{10} - \frac{1}{2}$

b $\frac{5}{7} + \frac{3}{7}$

c $\frac{3}{4} - \frac{1}{3}$

d $1\frac{2}{3} + \frac{1}{3}$

e $2\frac{1}{4} - 1\frac{1}{2}$

f $6 - 3\frac{2}{5}$

g $\frac{2}{5} + \frac{1}{3} + \frac{3}{4}$

h $6\frac{1}{4} + 3\frac{2}{3} - 5\frac{1}{2}$

**4** Evaluate:

a $\frac{4}{5} \times \frac{3}{5}$

b $\frac{3}{4} \times \frac{3}{8}$

c $\frac{7}{9} \times \frac{3}{4}$

d $\frac{5}{8} \times 1\frac{1}{3}$

e $1\frac{1}{2} \times \frac{3}{4}$

f $2\frac{1}{3} \times \frac{6}{7}$

g $8 \times 2\frac{2}{3} \times 2\frac{1}{4}$

h $(\frac{1}{2} + \frac{1}{3}) \times (\frac{1}{2} - \frac{1}{3})$

i $\frac{7}{12} \div \frac{3}{4}$

j $20 \div \frac{5}{8}$

k $6\frac{1}{4} \div 1\frac{7}{8}$

l $\frac{3}{4} \times 1\frac{3}{5} \div \frac{7}{10}$

m $\frac{2}{3} + \frac{3}{4} \times \frac{8}{9}$

n $10 - 3\frac{1}{3} \times 1\frac{4}{5}$

o $\dfrac{10}{\frac{3}{4} - \frac{1}{2}}$

p $\dfrac{\frac{2}{3} + \frac{1}{4}}{\frac{2}{3} - \frac{1}{4}}$

q $6\frac{1}{4} \times 1\frac{1}{5} + 3\frac{3}{4} \times 1\frac{1}{5}$

**5** Given that $a = \frac{2}{3}$, $b = \frac{1}{2}$, $c = \frac{2}{5}$ evaluate:

a $a + b$

b $a + b + c$

c $a + b - c$

d $ab$

e $abc$

f $a^2$

g $a^2 - b^2$

h $(a + b)(a - b)$

i $a + bc$

j $c^2 + bc$

k $c(b + c)$

l $\dfrac{a + b}{a - b}$

**6** $\frac{3}{4}$ of a tonne of sand costs \$45. How much will 1 tonne cost?

**7** Simone's salary was increased by $\frac{1}{5}$ to \$240 per week. Find Simone's salary prior to the increase and by how much her salary has increased.

**8** Jenny was given a $\frac{1}{3}$ discount on an engagement ring. The ring was originally marked at \$2400. How much does Jenny pay for the ring?

9 Pavlov is paid $2\frac{1}{2}$ times the normal hourly wage rate for training dogs to ring bells. If the normal rate is $8.64 per hour, how much would Pavlov be paid for 6 hours' training?

10 Complete the following statements:

a $\frac{7}{10} + \triangle = 1\frac{1}{2}$

b $4 - * = 2\frac{1}{4}$

c $\frac{4}{5} + * = 1\frac{1}{3}$

d $\frac{7}{10} + \frac{3}{10} \times 1\frac{1}{4} = \square$

e $\frac{3}{4} \times \square = 2$

## Decimals

1 Write as a fraction in simplest terms:

a 0.4
b 0.44
c 0.04
d 1.6
e 2.5
f 3.05
g 2.125
h 1.005

2 Change to decimal fractions:

a $\frac{7}{10}$

b $\frac{17}{100}$

c $\frac{7}{100}$

d $\frac{7}{10} + \frac{7}{100}$

e $3 + \frac{9}{10} + \frac{9}{100} + \frac{9}{1000}$

f $\frac{179}{100}$

g $\frac{424}{10}$

3 Evaluate:

a $7.2 + 10 + 6.44$
b $0.9 + 0.7 + 0.4$
c $6 - 3.4$
d $2.1 - 0.945$
e $100 \times 4.7$
f $10 \times 7.34$
g $6 \times 0.95$
h $4.5 \times 0.26$
i $4.8 \div 6$
j $3.2 \div 0.8$
k $42 \div 0.7$
l $540 \div 0.9$
m $25.84 \div 3.4$
n $16 \div 2.4$ (correct to 2 decimal places)

4 Given that $x = 1.5$, $y = 0.8$ and $z = 1.2$, evaluate:

a $x + y + z$
b $x + y - z$
c $xy$
d $xy \div z$
e $\frac{y + z}{x}$
f $\frac{z + y}{z - y}$
g $x + yz$
h $x^2$
i $4x^2$
j $x^2 - z^2$

5 Mitchell collected pieces of string. He had a continuous length of 19.45 metres and found a piece 3.9 metres long. What length of string does Mitchell have altogether?

6 Lyn has to cut a 15 m length of timber into 45 cm lengths. How many pieces will she cut and what length of timber is left over?

7 Allyn bought eight bags of ready-mixed concrete. Each bag weighed 4.5 kg. Calculate the total weight of the bags.

8 Maria cut a 3.28-metre length from a roll of cloth 10 metres long. How much cloth remained on the roll?

9 Brenda is paid $14.20 an hour for shaving poodles. Any overtime is paid at 1.5 times the normal rate of $14.20. If on Tuesday Brenda worked her normal 6 hours plus 3 hours overtime, how much was Brenda paid?

10 Calculate the missing number in the following statements:

a $6.7 + * = 9.34$

b $12 - \triangle = 3.76$

c $9.4 \times \square = 75.2$

d $16 \times \square = 92.8$

e $0.95 + 0.05 \times 7 = *$

f $4.7 \times 6.7 + \triangle \times 6.7 = 67$

## Percentages

1 Rewrite as simplified fractions:

a 17%

b 73%

c 7%

d 80%

e 6%

f 25%

g $7\frac{1}{2}\%$

h $12\frac{1}{4}\%$

2 Change the following to decimals:

a 69%

b 8%

c 80%

d 45%

e $6\frac{1}{2}\%$

f $17\frac{1}{4}\%$

g 140%

h 300%

3 Rewrite as percentages:

a $\frac{3}{4}$

b $\frac{7}{10}$

c $\frac{81}{100}$

d $\frac{4}{5}$

e $\frac{31}{50}$

f $1\frac{1}{4}$

g 0.63

h 0.7

i 0.05

j 1.2

k 0.375

l 1.205

4 Find:

a 45% of $600

b 28% of $7000

c 7% of 6500

d 4% of 600

e 35% of 2 metres

f 5% of 4 kg

5 What percentage is:

a 4 out of 16?

b 27 out of 30?

c $12 out of $120?

d 10c out of $1?

e 12 minutes out of an hour?

## Basic Algebra

1 Simplify these expressions:

a $7a + a$

b $11y - y$

c $4t + 3t$

d $9t - 4t$

e $a + a + 2a$

f $12y - 4y - y$

g $0.6x + 1.2x$

h $\frac{1}{3}y + \frac{1}{4}y$

i $-6y + 4y$

j $5a - 8a$

k $-3m - 4m$

l $1.5n - 0.8n$

2 Simplify these expressions:

a $4 \times y$

b $7 \times a \times 2$

c $a \times b$

d $5 \times 3t$

e $5y \times 7$

f $4a \times 7a$

g $6x \times 9y$

h $-6y \times 3$

i $-4t \times -7t$

j $24y \div 8$

k $18a \div 6a$

l $18xy \div 6x$

m $\dfrac{42a}{6a}$

n $-54pq \div 6p$

3 Simplify the following:

a $6x + y + 4x$

b $7y + 2x + 9y + 4x$

c $20a + 3 + 17a + 7$

d $20t + 15x - 5t + 2x$

e $14y - 3x + 2y + 5x$

f $a + b - 2a - 3b + a + 2b$

4 Write down the value of the pronumeral in the following equations:

a $x + 8 = 14$

b $a - 7 = 14$

c $17 + y = 31$

d $4a = 28$

e $9t = 72$

f $\dfrac{a}{3} = 9$

g $54 = 6y$

h $15 - m = 8$

i $x + 17 = 22$

j $3x + 11 = 17$

5 Calculate the perimeter of these shapes:

a

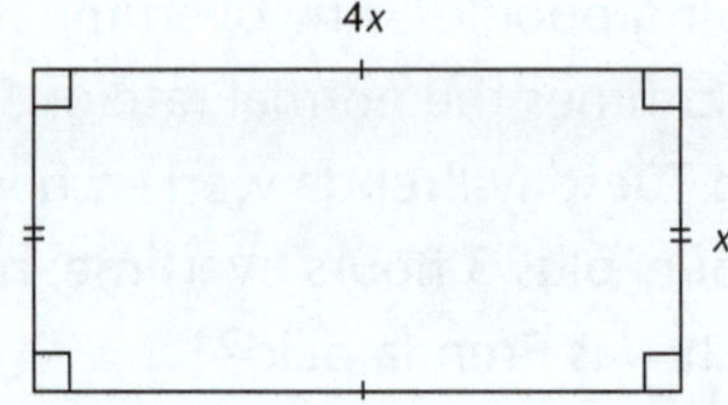


b

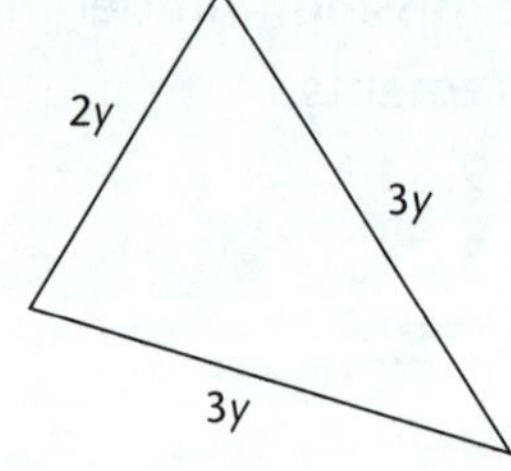


c

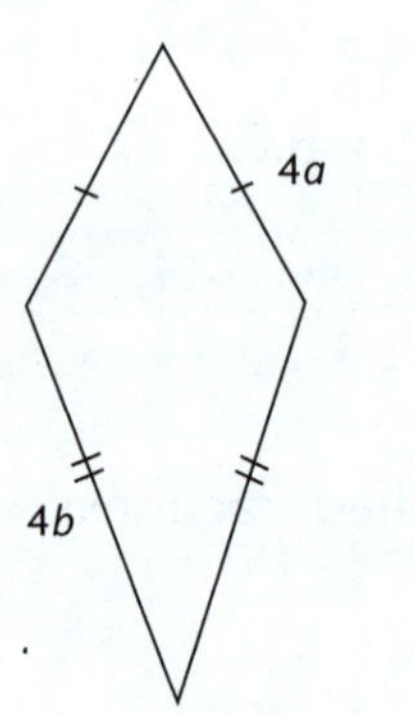


d

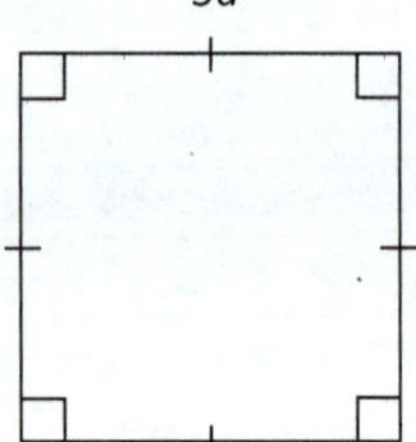


e

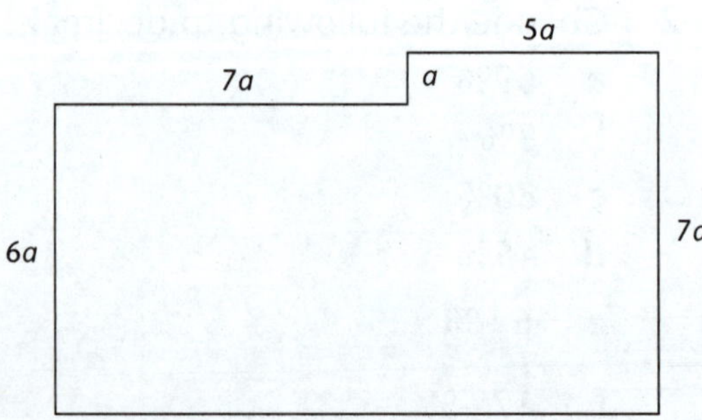

## Basic Geometry

Choose the correct name for each shape from the list. Answers are worth 1 mark each.

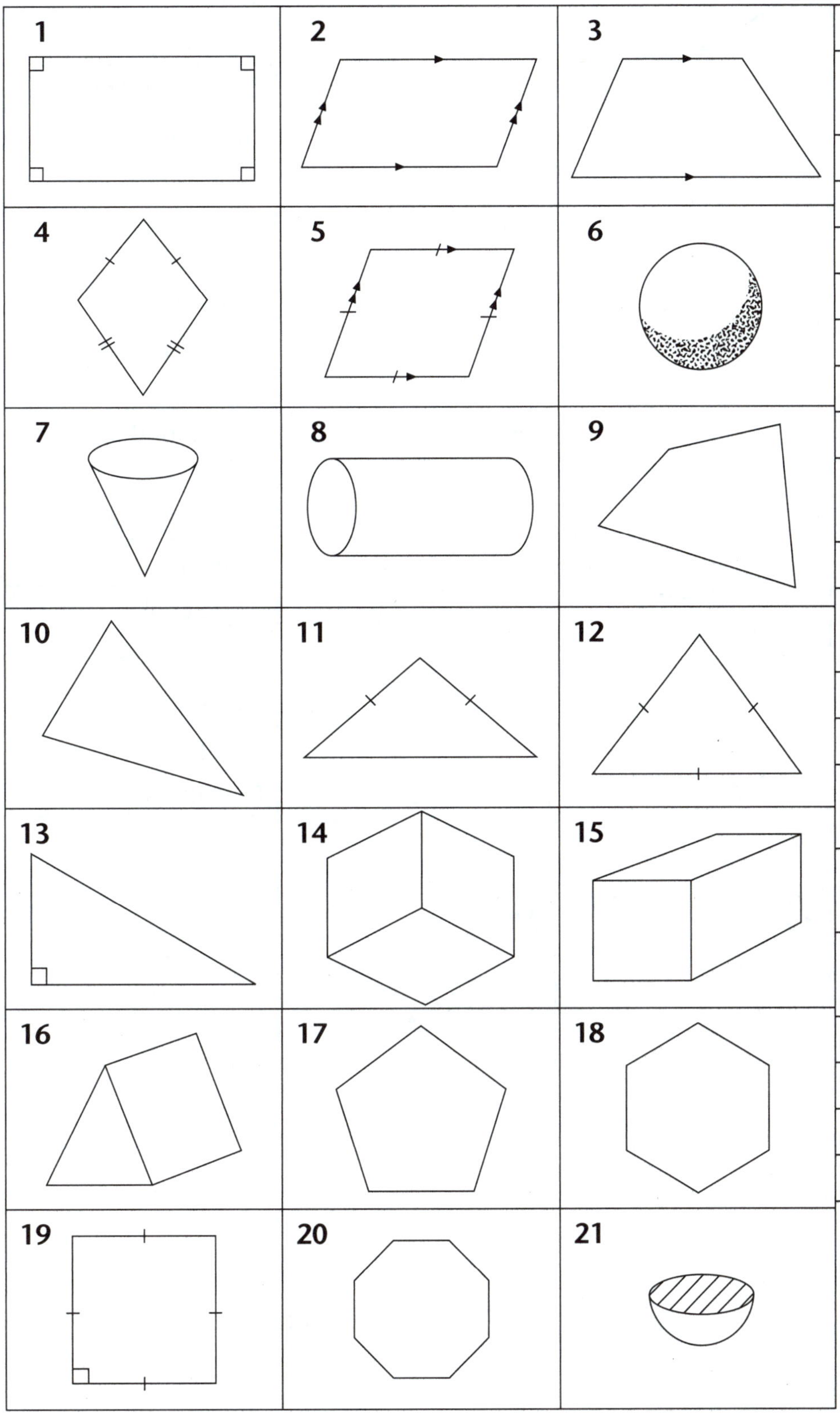

| | | | |
|---|---|---|---|
| 1 | 2 | 3 | Cube |
| | | | Square prism |
| | | | Cone |
| 4 | 5 | 6 | Cylinder |
| | | | Sphere |
| | | | Square |
| | | | Rectangle |
| 7 | 8 | 9 | Parallelogram |
| | | | Kite |
| | | | Isosceles triangle |
| 10 | 11 | 12 | Trapezium |
| | | | Equilateral triangle |
| | | | Rhombus |
| 13 | 14 | 15 | Quadrilateral |
| | | | Scalene triangle |
| | | | Triangular prism |
| 16 | 17 | 18 | Right-angled triangle |
| | | | Pentagon |
| | | | Hemisphere |
| | | | Octagon |
| 19 | 20 | 21 | Hexagon |

Name the important geometric property illustrated. Choose from the list.

| | | | |
|---|---|---|---|
| **22** | **23** | **24** | Complementary angles |
| | | | Collinear points |
| | | | Concurrent lines |
| | | | Parallel lines |
| **25** x | **26** x x | **27** x | Vertically opposite angles |
| | | | Co-interior angles |
| | | | Supplementary angles |
| **28** | **29** | **30** x | Corresponding angles |
| | | | Alternate angles |
| | | | Exterior angle |
| | | | Acute angle |
| **31** | **32** | **33** | Obtuse angle |
| | | | Reflex angle |
| | | | Straight angle |
| | | | Revolution |
| **34** | **35** | **36** | |

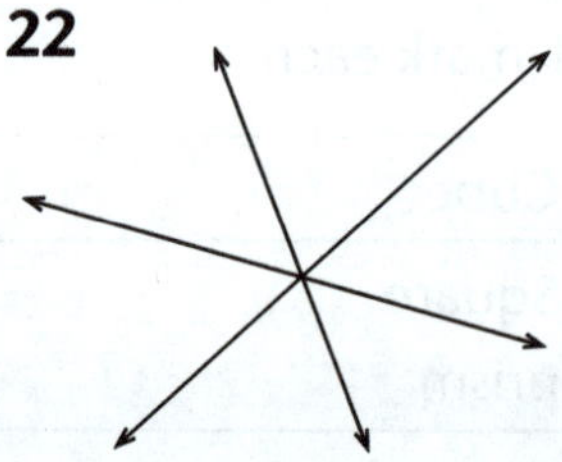

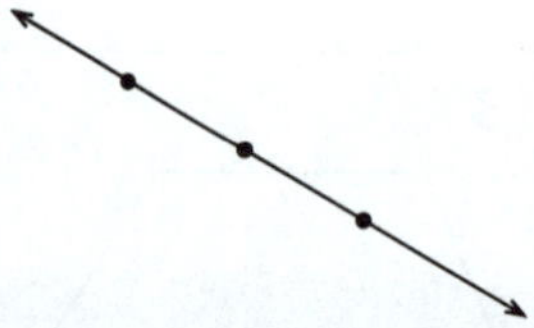

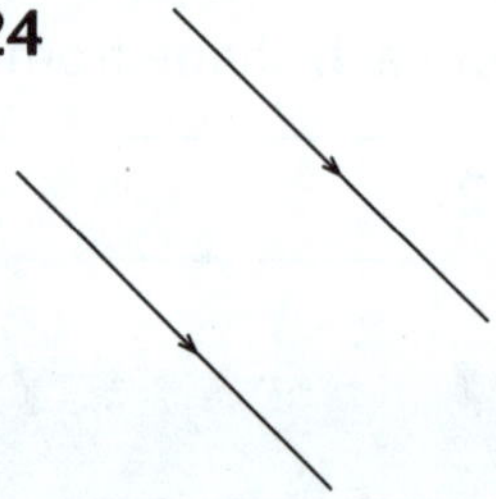

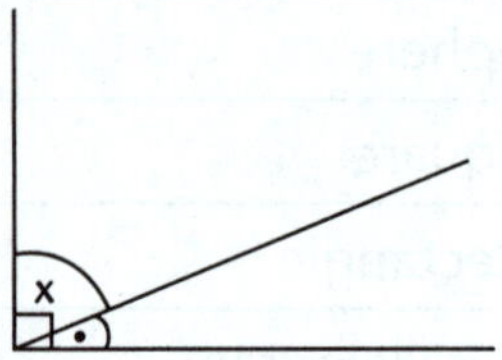

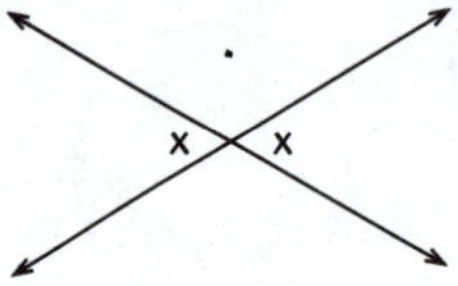

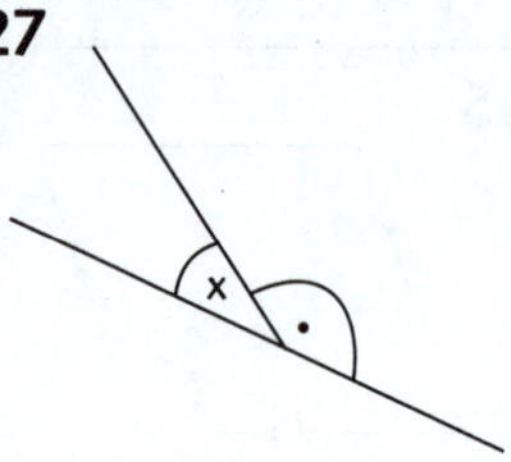

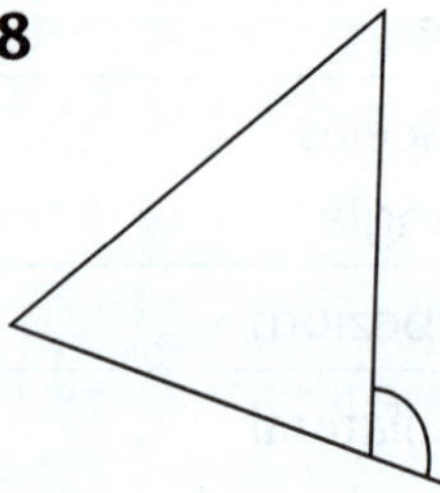

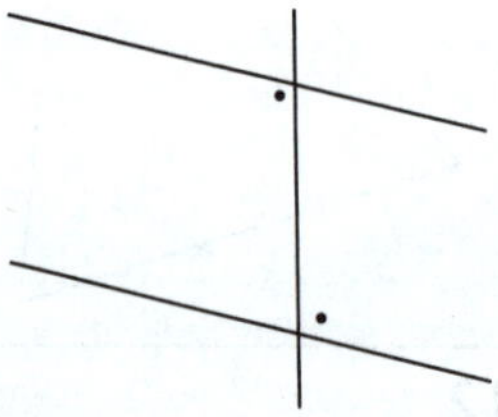

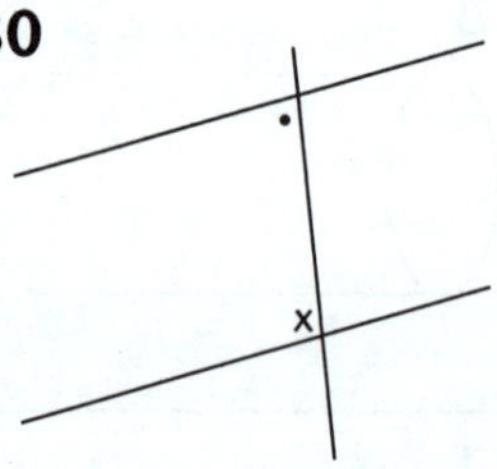

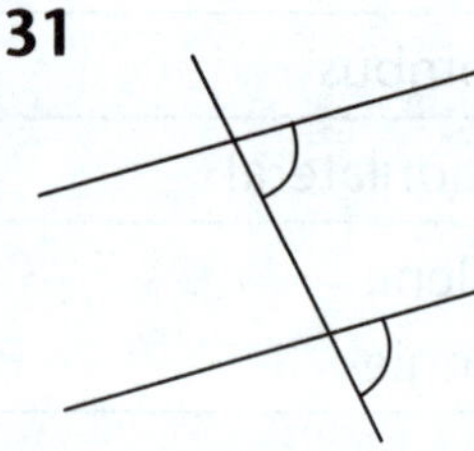

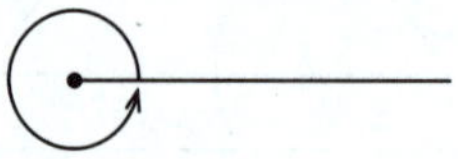

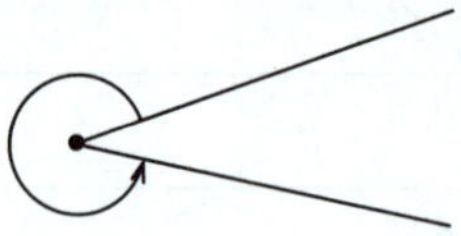

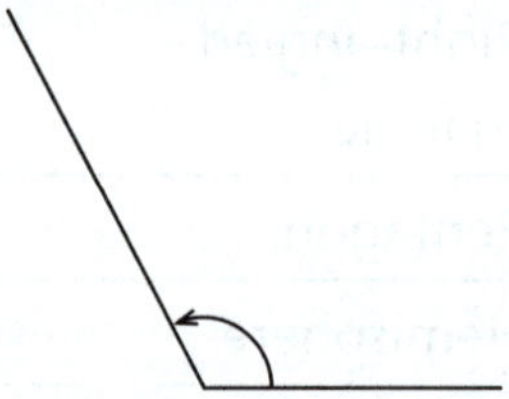

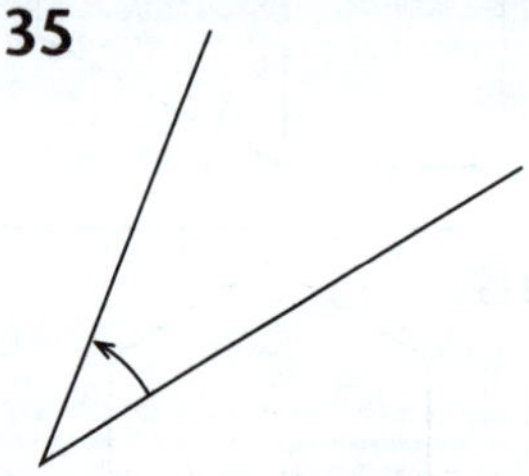

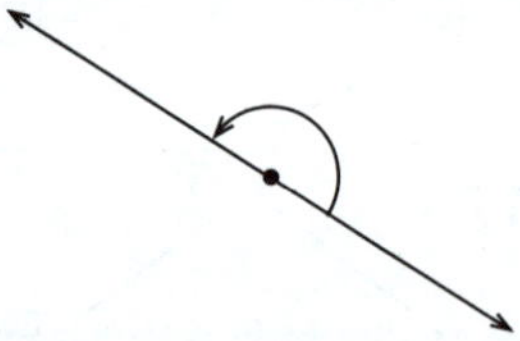

**37** Find the complement of:

**a** 67° **b** 48° **c** $x°$

**38** Find the supplement of:

**a** 67° **b** 48° **c** $x°$ **d** 148°

**39** Complete the following statements. Choose your answers from the table (some answers may be used more than once).

| Sum | Equal | 90° | 180° | 360° |
|---|---|---|---|---|
| Interior | Supplementary | Corresponding | Transversal | |

**a** Vertically opposite angles are ______________.

**b** Complementary angles have a sum of ______________.

**c** Supplementary angles have a sum of ______________.

**d** The base angles of an isosceles triangle are ______________.

**e** Angles at a point have a sum of ______________.

**f** The exterior angle of a triangle is equal to the ______________ of the two ______________ opposite angles.

**g** When a system of parallel lines is cut by a ______________ then the:

**i** alternate angles formed are ______________;

**ii** ______________ angles formed are equal;

**iii** co-interior angles are ______________.

Find the value of the pronumeral(s) in the following diagrams. Answers are worth 1 mark for each pronumeral value found.

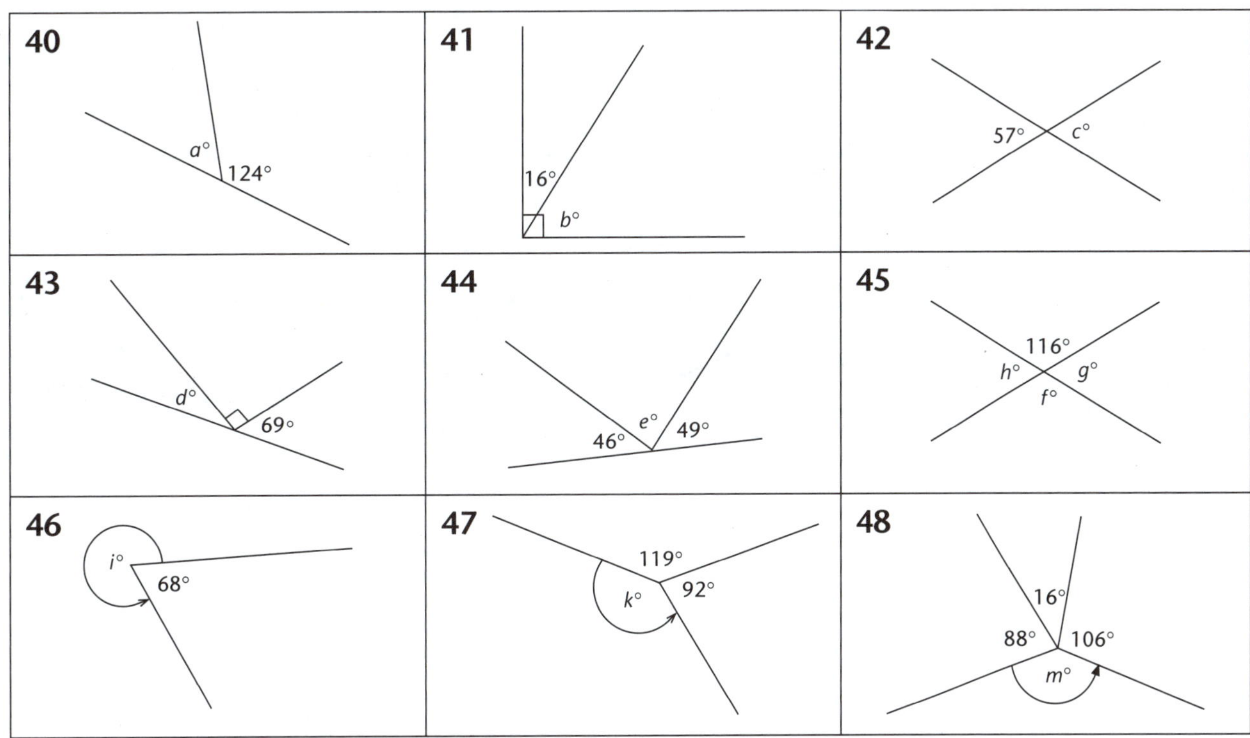

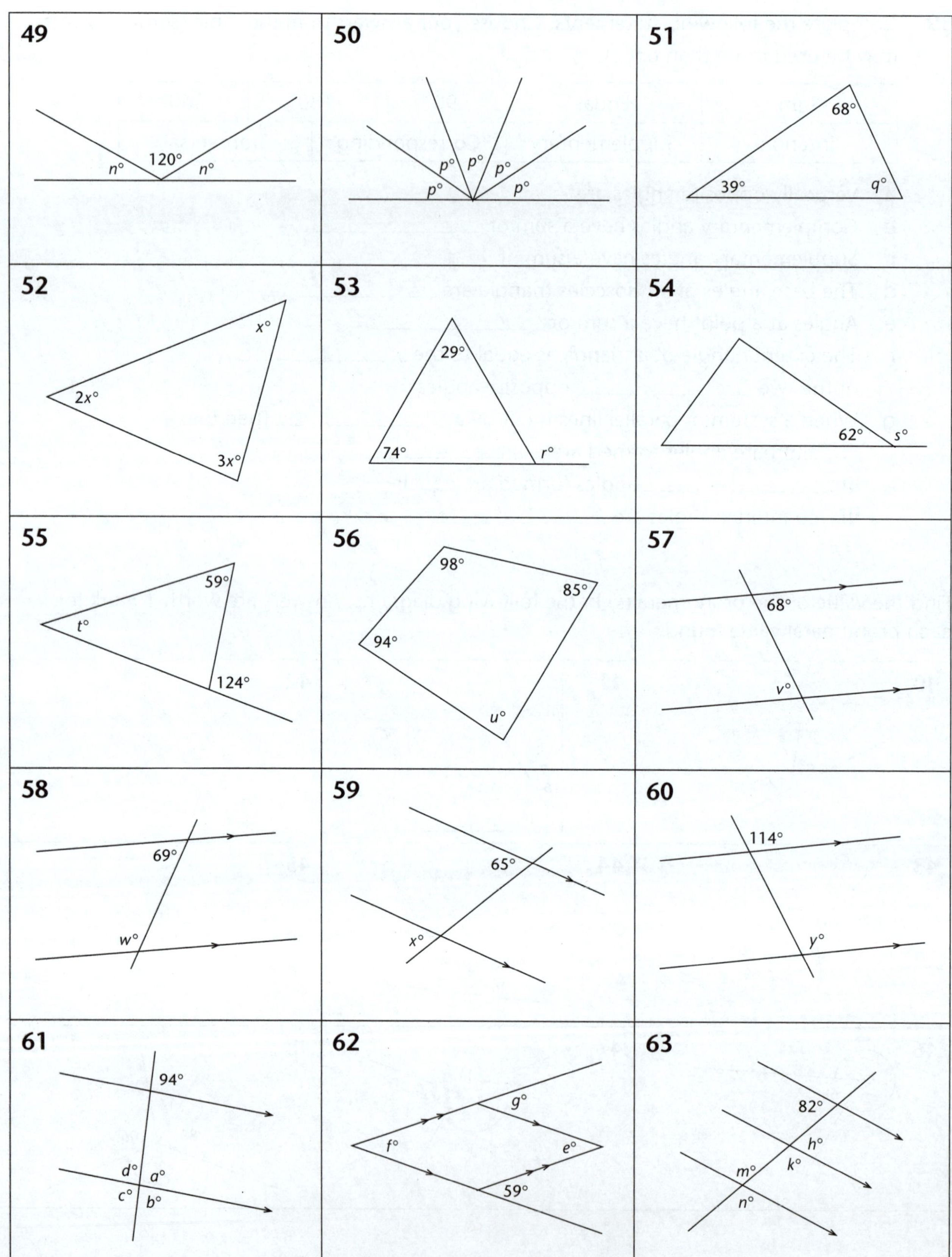
49
$n°$
120°
$n°$
50
$p°$
$p°$
$p°$
$p°$
$p°$
51
68°
39°
$q°$
52
$x°$
$2x°$
$3x°$
53
29°
74°
$r°$
54
62°
$s°$
55
59°
$t°$
124°
56
98°
85°
94°
$u°$
57
68°
$v°$
58
69°
$w°$
59
65°
$x°$
60
114°
$y°$
61
94°
$d°$
$a°$
$c°$
$b°$
62
$g°$
$f°$
$e°$
59°
63
82°
$h°$
$k°$
$m°$
$n°$

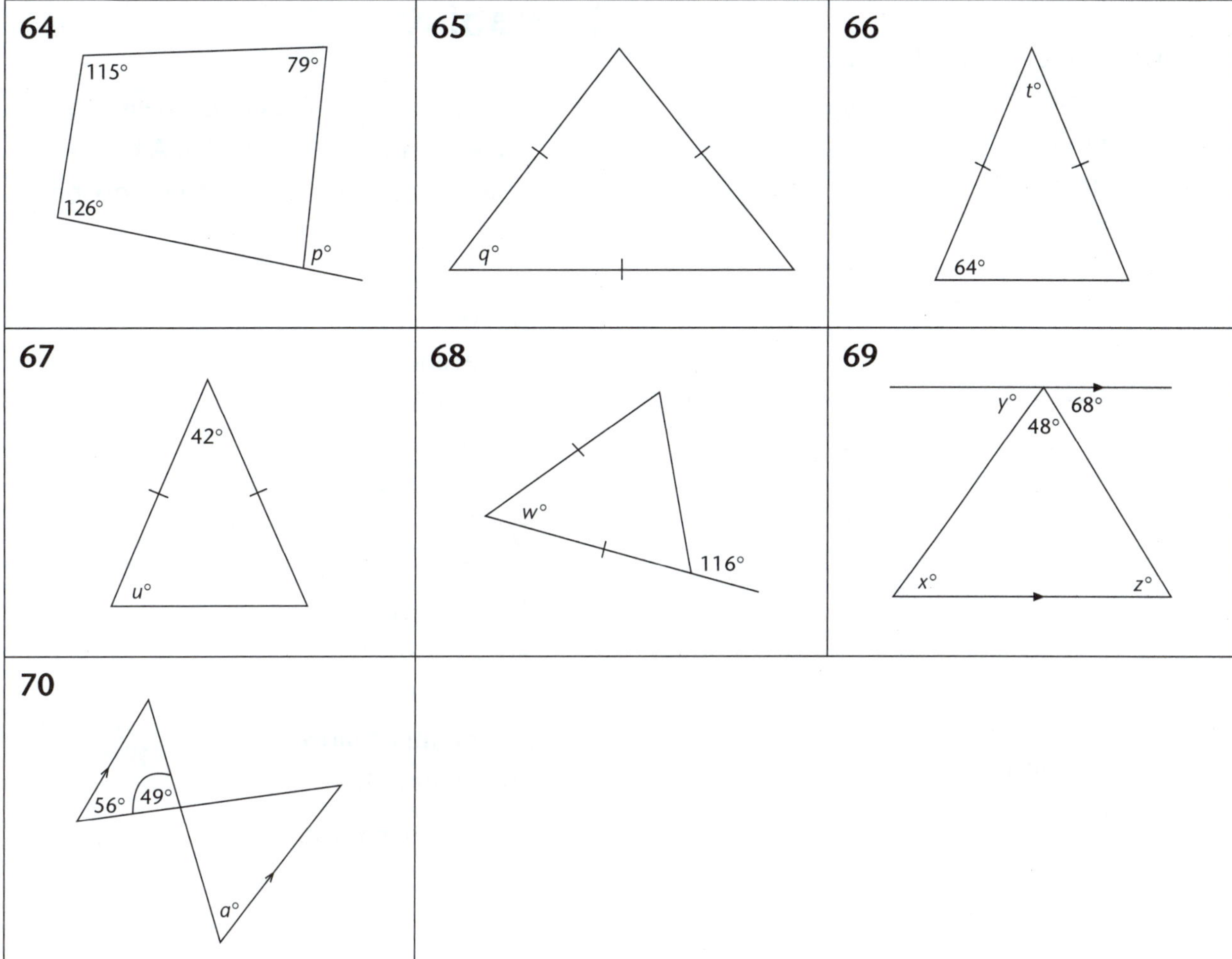


## Statistics

1 Over a 20-day period Tess recorded the number of SMSs she received:

| | | | | |
|---|---|---|---|---|
| 4 | 8 | 6 | 7 | 5 |
| 11 | 7 | 8 | 7 | 8 |
| 9 | 10 | 8 | 7 | 7 |
| 11 | 6 | 9 | 10 | 9 |

a Record the data in a frequency table.

b Use the table to draw a frequency histogram and polygon.

2 Draw a stem-and-leaf plot for the data:

| | | | | | |
|---|---|---|---|---|---|
| 63 | 47 | 58 | 41 | 40 | 57 |
| 73 | 70 | 48 | 40 | 68 | 63 |
| 52 | 59 | 61 | 53 | 49 | 47 |

3 For the scores

4, 7, 8, 3, 5, 8, 6, find the:

a range

b mode

c median

d mean.

4 For the scores
16, 4, 8, 8, 10, 12, find the:

a range
b mode
c median
d mean.

5

| Score ($x$) | Frequency ($f$) | $f \times x$ |
|---|---|---|
| 4 | 2 | |
| 5 | 4 | |
| 6 | 7 | |
| 7 | 5 | |
| 8 | 2 | |
| | ______ | ______ |

a Complete the table.
b Use the table to find the:
  i mode
  ii range
  iii mean.

6 A survey of favourite colours was conducted and the results recorded in the dot plot:

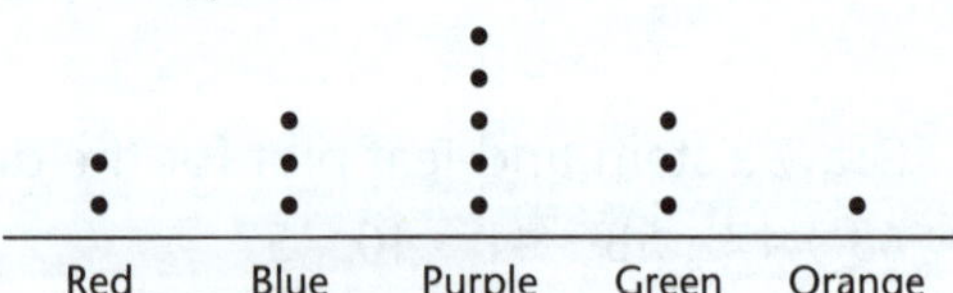


What is the mode?

## Probability

1 A bag contains 10 identical balls. The balls are numbered 1 to 10. A ball is selected at random. Find the probability that it is:

a 7
b odd
c even
d not 3
e prime
f composite
g 13
h 0
i a factor of 12
j divisible by 3
k less than 4
l greater than 7
m at least 4.

2 A die is rolled. Find the probability that it is:

a 4
b even
c not even
d 2 or 4
e a multiple of 2
f a square number
g a cube number.

☞ **Quick answers on pages 283-284**
☞ **Worked solutions on pages 296-303**

# NUMBER 2

- Factors and Highest Common Factor
- Multiples and Lowest Common Multiple
- Primes and Composities
- Index Notation
- Integers
- Order of Operation
- Conversions

## KEYWORDS

**Common**
**Composites**
**Conversions**
**Factors**
**Index**
**Integers**
**Multiples**
**Order of operations**
**Primes**

## Factors and Highest Common Factor

Any whole number that divides exactly into another number is called a **factor** of that number.

The factors of 12 are 1, 2, 3, 4, 6, 12.

The Highest Common Factor (HCF) of two or more numbers is the *largest number* that divides exactly into the numbers.

### For Example

1 List the factors of:

a 30 b 18

2 Find the Highest Common Factor (HCF) of the following pairs of numbers:

a 12 and 20 b 48 and 60

3 Jocyln wants to make bouquets using 16 red flowers and 28 white flowers. She wants to use the same number of flowers from each colour in all of her bouquets. What is the greatest number of bouquets Jocyln can make?

1 a Factors of 30 = 1, 2, 3, 5, 6, 10, 15, 30

b Factors of 18 = 1, 2, 3, 6, 9, 18

2 a 12: 1, 2, 3, (4), 6, 12
20: 1, 2, (4), 5, 10, 20
∴ HCF is 4

b 48: 1, 2, 3, 4, 6, 8, (12), 16, 24, 48
60: 1, 2, 3, 4, 5, 6, 10, (12), 15, 20, 30, 60
∴ HCF is 12

3 16: 1, 2, (4), 8, 16
28: 1, 2, (4), 7, 14, 28
∴ HCF is 4
∴ Jocyln can make 4 bouquets.

## Multiples and Lowest Common Multiple

A multiple is the result of multiplying a number by a whole number. The multiples of 7 are 7, 14, 21, …

The Lowest Common Multiple (LCM) is the *smallest number* that is a common multiple of two or more numbers.

### For Example

1 List the first six multiples of:

a 9 b 15

2 Find the Lowest Common Multiple (LCM) of the following pairs of numbers:

a 3 and 8 b 10 and 25

3 Two taps drip at the same instant. One tap drips every 4 seconds while the other drips every 10 seconds. How long will it be until they drip together again?

1 a 9: 9, 18, 27, 36, 45, 54

b 15: 15, 30, 45, 60, 75, 90

2 a 3: 3, 6, 9, 12, 15, 18, 21, (24), …
8: 8, 16, (24), …
∴ LCM is 24

b 10: 10, 20, 30, 40, (50), …
25: 25, (50), …
∴ LCM is 50

3 4: 4, 8, 12, 16, (20), …
10: 10, (20), …
∴ LCM is 20
∴ They will drip again together after 20 seconds.

## Primes and Composites

A **prime** is a number with only two factors. A **composite** is a number with more than two factors. Primes are 2, 3, 5, 7, 11, 13, …, while composites are 4, 6, 8, 9, 10, 12, 14, 15, …

The number 1 is neither prime nor composite.

A factor tree can be used to express a number as a product of its prime factors.

**For Example**

1 Use a factor tree to express as a product of prime factors:

a 60 b 144

1 a

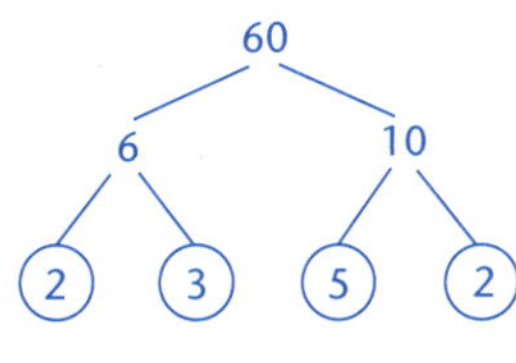


$\therefore 60 = 2 \times 2 \times 3 \times 5$

b

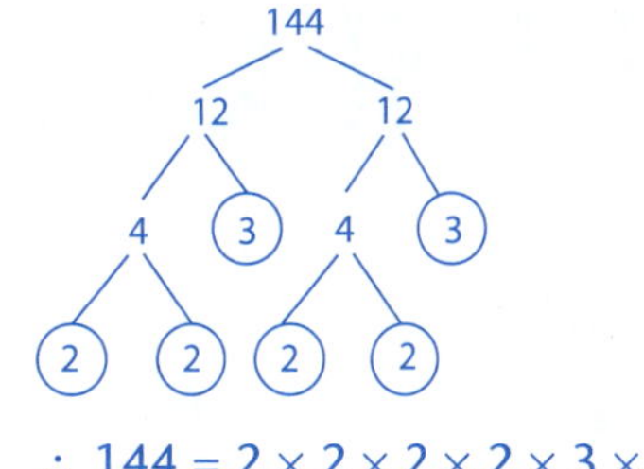


$\therefore 144 = 2 \times 2 \times 2 \times 2 \times 3 \times 3$

## Index Notation

This is a method of recording products containing the same number in a concise manner.

**For Example**

1 Express in index form:

a $4 \times 4 \times 4$

b $2 \times 2 \times 2 \times 2 \times 3 \times 3$

1 a $4 \times 4 \times 4 = 4^3$

b $2 \times 2 \times 2 \times 2 \times 3 \times 3 = 2^4 \times 3^2$

We can use these rules:

- $3^2 \times 3^4 = 3 \times 3 \times 3 \times 3 \times 3 \times 3 = 3^6$

  [When we multiply, we add the powers.]

- $3^5 \div 3^2 = \dfrac{\not{3} \times \not{3} \times 3 \times 3 \times 3}{\not{3} \times \not{3}} = 3^3$

  [When we divide, we subtract the powers.]

- $(3^2)^3 = 3^2 \times 3^2 \times 3^2$
  $= 3 \times 3 \times 3 \times 3 \times 3 \times 3$
  $= 3^6$

  [When we use the brackets, we multiply the powers.]

- $3^0 = 1$

  [Anything to the power of 0 is 1.]

1 Simplify:

a $5^2 \times 5^3$  b $7^4 \div 7^2$

c $3^6 \times 3^2 \div 3^5$  d $4^0 + 2^0$

e $(5 \times 2)^0$  f $(6^2)^4 \div (6^3)^2$

1 a $5^2 \times 5^3 = 5^5$

b $7^4 \div 7^2 = 7^2$

c $3^6 \times 3^2 \div 3^5 = 3^8 \div 3^5$
$= 3^3$

d $4^0 + 2^0 = 1 + 1$
$= 2$

e $(5 \times 2)^0 = 10^0$
$= 1$

f $(6^2)^4 \div (6^3)^2 = 6^8 \div 6^6$
$= 6^2$

## Integers

An integer is a whole number that can be positive, negative, or zero. There are rules to use when evaluating expressions involving positive and negative integers:

| Adding/ Subtracting | Multiplying | Dividing |
|---|---|---|
| + + = + | + × + = + | + ÷ + = + |
| + − = − | + × − = − | + ÷ − = − |
| − + = − | − × + = − | − ÷ + = − |
| − − = + | − × − = + | − ÷ − = + |

1 Evaluate:

a $4 - (-2)$  b $-3 + 7$

c $-4 - 6$  d $-3 - (-5)$

e $6 + (-2)$

2 Evaluate:

a $-3 \times 2$  b $-6 \times -5$

c $(-2) \times (-7)$  d $4 \times -8$

e $(-3)^2$

3 Evaluate:

a $6 \div -2$  b $-12 \div -4$

c $(-18) \div 6$  d $\frac{-10}{-2}$

e $\frac{-16}{4}$

1 a $4 - (-2) = 4 + 2 = 6$

b $-3 + 7 = 4$

c $-4 - 6 = -10$

d $-3 - (-5) = -3 + 5$
$= 2$

e $6 + (-2) = 6 - 2$
$= 4$

2 a $-3 \times 2 = -6$

b $-6 \times -5 = 30$

c $(-2) \times (-7) = 14$

d $4 \times -8 = -32$

e $(-3)^2 = -3 \times -3$
$= 9$

3 a $6 \div -2 = -3$

b $-12 \div -4 = 3$

c $(-18) \div 6 = -3$

d $\frac{-10}{-2} = 5$

e $\frac{-16}{4} = -4$

## Order of Operation

We can use the acronym BODMAS to help us remember the order of operation:

Brackets, Order, Division, Multiplication, Addition and Subtraction

[This is also remembered as BIDMAS and BEDMAS.]

For Example

1 Evaluate:

a $3 - 4 \times 2$

b $5 \times 4 - 10 \times 3$

c $(-4) \times 2 - 3 \times (-4)$

d $\frac{10 - 16}{3 \times 2}$

e $6(3 - 5)$

f $-2(4 - 6)$

g $(-4 + 2)(3 - 7)$

h $\frac{12 - 4 \times 5}{2 - 4}$

i $12 - [2(3 - 6)]$

j $(-3)^3 \div (-1)^5$

1 a $3 - 4 \times 2 = 3 - 8$
$= -5$

b $5 \times 4 - 10 \times 3 = 20 - 30$
$= -10$

c $(-4) \times 2 - 3 \times (-4) = -8 + 12$
$= 4$

d $\frac{10 - 16}{3 \times 2} = \frac{-6}{6}$
$= -1$

e $6(3 - 5) = 6 \times -2$
$= -12$

f $-2(4 - 6) = -2 \times -2$
$= 4$

g $(-4 + 2)(3 - 7) = -2 \times -4$
$= 8$

h $\frac{12 - 4 \times 5}{2 - 4} = \frac{12 - 20}{-2}$
$= \frac{-8}{-2}$
$= 4$

i $12 - [2(3 - 6)] = 12 - 2(-3)$
$= 12 + 6$
$= 18$

j $(-3)^3 \div (-1)^5 = -27 \div -1$
$= 27$

## Conversions

### Changing Fractions to Decimals

To change a fraction to a decimal we divide the numerator by the denominator. This can result in a terminating decimal, e.g. $\frac{3}{8} = 0.375$, or a repeating (recurring) decimal, e.g. $\frac{3}{11} = 0.272727\ldots = 0.\dot{2}\dot{7}$.

For Example

1 Convert to decimals:

a $\frac{3}{4}$ b $\frac{7}{8}$ c $\frac{1}{3}$

d $\frac{7}{11}$ e $\frac{4}{7}$

1 a $\frac{3}{4}$ $\quad 4\overline{)3.00}$ = 0.75

$\therefore \frac{3}{4} = 0.75$

b $\frac{7}{8}$ $\quad 8\overline{)7.000}$ = 0.875

$\therefore \frac{7}{8} = 0.875$

c $\frac{1}{3}$ $\quad 3\overline{)1.000}$ = 0.333...

$\therefore \frac{1}{3} = 0.\dot{3}$

**d** $\frac{7}{11}$

$$11\overline{)7.0000}\quad 0.6363\ldots$$

$\therefore \frac{7}{11} = 0.\dot{6}\dot{3}$

**e** $\frac{4}{7}$

$$7\overline{)4.00000000}\quad 0.57142857\ldots$$

$\therefore \frac{4}{7} = 0.\dot{5}7142\dot{8}$

## Changing Decimals to Fractions

The number of digits after the decimal point (the number of decimal places) equals the number of zeros in the denominator.

For Example

**1** Rewrite as simplified fractions:

**a** 0.7 **b** 0.03

**c** 0.84 **d** 0.083

**e** 0.04 **f** 0.075

**1** **a** $0.7 = \frac{7}{10}$

**b** $0.03 = \frac{3}{100}$

**c** $0.84 = \frac{84}{100}$

$= \frac{21}{25}$

**d** $0.083 = \frac{83}{1000}$

**e** $0.04 = \frac{4}{100}$

$= \frac{1}{25}$

**f** $0.075 = \frac{75}{1000}$

$= \frac{3}{40}$

Go to p. 284 for quick answers, or to pp. 303–305 for worked solutions

1 Find the Highest Common Factor (HCF) of: p. 14
a 16 and 24 b 45 and 75 c 24 and 36

2 A youth group is attended by 48 girls and 36 boys. The leaders need to form groups with the same number of boys and the same number of girls in each group. What is the greatest number of groups that can be formed? p. 14

3 The school choir has 20 altos and 16 sopranos. The choirmaster needs to form identical practise groups with the same number of altos and sopranos. What is the greatest number of groups that can be formed? p. 14

4 Find the Lowest Common Multiple (LCM) of: p. 14
a 4 and 6 b 6 and 9 c 12 and 16

5 Two lights are flashing. One light flashes every 6 seconds and the other every 4 seconds. If they flash at the same moment, when is the next time they flash at the same time? p. 14

6 Jack and Jill are training for a triathlon. Jack rides his bike every third morning while Jill rides her bike every seventh morning. If they ride together on the 1st of June, when will they ride again? p. 14

7 Write the positive numbers less than 30 that are: p. 15
a prime b composite.

8 How many primes lie between 50 and 60? p. 15

9 How many composites lie between 61 and 71? p. 15

10 Use a factor tree to rewrite the following as products of their prime factors: p. 15
a 72 b 96 c 250

11 Rewrite the following in index form: p. 15
a $4 \times 4 \times 4 \times 4 \times 4$ b $3 \times 3 \times 5 \times 5 \times 5$
c $10 \times 10 \times 10 \times 10 \times 10$

12 Use a factor tree to rewrite 216 as a product of its prime factors expressed in index form. p. 15

13 Rewrite the following in index form: pp. 15–16
a $2^4 \times 2^2$ b $3^5 \div 3^2$ c $6^4 \times 6 \times 6^2$
d $7^4 \div 7^3$ e $4^3 \div 4$ f $(2^4)^3$
g $(6^3)^2$

14 Evaluate: pp. 15–16
a $3^0$ b $(4 \times 2)^0$ c $5^0 - 2^0$ d $(5 - 2)^0$

15 Evaluate: p. 16
a $-2 - 3$ b $6 - (-2)$ c $-5 + (-1)$
d $10 - (-1)$ e $-5 - 7$ f $-6 + 4$

**16** Evaluate: p. 16

a $-2 \times 3$ b $-6 \times -2$ c $3 \times -4$

d $-5 \times -2 \times -1$ e $-4 \times -3 \times -2$ f $6 \times -1 \times -1$

g $(-3)^2$ h $(-5)^2$ i $(-2-1)^2$

**17** Evaluate: p. 16

a $-12 \div 6$ b $-16 \div -4$ c $15 \div -3$

d $-18 \div -2$ e $\frac{-12}{-4}$ f $\frac{-16}{8}$

g $\frac{-120}{-10}$ h $\frac{16}{-2}$ i $\frac{15}{-5}$

**18** Evaluate: p. 17

a $3 - 4 \times 2$ b $12 - 5 \times 3$ c $-6 + 4 \times 2$

d $-15 - 4 \div 2$ e $16 \div 4 - 10 \div 1$ f $\frac{30}{6} - \frac{20}{2}$

g $\frac{6+4}{-2}$ h $\frac{10-14}{-4}$ i $\frac{12-15}{6-9}$

j $\frac{10-14}{6-8}$ k $(-3-2)^2$ l $(10 - 3 \times 4)^2$

m $4(2-7)$ n $-2(-3-4)$ o $(4-6)(10-14)$

p $3(2 - 4 \times 2)$ q $-5(3 - 8 \times 2)$ r $\frac{12 \div 6 - 8 \times 2}{3 - 5 \times 2}$

**19** Convert to decimals: pp. 17–18

a $\frac{3}{5}$ b $\frac{11}{20}$ c $\frac{17}{100}$

d $\frac{2}{3}$ e $\frac{4}{11}$ f $\frac{6}{7}$

**20** Convert to simplified fractions: p. 18

a 0.7 b 0.07 c 0.08

d 0.31 e 0.145 f 0.072

Go to p. 284 for quick answers, or to pp. 303–305 for worked solutions

## YOUR CHECKLIST

**For a complete understanding of this topic you must be able to:**

| | | |
|---|---|---|
| ✓ | Determine the Highest Common Factor of two or more numbers | p. 14 |
| ✓ | Determine the Lowest Common Multiple of two or more numbers | pp. 14–15 |
| ✓ | Understand the difference between primes and composites | p. 15 |
| ✓ | Rewrite numbers in index form | p. 15 |
| ✓ | Use the index rules | pp. 15–16 |
| ✓ | Use the four operations involving positive and negative integers | p. 16 |
| ✓ | Use the rules for order of operations | p. 17 |
| ✓ | Change fractions to decimals involving terminating and repeating decimals | pp. 18–19 |
| ✓ | Change decimals to simplified fractions. | p. 19 |

**Now you are ready to do the tests!**

# Level 1 Test

**(35 marks)**

1. List the factors of:
   a 16 b 35 c 50 (3 marks)
2. Find the Highest Common Factor of:
   a 12 and 26 b 20 and 32 c 18 and 30 (6 marks)
3. Write the first five multiples of:
   a 3 b 5 c 7 (3 marks)
4. What is the Lowest Common Multiple of:
   a 4 and 6? b 3 and 5? c 6 and 8? (6 marks)
5. Using the numbers 1 to 10, list the:
   a composites b primes. (2 marks)
6. Rewrite in index form:
   a $2 \times 2 \times 2 \times 2$ b $6 \times 6 \times 7 \times 7 \times 7$ (2 marks)
7. Evaluate:
   a $4 - (-2)$ b $-6 + (-2)$ c $3 - 7$ (3 marks)
8. Evaluate:
   a $-3 \times 6$ b $2 \times -4$ c $5 \times -3$ (3 marks)
9. Find the value of:
   a $-12 \div 3$ b $18 \div -6$ c $-10 \div -2$ (3 marks)
10. Rewrite as decimals:
    a $\frac{4}{5}$ b $\frac{1}{3}$ (2 marks)
11. Change to simplified fractions:
    a 0.03 b 0.8 (2 marks)

☞ Quick answers on page 290
☞ Worked solutions on page 369

**Your Feedback** $\frac{\square}{35} \times 100\% = \square\%$

**(35 marks)**

1 Find the Highest Common Factor of:
a 20 and 24 b 25 and 45 c 60 and 80 (3 marks)

2 Find the Lowest Common Multiple of:
a 4 and 10 b 15 and 25 c 12 and 18 (3 marks)

3 Use a factor tree to express the following as products of their prime factors:
a 80 b 72 c 200 (6 marks)

4 Rewrite, leaving your answer in index form:
a $3^4 \times 3^2$ b $2^7 \div 2^5$ c $(5^2)^3$ (3 marks)

5 Evaluate:
a $4^0$ b $(6 - 2)^0$ c $3^0 \times 5^0$ (3 marks)

6 Evaluate:
a $-2 \times 4$ b $\frac{10}{-5}$ c $4 - (-2)$ (3 marks)

7 Evaluate:
a $12 - 4 \times 7$ b $16 \div (-4) - 3 \times (-5)$
c $\frac{6 - 12}{3 - 2 \times 3}$ d $(-4 - 2)(3 - 5)$ (8 marks)

8 Express as decimals:
a $\frac{1}{7}$ b $\frac{3}{11}$ c $\frac{3}{20}$ (3 marks)

9 Rewrite as simplified fractions:
a 0.03 b 0.033 c 0.3003 (3 marks)

☞ Quick answers on page 290
☞ Worked solutions on page 369

**Your Feedback** $\frac{\square}{35} \times 100\% = \square\%$

**(35 marks)**

1 Find the Highest Common Factor of:
a 60 and 100 b 75 and 40 (2 marks)

2 Office staff are making stationery packs. They have a total of 60 pens and 80 pencils. What is the maximum number of packs that can be made if they have the same number of pens and the same number of pencils? (1 mark)

3 Find the Lowest Common Multiple of:
a 30 and 40 b 16 and 24 (2 marks)

4 A box of oranges can be shared equally between four families, or five families or six families. What is the smallest number of oranges in the box? (2 marks)

5 A bag contains 10 cards with the numbers 1 to 10 written on the cards. What percentage of the cards are:
a prime? b composite? (2 marks)

6 Simplify, writing in index form:
a $4^3 \times 4^2 \div 4$ b $\frac{10^4}{10^3 \times 10}$ (2 marks)

7 Evaluate:
a $4^0 - 3^0$ b $(4 - 3)^0$ c $\frac{15^0 - 10^0}{5^0}$ (3 marks)

8 Simplify:
a $3(-2 - 5)$ b $(-16) \div (-4) + 3 \times (-2)$
c $\frac{15 - 4 \times 6}{3 - 6}$ d $25 - (4 \times 3 - 6 \times 5)$ (8 marks)

9 Evaluate:
a $(-1)^{2013}$ b $(-1)^{2014}$ (2 marks)

10 Use factor trees to express the following as products of their prime factors, written in index form:
a 320 b 256 c 1444 (6 marks)

11 If $a = -3$, $b = -4$, $c = -5$, evaluate:
a $2a - 3b + 4c$ b $\frac{bc + 1}{a}$ (4 marks)

12 Express $\frac{1}{99}$ as a decimal. (1 mark)

☞ Quick answers on page 290
☞ Worked solutions on page 370

**Your Feedback** $\frac{\square}{35} \times 100\% = \square\%$

# 3 PERCENTAGES AND APPLICATIONS

- Conversions
- Finding the Percentage of a Quantity
- Increasing/Decreasing Quantities by Percentages
- Expressing One Quantity as a Percentage of Another
- Unitary Method and Percentages
- Special Applications of Percentages

## KEYWORDS

**Conversion** **Loss**
**Convert** **Percentage**
**Discount** **Profit**

A percentage is a special fraction that has a denominator of 100. (Note that the symbol, %, is made up of two zeros.)

# Conversions

### Converting Fractions to Percentages

We multiply the fraction by 100% (i.e. by 1).

**For Example**

Convert to percentages:

**1** $\frac{3}{4}$ **2** $\frac{2}{3}$

**3** $\frac{7}{8}$ **4** $\frac{41}{1000}$

**5** $\frac{3}{40}$ **6** $2\frac{1}{2}$

**1** $\frac{3}{4} = \frac{3}{4} \times 100\%$
$= 75\%$

**2** $\frac{2}{3} = \frac{2}{3} \times 100\%$
$= 66\frac{2}{3}\%$ (or $66.\dot{6}\%$)

**3** $\frac{7}{8} = \frac{7}{8} \times 100\%$
$= \frac{700}{8}\%$
$= 87\frac{1}{2}\%$

**4** $\frac{41}{1000} = \frac{41}{1000} \times 100\%$
$= \frac{41}{10}\%$
$= 4.1\%$

**5** $\frac{3}{40} = \frac{3}{40} \times 100\%$
$= \frac{30}{4}\%$
$= 7\frac{1}{2}\%$

**6** $2\frac{1}{2} = 2\frac{1}{2} \times 100\%$
$= \frac{500}{2}\%$
$= 250\%$

### Converting Decimals to Percentages

We multiply the decimal by 100%.

**For Example**

Convert to percentages:

**1** 0.7 **2** 0.05

**3** 0.27 **4** 0.407

**5** 3.41

**1** $0.7 = 0.7 \times 100\%$
$= 70\%$

**2** $0.05 = 0.05 \times 100\%$
$= 5\%$

**3** $0.27 = 0.27 \times 100\%$
$= 27\%$

**4** $0.407 = 0.407 \times 100\%$
$= 40.7\%$

**5** $3.41 = 3.41 \times 100\%$
$= 341\%$

### Converting Percentages to Fractions

Remember that writing % is another way of writing a fraction with a denominator of 100.

Convert to fractions in their simplest form:

**1** 7% **2** 70%

**3** $4\frac{1}{2}\%$ **4** $7\frac{1}{4}\%$

**5** 120% **6** 600%

**1** $7\% = \frac{7}{100}$

**2** $70\% = \frac{70}{100}$
$= \frac{7}{10}$

**3** $4\frac{1}{2}\% = \frac{4\frac{1}{2}}{100}$
$= \frac{\frac{9}{2}}{100}$ [Write the numerator as an improper fraction.]
$= \frac{9}{200}$ [Multiply the numerator and denominator by 2.]

**4** $7\frac{1}{4}\% = \frac{7\frac{1}{4}}{100}$
$= \frac{\frac{29}{4}}{100}$ [Write the numerator as an improper fraction.]
$= \frac{29}{400}$ [Multiply the numerator and denominator by 4.]

**5** $120\% = \frac{120}{100}$
$= \frac{6}{5}$
$= 1\frac{1}{5}$

**6** $600\% = \frac{600}{100}$
$= 6$

## Converting Percentages to Decimals

We divide the percentage by 100. (This is equivalent to putting percentage over 100, e.g. $70\% = \frac{70}{100} = 70 \div 100 = 0.7$.)

Convert to decimals:

**1** 40% **2** 4%

**3** $18\frac{1}{2}\%$ **4** $9\frac{3}{4}\%$

**5** $66\frac{2}{3}\%$ **6** 247%

**1** $40\% = 40 \div 100$
$= 0.4$

**2** $4\% = 4 \div 100$
$= 0.04$

**3** $18\frac{1}{2}\% = 18\frac{1}{2} \div 100$
$= 0.185$ [The $\frac{1}{2}$ puts a 5 on the end.]

**4** $9\frac{3}{4}\% = 9\frac{3}{4} \div 100$
$= 0.0975$ [The $\frac{3}{4}$ puts a 75 on the end.]

**5** $66\frac{2}{3}\% = 66\frac{2}{3} \div 100$
$= 0.666\ldots$
$= 0.\dot{6}$ [The $\frac{2}{3}$ puts a $0.\dot{6}$ on the end.]

**6** $247\% = 247 \div 100$
$= 2.47$

## The Top Ten Conversions

Students should be familiar with these conversions.

| Fractions | Decimals | Percentages |
|---|---|---|
| 1 | 1 | 100% |
| $\frac{1}{2}$ | 0.5 | 50% |
| $\frac{1}{3}$ | $0.\dot{3}$ | $33\frac{1}{3}\%$ |
| $\frac{2}{3}$ | $0.\dot{6}$ | $66\frac{2}{3}\%$ |
| $\frac{1}{4}$ | 0.25 | 25% |
| $\frac{3}{4}$ | 0.75 | 75% |
| $\frac{1}{8}$ | 0.125 | 12.5% |
| $\frac{3}{8}$ | 0.375 | 37.5% |
| $\frac{5}{8}$ | 0.625 | 62.5% |
| $\frac{7}{8}$ | 0.875 | 87.5% |

## Finding the Percentage of a Quantity

Remember that 'of' means multiplication.

For Example

Find:

1. 16% of 300
2. 8% of 76
3. $4\frac{1}{2}$% of 350
4. 32% of \$15
5. 12.5% of 72 days
6. $33\frac{1}{3}$% of \$15.60
7. $\frac{3}{4}$% of \$70
8. 145% of \$60

Alternative sets of solutions are offered for these examples.

The first set of solutions converts the percentage to a decimal as the first step in the solution (i.e. 75% = 0.75).

The second set of solutions uses the conversion of the percentage to a fraction over 100 (i.e. $75\% = \frac{75}{100}$).

### Solutions 1

**1** 16% of 300
$0.16 \times 300 = 48$

**2** 8% of 76
$0.08 \times 76 = 6.08$

**3** $4\frac{1}{2}$% of 350
$0.045 \times 350 = 15.75$

**4** 32% of \$15
$0.32 \times 15 = 4.8$
$\therefore$ \$4.80

**5** 12.5% of 72 days
$0.125 \times 72 = 9$
$\therefore$ 9 days

**6** $33\frac{1}{3}$% of \$15.60
$0.333\ldots \times 1560 \approx 520$
$\therefore$ \$5.20

**7** $\frac{3}{4}$% of \$70
$0.0075 \times 70 = 0.525$
$\therefore$ \$0.53 (to nearest cent)

**8** 145% of \$60
$1.45 \times 60 = 87$
$\therefore$ \$87

### Solutions 2

**1** 16% of 300
$= \frac{16}{100} \times 300$
$= 48$

[Calculator:
16 ÷ 100 × 300 =
OR using fractions:
$\frac{16}{\cancel{100}_1} \times \frac{\cancel{300}^3}{1} = 48$]

**2** 8% of 76
$= \frac{8}{100} \times 76$
$= 6.08$

[8 ÷ 100 × 76]

**3** $4\frac{1}{2}$% of 350
$= \frac{4.5}{100} \times 350$
$= 15.75$

**4** 32% of \$15
$= \frac{32}{100} \times 15$
$=$ \$4.80

**5** 12.5% of 72 days
$= \frac{12.5}{100} \times 72$
$= 9$ days

**6** $33\frac{1}{3}$% of \$15.60
$= \frac{1}{3} \times 15.6$
$=$ \$5.20

[Note: $\frac{1}{3} = 33\frac{1}{3}\%$]

7 $\frac{3}{4}$% of \$70

$\frac{0.75}{100} \times 70$

$= 0.525$

= \$0.53 (to nearest cent)

8 145% of \$60

$= \frac{145}{100} \times 60 = \$87$

## Increasing/Decreasing Quantities by Percentages

- To increase an amount by 20%, we find 120% (100 + 20) of the amount.
- To decrease an amount by 20%, we find 80% (100 − 20) of the amount.

**For Example**

1 Increase \$400 by 20%.

2 Increase \$75 by 15%.

3 Decrease 200 litres by 12%.

4 Decrease 400 by 17%.

5 Increase \$200 by 30% and then decrease the amount by 30%.

6 \$600 is to be increased by 12% each month for the next three months. What will be the final amount?

1 Find 120% of \$400

i.e. 120% × 400

$= 1.2 \times 400$ [100 + 20]

$= 480$

∴ \$480

2 Find 115% of \$75

i.e. 115% × 75

$= 1.15 \times 75$ [100 + 15]

$= 86.25$

∴ \$86.25

3 Find 88% of 200 L

i.e. $0.88 \times 200$ [100 − 12]

$= 176$

∴ 176 L

4 Find 83% of 400

i.e. $0.83 \times 400$ [100 − 17]

$= 332$

∴ 332

5 ↑ by 30%: 130%; ↓ by 30%: 70%

∴ 130% × 200 × 70%

$= 1.3 \times 200 \times 0.7$

$= 182$

[Or 130% × 200
= 1.3 × 200
= 260
Then 70% of 260
= 0.7 × 260
= 182]

6 112% of 112% of 112% of \$600

$1.12 \times 1.12 \times 1.12 \times 600$

$= 842.9568$

= \$842.96 (to nearest cent)

**Alternatively**, we could have found the percentage of the amount and then added (for increasing) or subtracted (for decreasing).

**For Example**

1 Increase \$270 by 15%.

2 Decrease \$760 by 18%.

1 15% of \$270

$= 0.15 \times 270$

$= 40.5$

i.e. \$40.50

∴ New amount = \$270 + \$40.50

= \$310.50

2 18% of \$760
= $0.18 \times 760$
= 136.8
i.e. \$136.80
$\therefore$ New amount = \$760 − \$136.80
= \$623.20

## Expressing One Quantity as a Percentage of Another

We express the first quantity as a fraction of the second using consistent units and then convert this fraction to a percentage (by multiplying by 100%).

### For Example

1 What percentage is:
 a 15 of 20?
 b 75c of \$1?

2 Express 30 seconds as a percentage of 2 minutes.

3 Rewrite 25 out of 40 as a percentage.

4 What percentage of each figure has been shaded?

a

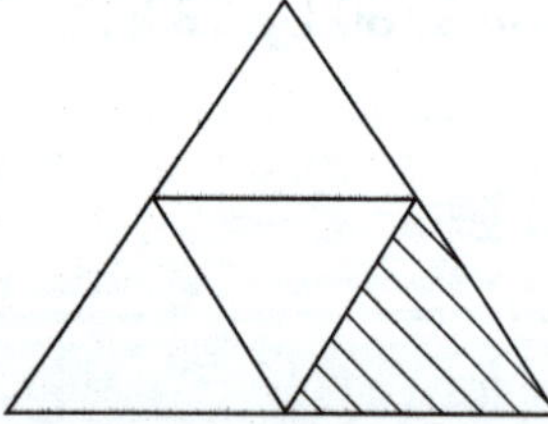

b

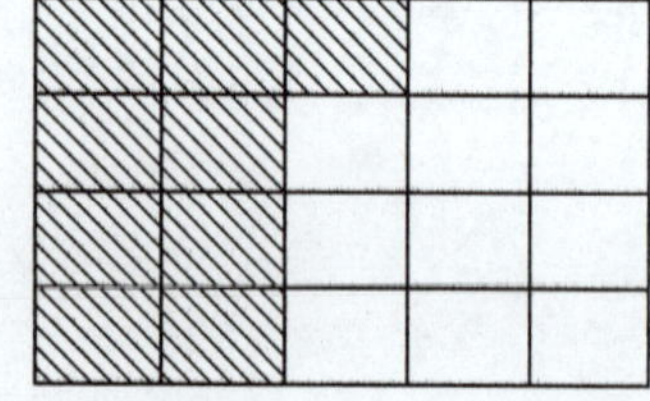

1 a $\frac{15}{20} \times 100\% = 75\%$ [Fraction is $\frac{15}{20}$.]
 b \$1 → 100c
 $\therefore \frac{75}{100} \times 100\% = 75\%$ [We must have the same units.]

2 2 minutes → 120 seconds
$\therefore \frac{30}{120} \times 100\% = 25\%$

3 $\frac{25}{40} \times 100\% = 62.5\%$

4 a $\frac{1}{4} \times 100\% = 25\%$
 b 9 shaded out of 20
 $\therefore \frac{9}{20} \times 100\% = 45\%$

## Unitary Method and Percentages

We can solve some problems by dividing to find 1%, then multiplying to find the whole amount (100%).

### For Example

1 If 7% of an amount is 56, find the amount.

2 Seventy-two per cent of a number is 360. What is the number?

3 15% of an amount is \$180. Find half of the amount.

1 7% of an amount = 56
$\therefore$ 1% of the amount = $56 \div 7$
= 8
100% of the amount = $8 \times 100$
= 800
$\therefore$ Amount is 800.

2 $72\%$ of a number $= 360$

$1\%$ of the number $= 360 \div 72$

$= 5$

$100\%$ of the number $= 5 \times 100$

$= 500$

$\therefore$ The number is 500.

3 $15\%$ of an amount $= 180$

$1\%$ of the amount $= 180 \div 15$

$= 12$

$50\%$ of an amount $= 12 \times 50$

$= 600$

$\therefore$ The amount is \$600.

[When convenient, we can drop to say 5% (by dividing by 3) and then 'build', rather than dropping to 1%.]

## Special Applications of Percentages

### Discount

Businesses often express reductions as a percentage discount.

**For Example**

1 How much is saved if a media player, valued at \$240, is discounted by 20%?

2 What will be paid for a pair of jeans, marked at \$80, but discounted by 30%?

3 A 'Seniors Card' entitles Bridget to a 15% discount off the price of a smorgasbord meal that costs \$13. What will Bridget pay for the meal?

4 A fridge is discounted by 20% and now costs \$960. What was its original price?

5 The price of a dress is cut by 40%. If this is a saving of \$96, what will the dress now cost?

6 Which is the better deal on a television marked at \$1200:

**a** a discount of 20%?

**b** a discount of 10% and then a further discount of 10%?

1 Savings $= 20\%$ of 240

$= 0.2 \times 240$

$= 48$

$\therefore$ \$48 is saved.

2 Amount of discount $= 30\%$ of 80

$= 24$

$\therefore$ new price $= 80 - 24$

$= 56$

$\therefore$ The new price is \$56.

Or alternatively:

Discounted price $= 70\%$ of 80

$= 0.7 \times 80$ [100 − 30]

$= 56$

$\therefore$ The new price is \$56.

[This second method is quicker.]

3 New discounted price $= 85\%$ of 13

$= 0.85 \times 13$

[100 − 15]

$= 11.05$

$\therefore$ The meal will cost \$11.05.

4 If already discounted,

$\therefore$ $80\%$ of original price $= 960$ [100 − 20]

$1\%$ of original price $= 960 \div 80$

$= 12$

$100\%$ of original price $= 12 \times 100$

$\therefore$ The original price was \$1200.

5 40% of original price = 96

1% of original price = 96 ÷ 40

= 2.4

∴ 60% of original price = 2.4 × 60 [after 40% discount]

= 144

∴ The dress (with 40% discount) will now cost $144.

6 a Discount of 20%

∴ 80% of $1200

= 0.8 × 1200

= 960

∴ The television to cost $960.

b Discount of 10%, then discount of 10%

∴ 90% of 1200

= 0.9 × 1200

= 1080

Now, 90% of 1080

= 0.9 × 1080

= 972

Note: a discount of 20% is NOT the same as two consecutive discounts of 10%.

∴ The television to cost $972.

∴ The better deal is (a).

## Profit and Loss

A profit occurs when the selling price > cost price.

A loss occurs when the selling price < cost price.

### For Example

1 Maria purchased a block of land for $495 000 and sold it later to make a 14% profit. How much did she sell the block for?

2 Cost price = $16; selling price = $20. Find the:

a profit

b profit as a percentage of the cost price

c profit as a percentage of the selling price.

3 Li bought some shares for $2.40 and sold them later for $1.80. Find her loss as a percentage of the cost price.

4 Jonty purchased a lawn mower for $40 and spent another $35 on spare parts, before selling it for $125. Find his percentage profit.

5 A fur coat was purchased in 1992 and sold 25 years later for $1400, which represented a loss of 20% on its purchase price. Find the purchase price.

6 An antique dresser was sold for $640, which represents a profit of 25% on the cost price. What was the cost price?

1 Profit = 14% of 495 000

= 69 300

∴ Profit was $69 300

∴ Block sold for $495 000 + $69 300; that is, for $564 300.

2 a Profit = Selling price – Cost price

= 20 – 16

= 4

∴ The profit is $4.

b $\frac{\text{Profit}}{\text{Cost price}} \times 100\% = \frac{4}{16} \times 100\%$

$= 25\%$

∴ The profit is 25% of cost price.

c $\frac{\text{Profit}}{\text{Selling price}} \times 100\% = \frac{4}{20} \times 100\%$

$= 20\%$

∴ The profit is 20% of selling price.

3 Loss = Cost price − Selling price

$= 2.40 - 1.80$

$= 0.60$

∴ Loss is \$0.60

$\therefore \frac{\text{Loss}}{\text{Cost price}} \times 100\% = \frac{0.60}{2.40} \times 100\%$

$\left[\frac{60}{240} \times 100\right]$

∴ Percentage loss is 25%.

4 Cost = 40 + 35

$= 75$

∴ Total cost is \$75

∴ Profit = 125 − 75

$= 50$

∴ Profit is \$50

$\therefore \text{Percentage profit} = \frac{\text{Profit}}{\text{Cost price}} \times 100\%$

$= \frac{50}{75} \times 100\%$

$= 66.\dot{6}\%$

∴ The percentage profit is $66.\dot{6}\%$.

5 \$1400 is 20% less than purchase price

∴ 80% of purchase price = \$1400

[100% is purchase price.]

1% of purchase price = \$1400 ÷ 80

= \$17.5

100% of purchase price = \$17.5 × 100

= \$1750

∴ The purchase price of the coat was \$1750.

6 \$640 is 25% more than cost price

∴ 125% of cost price = 640

[100% is cost price.]

1% of cost price = 640 ÷ 125

= 5.12

100% of cost price = 5.12 × 100

= 512

∴ The dresser's cost price was \$512.

Let's spend time checking this:
Dresser's cost = \$512,
and increased in price by 25%
i.e. 25% of 512 = 0.25 × 512 = 128
∴ 512 + 128 = 640
That is, the sale price is \$640.

## Goods and Services Tax

The Goods and Services Tax (GST) is a tax applied to most goods sold and services provided in Australia. Currently, the GST is set at 10%.

### For Example

1 The pre-GST price of a computer is \$800. Find the GST-inclusive price.

2 A doctor charged \$77 for a consultation. What amount of GST is included in the price?

3 A pair of jeans is priced at \$132, including GST. Find the pre-GST price of the jeans.

1 GST = 10% of \$800

= 0.1 × \$800

= \$80

New price = \$800 + \$80

= \$880

∴ The GST-inclusive price is \$880.

2 Amount of GST = \$77 ÷ 11

=\$7

∴ The GST was \$7.

3 Amount of GST = \$132 ÷ 11

= \$12

∴ The GST was \$12.

Pre-GST price = \$132 − \$12

= \$120

∴ Pre-GST price was \$120.

Go to p. 284 for quick answers, or to pp. 305–309 for worked solutions

1 Express each fraction as a percentage: p. 26

a $\frac{2}{5}$ b $\frac{9}{20}$ c $\frac{13}{25}$
d $\frac{7}{40}$ e $\frac{41}{100}$ f $\frac{1}{3}$
g $\frac{5}{6}$ h $3\frac{1}{4}$

2 Convert the following to percentages, correct to one decimal place: p. 26

a $\frac{2}{7}$ b $\frac{5}{11}$ c $2\frac{4}{15}$

3 Rewrite the following decimals as percentages: p. 26

a 0.3 b 0.03 c 0.003
d 0.19 e 0.576 f 1.12
g 4.7 h 0.075

4 Express as fractions, whole numbers or mixed numerals: pp. 26–27

a 60% b 3% c 95%
d 127% e 300% f $5\frac{1}{2}$%
g $12\frac{1}{4}$% h $3\frac{4}{5}$%

5 Convert these percentages to decimals: p. 27

a 16% b 7% c 60%
d 7.6% e 12.5% f 176%
g 104.6% h $\frac{1}{4}$% i $7\frac{3}{4}$%
j $33\frac{1}{3}$%

6 Find: pp. 28–29

a 27% of 600 b 16% of \$140 c 60% of \$42
d 7% of 6 km e $4\frac{1}{2}$% of 8 L f $12\frac{1}{2}$% of 4 hours
g $\frac{1}{4}$% of 6 kg h 145% of \$300 i 35.2% of \$100 000
j $15\frac{7}{8}$% of 4000

7 Increase the following by the given percentage: pp. 29–30

a \$20 by 10% b 700 by 18% c 45 L by 30%
d 3 kg by 8%

8 Decrease the following by the given quantity: pp. 29–30

a 260 by 20% b \$65 by 5% c 3000 by $12\frac{1}{2}$%
d \$60 by 42%

9 One-hundred dollars is increased by 30% and the result decreased by 30%. What is the result? pp. 29–30

**10** What percentage is: p. 30

**a** 12 of 25? **b** 20 cents of $2?
**c** 4 cm of 5 m? **d** 250 mL of 2 L?
**e** 20 seconds of 4 minutes? **f** 45c of $9?

**11** Express the first amount as a percentage of the second amount: p. 30

**a** 4 cm, 40 cm **b** 3.5 g, 2 kg
**c** 400 cm, 2 m **d** 3 seconds, 3 minutes

**12** **a** If 20% of an amount is 700, find the amount. pp. 30–31
**b** If 9% of an amount is 45, find the amount.

**13** We know that 15% of an object weighs 525 g. How much will the whole object weigh? pp. 30–31

**14** A store had a sale offering 25% off all its merchandise. Find the discounted price of a product that was originally marked: pp. 31–32

| 25% OFF EVERYTHING |
|---|

**a** $95 **b** $38

**15** A plumber received a discount of $12\frac{1}{2}$% on a shovel that was priced at $22. pp. 31–32

**a** How much will the plumber save?
**b** What will the shovel cost the plumber?

**16** The price of a television set was reduced from $800 to $700 as part of a discount sale. What was the percentage discount? pp. 31–32

**17** A dress marked at $84 was reduced in price by 35%. What was the new price for the dress? pp. 31–32

**18** As part of its 'Runout Sale' Honest Joe's Car Yard dropped its prices by 15%. How much will be saved on a car marked at $16 999? pp. 31–32

**19** A comic was purchased for 80 cents and later sold for $1.40. Find the: pp. 32–33

**a** profit
**b** profit as a percentage of the cost price.

**20** Allyn purchased a piano for $3400 and later sold it for $1000. Find his: pp. 32–33

**a** loss
**b** loss as a percentage of the cost price, correct to one decimal place.

**21** John purchased a case of oranges for $4.80 and planned to make a 20% profit when he sold it. pp. 32–33

**a** How much will he make as profit?
**b** How much will he be selling the box of oranges for?
**c** If the box contains 60 oranges, what will the profit be for each orange?

**22** A lounge suite is purchased for $1200 and six years later is sold for $240. What is the percentage loss (calculated on the cost price)? pp. 32–33

**23** When Maria sold her house for $480 000 she had made a profit of 25% on the price she had paid when she purchased the house six years earlier. What price had she paid when she bought the house? pp. 32–33

**24** A washing machine is priced at $473. What is the amount of GST included in the price? p. 33

**25** A dentist charges $71.50 for a check-up and clean. How much GST is included? p. 33

**26** A handbag has a pre-GST price of $54. When GST is added, what is the new price? p. 33

**27** A set of headphones is priced at $35.20. What would be the price without GST included? p. 33

**28** A jewellery store has a watch priced at $264. Find the price without GST. p. 33

**29** A car is priced at $9500, which includes GST. How much GST is included in the price, to the nearest cent? p. 33

**30** The price of a pair of shoes is $69, which includes GST. Find the price without GST, to the nearest cent. p. 33

**31** Adam's basketball team played 20 games during the season. If the team won 12 of the games, what percentage of games did they lose if there were no draws? p. 30

**32** A group of 50 students in Year 8 were surveyed regarding the number of televisions in their homes. The results are recorded below: p. 30

| Number of Televisions | Number of Students |
|---|---|
| 0 | 1 |
| 1 | 14 |
| 2 | 18 |
| 3 | 7 |
| 4 | 6 |
| 5 | 4 |

What percentage of the students had:

**a** two televisions at home?
**b** more than four televisions at home?

**33** In a bag of 40 jelly beans, Kathy counts eight black, seven red, seven green, four pink, six white and the rest yellow. p. 30

**a** What percentage of the jelly beans are yellow?
**b** If Kathy decides to eat all the black jelly beans first, what percentage of the remaining jelly beans are yellow?

**34** Jacylyn is presently being paid $602 per week. If she were to receive a pay rise of 4%, what will be her new weekly pay? pp. 29–30

**35** The Cessnock to Coonabarabran car rally attracted a large field of 80 cars. By the time the cars arrived at their destination only 16 cars remained. What percentage of the original number failed to finish? p. 30

**36** The enrolment of Year 8 at a particular school is 198. If this represents 22% of the school's total enrolment, find the number of students at the school. pp. 30–31

**37** Ben, Ken and Len combined their savings to purchase the complete set of 1994 All-Stars basketball cards. Ben's contribution of $68 represents 40% of the total cost of the cards. pp. 30–31

**a** Find the cost of the complete set.

**b** Ken contributed 20% of the total cost. How much was his contribution?

**38** Glendon Brook had a population of 240 in 2000. By 2014 it had increased to 270. Find the population increase as a percentage. p. 30

**39** During the big drought of 2008, Trevor reduced the number of sheep on his property from 4500 to 1400. p. 30

**a** What was the percentage decrease, to two decimal places?

**b** By the end of 2010 the better season enabled Trevor to increase his number of sheep by 30% from their lowest level in 2008. How many sheep did he have by the end of 2010?

Go to p. 284 for quick answers, or to pp. 305–309 for worked solutions

## YOUR CHECKLIST

**For a complete understanding of this topic you must be able to:**

| | | |
|---|---|---|
| ✓ | Convert between fractions, decimals and percentages | pp. 26–27 |
| ✓ | Find the percentage of a quantity | p. 28 |
| ✓ | Increase and decrease a quantity by a percentage | pp. 29–30 |
| ✓ | Express one quantity as a percentage of another | p. 30 |
| ✓ | Use the unitary method approach | pp. 30–31 |
| ✓ | Find discounts | pp. 31–32 |
| ✓ | Express profit or loss as a percentage of cost/selling price | pp. 32–33 |
| ✓ | Find the Goods and Services Tax (GST). | p. 33 |

**Now you are ready to do the tests!**

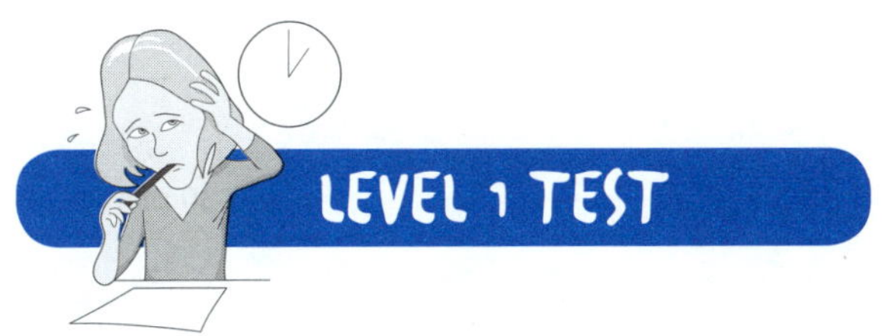

**(25 marks)**

1 Convert to a percentage:

a $\frac{3}{5}$ b $1\frac{1}{2}$ c 0.45

d 0.8 e 0.06 f 0.355 (6 marks)

2 Convert to a fraction:

a 27% b 8% c 40% (3 marks)

3 Convert to a decimal:

a 4% b $7\frac{1}{2}\%$ c 215% (3 marks)

4 What percentage is:

a 12 of 48? b \$3.20 of \$4? (2 marks)

5 Find:

a 15% of \$20 b 8% of \$600. (2 marks)

6 A hair straightener has a pre-GST price of \$88.

a Find the amount of GST.

b Find the selling price, including GST. (2 marks)

7 A motorbike is bought for \$2400 and sold again for \$3000. Find the:

a profit

b profit expressed as a percentage of cost price. (2 marks)

8 A tank is 60% full and currently holds 3600 L. What is the capacity of the tank when full? (2 marks)

9 A pair of jeans originally marked at \$95 is reduced in price by 15%. Find the new price. (1 mark)

10 A diamond ring increases in value by 20%. If it was previously worth \$6300, find the new value. (2 marks)

☞ **Quick answers on page 290**
☞ **Worked solutions on page 371**

**Your Feedback** $\frac{\square}{25} \times 100\% = \square\%$

**(30 marks)**

1 Rearrange in ascending order:

a $41\%, \frac{2}{5}, 0.42$ b $37\%, \frac{3}{8}, 0.4$ (2 marks)

2 Find:

a 4% of \$290 b 16% of \$360

c $8\frac{1}{2}\%$ of \$7000 d $5\frac{1}{4}\%$ of 4 metres. (4 marks)

3 What percentage is:

a 30c of \$3? b 5 mL of 2 L?

c 6 minutes of 2 hours? d 35 mg of 2 g? (4 marks)

4 Increase:

a \$25 by 10% b \$650 by 8%. (4 marks)

5 Decrease:

a \$650 by 8% b \$4000 by $9\frac{1}{2}\%$. (4 marks)

6 Find the discount on the following items if they originally cost:

a \$45 with a discount of 15%

b \$2480 with a discount of 20%. (4 marks)

7 A hardware store has a discount sale of 15% off all products. Find the new price of a:

a wheelbarrow marked at \$80

b bag of potting mix valued at \$18. (4 marks)

8 A rare book was purchased for \$1200 and later sold for \$1500. Find the profit expressed as a percentage of the cost price. (2 marks)

9 A shop owner bought 30 pairs of jeans for \$1950 and sold each pair for \$90 each. Find the profit as a percentage of the cost price. (2 marks)

☞ Quick answers on page 290
☞ Worked solutions on page 372

**Your Feedback** $\frac{\square}{30} \times 100\% = \square\%$

# LEVEL 3 TEST

**(25 marks)**

1 Rewrite as a percentage:

a $\frac{2}{3}$  b $\frac{7}{8}$ (2 marks)

2 Convert the following to decimals:

a $8\frac{1}{2}\%$  b 3.4%  c $12\frac{3}{4}\%$ (3 marks)

3 Find:

a $12\frac{1}{2}\%$ of \$70  b $\frac{4}{5}\%$ of \$50  c 115% of 300. (3 marks)

4 Increase 120 by 10% and then decrease the result by 10%. (2 marks)

5 An accountant charged \$120 for a consultation. The price included 10% GST. What is the cost of consultation without GST? Answer to the nearest dollar. (2 marks)

6 If a tank is 35% full and presently holds 2800 L, how much will the tank hold when $\frac{3}{4}$ full? (2 marks)

7 Jack sells a rare stamp for \$9600, making a profit of 20% on its cost price. How much had Jack bought the stamp for earlier? (3 marks)

8 Grant bought a phone for \$280. He was given a discount of 15% for being a previous customer. He then received a further discount of 5% for paying cash. What amount did Grant pay for the phone? (2 marks)

9 Liam bought a set of headphones for \$180 and sold them 18 months later to a friend for \$50. Find the loss as a percentage of the cost price. (3 marks)

10 The government plans to increase GST from 10% to 15%. The price of a bottle of perfume is currently \$75 (including GST). What will be the new price of the perfume? (3 marks)

☞ Quick answers on page 290
☞ Worked solutions on page 373

**Your Feedback** $\frac{\square}{25} \times 100\% = \square\%$

# ALGEBRA 4

- Revision of Operations with Pronumerals
- Substitution
- Like Terms
- Indices
- Removing Grouping Symbols
- Common Factors
- Generalised Arithmetic
- Generating Number Patterns

## KEYWORDS

**Algebra**
**Algebraic expression**
**Difference**
**Expanded form**
**Factorise**
**Factors**
**Generalised arithmetic**
**Grouping**
**Index notation**
**Indices**
**Power**
**Product**
**Pronumerals**
**Quotient**
**Simplify**
**Substitution**
**Sum**
**Terms**

# Revision of Operations with Pronumerals

The four operations of arithmetic (+, −, ×, ÷) have the same meaning in algebra as they have in arithmetic.

- $a + b$ means the **sum** of the numbers represented by the pronumerals $a$ and $b$.
- $a - b$ means the **difference** of the numbers represented by the pronumerals $a$ and $b$.
- $ab$ means the **product** ($a \times b$) of the numbers represented by the pronumerals $a$ and $b$.
- $\frac{a}{b}$ means the quotient $\left(\frac{a}{b} = a \div b\right)$ of the numbers represented by the pronumerals $a$ and $b$.

## Multiplying Pronumerals

When multiplying pronumerals, the multiplication sign, ×, is omitted between the letters.

For example: $m \times n$ is written $mn$
$3 \times a = 3a$
$2 \times b \times c = 2bc$
$5a \times 2b \times c = 10abc$

### For Example

Simplify the following algebraic expressions:

| | | | |
|---|---|---|---|
| **1** | $5 \times a$ | **2** | $m \times 3$ |
| **3** | $5 \times b \times 2$ | **4** | $3a \times 2b$ |
| **5** | $b \times b$ | **6** | $m \times 3n \times 2$ |
| **7** | $5 \times 2a \times 3a$ | | |

**1** $5 \times a = 5a$ — [Omit the multiplication sign. Always write the number first.]

**2** $m \times 3 = 3m$ — [Note the number is always written first.]

**3** $5 \times b \times 2 = 10b$ — [$5 \times 2 = 10$, $10 \times b = 10b$]

**4** $3a \times 2b = 6ab$ — [$3 \times 2 = 6$, $a \times b = ab$. Multiply the numbers first and then write the letters.]

**5** $b \times b = b^2$ — [$b \times b$ is not written as $bb$ but as $b^2$.]

**6** $m \times 3n \times 2 = 6mn$ — [$m \times n = mn$, $3 \times 2 = 6$. Multiply the numbers first then write the letters. Remember this expression could be written as $6nm$ since $mn = nm$.]

**7** $5 \times 2a \times 3a = 30a^2$ — [Remember: multiply numbers first.]

**Note:**

$2mn$ = $2 \times m \times n$

$2mn$ ↑ simplified form of the expression

$2 \times m \times n$ ↑ expanded form of the expression

## For Example

Write the following expressions in expanded form:

**1** $5a$ **2** $2abc$ **3** $2a^2$

**4** $ab^2c$ **5** $(3a)^2$

1 $5a = 5 \times a$

2 $2abc = 2 \times a \times b \times c$

3 $2a^2 = 2 \times a \times a$

4 $ab^2c = a \times b \times b \times c$

5 $(3a)^2 = 3a \times 3a = 3 \times a \times 3 \times a$

### Dividing Pronumerals

In some instances the question should be rewritten and then terms cancelled.

## For Example

Simplify:

**1** $15a \div 3$ **2** $6a \div 4a$

**3** $9ab^2c \div 3ab$ **4** $14ab \div 7a$

**5** $8a^2bc^2 \div 3abc^2$

1 $15a \div 3 = \frac{15a}{3}$ [Write in fractional form. Note: $a \div b = \frac{a}{b}$]

$= \frac{\cancel{15}^5 \times a}{\cancel{3}_1}$ [Write in expanded form then cancel common factors.]

$= \frac{5a}{1}$

$= 5a$ [Note: $\frac{5a}{1}$ is written as $5a$.]

2 $6a \div 4a = \frac{6a}{4a}$ [Write in fractional form.]

$= \frac{\cancel{6}^3 \times \cancel{a}^1}{\cancel{4}_2 \times \cancel{a}_1}$ [Write in expanded form.]

$= \frac{3}{2}$ [Cancel common factors.]

3 $9ab^2c \div 3ab = \frac{9ab^2c}{3ab}$

$= \frac{\cancel{9}^3 \times \cancel{a}^1 \times \cancel{b}^1 \times b \times c}{\cancel{3}_1 \times \cancel{a}_1 \times \cancel{b}_1}$

$= \frac{3bc}{1}$

$= 3bc$

4 $14ab \div 7a = \frac{14ab}{7a}$

$= \frac{\cancel{14}^2 \times \cancel{a}^1 \times b}{\cancel{7}_1 \times \cancel{a}_1}$

$= \frac{2b}{1}$

$= 2b$

5 $8a^2bc^2 \div 3abc^2 = \frac{8a^2bc^2}{3abc^2}$

$= \frac{8 \times \cancel{a}^1 \times a \times \cancel{b}^1 \times \cancel{c}^1 \times \cancel{c}^1}{3 \times \cancel{a}_1 \times \cancel{b}_1 \times \cancel{c}_1 \times \cancel{c}_1}$

$= \frac{8a}{3}$

# Substitution

### Substitution in Algebraic Expressions

In substitution we replace the pronumeral with its numerical value (which is given) and find the value of the resulting arithmetic expression.

## For Example

If $a = 2$, $b = 3$ and $c = -4$ evaluate:

**1** $ab$ **2** $a + b$

**3** $c + a$ **4** $\frac{c}{a}$

**5** $2b + c$ **6** $2c^2$

**7** $\frac{2b + 4}{a}$ **8** $abc$

**9** $a(b + a)$ **10** $a - c$

**11** $2c + b$

[Note: 'evaluate' means 'find the value' of the expression.]

**1** $ab = a \times b$
$= 2 \times 3$
$= 6$

**2** $a + b = 2 + 3$
$= 5$

**3** $c + a = -4 + 2$ [Remember, $c = -4$, $a = 2$.]
$= -2$

**4** $\frac{c}{a} = \frac{-4}{2}$ [$\frac{c}{a} = c \div a$]
$= -4 \div 2$
$= -2$ [$- \div + = -$]

**5** $2b + c = 2 \times b + c$ [Order of operations. Do multiplication first.]
$= 2 \times 3 + -4$
$= 6 - 4$ [Adding a negative is the same as subtracting.]
$= 2$

**6** $2c^2 = 2 \times c \times c$ [$2c^2 \neq (2c)^2$ but $2c^2 = 2 \times c \times c$]
$= 2 \times -4 \times -4$
$= 32$ [$- \times - = +$]

**7** $\frac{2b + 4}{a} = \frac{2 \times b + 4}{a}$
$= \frac{2 \times 3 + 4}{2}$
$= \frac{10}{2}$ [$\frac{10}{2}$ means $10 \div 2$]
$= 5$

**8** $abc = a \times b \times c$
$= 2 \times 3 \times -4$ [$+ \times - = -$]
$= -24$

**9** $a(b + a) = a \times (b + a)$ [Order of operations. Do brackets first.]
$= 2 \times (3 + 2)$
$= 2 \times 5$
$= 10$

**10** $a - c = 2 - -4$ [$2 - -4 = 2 + 4$]
$= 2 + 4$
$= 6$

**11** $2c + b = 2 \times c + b$ [Do multiplication first.]
$= 2 \times -4 + 3$
$= -8 + 3$
$= -5$

## Substitution into Formulae

When substituting, pronumerals are replaced with values.

### For Example

**1** If $M = 4n - 2$, find the value of $M$ when $n = 5$.

**1** $M = 4n - 2$
$= 4 \times n - 2$ [Replace $n$ with 5.]
$= 4 \times 5 - 2$
$= 20 - 2$
$= 18$

### For Example

**1** If $D = \frac{M}{V}$, find $D$ when $M = 3\frac{1}{4}$ and $V = 1\frac{1}{2}$.

**1** $D = \frac{M}{V}$ [Replace $M$ by $3\frac{1}{4}$ and $V$ by $1\frac{1}{2}$.]
$= \frac{3\frac{1}{4}}{1\frac{1}{2}}$
$= 3\frac{1}{4} \div 1\frac{1}{2}$
$= 2\frac{1}{6}$

Or by using the calculation → [$\frac{13}{4} \div \frac{3}{2} = \frac{13}{\cancel{4}_2} \times \frac{\cancel{2}^1}{3}$
$= \frac{13}{6}$
$= 2\frac{1}{6}$]

[*Calculator steps:*
3 [$a^b/_c$] 1 [$a^b/_c$] 4 [÷] 1 [$a^b/_c$] 1 [$a^b/_c$] 2 [=]]

1 If $A = \pi r^2$, find $A$ when $\pi = 3.14$ and $r = 10$.

1 $A = \pi r^2$
$= \pi \times r \times r$ [Write in expanded form.]
$= 3.14 \times 10 \times 10$ [Substitute $\pi = 3.14$ and $r = 10$.]
$= 314$

1 If $V = u + at$, find $V$ when $u = -12$, $a = 6$ and $t = \frac{1}{2}$.

1 $V = u + at$
$= -12 + 6 \times \frac{1}{2}$ [Do multiplication first.]
$= -12 + 3$
$= -9$

## Like Terms

Like terms are those terms of an expression that have the same pronumeral or pronumeral parts. For example: $2x$, $-5x$, $4x$ and $18x$ are like terms. $4x$, $3y$, $-5z$ are unlike terms. (Note: $xy$ is the same as $yx$.)

### Collecting Like Terms

When adding or subtracting pronumerals, only like terms can be added or subtracted.

Simplify the following:

1 $5x + 2x + x$
2 $5b + 2a + 3b + 5a$
3 $8y - y$
4 $2ab + 3ba$
5 $5x^2 + 3x - 2x^2 + x$
6 $4a - b - 3a + 5b$
7 $5b + 3a + 2b - 6a$
8 $2x - 6 + 3x - 2$
9 $2a - 6b - 3a - 2b$

Note: when considering terms of an expression, the sign preceding a term belongs to that term, for example:

$4a - b - 3a + 5b$ contains four terms $4a$, $-b$, $-3a$ and $+5b$

When moving any of the terms, the sign must move also:

$4a - b - 3a + 5b = 4a - 3a - b + 5b$

1 $5x + 2x + x = 8x$ [Note: $x = 1x$ $(5 + 2 + 1)x$]

2 $5b + 2a + 3b + 5a = 8b + 7a$

[This expression could be rearranged so that like terms are grouped together:
$5b + 2a + 3b + 5a = (5b + 3b) + (2a + 5a)$
$= 8b + 7a$]

3 $8y - y = 8y - 1y$ [Note: $y = 1y$]
$= 7y$

4 $2ab + 3ba = 5ab$

[Note:
$ab = ba$, therefore $2ab$ and $3ba$ are like terms.
The answer could also be written as $5ba$.]

5 $5x^2 + 3x - 2x^2 + x = (5x^2 - 2x^2) + (3x + x)$

[Grouping like terms.]

$= 3x^2 + 4x$

[Note: $x^2$ and $x$ are unlike terms.]

6 $4a - b - 3a + 5b = (4a - 3a) + (-b + 5b)$

[Grouping like terms.]

$= 1a + 4b$

$= a + 4b$

[Note: $-b + 5b = -1b + 5b = 4b$]

7 $5b + 3a + 2b - 6a = (5b + 2b) + (3a - 6a)$

[Grouping like terms.]

$= 7b + -3a$

$= 7b - 3a$

[When adding a negative, it is the same as subtracting.]

8 $2x - 6 + 3x - 2 = (2x + 3x) + (-6 - 2)$

[Grouping like terms.]

$= 5x + -8$

$= 5x - 8$

[Note: $-6 - 2 = -8$]

9 $2a - 6b - 3a - 2b = (2a - 3a) + (-6b - 2b)$

[Grouping like terms.]

$= -a + -8b$

$= -a - 8b$

[Note: $2a - 3a = -1a = -a$ $-6b - 2b = -8b$]

# Indices

## Index Notation

$a \times a$ is written as $a^2$
$a \times a \times a$ is written as $a^3$
$a \times a \times a \times a$ is written as $a^4$

Note:

$\underbrace{a \times a \times \ldots \times a}_{m \text{ times}} = a^m$

[It is called the $m$th power of $a$.]

Note that in the expression $a^m$ the $m$ is called the **index** or **power**. The plural of index is indices. Anything to the power of 0 is 1. This means $a^0 = 1$.

**For Example**

Evaluate:

**1** $4^3$ **2** $10^2$ **3** $7^0 + 2^0$

[Note: 'evaluate' means 'find the value of'.]

1 $4^3 = 4 \times 4 \times 4$
$= 64$

2 $10^2 = 10 \times 10$
$= 100$

3 $7^0 + 2^0 = 1 + 1$
$= 2$

**For Example**

Write the following expressions in index form:

**1** $2 \times 2 \times 2 \times 2 \times 2$

**2** $a \times a \times b \times b \times b$

**3** $5 \times a \times a \times a \times a \times b \times b \times b \times b \times b$

1 $2 \times 2 \times 2 \times 2 \times 2 = 2^5$

2 $a \times a \times b \times b \times b = a^2b^3$

3 $5 \times a \times a \times a \times a \times b \times b \times b \times b \times b = 5a^4b^5$

## Laws of Indices

- $a^m \times a^n = a^{m+n}$ [When multiplying, **add** indices.]
- $a^m \div a^n = a^{m-n}$ [When dividing, **subtract** indices.]
- $(a^m)^n = a^{mn}$
- $(ab)^m = a^m b^m$

For Example

Simplify:

| | | | |
|---|---|---|---|
| **1** | $a^6 \times a^2$ | **2** | $a^{10} \div a^2$ |
| **3** | $(b^3)^4$ | **4** | $(bd)^7$ |
| **5** | $(2a)^3$ | **6** | $(5x^7)^2$ |
| **7** | $4a^3 \times 2a$ | **8** | $\dfrac{x^5 \times x^4}{x^3}$ |
| **9** | $15y^5x^6 \div 3y^3x^2$ | **10** | $2^9 \times 2^5$ |
| **11** | $3^{12} \div 3^4$ | | |

1 $a^6 \times a^2 = a^8$

[When multiplying, add indices: $6 + 2 = 8$]

2 $a^{10} \div a^2 = a^8$

[When dividing, subtract indices: $10 - 2 = 8$]

3 $(b^3)^4 = b^{12}$ [$3 \times 4 = 12$]

4 $(bd)^7 = b^7d^7$

5 $(2a)^3 = 2^3 \times a^3$
$= 8a^3$

6 $(5x^7) = 5^2 \times (x^7)^2$
$= 25x^{14}$

7 $4a^3 \times 2a = (4 \times 2)a^{3+1}$
$= 8a^4$
[Multiply numbers first. $a = a^1$]

8 $\dfrac{x^5 \times x^4}{x^3} = \dfrac{x^9}{x^3}$
$= x^6$
[$\dfrac{x^9}{x^3}$ means $x^9 \div x^3 = x^6$]
[When dividing, subtract indices.]

9 $15y^5x^6 \div 3y^3x^2 = 5y^2x^4$ [Divide numbers first. Subtract indices.]

10 $2^9 \times 2^5 = 2^{14}$ [Add indices.]

11 $3^{12} \div 3^4 = 3^8$ [Subtract indices.]

## Removing Grouping Symbols

When removing brackets (expanding), multiply each term **inside** the brackets by the term **outside**.

For Example

Expand:

| | | | |
|---|---|---|---|
| **1** | $5(a + 3)$ | **2** | $4(2x - 3)$ |
| **3** | $3(2a - 5b)$ | **4** | $b(3b - a)$ |
| **5** | $-3(x - 2)$ | **6** | $-(x - 5)$ |

1 $5(a + 3) = (5 \times a) + (5 \times 3)$
$= 5a + 15$

2 $4(2x - 3) = (4 \times 2x) - (4 \times 3)$
$= 8x - 12$

3 $3(2a - 5b) = 6a - 15b$

4 $b(3b - a) = 3b^2 - ba$

5 $-3(x - 2) = -3x + 6$ [$-3 \times -2 = +6$]

6 $-(x - 5) = -1(x - 5)$
$= -x + 5$
$= 5 - x$ [$-x + 5 = 5 - x$]

Expand and simplify:

1 $2(x + 3) + 4x$

2 $4 + 2(a + 3)$

3 $3(x + 4) + 2(x - 3)$

4 $4(2a - 3) - 3(2 - a)$

5 $5 - 4(x - 1)$

6 $4(2x - 3) + 3(x - 1)$

1 $2(x + 3) + 4x = 2x + 6 + 4x$

[Remove brackets first. Collect like terms.]

$= 6x + 6$

2 $4 + 2(a + 3) = 4 + 2a + 6$

$= 10 + 2a$

3 $3(x + 4) + 2(x - 3) = 3x + 12 + 2x - 6$

$= 5x + 6$

4 $4(2a - 3) - 3(2 - a) = 8a - 12 - 6 + 3a$

[Note that second expansion is:
$-3(2 - a) = -3 \times 2 - 3 \times -a$
$= -6 + 3a$]

$= 11a - 18$

[Note: $-3 \times -a = +3a$]

5 $5 - 4(x - 1) = 5 - 4x + 4$

$= 5 + 4 - 4x$

$= 9 - 4x$

6 $4(2x - 3) + 3(x - 1) = 8x - 12 + 3x - 3$

$= 11x - 15$

## Common Factors

Factorising is the process by which an expression in an expanded form is changed to a factored form.

Factorise:

1 $6a + 12$

2 $3m - 3n$

3 $ab + ac$

4 $2x^2 + 4x$

5 $9a - 3a^2$

1 $6a + 12 = 6(a + 2)$

[Common factor was 6. Check by expansion.]

2 $3m - 3n = 3(m - n)$

[Common factor was 3.]

3 $ab + ac = a(b + c)$

4 $2x^2 + 4x = 2x(x + 2)$

[Take both 2 and $x$ as common factors; that is, $2 \times x = 2x$.]

5 $9a - 3a^2 = 3a(3 - a)$

[Highest common factor is $3a$.]

## Generalised Arithmetic

Note that **sum** means add. **Difference** means subtract. **Product** means multiply. **Quotient** means divide.

1 Find the product of $a$ and $b$.

2 Find the sum of $x$ and 4.

3 Find the number 8 less than $2a$.

4 Eleni had \$$a$. She spends \$10. Write an expression for the amount of money she has left.

5 Convert $y$ metres to centimetres.

6 Find the average of $2a$, $b$ and $c$.

7 If $b$ is an even number, find the next two consecutive even numbers.

8 Write expressions for the area and perimeter of the rectangle:

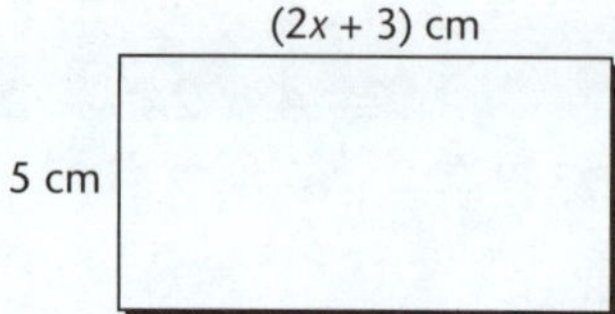


9 Find an expression for the volume of the rectangular prism:

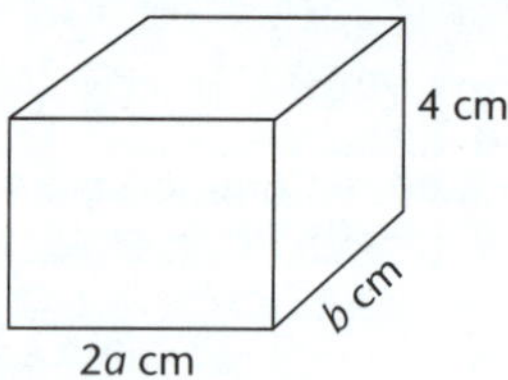


1 $a \times b = ab$ [Product means multiply.]

2 $x + 4$ [Sum means add. $x$ and 4 are unlike terms; therefore, $x + 4$ cannot be simplified any further.]

3 $2a - 8$

4 Amount left $= \$a - \$10$
$= \$(a - 10)$

5 $y$ metres $= 100 \times y$ centimetres
$= 100y$ centimetres

[1 m = 100 cm
2 m = $2 \times 100$ cm
$y$ m = $y \times 100$ cm]

6 Average of $2a$, $b$ and $c = \dfrac{2a + b + c}{3}$

[Average $= \dfrac{\text{Sum of scores}}{\text{Number of scores}}$]

7 $b + 2$, $b + 4$ [Note: all **even** numbers and **odd** numbers are separated by 2.]

8 $A = 5 \times (2x + 3)$
$= 5(2x + 3)$
$= 10x + 15$
$\therefore$ Area is $(10x + 15)\ \text{cm}^2$
$P = 2x + 3 + 5 + 2x + 3 + 5$
$= 4x + 16$
$\therefore$ Perimeter is $(4x + 16)$ cm

9 $V = lbh$
$= 2a \times b \times 4 = 8ab$
$\therefore$ Volume is $8ab\ \text{cm}^3$

## Generating Number Patterns

Algebraic expressions can be used to describe number patterns. To generate a number pattern we substitute different values for the pronumeral used.

### For Example

1 If $y = x + 2$, substitute the values 1, 2, 3, 4, 5, 6 for $x$.

2 Complete the table for the rule $y = 3x - 2$.

| $x$ | −2 | −1 | 0 | 1 | 2 | 3 |
|---|---|---|---|---|---|---|
| $y$ | | | | | | |

1

| $x$ | 1 | 2 | 3 | 4 | 5 | 6 |
|---|---|---|---|---|---|---|
| $y$ | 3 | 4 | 5 | 6 | 7 | 8 |

2

| $x$ | −2 | −1 | 0 | 1 | 2 | 3 |
|---|---|---|---|---|---|---|
| $y$ | −8 | −5 | −2 | 1 | 4 | 7 |

Go to p. 284 for quick answers, or to pp. 309–315 for worked solutions

**1** Write the following expressions in simplest form: p. 43

**a** $2 \times b \times c$ **b** $3a \times 4$ **c** $a \times b$
**d** $b \times 3$ **e** $a \times a$ **f** $5 \times t \times 2$
**g** $3 \times b \times b$ **h** $2a \times 3a$ **i** $5 \times a \times 2a \times 3$
**j** $3a \times a$ **k** $7a \times 3b$ **l** $2a \times a \times a$
**m** $3y \times 4xy$ **n** $4x \times 3 \times y$ **o** $(2x)^2$

**2** Write the following algebraic expressions in expanded form: p. 44

**a** $ab$ **b** $2xy$ **c** $3a^2$
**d** $5abc$ **e** $2a^2bc^2$ **f** $(5x)^2$

**3** If $a = 5$, $b = 2$, $c = 3$, evaluate: pp. 44–45

**a** $bc$ **b** $3a$ **c** $a + b$
**d** $c - b + a$ **e** $b^2$ **f** $3b - c$
**g** $abc$ **h** $a^2 - c^2$ **i** $\sqrt{5a}$
**j** $a(b + c)$ **k** $4c - a + b$ **l** $2a^2$
**m** $3b^2$ **n** $(2a)^2$ **o** $\frac{4c}{6}$
**p** $\frac{abc}{a + b + c}$ **q** $bc - b^2$ **r** $\frac{3a^2}{c}$
**s** $\sqrt{a^2 - c^2}$

**4** If $a = 0.72$, $b = 0.6$, evaluate: pp. 44–45

**a** $a + b$ **b** $\frac{a}{b}$ **c** $ab$
**d** $5b$ **e** $4a - b$ **f** $a^2$
**g** $3b - 2a$ **h** $2b^2$

**5** If $m = \frac{4}{5}$ and $n = \frac{2}{3}$, evaluate: pp. 44–45

**a** $m + n$ **b** $\frac{m}{n}$ **c** $mn$
**d** $3n$ **e** $2m$ **f** $m^2$
**g** $3n^2$ **h** $5 - mn$ **i** $2m + 3n$
**j** $\frac{m + n}{m - n}$

**6** If $x = -4$, $y = 3$ and $z = 2$, evaluate: pp. 44–45

**a** $xy$ **b** $x^2$ **c** $3x + z$
**d** $x - z$ **e** $y - x$ **f** $y + 2x$
**g** $\frac{x}{z}$ **h** $x(y + z)$ **i** $(3x)^2$
**j** $-y - z - x$ **k** $y^2 - x^2$

7 **a** If $V = lbh$, find $V$ when $l = 4$, $b = 3$ and $h = \frac{1}{2}$. pp. 45–46
**b** If $T = 2R^2$, find $T$ when $R = -5$.
**c** If $V = u + at$, find $V$ when $u = -4$, $a = -2$ and $t = 3$.
**d** If $A = \frac{22}{7}R^2$, find $A$ when $R = 7$.
**e** If $A = 4xy + x^2$, find $A$ when $x = 4$, $y = 3$.
**f** If $D = \dfrac{M}{V}$, find $D$ when $M = 2\frac{3}{4}$, $V = 1\frac{1}{2}$.
**g** If $T = \dfrac{10}{A + B}$, find $T$ when $A = 4$ and $B = 11$.
**h** If $P = 2(l + b)$, find $P$ when $l = 1.72$ and $b = 0.8$.
**i** If $d = \sqrt{a^2 - b^2}$, find $d$ when $a = -5$, $b = 4$.
**j** If $N = \dfrac{4B}{A + C}$, find $N$ when $B = \frac{1}{2}$, $A = \frac{1}{3}$, $C = \frac{1}{4}$.

**8** Simplify the following by collecting like terms: pp. 46–47
**a** $6a + 4a - 3a$
**b** $7a - a$
**c** $3a + 4b + 2a - 2b$
**d** $2a + a$
**e** $7b - 2 + 3b$
**f** $8x - 4 - 7x$
**g** $3a^2 + 4a + 2a^2 - 2a$
**h** $a + 6b - a + 2b$
**i** $7p + 3 - 4p - 7$
**j** $7x - 3x - 2x + y$
**k** $5ab - 2ba$
**l** $m^2 + m + 3m^2 + 4m$
**m** $5b + 6a - 2b + a$

**9** Simplify: p. 44
**a** $15b \div 3$
**b** $15a \div 3a$
**c** $12ab \div 4a$
**d** $20ab \div 4ac$
**e** $14xy^2t \div 7xyt$
**f** $12a^2b \div 8ab$
**g** $10mn^2 \div 15m^2n$
**h** $35x^8y^4 \div 5x^2y^3$
**i** $3ab \div 7a$
**j** $6a \div 3a$

**10** Simplify: pp. 47–48
**a** $a^5 \times a^3$
**b** $a^2 \times a^6$
**c** $a \times a^9$
**d** $a^{12} \times a^2$
**e** $a^{15} \times a^2$
**f** $a^{16} \div a^4$
**g** $a^{15} \div a^5$
**h** $a^{12} \div a^3$
**i** $(a^4)^3$
**j** $(a^7)^4$
**k** $(a^5)^2$
**l** $(ab)^3$
**m** $(3x)^2$
**n** $(4ab)^2$
**o** $(4a)^3$
**p** $(5a^4)^2$
**q** $(7x)^2$
**r** $(a^4b^2)^3$
**s** $y \times y^5 \times y^2$
**t** $\dfrac{a^5 \times a^6}{a^3}$
**u** $15a^6b^3 \div 5a^3b^2$
**v** $12x^2y \div 4xy$
**w** $\dfrac{4 \times a^2 \times a^4}{a \times 2 \times a^3}$
**x** $(2x)^3 \div x^2$
**y** $\dfrac{15a^4b^6}{5a^2b^2}$
**z** $\dfrac{(3a^4)^2}{3a^4}$
**aa** $2^7 \times 2^3$
**bb** $5 \times 5^4$
**cc** $10^{14} \div 10^7$

**11** Expand: pp. 48–49

**a** $5(a + 3)$ **b** $4(b - 2)$ **c** $3(a - b)$
**d** $4(2x - 3)$ **e** $a(b - 2)$ **f** $-4(x + 2)$
**g** $-3(x - 2)$ **h** $4(2a - 3b)$ **i** $-2(2a - 3b)$
**j** $a(2a - b)$ **k** $-4(a - 3)$ **l** $-(a - 2)$
**m** $-(2 - 3a)$ **n** $3(2 - 4a)$ **o** $-3(2 + a)$

**12** Simplify by collecting like terms: p. 46

**a** $7a - 2a - 3a - 6a$ **b** $-8t + 2t$
**c** $-3a - 4a$ **d** $-2a - 2a$
**e** $-5a + 5a$ **f** $2x - 7y + x + 4y$
**g** $7x - 3y - x - y$ **h** $4a^2 + 6a - 3a^2 - 10a$
**i** $6x - 5 - 2x - 4$ **j** $5 - x^2 - 3$
**k** $2x - 8 - 4x + 3$ **l** $8x - 12 - 6 + 3x$
**m** $8b - 12 + 3b + 6$ **n** $a - 7b + a - 3b$
**o** $a^2 - 6b - a^2 + 2b$ **p** $a - 3b + a - b$

**13** Expand and simplify: pp. 48–49

**a** $5(x + 2) + 3x$ **b** $4(a + b) + 3b$
**c** $2 + 3(a + 4)$ **d** $5(a + 3) + 2(a + 1)$
**e** $4(x + 3) + 2(x - 1)$ **f** $3(x - 2) + 2(x - 1)$
**g** $7 - 2(x - 3)$ **h** $4 - (x - 2)$
**i** $3(x - 2) - 4(x - 1)$ **j** $4x - 3(2 - 3x)$
**k** $5(2x - 3) - 2(x - 2)$ **l** $4(1 - x) - 2(3 - x)$
**m** $3 - 2(1 - a)$ **n** $a(a + 2) + 3(a + 1)$
**o** $2a(a - 4) - a(a - 2)$ **p** $2b(3b - 1) - b(b + 2)$
**q** $5p(2p - 1) - 3(p - 2)$

**14** Simplify: pp. 43–44

**a** $-4 \times 2a$ **b** $-3a \times 2b$
**c** $-2a \times -3b$ **d** $4 \times -b \times -c$
**e** $-2x \times x$ **f** $5b \times -3b$
**g** $-12a \div 3$ **h** $-12b \div -4$
**i** $5 \times -2p \times -p$ **j** $-8x^2y \div 4xy$
**k** $-12b \div -b$ **l** $4t \div -2t$

**15** Simplify: pp. 43–44

**a** $6a - 4 + 3a + 2$ **b** $2t \times -4$
**c** $18a \div 3$ **d** $25a \div 5a$
**e** $12p^2 \div 6p$ **f** $-2a \times -3a \times b$
**g** $a + a + a$ **h** $a \times a \times a$
**i** $5a - a$ **j** $(4x)^2$
**k** $(2a^4)^3$ **l** $5a - 12a$
**m** $5a \times -4a$ **n** $(-2ab)^2$
**o** $5x^2 - 6x + x^2 + x$ **p** $18x^2y \div 3xy$
**q** $2x^4y \div 3x^3y^2$

**16** Factorise: p. 49

a $4a + 20$
b $5a + 5b$
c $ab - ac$
d $6a + 4b$
e $3b^2 - 6$
f $5x^2 + 25x$
g $4a^2 - 2a$
h $5x^2 + 10y + 20$
i $12 - 6a$
j $5x^2 - 6x$
k $35x^2y - 7xy$
l $4ab - 2b$
m $8a - 4a^2$
n $12b - 8b^2$

**17** Find the: pp. 49–50

a sum of $2a$ and $c$
b product of $m$ and $n$
c average of $5x$, $7x$ and $3x$
d number $a$ more than $2b$
e number $2x$ less than $5x + 8$
f cost of $y$ items at 12 cents per item
g change from $\$x$ if \$6 was spent
h next three consecutive whole numbers after $a$
i next two consecutive even numbers after $a$, if $a$ is even
j area of a square with length $3b$ cm.

**18** a Convert: pp. 49–50

i $b$ metres into centimetres
ii $a$ minutes to seconds
iii $2a$ litres to millilitres
iv $3a$ dollars to cents
v $120t$ seconds to minutes.

b Find the perimeter of the following (all measurements in cm):

i
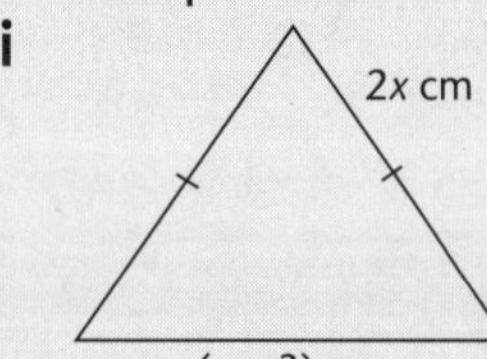


ii
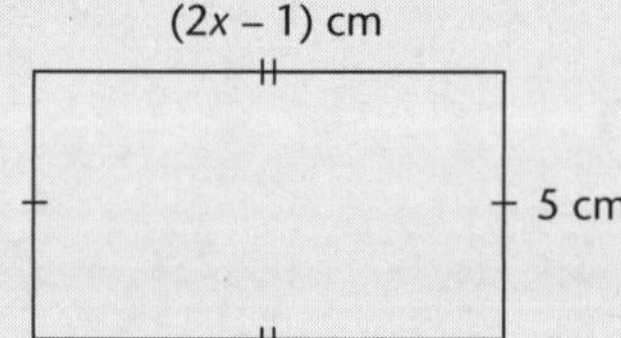


iii
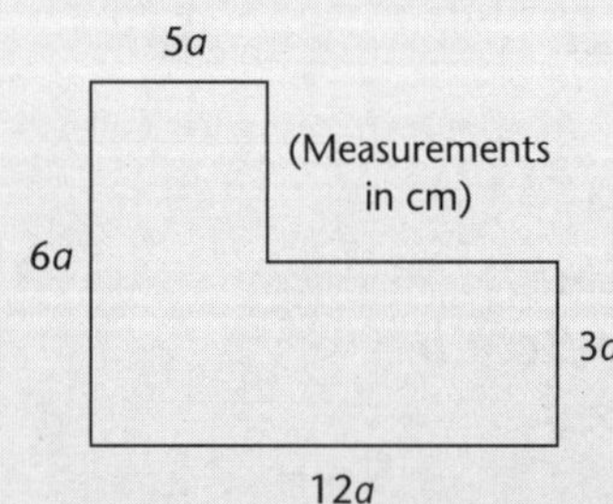


c Find the area of the following (all measurements in cm):

i
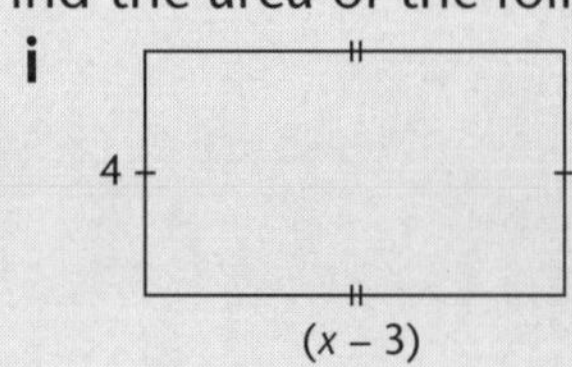


ii
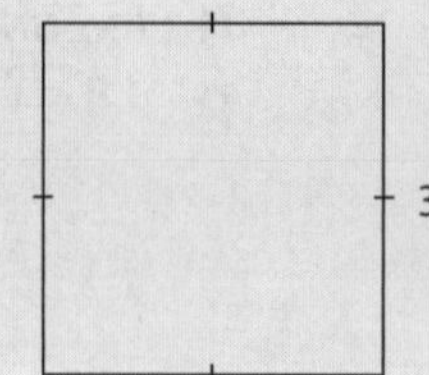


iii
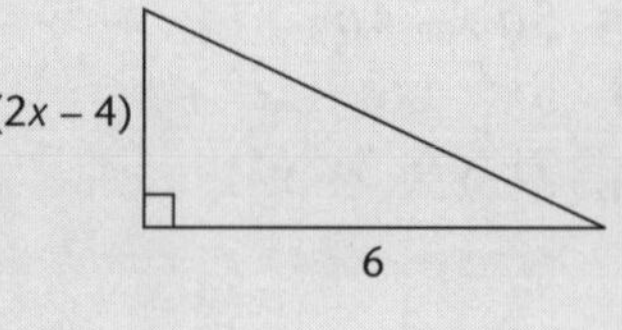

pp. 43–49

**19** **a** Increase $2a$ by 3.
**b** Expand $4(2a - 3)$.
**c** Simplify $15ab \div 3a$.
**d** Simplify $3c^2 - 2c + 5c^2$.
**e** Simplify $3y^4 \times 2y^3$.
**f** If $d = -4$, $d^2 = ?$
**g** Simplify $-3a \times -2a$.
**h** Simplify $4m - 8n + 3m + 2n$.
**i** If $a = -4$, find the value of $5 - a$.
**j** If $a = -2$, $b = 7$, find the value of $4a - b$.
**k** Expand and simplify $3(2x - 5) + 2(x + 3)$.
**l** If $y = -4$, evaluate $2y^2 - (2y)^2$.
**m** Given $D = \frac{M}{V}$, find $D$ if $M = -42$ and $V = 3$.
**n** If $M = \frac{x}{y}$ and $x = \frac{3}{4}$, $y = \frac{5}{8}$, find the value of $\frac{1}{M}$.
**o** Simplify $2ab^2 \times 3a^2$.
**p** Simplify $3ab + 2a - 4ab + 6a$.
**q** Evaluate $a^2 - b$ if $a = 2$ and $b = -4$.
**r** Simplify $3 - 2(a - 4)$.
**s** Factorise $8b - 2a^2$.
**t** Simplify $12x^4y^6 \div 6y^3x$.
**u** Factorise fully $4x^2 + 16x$.
**v** Expand and simplify $4x(x - 2) - x(3 - x)$.
**w** If $x$ is an odd number, find the next two consecutive odd numbers.
**x** If $a = 0.4$, evaluate $2a^2$.
**y**

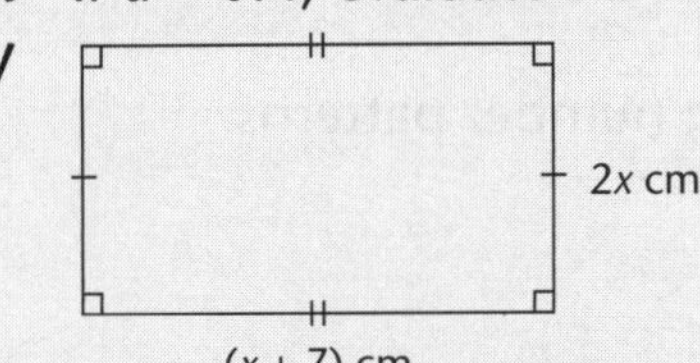


Write an expression for the perimeter and area of the rectangle.

p. 50

**20** Complete the following tables:

**a** $y = x + 3$

| $x$ | 1 | 2 | 3 | 4 | 10 | 100 |
|---|---|---|---|---|---|---|
| $y$ | | | | | | |

**b** $y = 2x - 3$

| $x$ | 2 | 3 | 4 | 10 | 40 | 100 |
|---|---|---|---|---|---|---|
| $y$ | | | | | | |

**c** $y = 3x + 1$

| $x$ | −2 | −1 | 0 | 1 | 20 | 50 |
|---|---|---|---|---|---|---|
| $y$ | | | | | | |

**d** $y = 5 - 2x$

| $x$ | −3 | −2 | −1 | 0 | 1 | 2 |
|---|---|---|---|---|---|---|
| $y$ | | | | | | |

**e** $y = x^2 + 1$

| $x$ | −3 | −2 | −1 | 0 | 1 | 2 |
|---|---|---|---|---|---|---|
| $y$ | | | | | | |

**f** $T = -6 + 3d$

| $d$ | 1 | 2 | 3 | 4 | 10 | 100 |
|---|---|---|---|---|---|---|
| $T$ | | | | | | |

**g** $p = 2q - 1$

| $q$ | 0 | $\frac{1}{4}$ | $\frac{1}{2}$ | 1 | $1\frac{1}{4}$ | $1\frac{1}{2}$ |
|---|---|---|---|---|---|---|
| $p$ | | | | | | |

**h** $S = 5t - 6$

| $t$ | 0 | 0.2 | 0.4 | 0.6 | 0.8 | 1 |
|---|---|---|---|---|---|---|
| $S$ | | | | | | |

**21** Find the relationship between $x$ and $y$ in the following number patterns: p. 50

**a**

| $x$ | 0 | 1 | 2 | 3 | 4 | 5 |
|---|---|---|---|---|---|---|
| $y$ | 0 | 4 | 8 | 12 | 16 | 20 |

**b**

| $x$ | 0 | 1 | 2 | 3 | 4 | 5 |
|---|---|---|---|---|---|---|
| $y$ | 1 | 3 | 5 | 7 | 9 | 11 |

**c**

| $x$ | 0 | 1 | 2 | 3 | 4 | 5 |
|---|---|---|---|---|---|---|
| $y$ | 1 | 4 | 7 | 10 | 13 | 16 |

**d**

| $x$ | 0 | 1 | 2 | 3 | 4 | 5 |
|---|---|---|---|---|---|---|
| $y$ | 4 | 3 | 2 | 1 | 0 | −1 |

**e**

| $x$ | –2 | –1 | 0 | 1 | 2 | 3 |
|---|---|---|---|---|---|---|
| $y$ | –3 | –1 | 1 | 3 | 5 | 7 |

**f**

| $x$ | 0 | 1 | 2 | 3 | 4 | 5 |
|---|---|---|---|---|---|---|
| $y$ | 0 | 1 | 4 | 9 | 16 | 25 |

**g**

| $x$ | –2 | –1 | 0 | 1 | 2 | 3 |
|---|---|---|---|---|---|---|
| $y$ | 5 | 2 | 1 | 2 | 5 | 10 |

**h**

| $x$ | 0 | 1 | 2 | 3 | 4 | 5 |
|---|---|---|---|---|---|---|
| $y$ | 3 | 1 | –1 | –3 | –5 | –7 |

Go to p. 284 for quick answers, or to pp. 309–315 for worked solutions

## YOUR CHECKLIST

**For a complete understanding of this topic you must be able to:**

| | | | |
|---|---|---|---|
| ✓ | Simplify algebraic expressions that involve multiplication and division | | pp. 43–44 |
| ✓ | Substitute in algebraic expressions and formulae | | pp. 44–45 |
| ✓ | Simplify algebraic expressions by collecting like terms | | pp. 46–47 |
| ✓ | Use the laws of indices to simplify algebraic expressions | | pp. 47–48 |
| ✓ | Expand algebraic expressions by removing grouping symbols | | pp. 48–49 |
| ✓ | Factorise algebraic expressions | | p. 49 |
| ✓ | Translate from everyday language to algebraic language and vice versa | | pp. 49–50 |
| ✓ | Generate number patterns from an algebraic expression. | | p. 50 |

**Now you are ready to do the tests!**

**(50 marks)**

**1** Simplify:

a $4 \times x$ b $n \times 2$ c $3 \times y \times 4$
d $2 \times 3a$ e $a \times b$ f $a \times a$
g $m \times 2n \times 4$ h $3 \times b \times 2 \times c$ i $2a \times a$
j $5a \times 2a$ (10 marks)

**2** Write the following expressions in expanded form:

a $4a$ b $2ab$ c $3a^2$ (3 marks)

**3** Simplify:

a $12a \div 4$ b $20t \div 10$ c $4t \div 2$ d $5t \div t$ (4 marks)

**4** If $a = 4$, $b = 2$ find the value of:

a $a + b$ b $a - b$ c $ab$
d $3a$ e $b^2$ (5 marks)

**5** Simplify:

a $5x + 3x$ b $7x - 2x$ c $6y - 3y + 2y$ (3 marks)

**6** Expand:

a $2(a + 3)$ b $5(x - 2)$ c $3(2a - 4)$ (3 marks)

**7** Write algebraic expressions for the:

a product of 3 and $b$
b sum of $a$ and $b$
c number $x$ less 2. (3 marks)

**8** Complete the following table:

$y = x + 2$

| $x$ | 0 | 1 | 2 | 3 | 4 | 5 |
|---|---|---|---|---|---|---|
| $y$ | | | | | | |

(2 marks)

*(cont.)*

9 Factorise the following:

a $2b + 6$ b $3x + 6y + 12$ c $12x + 8$ (3 marks)

10 Simplify:

a $4a + 3b + 2a$ b $4x + 3y + 2x + 2y$ c $5x + y - x + 3y$ (6 marks)

11 a If $a = 4$ and $b = 2$, find the value of $a^2 - b$.

b If $M = 3t - 1$, find $M$ when $t = 3$. (4 marks)

12 Expand and simplify:

a $2(x + 3) + 4x$ b $3(x + 1) + 2(x + 3)$ (4 marks)

☞ Quick answers on page 290
☞ Worked solutions on page 374

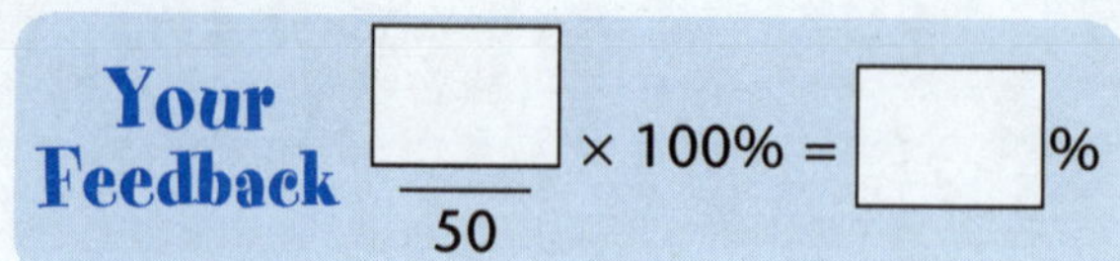

(50 marks)

## Part A

1 Simplify $5x - 3 + 2x$. (1 mark)

2 Write $3ab^2$ in expanded form. (1 mark)

3 The product of $2m$ and $n$ is? (1 mark)

4 If $x = 5$, $4x - 2 = ?$ (1 mark)

5 If $a = 4$, $2a^2 = ?$ (1 mark)

6 If $x - 7 = 10$, find the value of $x$. (1 mark)

7 $-2 \times a \times b \times 3 = ?$ (1 mark)

8 Evaluate $4^0 + 2^0 - 3^0$. (1 mark)

9 If $x = 4$, $y = 3$, evaluate $(xy)^0$. (1 mark)

10 Factorise $3a - 12b$. (1 mark)

11 Expand $-3(x + 2)$. (1 mark)

12 Simplify $2x - 6x$. (1 mark)

13 If $F = ma$ and $m = 10$, $a = 3$, find the value of $F$. (1 mark)

14 If $a = 5$, the value of $(2a)^2 = ?$ (1 mark)

15 Simplify $3a \times 5a$. (1 mark)

## Part B

1 Simplify:

a $2b \times 3a$ b $4 \times b \times 3b$ c $-3 \times 2y$ (3 marks)

2 Simplify:

a $5x - 2 + 3x$ b $6a + 2b + 3a + b$ c $9x + 4y - x + 2y$ (3 marks)

3 Simplify:

a $8b \div 4$ b $6ab \div 2b$ c $\frac{8x}{6y}$ (3 marks)

*(cont.)*

4 a

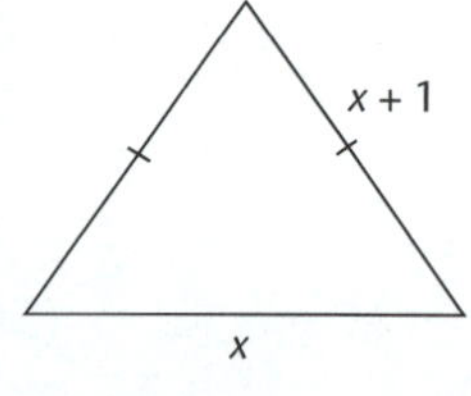


Perimeter =

b

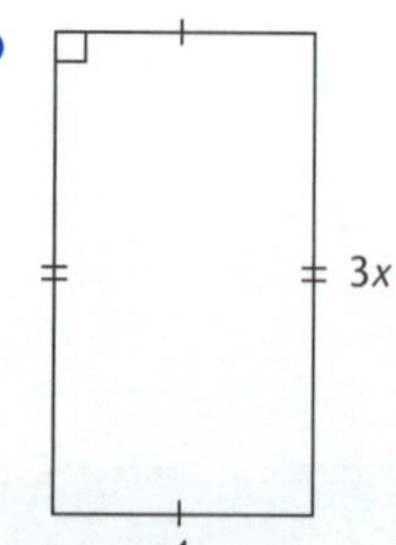


Perimeter =

c

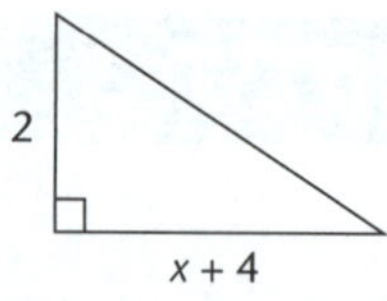


Area =

(3 marks)

5 a If $a = 14$ and $b = 2$, find the value of $\frac{a}{b}$.

b If $x = 5$ and $y = -2$, find the value of $y^2 + x$.

c If $k = \frac{3b - a}{c}$, find the value of $k$ when $b = 7$, $a = 5$ and $c = 4$. (4 marks)

6 a Paul had \$$a$ and spent \$$b$. How many dollars does Paul have left?

b Write an algebraic expression for the perimeter of the following rectangle:

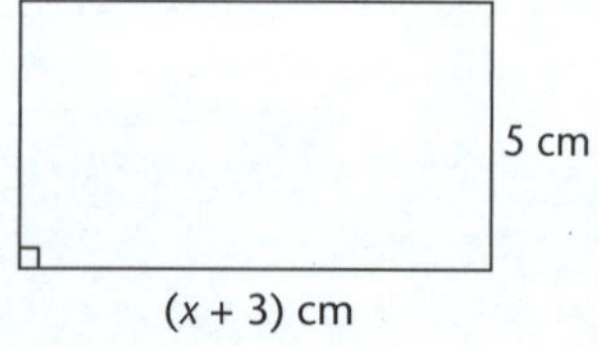


(3 marks)

7 Factorise:

a $6a - 4$ b $a^2 - 4a$ c $12xy - 4x$ (3 marks)

8 Expand:

a $x(x + 7)$ b $4a(3 - a)$ c $-(2y - 7)$ (3 marks)

9 Expand and simplify:

a $5(x + 2) - 3$ b $5(x + 1) + 2(x - 1)$ c $2x + 3(x + 5)$ (6 marks)

10 Complete the following table:

$y = 3x - 2$

| $x$ | –1 | 0 | 1 | 2 |
|---|---|---|---|---|
| $y$ | | | | |

(2 marks)

11 Study the pattern below and write a formula that relates $x$ and $y$:

| $x$ | 0 | 1 | 2 | 3 |
|---|---|---|---|---|
| $y$ | 2 | 3 | 4 | 5 |

(2 marks)

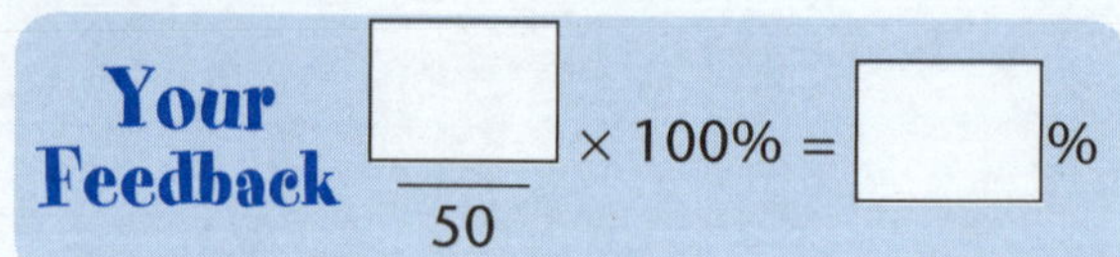


☞ **Quick answers on page 291**
☞ **Worked solutions on page 374**

**(50 marks)**

## Part A

1 Simplify $-4a \times -2a$. (1 mark)

2 Increase $3a$ by 5. (1 mark)

3 If $d = -5$, find the value of $2d^2 - 3$. (1 mark)

4 Simplify $3x - 4y + x - y$. (1 mark)

5 If $x = -3$ and $y = 6$, find the value of $y - x$. (1 mark)

6 Simplify $\frac{x^2 \times x^3}{x^4}$ (1 mark)

7 Simplify $16ab \div 4b$. (1 mark)

8 Expand $3x(2 - x)$. (1 mark)

9 Factorise $5t^2 - 30t$. (1 mark)

10 Given $a = \frac{F}{m}$, find $a$ when $F = 3.2$ and $m = 0.2$. (1 mark)

11 If $x$ is an even number, find the next two consecutive odd numbers. (1 mark)

12 Simplify $(5x)^2$. (1 mark)

13 Simplify $8x \div 6y^2$. (1 mark)

14 Evaluate $\frac{1}{a}$ if $a = 1\frac{1}{4}$. (1 mark)

15 Simplify: $(2y)^3 \div 4y$ (1 mark)

## Part B

1 Simplify:

a $2a \times (-a) - 3a^2 \times (-2)$

b $2 - (a - 1)$

c $15xy^2 \div 3xy$ (6 marks)

2 Find the value of $2y^2 - 3y$ when $y = -2$. (2 marks)

3 Factorise:

a $-5x - 25$ b $3ay^2 - 6a^2$ c $t^3 - 2t^2$ (3 marks)

4 Simplify:

a $(x^2y^3)^2 \div xy$ b $\frac{x^2y^3}{(xy)^0}$ c $\frac{(x^2)^0y^3}{(x^2y^3)^0}$ (6 marks)

*(cont.)*

5 Expand and simplify:

a $3(x-2)+4(x+1)$ b $2x-(x-2)+4x$ (4 marks)

6 Complete the following table:

| $a$ | $-1$ | $0$ | $1$ | $2$ |
|---|---|---|---|---|
| $a^2-2$ | | | | |

(2 marks)

7 Given that $C=\frac{5}{9}(F-32)$, find the value of $C$ when $F=104$. (2 marks)

8 Write algebraic expressions for the following statements:

a 7 more than 3 times $x$

b Athena is $x$ years old. Her sister Lucy is 2 years younger. How old is Lucy? (2 marks)

9 Simplify:

a $(6y)^2$ b $\sqrt{81y^2}$ (2 marks)

10 Simplify:

a $a^2-4a+a^2-a$ b $5xy-4x+2yx$ (2 marks)

11 Study the patterns below and write a formula that relates $x$ and $y$:

a

| $x$ | 0 | 1 | 2 | 3 |
|---|---|---|---|---|
| $y$ | 1 | 4 | 7 | 10 |

b

| $x$ | 0 | 1 | 2 | 3 |
|---|---|---|---|---|
| $y$ | 3 | 1 | $-1$ | $-3$ |

(4 marks)

☞ Quick answers on page 291
☞ Worked solutions on page 376

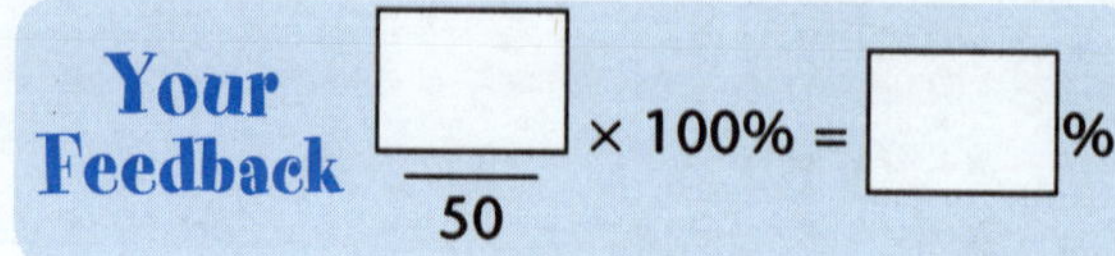

# 5 PATTERNS AND LINEAR RELATIONSHIPS

- Graphing Ordered Pairs on the Number Plane
- Number Patterns
- Graphing Lines on the Number Plane
- Points on Lines
- Horizontal and Vertical Lines
- Points of Intersection
- Parallel and Perpendicular Lines
- Using the Linear Relationships to Solve Problems

## KEYWORDS

| | |
|---|---|
| **Abscissa** | **Horizontal** |
| **Axes** | **Intersection** |
| **Axis** | **Number plane** |
| **Coefficient** | **Ordered pair** |
| **Collinear** | **Ordinate** |
| **Coordinates** | **Origin** |
| **Cubic** | **Parabola** |
| **Domain** | **Parallel** |
| **Function** | **Perpendicular** |
| **Gradient** | **Range** |
| **Graph** | **Vertical** |

# Graphing Ordered Pairs on the Number Plane

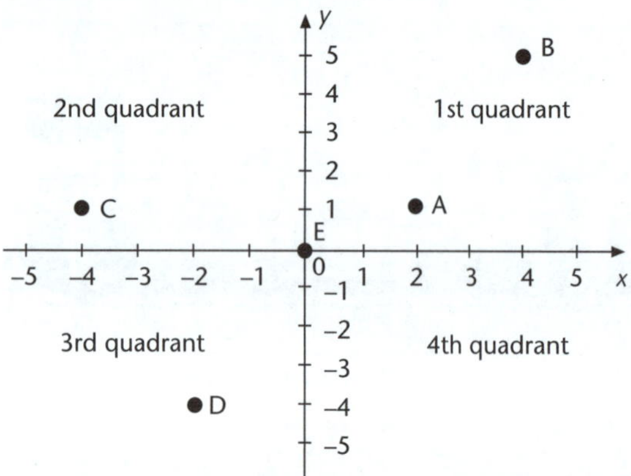


A(2, 1): From origin, across 2 and up 1

### Important Features

- Horizontal axis = ***x*-axis**; vertical axis = ***y*-axis**
- Every point has an 'address' called its **coordinates** (*x*, *y*):
  – note that *x* comes before *y*
  – go across, then up/down.
- *x* coordinate is called the **abscissa**.
- *y* coordinate is called the **ordinate**.
- Intersection of axes called the **origin** (0, 0).
- Coordinates of points are sometimes called **ordered pairs**.

## For Example

1 Use the number plane shown to determine the coordinates of the points B, C, D, E.

2 Mark the following points on a number plane:
P(4, 1), Q(–3, 2), R(–2, –5), S(2, 0)

3 In which quadrant would the following points belong:

a (3, 9) b (–2, 1)

c (–4, –2) d (3, –5)

4 On a number plane, plot the points F(4, 0), G(4, 3), H(–2, 3), I(–2, 0). Join the points to form a quadrilateral.

a Name the type of quadrilateral.

b Find the perimeter of FGHI.

c Find the area of FGHI.

5 Plot the points X(–3, –4), Y(3, –4) and Z(0, 3) on a number plane. Join the points to form the triangle XYZ.

a Find the length of the base XY.

b Find the perpendicular height of the triangle.

c Find the area of triangle XYZ.

6 If A(2, 3), B(2, –2), C(–3, –2) and D are the vertices of a rectangle, find the coordinates of the point D.

7 On a number plane, plot the four points, P(2, 1), Q(–1, 4), R(0, 3) and S(4, 3). Three of the points lie on a straight line. Which point does not lie on the line?

1 B(4, 5), C(–4, 1), D(–2, –4), E(0, 0).

2

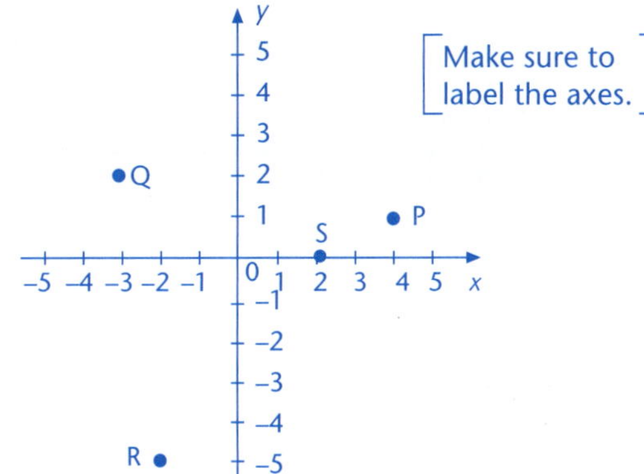


3 a (3, 9): 1st Q

b (–2, 1): 2nd Q

c (–4, –2): 3rd Q

d (3, –5): 4th Q

4

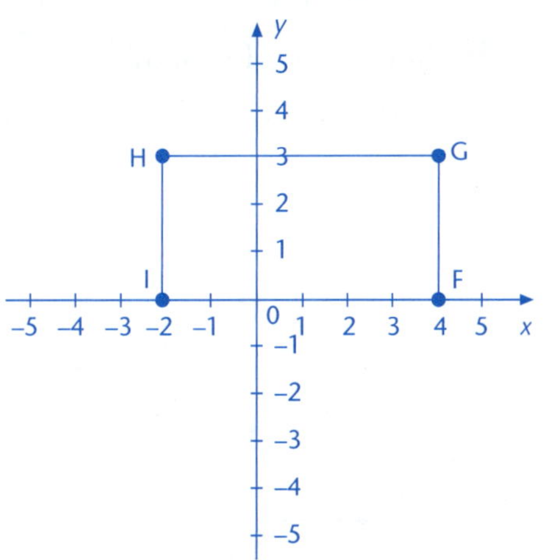


a Rectangle

b HG = 6 units

HI = 3 units

∴ perimeter = 2(6 + 3)

= 2(9)

= 18

∴ The perimeter is 18 units.

c Area = 6 × 3

= 18

∴ Area is 18 units$^2$.

5

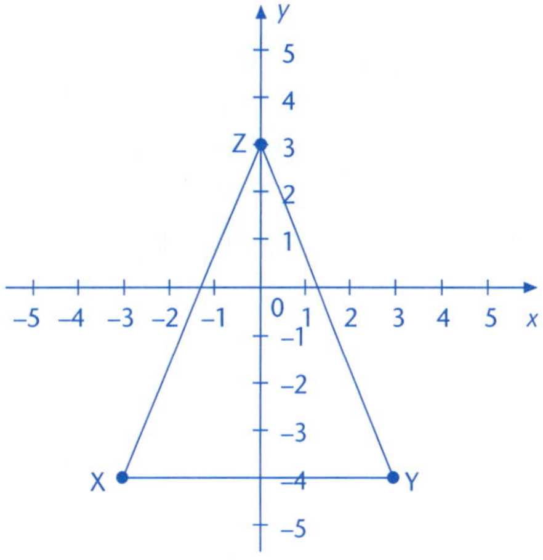


a XY = 6 units

b Height = 7 units

c Area = $\frac{1}{2}$ × base × height

= $\frac{1}{2}$ × 6 × 7

= 21

∴ The area is 21 units$^2$.

6 From the diagram, we can see that D must have coordinates (–3, 3).

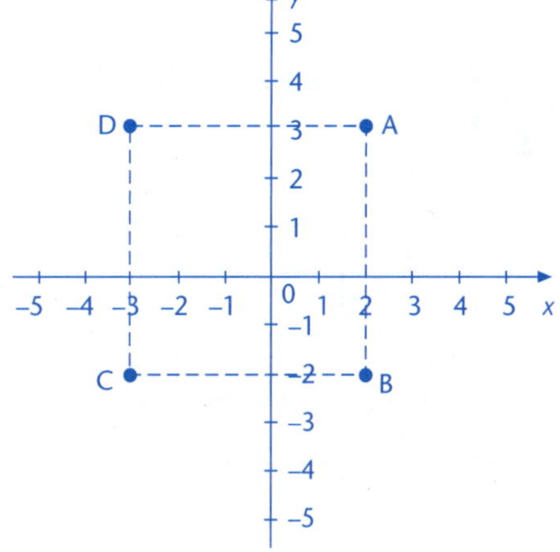

7

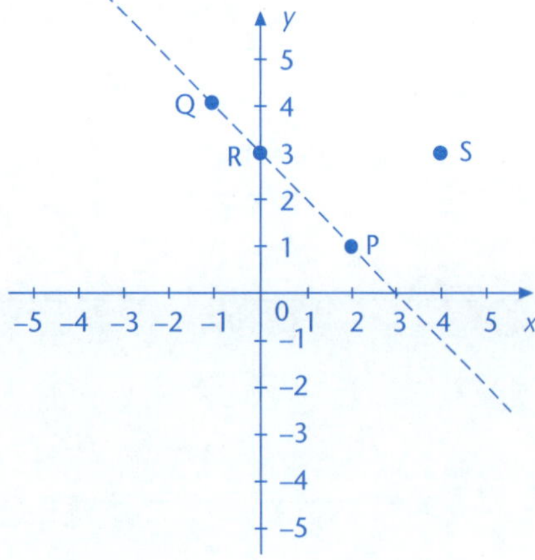


P, Q, R are points on a line (shown on diagram).
∴ S is not on the line.

## Number Patterns

Number patterns, or pattern rules, work like input/output machines. Sometimes we are given the rule and the input and we have to find the output; sometimes we are given the input and output and we have to find the rule.

### For Example

1

| $y = 2x + 1$ | |
|---|---|
| $x$ | $y$ |
| 0 | |
| 1 | |
| 2 | |
| 3 | |

**a** Copy and complete the table for the rule $y = 2x + 1$.

**b** List the ordered pairs formed in the table.

**c** On a number plane, graph the ordered pairs calculated.

**d** Are the points collinear?

2

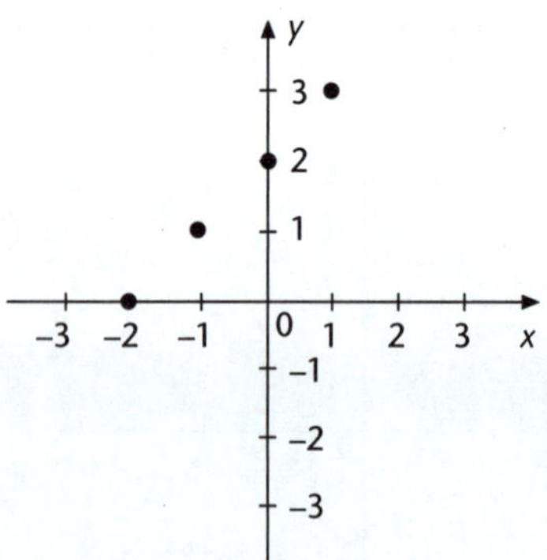


| $x$ | $y$ |
|---|---|
| −2 | |
| −1 | |
| 0 | |
| 1 | |

**a** Complete the table of values from the number plane.

**b** State the rule used.

**3** State the rule or number pattern used to obtain this table:

| $x$ | $y$ |
|---|---|
| 0 | 2 |
| 1 | 4 |
| 2 | 6 |
| 3 | 8 |
| 4 | 10 |

**4** State the rule or number pattern used to determine this set of points:

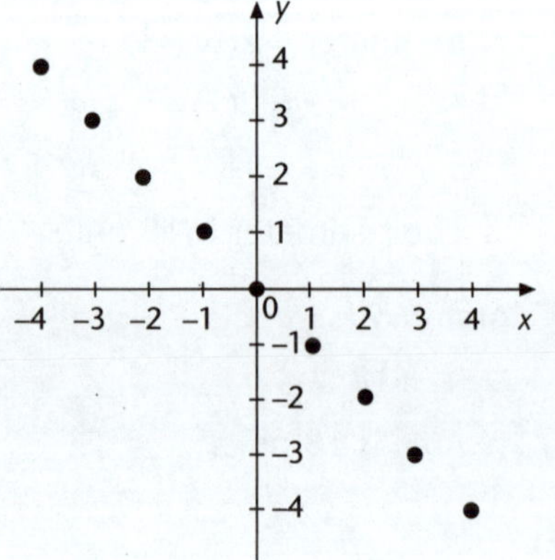

**1** **a**

| $y = 2x + 1$ | |
|---|---|
| $x$ | $y$ |
| 0 | 1 |
| 1 | 3 |
| 2 | 5 |
| 3 | 7 |

**b** (0, 1), (1, 3), (2, 5), (3, 7)

**c**

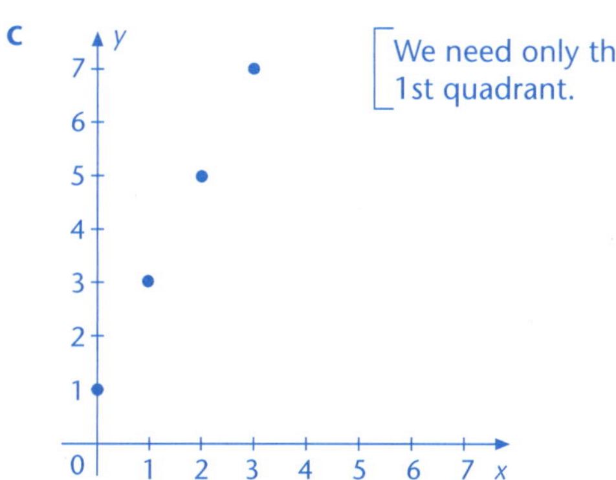

[We need only the 1st quadrant.]

**d** The points are collinear.

[Collinear points lie on a straight line.]

**2** **a**

| $x$ | $y$ |
|---|---|
| –2 | 0 |
| –1 | 1 |
| 0 | 2 |
| 1 | 3 |

**b** $y$ is 2 added onto $x$

$\therefore y = x + 2$

[Check by substitution of the points.]

**3** $y$ is 2 added onto twice $x$

$\therefore y = 2 + 2x$

or $y = 2x + 2$

**4** The ordered pairs are (–4, 4), (–3, 3), (–2, 2), (–1, 1), (0, 0), (1, –1), (2, –2), (3, –3), (4, –4)

$\therefore y$ is the negative of $x$

$\therefore y = -x$

or $x + y = 0$

# Graphing Lines on the Number Plane

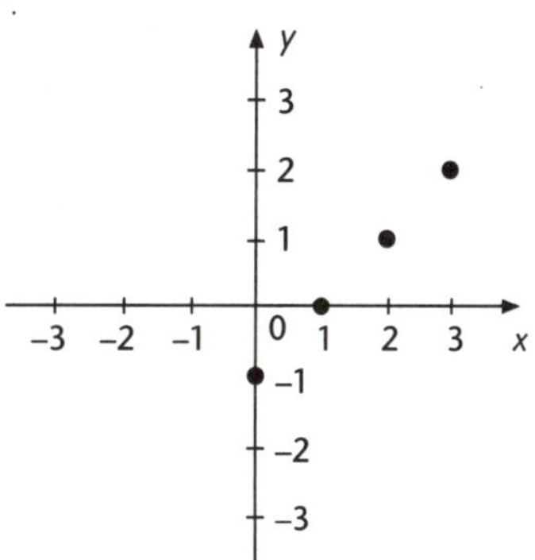

The set of coordinates (0, –1), (1, 0), (2, 1) and (3, 2) lie in a straight line and the coordinates are related in a mathematical sense. This is called a **relation**.

The set of $x$ values (0, 1, 2, 3) is called the **domain**.

The set of $y$ values (–1, 0, 1, 2) is called the **range**.

The number pattern used here is $y = x - 1$. This is called a **function**.

Functions can be graphed by choosing suitable values for $x$. Values like 0, 1 and 2 are easy to substitute into our function (or number pattern or equation).

(You will learn more about functions and relations and the distinction between them in later years.)

## For Example

**1** Graph the function $y = 2x - 3$ by first completing the table given:

| $y = 2x - 3$ | |
|---|---|
| $x$ | $y$ |
| 0 | |
| 1 | |
| 2 | |

**2** Graph the function $y = 3 - x$ by first completing the table of values given:

| $x$ | 0 | 1 | 2 |
|---|---|---|---|
| $y$ | | | |

**3** Graph the function $y = 1 - 2x$, by first selecting suitable values for $x$ and completing a table of values.

**1**

| $y = 2x - 3$ | |
|---|---|
| $x$ | $y$ |
| 0 | −3 |
| 1 | −1 |
| 2 | 1 |

This has given us (0, −3), (1, −1) and (2, 1), which are three points that 'satisfy' $y = 2x - 3$ and lie on its graph.

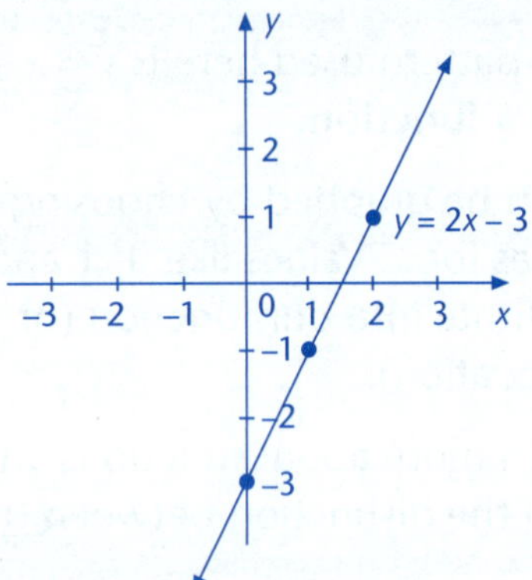


Note:
- points plotted
- joined by a straight line
- line extended and arrows used because it continues on and on (many other $x$ values could be chosen)
- the function name $y = 2x - 3$ is used to label graph.

**2** $y = 3 - x$

| $x$ | 0 | 1 | 2 |
|---|---|---|---|
| $y$ | 3 | 2 | 1 |

∴ Plot (0, 3), (1, 2) (2, 1).

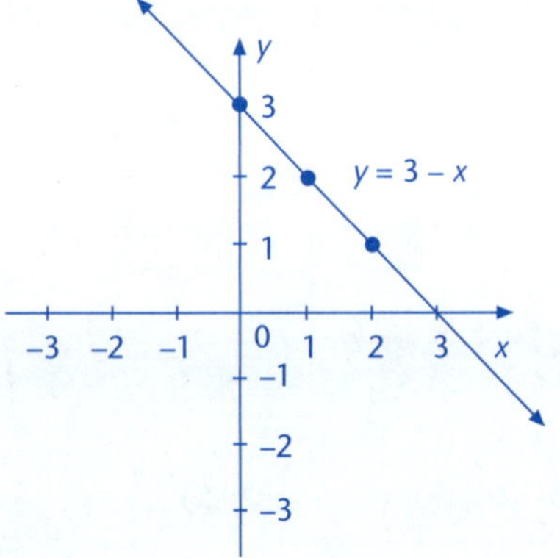


Remember we can choose other $x$ values but:
- 0, 1, 2 are easy to substitute
- two values would establish a line but a third is used as check.

**3** $y = 1 - 2x$

| $x$ | 0 | 1 | 2 |
|---|---|---|---|
| $y$ | 1 | −1 | −3 |

∴ Plot (0, 1), (1, −1) (2, −3).

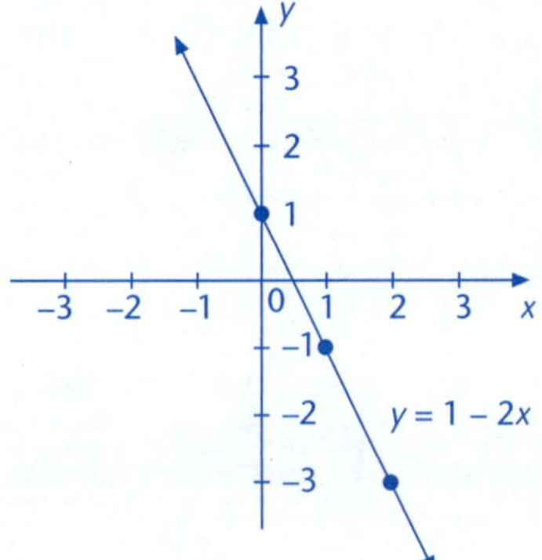


## Points on Lines

In the examples above, the three points plotted showed us the position of the straight line. But each function's graph is made up of an infinite number of points (or ordered pairs).

We can test whether a line will pass through a specific point, or whether a point is on a line, by substituting values of $x$ and $y$ into the function.

## For Example

1 Does the point (2, 4) lie on the line $y = 3x - 2$?

2 Will $y = 2x + 1$ pass through:

a (7, 15)?

b (5, 14)?

3 Does $y = 7x$ pass through the origin?

4 Show that A(3, 2) and B(7, 6) are on the line $y = x - 1$. If C has the coordinates (9, 8), determine whether A, B and C are collinear points.

1 If (2, 4) lies on $y = 3x - 2$, it means that by substituting $x = 2$ into the function, then $y$ will equal 4:

$y = 3x - 2$

$4 = 3(2) - 2$ [Substitute $x = 2$, $y = 4$]

$4 = 6 - 2$

$4 = 4$

This is a true statement.
$\therefore$ (2, 4) lies on $y = 3x - 2$.

2 a Again we substitute $x = 7$ and $y = 15$ into $y = 2x + 1$

$\therefore y = 2x + 1$

$15 = 2(7) + 1$

$15 = 14 + 1$

$15 = 15$

$\therefore y = 2x + 1$ passes through (7, 15).

b Substitute $x = 5$, $y = 14$ in $y = 2x + 1$

$y = 2x + 1$

$14 = 2(5) + 1$

$14 = 11$ No!

$\therefore y = 2x + 1$ does *not* pass through (5, 14).

3 Origin is (0, 0)

$\therefore$ Substitute $x = 0$, $y = 0$ in $y = 7x$

$y = 7x$

$0 = 7(0)$

$0 = 0$

$\therefore y = 7x$ passes through origin.

4 Substitute both points into $y = x - 1$.

That is, for A(3, 2): $y = x - 1$

$2 = 3 - 1$

$2 = 2$

$\therefore$ A is on the line

For B(7, 6): $y = x - 1$

$6 = 7 - 1$

$6 = 6$

$\therefore$ B is on the line

Now, for C(9, 8): $y = x - 1$

$8 = 9 - 1$

$8 = 8$

$\therefore$ C is also on the line

$\therefore$ As A, B, C are all on $y = x - 1$, then A, B, C are collinear points.

[Collinear points lie on a straight line.]

## Horizontal and Vertical Lines

- To graph $y = 2$ let's look at the table of values:

| $x$ | 0 | 1 | 2 |
|---|---|---|---|
| $y$ | 2 | 2 | 2 |

$y$ has to be always 2 no matter the value of $x$.

i.e. (0, 2), (1, 2), (2, 2), (–2, 2), etc.

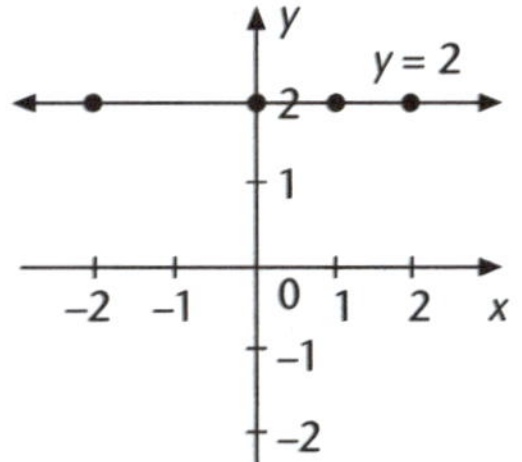


[The graph shows all points with a $y$ value of 2.]

- To graph $x = 3$, our table of values can have only $x = 3$:

| $x$ | 3 | 3 | 3 |
|---|---|---|---|
| $y$ | 2 | 0 | −3 |

Thus $y$ can be *anything*.

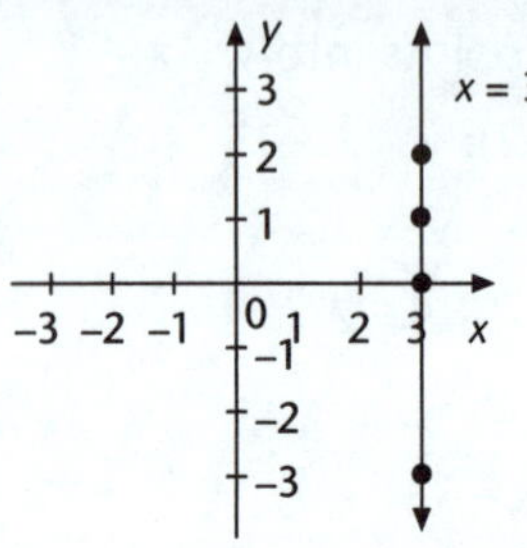


i.e. (3, 2), (3, 0), (3, −3), (3, 1)

[The graph shows all points with $x$ value of 3.]

[$y = b$ parallel to $x$-axis; $x = a$ parallel to $y$-axis.]

## For Example

1 On the same number plane, graph $y = -2$ and $y = 4$.

2 On the same number plane, graph $x = -5$ and $x = 8$.

3 What function has as its graph the:

a $x$-axis?

b $y$-axis?

4 What functions have been graphed:

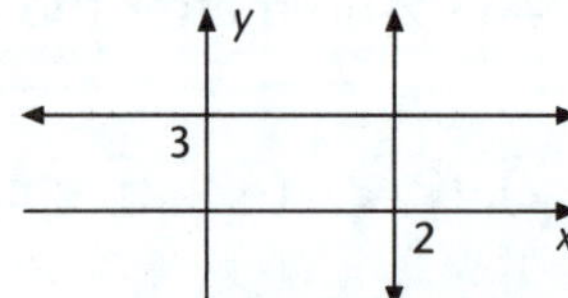


1

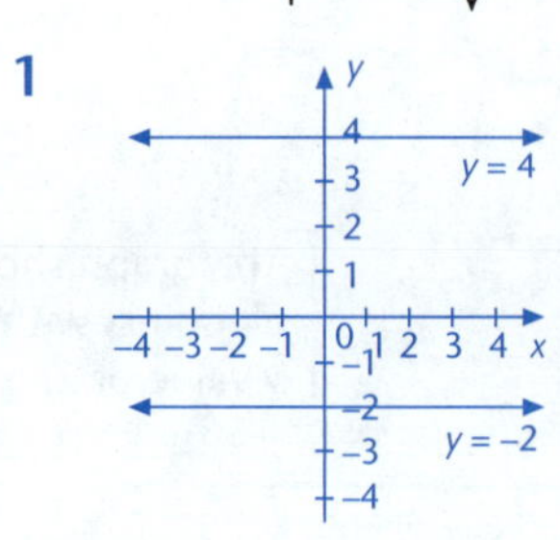


For $y = -2$ ∴ points like (0, −2), (3, −2), (2, −2) …

For $y = 4$ ∴ points like (1, 4), (3, 4), (−2, 4) …

2

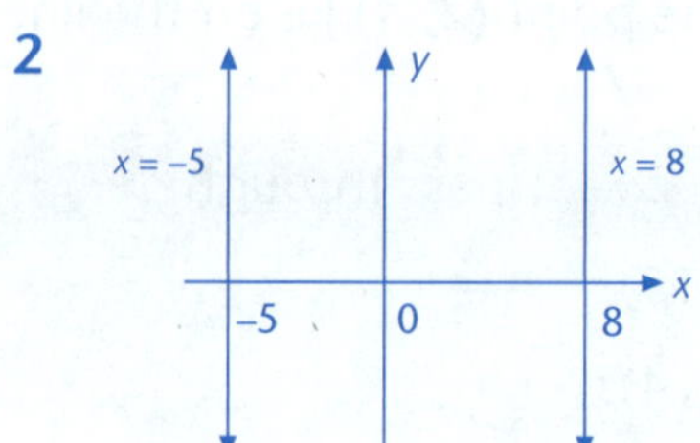


For $x = -5$, points like (−5, 0), (−5, 7), (−5, 1), etc.

For $x = 8$, points like (8, 3), (8, 7), (8, −4), etc.

3 a the $x$-axis is $y = 0$

b the $y$-axis is $x = 0$

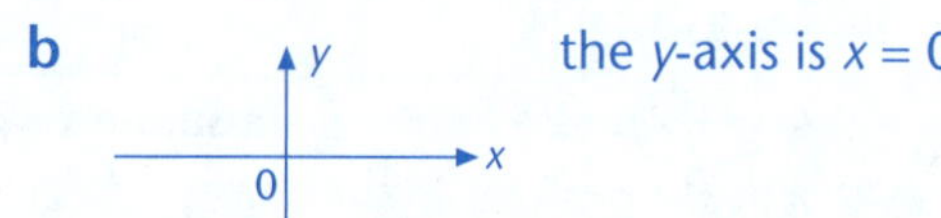


4

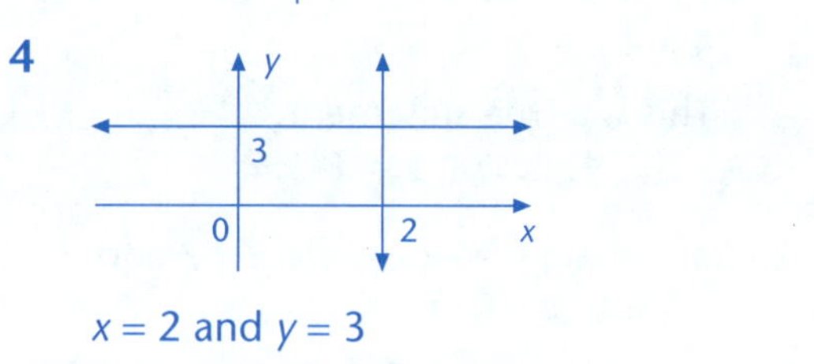


$x = 2$ and $y = 3$

## Points of Intersection

When a pair of equations are graphed, the point of intersection can be read from the graph. This is the only common point between two lines.

## For Example

Find the point of intersection when the following pairs of equations are graphed on the same number plane:

**1** $y = 1$ and $x = -3$

**2** $y = 2x - 3$ and $y = 3 - x$

**1** $x = -3$

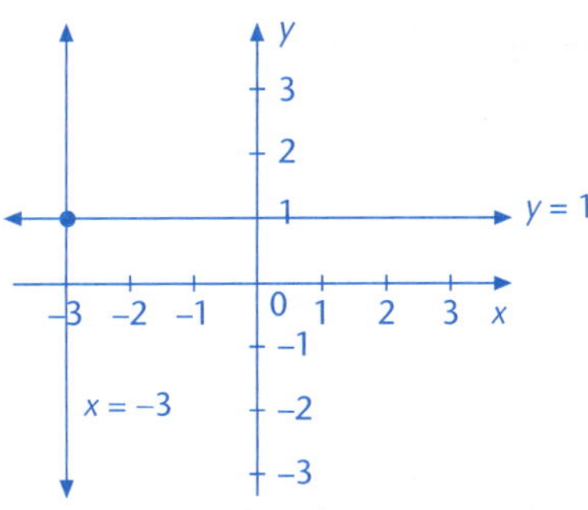


$y = 1 \therefore (0, 1), (3, 1)$, etc.

$x = -3 \therefore (-3, 2), (-3, 0)$, etc.

$\therefore$ Point of intersection is $(-3, 1)$.

**2** $y = 2x - 3$

| $x$ | 0 | 1 | 2 |
|---|---|---|---|
| $y$ | –3 | –1 | 1 |

$y = 3 - x$

| $x$ | 0 | 1 | 2 |
|---|---|---|---|
| $y$ | 3 | 2 | 1 |

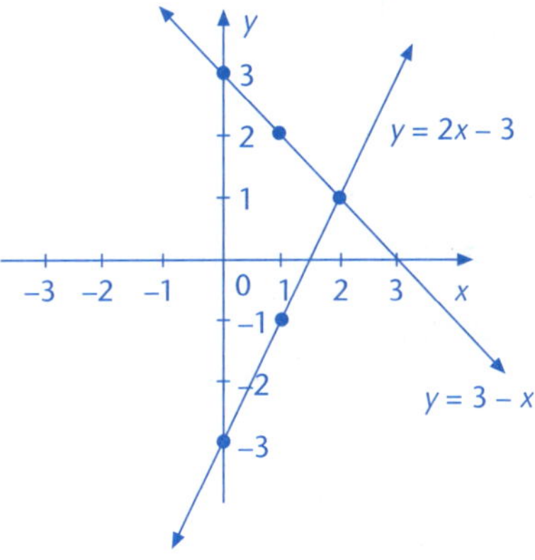


$\therefore$ Point of intersection is $(2, 1)$.

In both tables of values, $(2, 1)$ is a common point $\therefore$ the point of intersection.

# Parallel and Perpendicular Lines

Parallel lines are the same distance apart for their entire length.

Perpendicular lines meet at right-angles.

For Example

**1** On the same set of axes sketch the lines $y = 2x + 1$ and $y = 2x - 3$.

**1** $y = 2x + 1$

| $x$ | 0 | 1 | 2 |
|---|---|---|---|
| $y$ | 1 | 3 | 5 |

$y = 2x - 3$

| $x$ | 0 | 1 | 2 |
|---|---|---|---|
| $y$ | –3 | –1 | 1 |

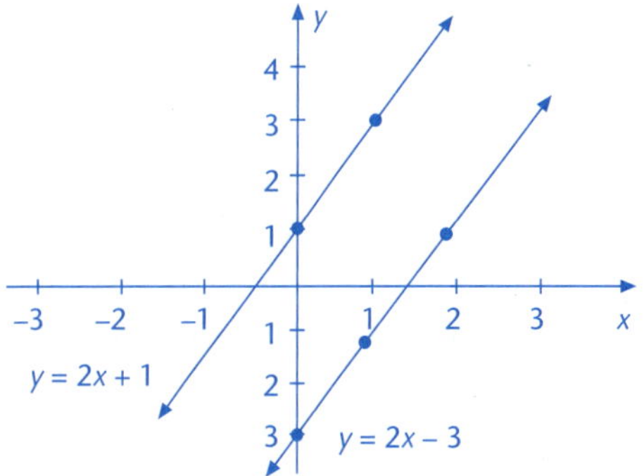


Notes: **1** The lines are parallel.

**2** The coefficient of $x$ for both equations is 2.

Parallel lines will have the same $x$ coefficient when the equation is written as $y = mx + b$, where $m$ is the gradient of the line.

The above lines have a gradient of 2.

1 On the same set of axes sketch the lines $y = 2x - 3$ and $y = -\frac{1}{2}x + 1$.

1 $y = 2x - 3$

| $x$ | 0 | 1 | 2 |
|---|---|---|---|
| $y$ | −3 | −1 | 1 |

$y = -\frac{1}{2}x + 1$

| $x$ | 0 | 1 | 2 |
|---|---|---|---|
| $y$ | 1 | $\frac{1}{2}$ | 0 |

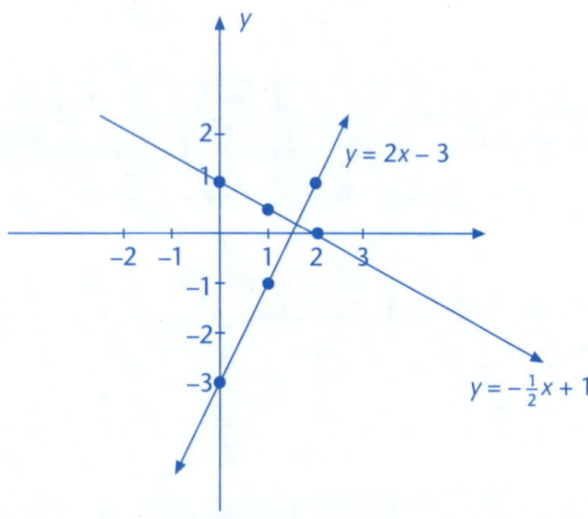


Note that the lines are perpendicular to each other (check with your protractor).

Note also that the $x$ coefficient for line one is 2, while that for line two is $-\frac{1}{2}$. They are negative reciprocals. (The reciprocal of 2 is $\frac{1}{2}$, the negative reciprocal of 2 is $-\frac{1}{2}$.)

Perpendicular lines have $x$-coefficients that are negative reciprocals of each other when equations are in the form $y + mx = b$ where $m$ is the gradient of the line.

## Using Linear Relationships to Solve Problems

In mathematical word problems that contain variables we use pronumerals to represent the variables. For example, cost = $c$ and time = $t$. Word problems that involve a constant rate have a linear relationship. This means that if a graph is drawn to represent this relationship the points are in a line.

1 Aaron earns $25 per hour as a sales assistant.

a Using total wage = $w$ and hours = $h$, complete the table.

| $h$ | 0 | 1 | 2 | 3 | 4 | 5 |
|---|---|---|---|---|---|---|
| $w$ | | | | | | |

b Plot the graph of this relationship.

c How much will Aaron earn when he works for 8 hours?

d How many hours will Aaron work if he earns $150?

2 An electrician charges a call out fee of $80 and $50 per hour.

a Using $c$ = cost in dollars and $t$ = time in hours, complete the table:

| $t$ | 1 | 2 | 3 | 4 | 5 | 6 |
|---|---|---|---|---|---|---|
| $c$ | | | | | | |

b Plot the graph of this relationship.

c What will the cost of employing the electrician for $1\frac{1}{2}$ hours?

d What is the value of $c$ when $t = 0$?

**3** Laura draws a conversion graph relating temperatures expressed in Celsius (C) and Fahrenheit (F). It is known that 0 °C = 32 °F and 38 °C = 100 °F.

**a** Draw a conversion graph, using Celsius on the horizontal axis and Fahrenheit on the vertical axis.

**b** Complete the table using the graph:

| Celsius (°C) | 0 | 20 | 38 |
|---|---|---|---|
| Fahrenheit (°F) | | | |

**1 a**

| $h$ | 0 | 1 | 2 | 3 | 4 | 5 |
|---|---|---|---|---|---|---|
| $w$ | 0 | 25 | 50 | 75 | 100 | 125 |

**b**

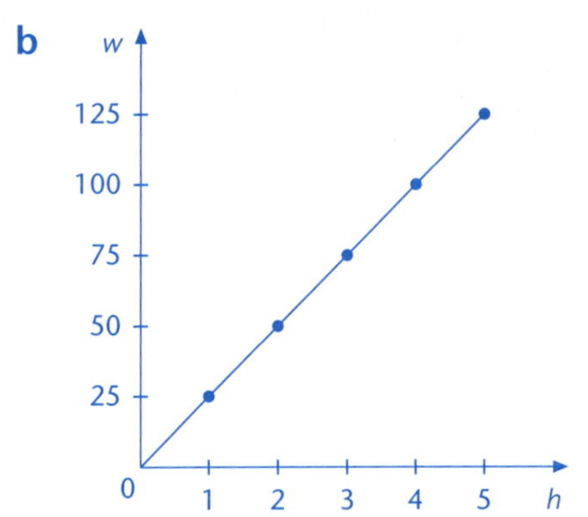


**c** $\$25 \times 8 = \$200$

**d** $\$150 \div \$25 = 6$

$\therefore$ 6 hours

**2 a**

| $t$ | 1 | 2 | 3 | 4 | 5 | 6 |
|---|---|---|---|---|---|---|
| $c$ | 130 | 180 | 230 | 280 | 330 | 380 |

**b**

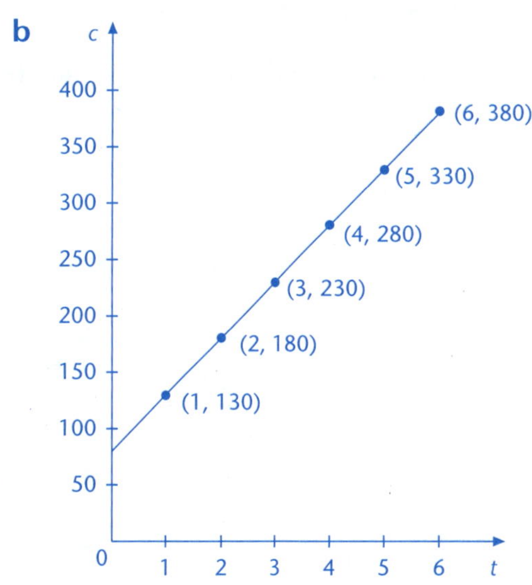


**c** In $1\frac{1}{2}$ h, the electrician charges \$155.

**d** When $t = 0$, $c = 80$.

**3 a**

°F
100
90
80
70
60
50
40
30
20
10
0
10 20 30 40 °C
(38, 100)
(20, 68)
(0, 32)

**b**

| Celsius (°C) | 0 | 20 | 38 |
|---|---|---|---|
| Fahrenheit (°F) | 32 | 68 | 100 |

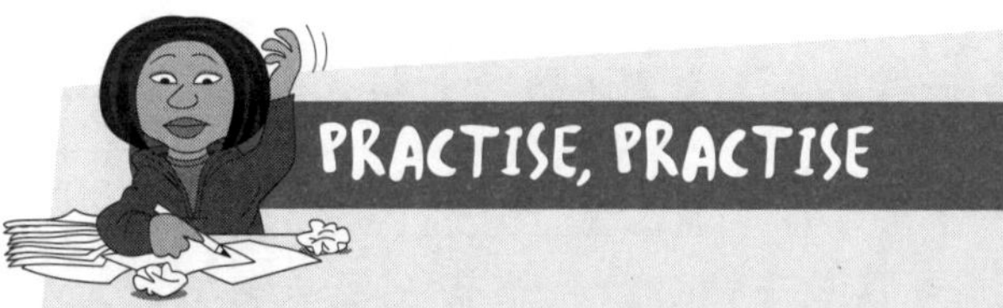


Go to p. 285 for quick answers, or to pp. 315–320 for worked solutions

**1** Write down the ordered pairs for the points A to J. pp. 66–68

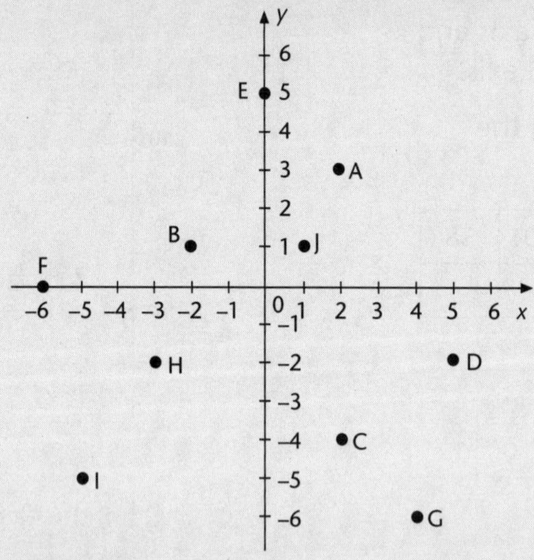


**2** Plot the points P(2, 1), Q(2, –5), R(–3, –5), S(–3, 1). Join the points to form a plane shape. pp. 66–68

- **a** What is the name of the shape?
- **b** Find the length of PQ and PS.
- **c** What is the perimeter?
- **d** What is the area?
- **e** In what quadrant is the point S?

**3** **a** Write the coordinates of K, L, M, N. pp. 66–68

- **b** What is the shape of KLMN?
- **c** Find the area of KLMN.

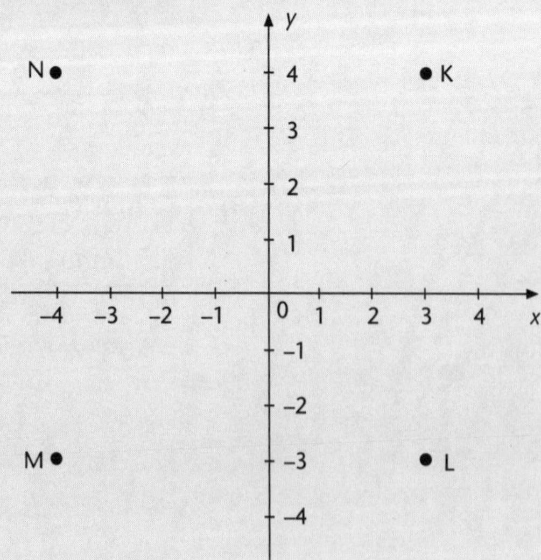

4 Plot the points, A(0, –4), B(0, 3) and C(4, 0). Join A to B, B to C and C to A. pp. 66–68

a What is the shape of ABC?

b Find the length of AB.

c Find the distance between the origin and C.

d Determine the area of ABC.

5 a Find the length of: pp. 66–68

i XY ii YZ

b From what has been learned in (a), what type of triangle is XYZ (besides right-angled)?

c Find the area of $\triangle XYZ$:

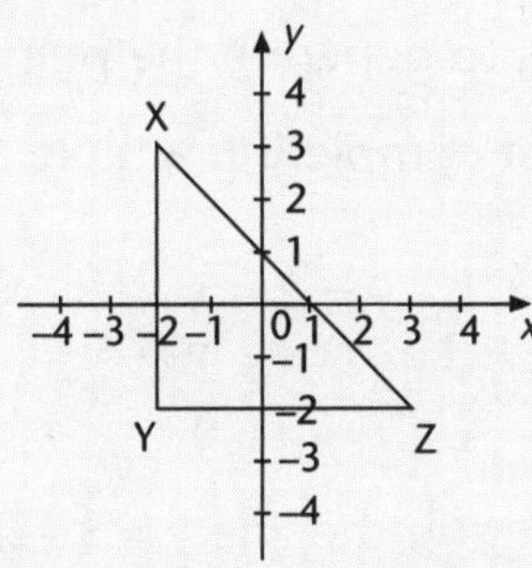


6 A(0, 0) and B(3, 4) are plotted on a number plane. C is plotted so that $\triangle ABC$ is right-angled at C. pp. 66–68

a What are the two possible locations of point C?

b Find the area of the triangle.

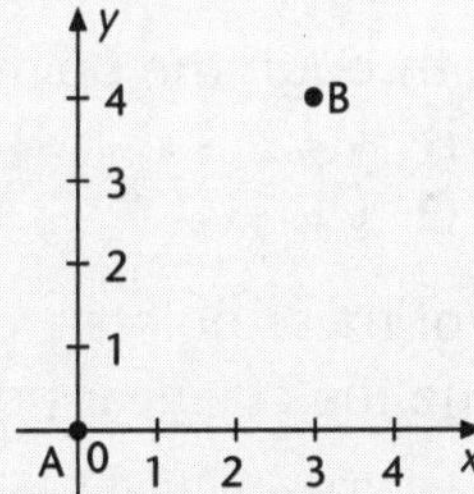


7 ABCD is a rectangle. If A(–3, –4), B(–3, 2), C(4, 2), find the coordinates of D. pp. 66–68

8 Copy and complete the following tables: pp. 68–69

a

| $y = 3x + 2$ | |
|---|---|
| $x$ | $y$ |
| 0 | |
| 1 | |
| 2 | |
| 3 | |

b

| $y = 4 - 2x$ | |
|---|---|
| $x$ | $y$ |
| 0 | |
| 1 | |
| 2 | |
| 3 | |

**9** State the equations used to generate these tables: pp. 68–69

**a**

| $x$ | 0 | 1 | 2 |
|---|---|---|---|
| $y$ | 0 | 4 | 8 |

**b**

| $x$ | 0 | 1 | 2 |
|---|---|---|---|
| $y$ | –2 | –1 | 0 |

**c**

| $x$ | 0 | 1 | 2 |
|---|---|---|---|
| $y$ | –1 | 2 | 5 |

**10** For the set of points (0, 0), (1, 2), (2, 4), (3, 6), find the: pp. 69–70
**a** set of $x$ values (domain)
**b** set of $y$ values (range)
**c** number pattern (or function) used to generate the points.

**11** Graph the following equations, by first completing a table of values using $x = 0, 1, 2$: pp. 69–70
**a** $y = 3x$ **b** $y = 2 - x$
**c** $y = x + 3$ **d** $y = 4 - 2x$
**e** $x + y = 3$ **f** $x - y = 0$

**12** Determine whether the point (3, 1) lies on the line $y = x - 2$. pp. 70–71

**13** Does the line $y = 3x + 1$ pass through the point (2, 4)? pp. 70–71

**14** Which one of the following points lies on the line $y = 3x - 2$? pp. 70–71
**A** (4, 1) **B** (2, 0)
**C** (–1, 1) **D** (3, 7)

**15** Which one of the following lines passes through the point (1, 5)? pp. 70–71
**A** $y = x + 2$ **B** $y = 3 - x$
**C** $y = 2x + 3$ **D** $y = 3x - 1$

**16** A(1, 2), B(2, 5), C(0, 1), D(1, 3). Three of these points are **collinear** because they lie on $y = 2x + 1$. Determine the collinear points. pp. 70–71

**17** $y = 3x + 2$, $y = 4x$, $y = 6 - x$ and $y = 6x - 1$. Three of these lines are **concurrent** because they pass through the point (1, 5). Name the three concurrent lines. pp. 70–71

**18** Graph the following lines on the same number plane: pp. 71–72
$x = -5$, $y = 3$

**19** The lines $x = 0$, $y = 0$, $x = 3$ and $y = 4$ are graphed on the same number plane. The four lines enclose a region of the number plane. pp. 71–72
**a** What is the shape of this enclosed region?
**b** Find the area of this enclosed region.

**20** Graph the following pairs of equations on the same number plane, and find the point of intersection: pp. 71–72
**a** $y = 3$ and $x = -1$ **b** $x = 2$ and $y = x + 1$
**c** $y = x$ and $y = 2x - 2$ **d** $y = 3x - 1$ and $y = 3 - x$

**21** Sketch the lines below on the same number plane and decide which are parallel, which are perpendicular, which are neither: pp. 73–74

**a** $y = 2x$ **b** $y = -\frac{1}{2}x$ **c** $y = \frac{1}{2}x$

**d** $y = 2x - 1$ **e** $y = -\frac{1}{2}x + 2$ **f** $2x - y = 2$

Write down the gradient of each line.

**22** Laura is saving for a trip to Europe. She plans to save $80 per week. pp. 74–75

**a** Using $n$ = number of weeks and $s$ = savings, complete the table:

| $n$ | 0 | 4 | 8 | 12 | 16 | 20 |
|---|---|---|---|---|---|---|
| $s$ | | | | | | |

**b** Draw a graph showing the linear relationship between $n$ and $s$.

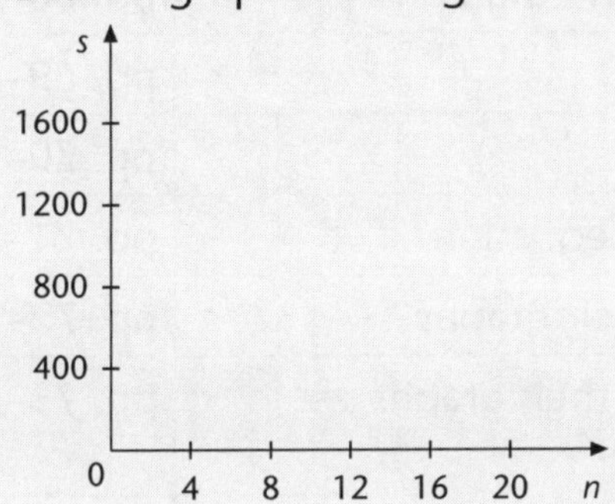


**23** A function centre charges $200 to hire the centre, plus $10 per person who attends. pp. 74–75

**a** Using $n$ = number of people and $c$ = cost, complete the table:

| $n$ | 0 | 10 | 20 | 30 | 40 |
|---|---|---|---|---|---|
| $c$ | | | | | |

**b** Draw the graph linking $n$ and $c$.

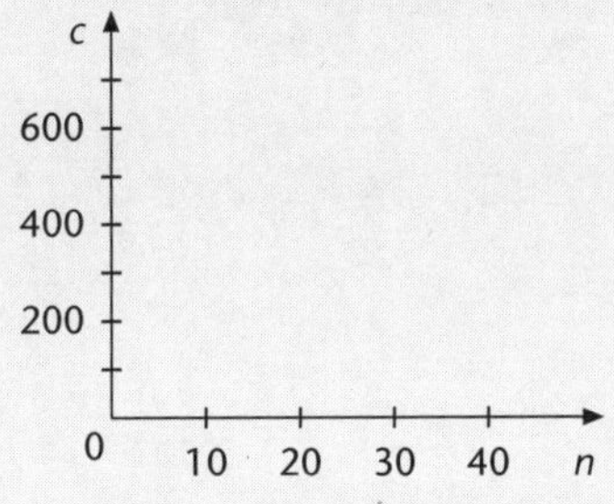


**c** Find the cost of hiring the centre for 25 people.

Go to p. 285 for quick answers, or to pp. 315–320 for worked solutions

## YOUR CHECKLIST

**For a complete understanding of this topic you must be able to:**

| | | |
|---|---|---|
| ✓ | Plot points on the number plane | pp. 66–68 |
| ✓ | Use and understand terminology associated with the number plane | pp. 66–68 |
| ✓ | Substitute into simple equations in *x* and *y* and plot the resulting ordered pairs | pp. 68–69 |
| ✓ | Deduce an equation in *x* and *y* from a table and a graph | pp. 68–69 |
| ✓ | Graph a straight line given its equation | pp. 69–70 |
| ✓ | Deduce whether a point lies on a line | pp. 70–71 |
| ✓ | Graph horizontal and vertical lines from their equation | pp. 71–72 |
| ✓ | Deduce points of intersection of lines from their graphs | pp. 72–73 |
| ✓ | Identify parallel and perpendicular lines from their graphs and/or equations | pp. 73–74 |
| ✓ | Use linear relationships to solve problems. | pp. 74–75 |

**Now you are ready to do the tests!**

**(40 marks)**

**1**

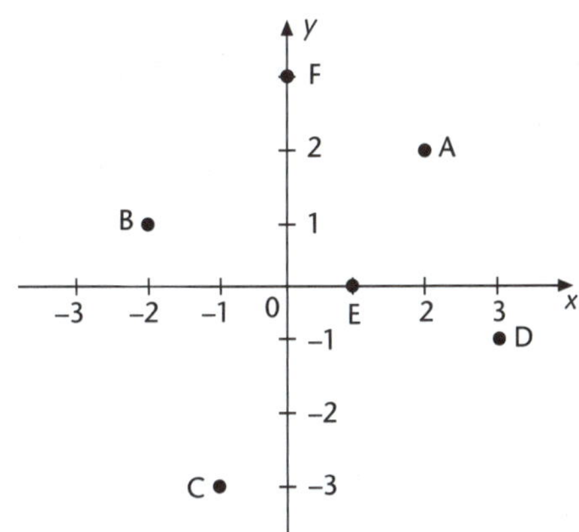


**a** Write down the ordered pair that represents A, B, C, D, E and F. (6 marks)

**b** On the number plane plot G(–3, 0), H(–1, 2) and I(2, –2). (3 marks)

**2** Complete the table of values for $y = 11 + 3x$: (3 marks)

| $y = 11 + 3x$ | |
|---|---|
| $x$ | $y$ |
| 0 | |
| 2 | |
| 4 | |

**3** Find the number pattern rule for this table: (1 mark)

| $x$ | $y$ |
|---|---|
| 0 | 0 |
| 1 | 3 |
| 2 | 6 |
| 3 | 9 |

**4** Complete a table of values and graph the resulting line: (8 marks)

**a**

| $y = x + 3$ | |
|---|---|
| $x$ | $y$ |
| 0 | |
| 1 | |
| 2 | |

**b**

| $y = \frac{1}{2}x + 2$ | |
|---|---|
| $x$ | $y$ |
| 0 | |
| 2 | |
| 4 | |

*(cont.)*

**5** Which of these ordered pairs lies on the line $y = 10 - x$?

a (4, 6) b (6, 4) c (10, –2)

d (11, –1) e (–3, 7) (5 marks)

**6** By first graphing the lines $x = -2$ and $y = 5$, find the point of intersection of the lines. (3 marks)

**7** By first graphing the lines $y = x + 2$ and $x - y = 3$, decide whether they are parallel or perpendicular or neither. (5 marks)

**8** At the local supermarket potatoes are priced at $3 per kilogram.

a Complete the table, using $n$ = number of kilograms and $p$ = price (in $): (2 marks)

| $n$ | 0 | 2 | 4 | 6 |
|---|---|---|---|---|
| $p$ | | | | |

b Use the table to graph the relationship between $n$ and $p$: (2 marks)

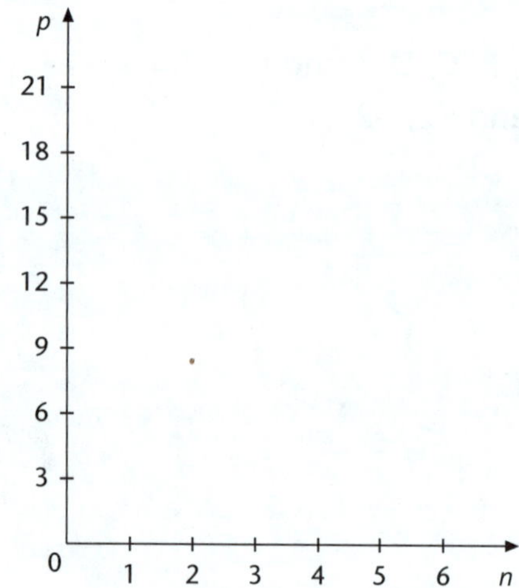


c Use the graph to estimate the cost of 5 kilograms of potatoes. (1 mark)

d Tess paid $21 for her potatoes. How many kilograms did she buy? (1 mark)

☞ Quick answers on page 291
☞ Worked solutions on page 377

**Your Feedback** $\frac{\square}{40} \times 100\% = \square\%$

**(40 marks)**

**1**

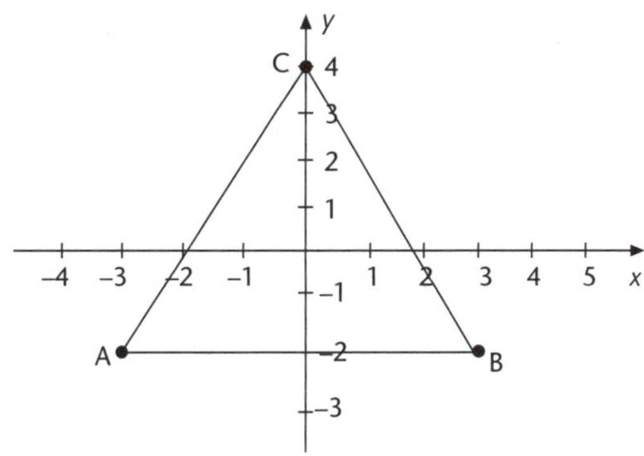


**a** Write down the coordinates of the points A, B and C. (3 marks)

**b** Name the shape formed by joining ABC. (1 mark)

**c** Calculate the area of ABC. (3 marks)

**2** Complete this pattern table for $y = 12 - 2x$: (4 marks)

| $y = 12 - 2x$ | |
|---|---|
| $x$ | $y$ |
| –1 | |
| 1 | |
| 3 | |
| 5 | |

**3** **a** Complete the table from this graph: (6 marks)

| $x$ | –3 | –2 | –1 | 0 | 1 | 2 |
|---|---|---|---|---|---|---|
| $y$ | | | | | | |

**b** Write down the rule for this pattern $y =$ ________________. (1 mark)

*(cont.)*

4 a By completing a table of values for the equations $y = x - 1$ and $y = 2x - 3$, sketch the lines on the same number plane. (6 marks)

$y = x - 1$

| $x$ | 0 | 1 | 2 |
|---|---|---|---|
| $y$ | | | |

$y = 2x - 3$

| $x$ | 0 | 1 | 2 |
|---|---|---|---|
| $y$ | | | |

b Write down the coordinates of the point of intersection. (2 marks)

5 Show that A(1, 5), B(–1, 9) lie on the line $y = 7 - 2x$.
If C has coordinates (5, –3), show that A, B and C are collinear. (4 marks)

6 Write down the gradients of:

a $y = 3x + 1$

b $y = -\frac{1}{3}x - 4$ (2 marks)

Are the lines parallel or perpendicular, and why? (2 marks)

7 A financial adviser charges a $70 consultation fee and $180 for every hour.

a Make a table using $c$ to represent the total cost and $n$ for the number of hours: (2 marks)

| $n$ | 0 | 2 | 4 | 6 | 8 |
|---|---|---|---|---|---|
| $c$ | | | | | |

b Draw a graph of $c$ against $n$. (2 marks)

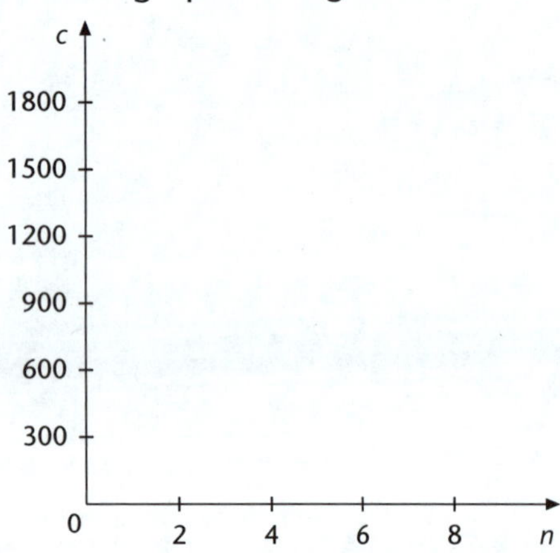


c Use the graph to estimate the:

i cost if the advisor worked for 5 hours

ii number of hours he or she worked if the cost was $1330. (2 marks)

**Your Feedback** $\frac{\square}{40} \times 100\% = \square\%$

☞ Quick answers on page 291
☞ Worked solutions on page 378

**(40 marks)**

1 The points A(–2, –3), B(–1, 2), C(4, 1) and D are the vertices of a parallelogram.

a Plot A, B and C on a number plane and write down the coordinates of D. (3 marks)

b Find the midpoints of the sides AB and DC and label them E and F on your diagram. (2 marks)

c Write down the ordered pairs for E and F. (4 marks)

d Are the lines AD, BC and EF parallel?
How can this be seen from the diagram? (2 marks)

2 Complete these pattern tables for:

a $2x + y = 7$

| $x$ | $y$ |
|---|---|
| –1 | |
| 0 | |
| 1 | |

b $y = \dfrac{4}{x+2}$ (5 marks)

| $x$ | $y$ |
|---|---|
| 0 | |
| 2 | |
| 4 | |

3 Write down the pattern rule from this table: (2 marks)

$y =$ ______________

| $x$ | –2 | 0 | 1 | 2 | 5 |
|---|---|---|---|---|---|
| $y$ | 7 | 3 | 4 | 7 | 28 |

4 a Complete a table of values for the equations $2x + y = 5$ and $2x - y = 7$ and sketch them on the number plane:

$2x + y = 5$

| $x$ | 0 | 1 | 2 | 3 |
|---|---|---|---|---|
| $y$ | | | | |

$2x - y = 7$ (6 marks)

| $x$ | 0 | 1 | 2 | 3 |
|---|---|---|---|---|
| $y$ | | | | |

b Write down the coordinates of the point of intersection. (2 marks)

c Show that the line $y = 3x - 10$ is concurrent with $2x + y = 5$ and $2x - y = 7$. (3 marks)

*(cont.)*

5 Show that C(2, –3), D(–2, –7) and E(4, –1) lie on the line $x - y = 5$. (3 marks)

6 Colin is preparing for a triathlon. Today he cycled for 30 km and ran for 6 km. Each day he intends to cycle 2 km less and run 1 km more than the previous day.

a Using $d$ = distance and $n$ = number of days, complete the tables: (2 marks)

i Cycling

| $n$ | 0 | 1 | 2 | 3 | 4 | 5 |
|---|---|---|---|---|---|---|
| $d$ | | | | | | |

ii Running

| $n$ | 0 | 1 | 2 | 3 | 4 | 5 |
|---|---|---|---|---|---|---|
| $d$ | | | | | | |

b Use the tables to draw the two graphs on the same set of axes. (2 marks)

c By extending the graphs, estimate the day when Colin will cycle and run identical distances. (2 marks)

d Estimate the day that Colin will not cycle. (2 marks)

**Your Feedback** $\frac{\square}{40} \times 100\% = \square\%$

☞ Quick answers on page 292
☞ Worked solutions on page 379

# 6 EQUATIONS AND FORMULAE

- Testing Solutions
- One-step Equations
- Two-step Equations
- Pronumerals on Both Sides of the Equation
- Equations with Grouping Symbols
- Equations with Fractions
- Problem Solving with Equations
- Formulae

## KEYWORDS

| | |
|---|---|
| **Consecutive** | **Pronumeral** |
| **Equation** | **Solution** |
| **Expression** | **Solve** |
| **Formula** | **Substitution** |
| **Formulae** | |

An equation is a mathematical sentence containing numbers and pronumerals (letters) separated by an equals sign. Examples of equations are:

$6x = 42$

$x + 7 = 90$

[We are asked to *solve* equations.]

$4a - 13 = 17$

$14 = 2y - 3$

$5(b + 6) = 3(b - 2)$

$$\frac{m-3}{7} - 4 = \frac{m+2}{5}$$

To **solve** an equation you are required to find the value of the pronumeral that makes the sentence true.

For example, if $6 + y = 17$, then $y = 11$ is the solution as $6 + 11 = 17$.

Remember: only the correct solution will satisfy an equation.

[*Satisfy* means the correct value of the letter makes the statement true.]

In this chapter only equations in one variable are considered.

[*Variable* means letter, pronumeral.]

## Testing Solutions

A possible solution for an equation can be tested by substituting that value for the pronumeral.

### For Example

1 Show that $x = 4$ is the correct solution of $5x - 12 = 8$.

1 Consider $5x - 12 = 8$ [$5x$ means $5 \times x$]

$LHS = 5x - 12$

$= 5 \times 4 - 12$

$= 20 - 12$ [LHS = left-hand side]

$= 8$

$RHS = 8$

As LHS = RHS, then $x = 4$ is a correct solution.

## One-step Equations

Remember, pronumerals must finish up on one side, with numbers on the other side. Changing sides means changing signs (to the opposite sign).

### For Example

Solve the following equations:

| | | | |
|---|---|---|---|
| 1 | $a + 11 = 19$ | 2 | $y + 15 = 3$ |
| 3 | $t - 7 = 15$ | 4 | $19 = t - 8$ |
| 5 | $6y = 42$ | 6 | $4t = -24$ |
| 7 | $8a = 20$ | 8 | $\frac{x}{9} = 3$ |
| 9 | $\frac{m}{4} = \frac{1}{2}$ | 10 | $\frac{y}{6} = -8$ |

**1** $a + 11 = 19$
$a = 19 - 11$
$= 8$

Move + 11 to RHS. It becomes – 11, i.e. the opposite sign.

**2** $y + 15 = 3$
$y = 3 - 15$
$= -12$

**3** $t - 7 = 15$
$t = 15 + 7$
$= 22$

– 7 is moved to the RHS. It becomes + 7, i.e. the opposite sign.

**4** $19 = t - 8$
$19 + 8 = t$
$27 = t$
or $t = 27$

Note that $27 = t$ means the same as $t = 27$.

**5** $6y = 42$
$y = \frac{42}{6}$
$= 7$

$6y$ means $6 \times y$. 6 moves to the RHS and becomes division, i.e. the opposite sign.

**6** $4t = -24$
$t = \frac{-24}{4}$
$= -6$

**7** $8a = 20$
$a = \frac{20}{8}$
$= 2\frac{4}{8}$
$= 2\frac{1}{2}$

**8** $\frac{x}{9} = 3$
$x = 3 \times 9$
$= 27$

$\frac{x}{9}$ means $x \div 9$. 9 moves to RHS and becomes multiplication, i.e. opposite sign.

**9** $\frac{m}{4} = \frac{1}{2}$
$m = \frac{1}{2} \times 4$
$= 2$

**10** $\frac{y}{6} = -8$
$y = -8 \times 6$
$= -48$

# Two-step Equations

Remember, pronumerals must finish up on one side, with numbers on the other side. Changing sides means changing signs (to the opposite sign).

## For Example

Solve the following equations:

| | | | |
|---|---|---|---|
| **1** | $4x + 11 = 31$ | **2** | $7x = 3x + 16$ |
| **3** | $8a - 19 = 17$ | **4** | $29 = 7x - 13$ |
| **5** | $47 = 32 - 3x$ | **6** | $\frac{x}{5} - 2 = 4$ |
| **7** | $11y = 7y - 24$ | **8** | $5t = 8t - 11$ |

**1** $4x + 11 = 31$
$4x = 31 - 11$
$4x = 20$
$x = \frac{20}{4}$
$x = 5$

+ 11 moves to RHS and becomes – 11.

× 4 moves to RHS and becomes ÷ 4.

**2** $7x = 3x + 16$
$7x - 3x = 16$
$4x = 16$
$x = \frac{16}{4}$
$x = 4$

Note $3x$ means $+ 3x$. When it moves to LHS it becomes $- 3x$.

× 4 moves to RHS and becomes ÷ 4.

**3** $8a - 19 = 17$
$8a = 17 + 19$
$8a = 36$
$a = \frac{36}{8}$
$a = 4\frac{4}{8}$
$a = 4\frac{1}{2}$

– 19 moves to RHS and becomes + 19.

× 8 moves to RHS and becomes ÷ 8.

**4** $29 = 7x - 13$

$29 + 13 = 7x$

$42 = 7x$

$\frac{42}{7} = x$

[× 7 moves to LHS and becomes ÷ 7.]

$6 = x$

$x = 6$

**5** $47 = 32 - 3x$

[Remember 32 is + 32. + 32 becomes − 32 as it moves to LHS.]

$47 - 32 = -3x$

$15 = -3x$

$\frac{15}{-3} = x$

[× −3 becomes ÷ −3 as it moves to LHS]

$-5 = x$

$x = -5$

**6** $\frac{x}{5} - 2 = 4$

[− 2 moves to RHS and becomes + 2.]

$\frac{x}{5} = 4 + 2$

$\frac{x}{5} = 6$

[÷ 5 moves to RHS and becomes × 5.]

$x = 6 \times 5$

$x = 30$

**7** $11y = 7y - 24$

[7*y* means + 7*y*. + 7*y* becomes − 7*y* when it moves to LHS.]

$11y - 7y = -24$

[Note that − 24 remains on RHS]

$4y = -24$

$y = \frac{-24}{4}$

[× 4 moves to RHS and becomes ÷ 4.]

$y = -6$

**8** $5t = 8t - 11$

[8*t* means + 8*t*. + 8*t* becomes − 8*t* when it moves to LHS.]

$5t - 8t = -11$

$-3t = -11$

$t = \frac{-11}{-3} = \frac{11}{3}$

[× −3 becomes ÷ −3 when it moves to RHS.]

$t = 3\frac{2}{3}$

## Pronumerals on Both Sides of the Equation

More complex equations have pronumerals and numbers on both sides.

### For Example

Solve the following equations:

1. $9t + 14 = 7t$
2. $5y + 17 = 3y + 29$
3. $11t - 12 = 7t + 8$
4. $5t + 13 = 9t + 25$
5. $9m - 15 = 5m - 3$
6. $4y + 11 = 19 - 2y$
7. $7 - 3t = 22 - 8t$

**1** $9t + 14 = 7t$

[+ 14 moves to RHS and becomes − 14.]

$9t = 7t - 14$

$9t - 7t = -14$

[+ 7*t* becomes − 7*t* when it moves to RHS.]

$2t = -14$

$t = \frac{-14}{2}$

[× 2 becomes ÷ 2 when it moves to RHS.]

$t = -7$

**2** $5y + 17 = 3y + 29$

[+ 17 moves to RHS and becomes − 17.]

$5y = 3y + 29 - 17$

$5y = 3y + 12$

[+ 3*y* moves to LHS and becomes − 3*y*.]

$5y - 3y = 12$

$2y = 12$

[× 2 becomes ÷ 2 when it moves to RHS.]

$y = \frac{12}{2} = 6$

**3** $11t - 12 = 7t + 8$

[− 12 moves to RHS and becomes + 12.]

$11t = 7t + 8 + 12$

$11t = 7t + 20$

[+ 7*t* becomes − 7*t* as it moves to RHS.]

$11t - 7t = 20$

$4t = 20$

[× 4 becomes ÷ 4 when it moves to RHS.]

$t = \frac{20}{4}$

$t = 5$

**4** $5t + 13 = 9t + 25$

[+ 13 becomes − 13 as it moves to LHS.]

$5t = 9t + 25 - 13$

$5t = 9t + 12$

[+ 9*t* becomes − 9*t* as it moves to LHS.]

$5t - 9t = 12$

$-4t = 12$

[× − 4 becomes ÷ − 4 as it moves to RHS.]

$t = \frac{12}{-4}$

$t = -3$

5 $9m - 15 = 5m - 3$

$9m = 5m - 3 + 15$

$9m = 5m + 12$

$9m - 5m = 12$

$4m = 12$

$m = \frac{12}{4}$

$m = 3$

[ $-15$ becomes $+15$ as it moves to RHS. ]

[ $+5m$ becomes $-5m$ as it moves to RHS. ]

[ $\times 4$ becomes $\div 4$ as it moves to RHS. ]

6 $4y + 11 = 19 - 2y$

$4y = 19 - 11 - 2y$

$4y = 8 - 2y$

$4y + 2y = 8$

$6y = 8$

$y = \frac{8}{6}$

$y = 1\frac{1}{3}$

[ $+11$ moves to RHS and becomes $-11$. ]

[ $-2y$ moves to LHS and becomes $+2y$. ]

[ $\times 6$ moves to RHS and becomes $\div 6$. ]

7 $7 - 3t = 22 - 8t$

$-3t = 22 - 7 - 8t$

$-3t = 15 - 8t$

$-3t + 8t = 15$

$5t = 15$

$t = \frac{15}{5}$

$t = 3$

[ $+7$ moves to RHS and becomes $-7$. ]

[ $-8t$ moves to LHS and becomes $+8t$. ]

[ $\times 5$ moves to RHS and becomes $\div 5$. ]

## Equations with Grouping Symbols

Remove grouping symbols before moving any terms.

### For Example

1 $5(x + 7) = 45$

2 $6(x - 3) = 12$

3 $8(a + 7) = 5(a + 4)$

4 $7(y - 3) + 18 = 11$

5 $5(2t + 7) + 3(t - 8) = 50$

6 $8(t + 7) - 5(t + 3) = 27$

7 $7(2y - 5) - 8(y - 9) = 19$

1 $5(x + 7) = 45$

$5x + 35 = 45$

$5x = 45 - 35$

$5x = 10$

$x = \frac{10}{5}$

$x = 2$

[ Remove grouping symbols. ]

2 $6(x - 3) = 12$

$6x - 18 = 12$

$6x = 12 + 18$

$6x = 30$

$x = \frac{30}{6}$

$x = 5$

3 $8(a + 7) = 5(a + 4)$

$8a + 56 = 5a + 20$

$8a = 5a + 20 - 56$

$8a = 5a - 36$

$8a - 5a = -36$

$3a = -36$

$a = \frac{-36}{3}$

$a = -12$

4 $7(y - 3) + 18 = 11$

$7y - 21 + 18 = 11$

$7y - 3 = 11$

$7y = 11 + 3$

$7y = 14$

$y = \frac{14}{7}$

$y = 2$

5 $5(2t + 7) + 3(t - 8) = 50$

$10t + 35 + 3t - 24 = 50$

$13t + 11 = 50$

$13t = 50 - 11$

$13t = 39$

$t = \frac{39}{3}$

$t = 3$

**6** $8(t+7) - 5(t+3) = 27$

$8t + 56 - 5t - 15 = 27$

$3t + 41 = 27$

$3t = 27 - 41$

$3t = -14$

$t = \frac{-14}{3}$

$t = -4\frac{2}{3}$

[Note that second expansion is $-5(t+3)$.]

**7** $7(2y-5) - 8(y-9) = 19$

$14y - 35 - 8y + 72 = 19$

$6y + 37 = 19$

$6y = 19 - 37$

$6y = -18$

$y = \frac{-18}{6}$

$y = -3$

[Second expansion is $-8(y-9)$.]

## Equations with Fractions

### Single Fraction

Multiply *both* sides of the equation by the denominator of the fraction.

**For Example**

**1** $\frac{x+7}{4} = 9$  **2** $\frac{3x-4}{5} = 2$

**3** $\frac{4(x+3)}{5} = 7$  **4** $\frac{3}{4}(x-2) = 9$

**1** $\cancel{4} \times \frac{x+7}{\cancel{4}} = 9 \times 4$

$x + 7 = 36$

$x = 36 - 7$

$x = 29$

[Multiply both sides by 4.]

**2** $\cancel{5} \times \frac{3x-4}{\cancel{5}} = 2 \times 5$

$3x - 4 = 10$

$3x = 10 + 4$

$3x = 14$

$x = \frac{14}{3}$

$x = 4\frac{2}{3}$

[Multiply both sides by 5.]

**3** $\cancel{5} \times \frac{4(x+3)}{\cancel{5}} = 7 \times 5$

$4(x+3) = 35$

$4x + 12 = 35$

$4x = 35 - 12$

$4x = 23$

$x = \frac{23}{4}$

$x = 5\frac{3}{4}$

[Multiply both sides by 5.]

**4** $4 \times \frac{3}{4}(x-2) = 9 \times 4$

$3(x-2) = 36$

$3x - 6 = 36$

$3x = 36 + 6$

$3x = 42$

$x = \frac{42}{3}$

$x = 14$

[Multiply both sides by 4.]

### More than One Fraction

Multiply all terms by Lowest Common Denominator (LCD).

Remember your work on simple fractions. The Lowest Common Denominator of the fractions:

LCD of $\frac{3}{4}$ and $\frac{2}{3}$ is 12

LCD of $\frac{3}{4}$ and $\frac{1}{2}$ is 4

LCD of $\frac{2}{5}$ and $\frac{1}{2}$ is 10.

Multiplying by the LCD will remove all fractions from the equation.

## For Example

Solve:

**1** $\frac{x}{2} - \frac{x}{3} = 7$

**2** $\frac{x}{3} + \frac{x}{5} = 4$

**3** $a - \frac{a}{5} = 4$

**4** $\frac{t+3}{4} + \frac{t}{8} = 3$

**5** $\frac{y-7}{5} - \frac{y}{3} = 2$

**6** $a + \frac{a+2}{5} = 7$

**7** $y - \frac{y-2}{4} = 3$

**8** $\frac{a+7}{3} + \frac{a-4}{2} = 5$

**9** $\frac{y-7}{3} - \frac{y-4}{5} = 2$

**10** $a - \frac{a-2}{2} - \frac{a-3}{3} = 5$

**1** $\not{6}^3 \times \frac{x}{\not{2}_1} - \not{6}^2 \times \frac{x}{\not{3}_1} = 7 \times 6$

[Multiply all terms by LCD, which is 6.]

$$3x - 2x = 42$$
$$x = 42$$

**2** $\not{15}^5 \times \frac{x}{\not{3}_1} + \not{15}^3 \times \frac{x}{\not{5}_1} = 4 \times 15$

[Multiply all terms by LCD = 15.]

$$5x + 3x = 60$$
$$8x = 60$$
$$x = \tfrac{60}{8}$$
$$x = 7\tfrac{1}{2}$$

**3** $5 \times a - \not{5}^1 \times \frac{a}{\not{5}_1} = 4 \times 5$

[Multiply all terms by LCD = 5.]

$$5a - a = 20$$
$$4a = 20$$
$$a = \tfrac{20}{4}$$
$$a = 5$$

**4** $\not{8}^2 \times \frac{(t+3)}{\not{4}_1} + \not{8}^1 \times \frac{t}{\not{8}_1} = 3 \times 8$

[Multiply all terms by LCD = 8.]

$$2(t+3) + t = 24$$
$$2t + 6 + t = 24$$
$$3t + 6 = 24$$
$$3t = 24 - 6$$
$$= 18$$
$$t = \tfrac{18}{3}$$
$$t = 6$$

[Remember, you can substitute your answer into the original equation to check that you are correct.]

**5** $\not{15}^3 \times \frac{(y-7)}{\not{5}_1} - \not{15}^5 \times \frac{y}{\not{3}_1} = 2 \times 15$

[Multiply both sides by 15.]

$$3(y-7) - 5y = 30$$
$$3y - 21 - 5y = 30$$
$$-2y - 21 = 30$$
$$-2y = 30 + 21$$
$$-2y = 51$$
$$y = \tfrac{51}{-2}$$
$$y = -25\tfrac{1}{2}$$

**6** $5 \times a + \not{5}^1 \times \frac{(a+2)}{\not{5}_1} = 7 \times 5$

[Multiply all terms by LCD = 5.]

$$5a + a + 2 = 35$$
$$6a + 2 = 35$$
$$6a = 35 - 2$$
$$6a = 33$$
$$a = \tfrac{33}{6}$$
$$a = 5\tfrac{1}{2}$$

**7** $4 \times y - \cancel{4}^{1} \times \dfrac{(y-2)}{\cancel{4}_{1}} = 3 \times 4$

$$4y - (y - 2) = 12$$
$$4y - y + 2 = 12$$
$$3y + 2 = 12$$
$$3y = 12 - 2$$
$$3y = 10$$
$$y = \tfrac{10}{3}$$
$$y = 3\tfrac{1}{3}$$

**8** $\cancel{6}^{2} \times \dfrac{(a+7)}{\cancel{3}_{1}} + \cancel{6}^{3} \times \dfrac{(a-4)}{\cancel{2}_{1}} = 5 \times 6$

[Multiply all terms by LCD = 6.]

$$2(a + 7) + 3(a - 4) = 30$$
$$2a + 14 + 3a - 12 = 30$$
$$5a + 2 = 30$$
$$5a = 30 - 2$$
$$5a = 28$$
$$a = \tfrac{28}{5}$$
$$= 5\tfrac{3}{5}$$

**9** $\dfrac{y-7}{3} - \dfrac{y-4}{5} = 2$

[Multiply by LCD = 15.]

$$\cancel{15}^{5} \times \frac{(y-7)}{\cancel{3}_{1}} - \cancel{15}^{3} \times \frac{(y-4)}{\cancel{5}_{1}} = 2 \times 15$$
$$5(y - 7) - 3(y - 4) = 30$$
$$5y - 35 - 3y + 12 = 30$$
$$2y - 23 = 30$$
$$2y = 30 + 23$$
$$= 53$$
$$y = \tfrac{53}{2}$$
$$y = 26\tfrac{1}{2}$$

**10** $a - \dfrac{a-2}{2} - \dfrac{a-3}{3} = 5$

[Multiply all terms by LCD = 6.]

$$6 \times a - \cancel{6}^{3} \times \frac{(a-2)}{\cancel{2}_{1}} - \cancel{6}^{2}\frac{(a-3)}{\cancel{3}_{1}} = 5 \times 6$$
$$6a - 3(a - 2) - 2(a - 3) = 30$$
$$6a - 3a + 6 - 2a + 6 = 30$$
$$a + 12 = 30$$
$$a = 30 - 12$$
$$a = 18$$

# Problem Solving with Equations

## Mathematical Expressions

A sentence can be rewritten in symbols by using sum (+), difference (–), product (×) and division (÷).

**For Example**

Write down mathematical expressions for:

1. The sum of $x$ and 4.
2. The product of $y$ and 21.
3. The sum of $N$ and the next two consecutive whole numbers.
4. The sum of $M$ and the next even number (assuming $M$ is even).
5. The number of dollars in $y$ cents.

**1** Sum $= x + 4$

**2** Product $= y \times 21$
$= 21y$

3 $\text{Sum} = N + (N + 1) + (N + 2)$

$= N + N + N + 1 + 2$

$= 3N + 3$

There is one difference between consecutive whole numbers, i.e. 4, 5, 6 … or $N, N + 1, N + 2, \ldots$

There are two differences between consecutive even (or odd) numbers, i.e. 4, 6, 8, …
3, 5, 7, …
or $N, N + 2, N + 4, \ldots$

4 $\text{Sum} = M + (M + 2)$

$= 2M + 2$

5 $\text{Dollars} = \$(y \div 100)$

$= \$\left(\frac{y}{100}\right)$

## Mathematical Equations

Many problems are best solved using algebraic equations. The unknown quantity in the problem is represented by a pronumeral in your statement of the problem.

For example, consider this problem.

Jack is 7 years younger than Jill and the sum of their ages is 53. Find the age of Jack.

**Step one:** Start with your pronumeral. Let Jack's age be '$a$' years.

**Step two:** Construct an equation.

Jill's age must be $(a + 7)$ years (Jill is 7 years older than Jack).

Then $a + (a + 7) = 53$

$\therefore 2a + 7 = 53$

**Step three:** Solve the equation.

$2a + 7 = 53$

$2a = 53 - 7$

$2a = 46$

$a = \frac{46}{2}$

$a = 23$

**Step four:** Answer the problem.

This must always be a specific, written statement.

Jack's age is 23.

In reality, when solving problems the headings in steps 1 to 4 are not used. The steps are:

- Start with your own pronumeral
- Construct your equation
- Solve your equation
- Write down your answer to the problem (in words).

For Example

1 Five times a certain number, less 13, is equal to 47. Find the number.

1 Let the number be $n$. [STEP 1]

Then $5n - 13 = 47$ [STEP 2]

$5n = 47 + 13$

$5n = 60$ [STEP 3]

$n = \frac{60}{5}$

$n = 12$

The number is 12. [STEP 4]

For Example

1 The sum of two consecutive odd numbers is 60. Find the value of the two numbers.

1 Let one number be $N$.

Then next odd number is $(N + 2)$.

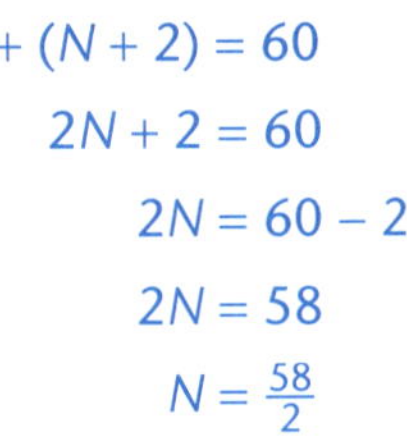


$N + (N + 2) = 60$

$2N + 2 = 60$

$2N = 60 - 2$

$2N = 58$

$N = \frac{58}{2}$

$N = 29$

The two numbers are 29 and 31.

## For Example

1 The length of a rectangle is three times the width of the rectangle. The perimeter of the rectangle is 144 cm. Find the dimensions of the rectangle.

[With geometric problems, always draw a diagram and mark the information on the diagram.]

1 Let the width be $w$ cm.

$\therefore$ Length = $3w$ cm

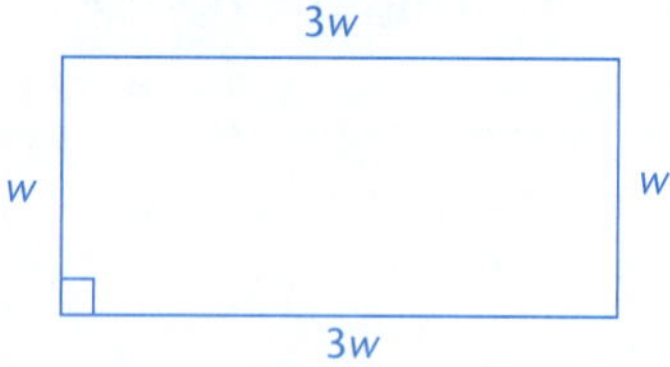


Perimeter $= w + 3w + w + 3w$

$= 8w$

Then $8w = 144$

$w = \frac{144}{8}$

$w = 18$

The width is 18 cm and length is 54 cm.

## For Example

1 Benny is twice as old as his son Lenny. Twelve years ago he was three times as old as Lenny. Calculate the present ages of Benny and Lenny.

1 Let Lenny's present age be $y$ years old.

$\therefore$ Benny's present age is $2y$ years old.

Twelve years ago Lenny's age was $(y - 12)$ years old.

Twelve years ago Benny's age was $(2y - 12)$ years old.

$2y - 12 = 3(y - 12)$

$2y - 12 = 3y - 36$

$2y = 3y - 36 + 12$

$2y = 3y - 24$ [Benny' s age was 3 times Lenny' s age.]

$2y - 3y = -24$

$-y = -24$

$y = 24$

Presently Lenny is 24 years old, while Benny is 48 years old. (This can be checked because 12 years ago, Lenny was 12, and Benny was 36.)

## For Example

1 At a supermarket Lim purchased $21 worth of groceries using only $2 and $1 coins. One and a half times as many $1 coins were used as $2 coins. How many $1 coins and $2 coins were used?

1 Let number of $2 coins be $x$.

Then number of $1 coins used is $\frac{3x}{2}$.

Value of $2 coins is $\$(2 \times x)$, i.e. $\$2x$

Value of $1 coins is $\$(1 \times \frac{3x}{2})$, i.e. $\$\frac{3x}{2}$

Then $1\frac{1}{2}x + 2x = 21$ [Total is $21.]

Then $2 \times \frac{3x}{2} + 2 \times 2x = 21 \times 2$

$3x + 4x = 42$

$7x = 42$

$x = \frac{42}{7}$

$x = 6$

The number of $2 coins is six, while there are nine $1 coins.

### For Example

**1** Anita and Lisa cycle towards each other between Muswellbrook and Wyong, a distance of 192 km. If they meet along the road after 8 hours and Anita is travelling three times as fast as Lisa, find the speed at which they are travelling.

**1** Let Lisa's speed be $v$ km/h

$\therefore$ Anita's speed is $3v$ km/h.

They travel for 8 hours.

Distance travelled = Speed × Time taken

For Lisa: Distance travelled $= v \times 8 = 8v$ km

For Anita: Distance travelled $= 3v \times 8$

$= 24v$ km

24$v$ km — 8$v$ km

Muswellbrook — Wyong

192 km

As they meet along the road:

$$24v + 8v = 192$$
$$32v = 192$$
$$v = \frac{192}{32}$$
$$v = 6$$

$\therefore$ Lisa's speed is 6 km/h while Anita's speed is 18 km/h.

## Formulae

A formula is a mathematical statement connecting more than one variable with an equals sign.

If you know the values of all but one of the variables, the unknown variable can be obtained by substituting for the known variables.

### For Example

**1** If $A = 2h(L + B)$, find the value of $A$ given that $h = 7$, $L = 19$ and $B = 11$.

**1**

| | |
|---|---|
| $A = 2h(L + B)$ | $h = 7$ |
| $= 2 \times 7 \times (19 + 11)$ | $L = 19$ |
| $= 14 \times 30$ | $B = 11$ |
| $= 420$ | $A = ?$ |

### For Example

**1** Given that $s = \frac{1}{2}gt^2$, calculate the value of $s$ if $g = -10$ and $t = 4$.

**1**

| | |
|---|---|
| $s = \frac{1}{2}gt^2$ | $g = -10$ |
| $= \frac{1}{2} \times -10 \times (4)^2$ | $t = 4$ |
| $= -5 \times 16$ | $s = ?$ |
| $= -80$ | |

## Equations Arising from Substitution in Formulae

### For Example

**1** If $P = 2L + 2B$ and $P = 75$, $L = 15$ find the value of $B$.

**1**

| | |
|---|---|
| $P = 2L + 2B$ | $P = 75$ |
| $75 = 2 \times 15 + 2B$ | $L = 15$ |
| $75 = 30 + 2B$ | $B = ?$ |
| $75 - 30 = 2B$ | |
| $45 = 2B$ | |
| $\frac{45}{2} = B$ | |
| $B = 22\frac{1}{2}$ | |

## For Example

**1** Given that $A = \frac{1}{2}xy$ and when $x = 12$, $A = 60$, find the value of $y$.

**2** Given that $p = a + (n - 1)b$ and that, when $p = 59$, $a = 17$ and $b = 6$, calculate the value of $n$.

**1**

$A = \frac{1}{2}xy$

$60 = \frac{1}{2} \times 12 \times y$

$60 = 6y$

$6y = 60$

$y = 10$

**2**

$p = a + (n - 1)b$

$59 = 17 + (n - 1)6$

$59 = 17 + 6n - 6$

$59 = 11 + 6n$

$\begin{bmatrix} p = 59 \\ a = 17 \\ b = 6 \\ n = ? \end{bmatrix}$

i.e. $6n + 11 = 59$

$6n = 59 - 11$

$6n = 48$

$n = \frac{48}{6}$

$n = 8$

Go to p. 286 for quick answers, or to pp. 320–331 for worked solutions.

**1** Show that $x = 3$ is a solution of these equations: p. 88

**a** $7x - 15 = 6$ **b** $11 - 2x = 7x - 16$

**2** Decide whether the given values are solutions of the given equation: p. 88

**a** $8y - 5 = 11$ $[y = 2]$ **b** $3(y + 2) = 27$ $[y = 7]$

**c** $30 - 4x = 8$ $[x = 6]$ **d** $\frac{x}{5} - 11 = -9$ $[x = 10]$

**3** Solve the following one-step equations: pp. 88–89

**a** $y + 7 = 11$ **b** $y + 11 = 7$ **c** $y - 11 = 7$
**d** $y - 7 = 11$ **e** $y + 17 = 4$ **f** $15 + t = 4$
**g** $19 = 11 + m$ **h** $t - 5 = -2$ **i** $p + 13 = 24$
**j** $m - 17 = 4$ **k** $n + 31 = 14$ **l** $11 - t = 4$
**m** $15 - y = 17$ **n** $4a = 24$ **o** $12t = 72$
**p** $6x = 48$ **q** $3a = -27$ **r** $7t = -35$
**s** $-4y = 16$ **t** $-8t = -64$ **u** $\frac{t}{6} = 12$
**v** $\frac{z}{3} = 9$ **w** $\frac{a}{5} = -6$ **x** $\frac{n}{-8} = 6$
**y** $\frac{m}{-7} = -4$ **z** $27 = 9x$

**4** Solve these equations. Write answers as fractions in simplest terms: pp. 88–89

**a** $4y = 9$ **b** $6a = 15$ **c** $9t = 21$
**d** $5a = 14$ **e** $5y = 8$ **f** $7m = 30$
**g** $9a = 24$ **h** $8a = 14$ **i** $4a = 3$
**j** $15t = 10$ **k** $12y = 30$ **l** $-8a = 10$
**m** $6y = -20$ **n** $-8t = -20$ **o** $24z = 30$
**p** $15m = 25$ **q** $3y = 0$

**5** Solve these simple two-step equations: pp. 88–89

**a** $8a + 3 = 27$ **b** $6x - 3 = 15$ **c** $4y + 13 = 29$
**d** $7t - 15 = 13$ **e** $11t - 3 = 30$ **f** $9y + 17 = 53$
**g** $21 = 2t + 7$ **h** $13 + 9y = 40$ **i** $11n - 31 = 13$
**j** $7a = 3a + 24$ **k** $5t = t + 36$ **l** $13y = 8y - 35$
**m** $3m = 7m + 32$ **n** $5y = 9y - 16$ **o** $\frac{a}{2} - 3 = 7$
**p** $7a + 21 = 14$ **q** $3t + 17 = 2$ **r** $11n + 47 = 3$
**s** $15 - 7t = 29$ **t** $11 - 3n = 2$

**6** Solve these equations, leaving your answers as mixed numbers in their simplest terms: pp. 89–91

| | | |
|---|---|---|
| **a** $5a-2=4$ | **b** $2y+7=12$ | **c** $7t+13=24$ |
| **d** $5a+11=17$ | **e** $4t-9=6$ | **f** $25=11+4t$ |
| **g** $27+2m=24$ | **h** $8a-13=4$ | **i** $6-3t=17$ |
| **j** $12y-15=7$ | **k** $30-2y=25$ | **l** $24t+15=47$ |
| **m** $13t=8t+7$ | **n** $5m=7m+9$ | |

**7** Solve the following equations. Where necessary, answers should be left as fractions or mixed numbers in their simplest terms: pp. 89–91

| | |
|---|---|
| **a** $5a+14=3a$ | **b** $3y-2=y+8$ |
| **c** $8y+13=5y+26$ | **d** $9n-4=5n+16$ |
| **e** $11a-17=6a+8$ | **f** $8y-19=5y-7$ |
| **g** $13t-19=7t-1$ | **h** $10y+21=6y+1$ |
| **i** $30y+27=15y+12$ | **j** $13m-17=6m+25$ |
| **k** $5+3t=17-t$ | **l** $3-5a=17-2a$ |
| **m** $5y-32=4-3y$ | **n** $25-7a=27-4a$ |

**8** Solve the following equations: pp. 91–92

| | |
|---|---|
| **a** $6(x-3)=12$ | **b** $5(a+2)=35$ |
| **c** $4(t-5)=24$ | **d** $32=4(a+3)$ |
| **e** $6(y-7)=12$ | **f** $3(2m+5)=33$ |
| **g** $5(t-6)=15$ | **h** $6(a+3)-11=5$ |
| **i** $8(y-2)=5(y+3)$ | **j** $6(t+3)+2(t-4)=26$ |
| **k** $9(t+3)-4(t-2)=6$ | **l** $8(y-6)-6(y-8)=24$ |
| **m** $7a-4(a-2)=0$ | **n** $15a-7(a-2)-5(a-3)=21$ |
| **o** $7a-(5a-2)=26$ | **p** $3a-(7-2a)=42$ |
| **q** $11-7(2-a)=18$ | **r** $3(3x+1)-2(x-3)=5(x+7)$ |

**s** $3(2x+5)-4(x-3)=3(5x+1)-2$

**t** $11(a-2)-4(2-a)-2(1-4a)=19(a-3)$

**9** Solve the following equations containing fractions: pp. 92–94

| | | |
|---|---|---|
| **a** $\dfrac{a-2}{5}=3$ | **b** $\dfrac{y+2}{3}=4$ | **c** $\dfrac{3-2a}{4}=5$ |
| **d** $\dfrac{2a+5}{3}=7$ | **e** $\dfrac{5(a-2)}{3}=2$ | **f** $\frac{4}{5}(x+3)=6$ |

**10** Solve these equations: pp. 92–94

| | |
|---|---|
| **a** $\dfrac{y}{3}+\dfrac{y}{5}=16$ | **b** $\dfrac{a}{3}-\dfrac{a}{5}=12$ |
| **c** $\dfrac{3y}{4}-\dfrac{y}{2}=5$ | **d** $t+\dfrac{t}{2}=6$ |
| **e** $a-\dfrac{a}{4}=9$ | **f** $\dfrac{a}{5}+\dfrac{a-2}{3}=4$ |
| **g** $\dfrac{m+2}{4}-\dfrac{m}{3}=2$ | **h** $t+\dfrac{t-3}{3}=5$ |

**i** $m - \dfrac{m-5}{3} = 5$ **j** $\dfrac{a+2}{3} + \dfrac{a+3}{4} = 8$

**k** $\dfrac{3y-2}{3} - \dfrac{y-3}{2} = 4$ **l** $\frac{2}{5}(x+4) - \frac{3}{4}(x-2) = 3$

**11** Try to solve these equations. They are much harder: pp. 92–94

**a** $7 - \dfrac{3x}{5} = 3 - \dfrac{2x}{7}$

**b** $x - 3 - \dfrac{x-2}{2} - \dfrac{x-4}{3} = 1$

**c** $\dfrac{x}{4} - \dfrac{x-2}{5} = 15 + \dfrac{7-x}{2}$

**d** $\frac{1}{3}(x-3) - \frac{1}{5}(x-5) = \frac{3}{4}(3x-4) + 5$

**e** $\dfrac{3x+2}{5} - 3(x-2) = 8 - \dfrac{2(3x+1)}{3} - \dfrac{5(4x-2)}{6}$

**12** Write an algebraic expression for each statement: pp. 94–95

**a** The sum of *a* and *b*.
**b** The product of *a* and *b*.
**c** The next even number after *y* (*y* is even).
**d** The next even number after *y* (*y* is odd).
**e** The sum of *p* and the next consecutive whole number.
**f** The number of kilometres in *x* metres.
**g** The number of metres in *x* kilometres.
**h** The difference between *a* and 2*b* ($2b > a$).

**13** Write an equation and then solve it to answer these simple problems: pp. 95–97

**a** If 19 is added to a number the result is 37. Find the number.
**b** The product of a number and 6 is 84. Find the number.
**c** When a number is doubled and then 7 is added the answer is 35. Find the number.
**d** Five times a certain number less 11 gives an answer of 64. Find the number.
**e** Three times a number is subtracted from 76. If the answer is 52, find the number.

**14** Form equations in order to answer these problems: pp. 95–97

**a** Six times the sum of a number and 3 is equal to twice the number plus 30. Find the number.
**b** The sum of three consecutive whole numbers is 42. Find the numbers.
**c** The sum of two consecutive odd numbers is 64. Find the numbers.
**d** Mai has three times as many teddy bears as Tony. Altogether they have 24 teddy bears. How many bears has Mai?
**e** Uncle Scrooge has made a strange bequest in his will. Huey is to receive three times as much as Duey, while Louie is to receive twice as much as Duey less one million dollars. If the total amount to be divided between them is 107 million dollars, calculate how much each is to inherit.

**15** Write an equation and then solve it for each of the following problems: pp. 95–97

**a**

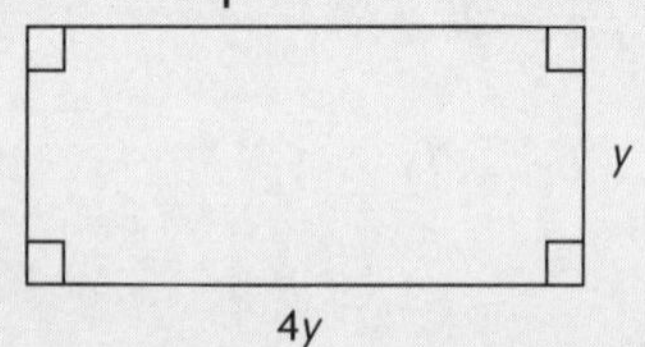


Perimeter = 105 metres
Find the value of $y$.

**b**

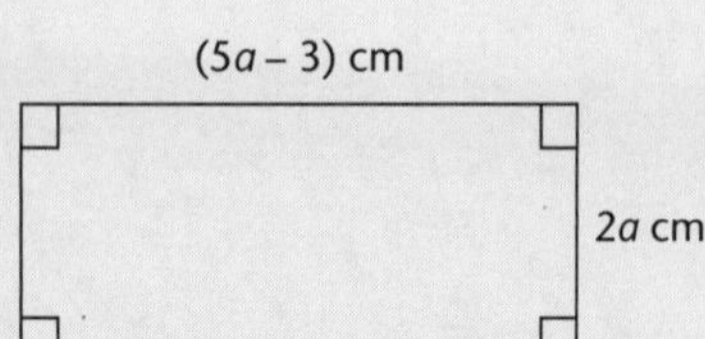


Perimeter = 78 cm
Calculate the value of $a$.

**c**

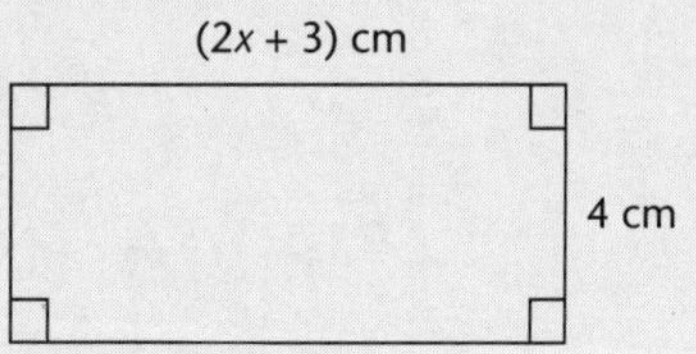


Area = 84 $cm^2$
Calculate the value of $x$.

**d**

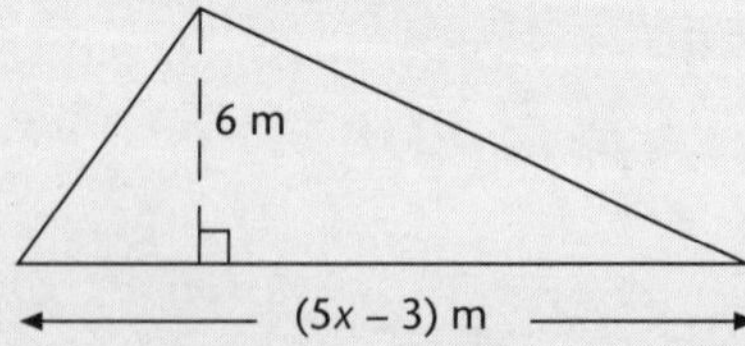


Area = 51 $m^2$
Calculate the length of the base.

**e**

Calculate the value of $x$.

**f**

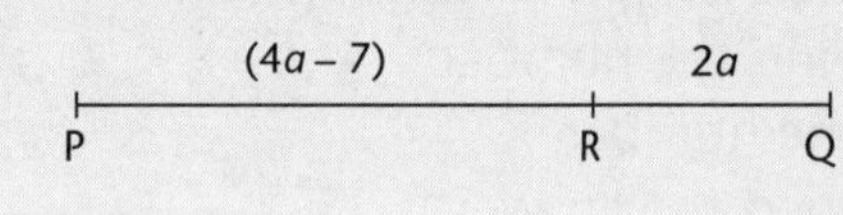


$PQ = 41$ cm
$PR = (4a - 7)$ cm
$RQ = 2a$ cm
Find the value of $a$.

**16** Use your knowledge of geometry to form equations. Solve the equations to calculate the value of these pronumerals: pp. 95–97

**a**

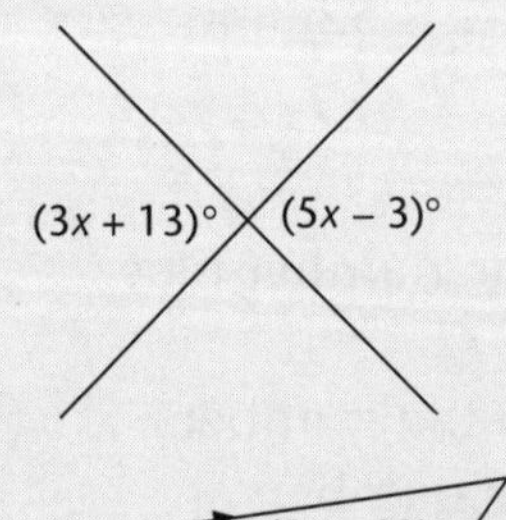


**b**

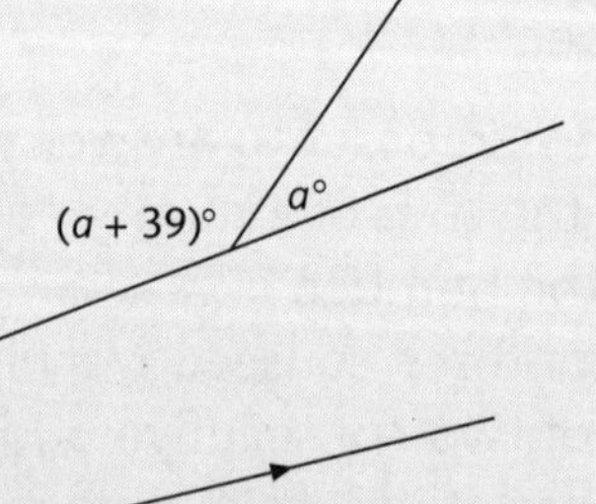


**c**

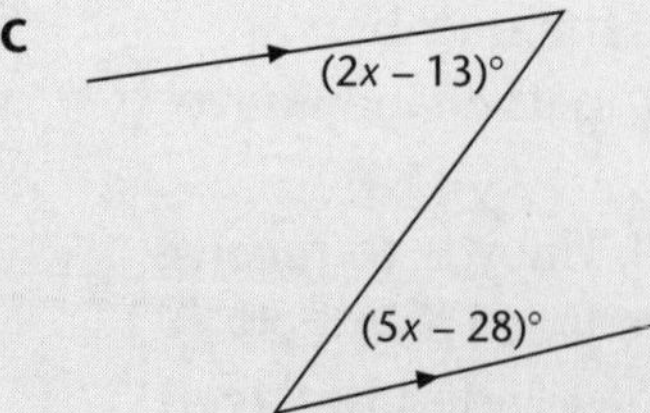


**d**

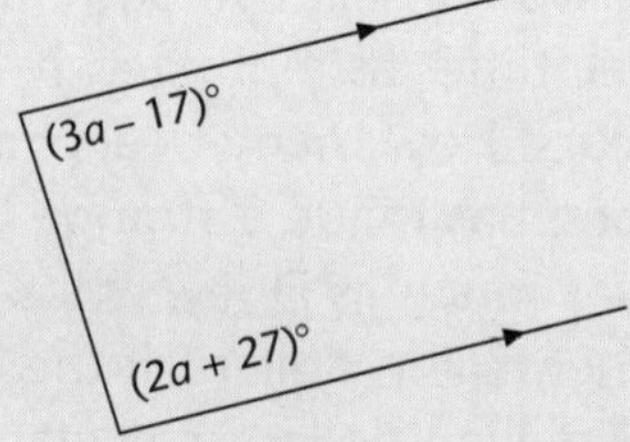

**17** Reis has $n$ apples. When he is given 30 more he has four times as many as he started with. Find the value of $n$. pp. 95–97

**18** Allyn is 31 years older than David. The sum of their ages is 49. How old is Allyn? pp. 95–97

**19** Aaron has an assortment of coins and notes in his money box. He has nine times as many \$1 coins as \$5 notes, while there are eight times as many \$2 coins as \$5 notes. After counting his money Aaron discovered he had exactly \$90. How many \$2 coins did Aaron have? pp. 95–97

**20** Brad kept pet ducks and sheep in his yard. He could never remember how many of each animal he had, but he always knew that the total number of legs was 56. If Brad had three times as many sheep as ducks, how many sheep did he have? pp. 95–97

**21** Lani walked a certain distance, cycled four times that distance, ran half the first distance and then swam one-quarter of the first distance. If the total distance covered was 46 km, how far did Lani swim? pp. 95–97

**22** In a zoo there are 17 more monkeys than lions and 30 more lizards than lions. If there were 161 monkeys, lions and lizards in total, how many of each kind are there? pp. 95–97

**23** Ben sails one-third of his trip at 4 km/h, the next third at 8 km/h and the last third at 6 km/h. The trip takes him 9 hours 45 minutes. How far was the complete trip? pp. 95–97

**24** For the given formulae, calculate the value of the unknown variable: pp. 97–98

**a** $v = u + at$, find $v$ given $u = 70$, $a = 8$, $t = 6$.
**b** $y = mx + b$, find $y$ given $m = 3$, $x = 5$, $b = -4$.
**c** $E = \frac{1}{2}mv^2$, find $E$ given $m = \frac{1}{2}$, $v = 8$.
**d** $D = \dfrac{M}{V}$, find $D$ when $M = 72$, $V = 16$.
**e** $I = \pi rs$, find $I$ when $\pi = 3.14$, $r = 10$, $s = 5$.
**f** $C = 2\pi r$, find $C$ when $\pi = 3\frac{1}{7}$, $r = 28$.
**g** $I = Prn$, find $I$ when $P = 500$, $r = 0.09$, $n = 3$.
**h** $S = \dfrac{n}{2}(a + l)$, find $S$ when $n = 31$, $a = 17$, $l = 35$.
**i** $T = a + (n - 1)d$, find $T$ when $a = -8$, $n = 14$, $d = 2$.
**j** $S = \dfrac{n}{2}(2a + [n - 1]d)$, find $S$ when $n = 16$, $a = 5$, $d = 4$.
**k** $C = \frac{5}{9}(F - 32)$, find $C$ when $F = 50$.
**l** $F = Ma$, find $F$ when $M = 0.25$, $a = 8$.
**m** $S = \dfrac{a}{1 - r}$, find $S$ when $a = 12$, $r = 0.5$.

**n** $F = 32 + \frac{9}{5}C$, find $F$ when $C = 10$.

**o** $V = \pi r^2 h$, find $V$ when $\pi = 3.14$, $r = 4$, $h = 8$.

**25** In these equations, find the value of the unknown letter: pp. 97–98

**a** $M = \dfrac{r}{r - w}$, find $M$ when $r = 15$, $w = 9$.

**b** $Y = \dfrac{ax}{x - p}$, find $Y$ when $a = \frac{1}{4}$, $x = 12$, $p = 6$.

**c** $A = lb$, find $b$ when $A = 36$, $l = 9$.

**d** $V = lbh$, find $b$ when $l = 3$, $h = 5$, $V = 90$.

**e** $I = Prn$, find $n$ when $P = 450$, $r = 0.1$, $I = 90$.

**f** $F = \dfrac{L}{L + S}$, find $S$ when $F = 4$, $L = 7$.

**g** $T = a + (n - 1)d$, find $d$ when $n = 7$, $a = 17$, $T = 71$.

**h** $A = \frac{1}{2}xy$, find $x$ when $A = 81$, $y = 18$.

Go to p. 286 for quick answers, or to pp. 320–331 for worked solutions.

## YOUR CHECKLIST

**For a complete understanding of this topic you must be able to:**

| | | | |
|---|---|---|---|
| ✓ | Test whether a solution satisfies an equation | | p. 88 |
| ✓ | Solve equations with one variable | | pp. 88–94 |
| ✓ | Use equations to solve word problems, geometric problems | | pp. 94–97 |
| ✓ | Evaluate a variable directly using a formula | | pp. 97–98 |
| ✓ | Solve equations arising from formulae. | | pp. 97–98 |

**Now you are ready to do the tests!**

**(40 marks)**

1 Show that $x = 5$ is a solution of $4x - 1 = 19$. (2 marks)

2 Solve the basic equations:

a $7x = 35$
b $4y = -12$
c $t + 11 = 19$
d $m - 7 = 13$ (4 marks)

3 Solve the equations:

a $3x - 5 = 16$
b $8y + 7 = 23$
c $4(m - 2) = 12$
d $8a = 3a + 30$
e $\frac{x - 3}{4} = 5$ (10 marks)

4 Solve the equations:

a $\frac{y}{2} + \frac{y}{3} = 10$
b $\frac{2t - 5}{3} = 1$ (6 marks)

5 Write mathematical statements for the:

a sum of $m$ and $n$
b number of cents in $\$y$
c next consecutive whole number after $a$. (6 marks)

6 Four times a certain number added to 15 gives a sum of 79. Find the number. (2 marks)

7 Given that $S = \frac{D}{t}$, evaluate $S$ when $D = 840$ and $t = 7$. (2 marks)

8 Solve the equations:

a $8y = -16$
b $-4y = -20$
c $\frac{x}{-3} = -2$ (3 marks)

9 Solve the equation $2 - 3x = 13$. (2 marks)

10 Solve and graph your solution on a number line:
$4x + 13 = 29$ (3 marks)

☞ Quick answers on page 292
☞ Worked solutions on page 381

**Your Feedback** $\frac{\square}{40} \times 100\% = \square\%$

# LEVEL 2 TEST

**(40 marks)**

1 Which of the given values is a solution of $3x - 12 = 7x$?

a $x = 3$ b $x = -3$ (2 marks)

2 Solve the one-step equations:

a $8x = 12$ b $y + 7 = 5$ c $m - 8 = -4$ d $\frac{t}{8} = 4$ (4 marks)

3 Solve the equations:

a $4y + 6 = 20$ b $7m - 3 = 39$ (2 marks)

c $6(m + 7) = 35$ (2 marks)

4 Solve:

a $5y - 3 = 3y + 15$ b $7(m + 2) = 3m + 30$ (4 marks)

c $\frac{m}{2} - \frac{m}{5} = 9$ d $\frac{m+2}{2} + \frac{m}{3} = 5$ (6 marks)

5 [Rectangle with length $3x - 1$ and width $x$]

The length of a rectangle is three times the width less one metre. The perimeter of the rectangle is 22 metres. Write an equation to show this information, and solve the equation to find the dimensions of the rectangle. (4 marks)

6 Given that $v = u + at$, and $u = 18$, $a = -3$ and $t = 5$, find the value of $v$. (3 marks)

7 a If $P = 2(l + b)$, find $b$ if $P = 46$ and $l = 15$. (2 marks)

b If $M = \frac{P}{x + y}$, find $P$ if $M = 8$, $x = 6$, $y = 2$. (2 marks)

c If $T = \frac{1}{2}p(a + b)$, find $p$ if $T = 12$, $a = 4$, $b = 2$. (3 marks)

8 Marcus is 6 years older than his sister Rebecca. The sum of their ages is 28. Let the age of Marcus be $x$. Solve an equation to find the age of both Marcus and Rebecca. (3 marks)

9 At a small school there are two classes in Year 8. One class has eight more students than the other class. Altogether there are 50 Year 8 students. How many students are in each class? (3 marks)

☞ **Quick answers on page 292**
☞ **Worked solutions on page 382**

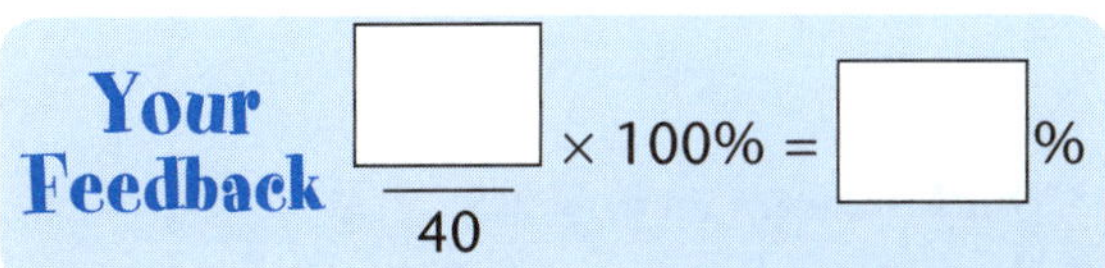

# LEVEL 3 TEST

**(40 marks)**

**1** Check whether these equations has a solution of $y = -2$:

**a** $7y + 9 = -5$ **b** $4 - 3y = -10$ **c** $\frac{9y}{2} - 4y = -1$ (3 marks)

**2** Solve the simple equations:

**a** $5x = -4$ **b** $y + 11 = 7$ **c** $\frac{t}{3} = -5$ (3 marks)

**3** Solve the equations:

**a** $3m - 11 = 5$ **b** $17 + 5y = 11$ **c** $3(2a - 5) = 12$ (6 marks)

**d** $5 - 7m = 12 - 2m$ **e** $7(2t - 5) - 4(t - 6) = 7$ (6 marks)

**4** Solve the equations:

**a** $4y + \frac{2y - 3}{2} = 8$ (3 marks)

**b** $\frac{x - 3}{2} - \frac{x - 2}{5} = 5$ (4 marks)

**5** Betty's age is 9 years more than twice the age of her cousin Brad. The sum of their ages is 30. Write an equation to show this information, and solve the equation to find Betty's age. (5 marks)

**6** Given that $v^2 = u^2 + 2as$ and that $v = 13$, $u = 11$ and $s = 3$, find the value of $a$. (3 marks)

**7** The size of an angle in a triangle is 20° more than one angle and 10° degrees more than the other. By solving an equation, find the size of each angle. (3 marks)

**8** The length of a rectangle is 4 cm less then twice its width. If the perimeter of the rectangle is 52 cm, solve an equation to find the dimensions of the rectangle. (4 marks)

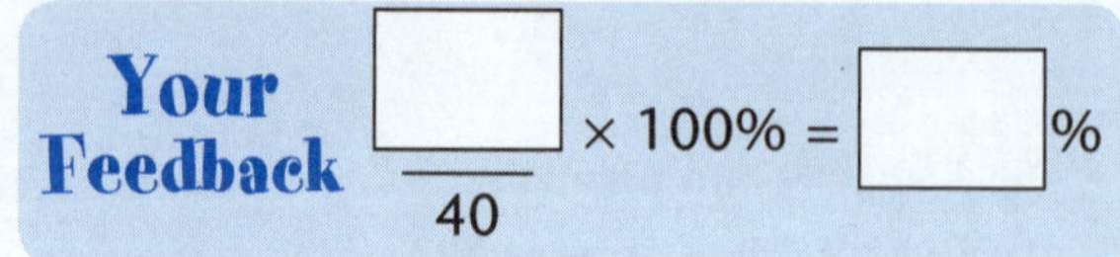


☞ **Quick answers on page 292**
☞ **Worked solutions on page 383**

# PYTHAGORAS' THEOREM

- Pythagoras' Theorem
- A Note on Squares and Square Roots
- Finding the Length of the Hypotenuse
- Finding the Length of a Side not the Hypotenuse
- Using Pythagoras' Theorem to Prove a Triangle is Right-Angled
- Pythagorean Triads (or Triples)
- Using the Calculator for Pythagoras' Theorem
- Problem Solving Using Pythagoras' Theorem

## KEYWORDS

| | |
|---|---|
| **Exact** | **Square root** |
| **Hypotenuse** | **Surd** |
| **Pythagoras** | **Theorem** |
| **Pythagorean** | **Triad** |
| **Right-angled** | **Triangle** |
| **Square** | |

## Pythagoras' Theorem

The hypotenuse is the longest side of the right-angled triangle. It is always the side opposite the right-angle.

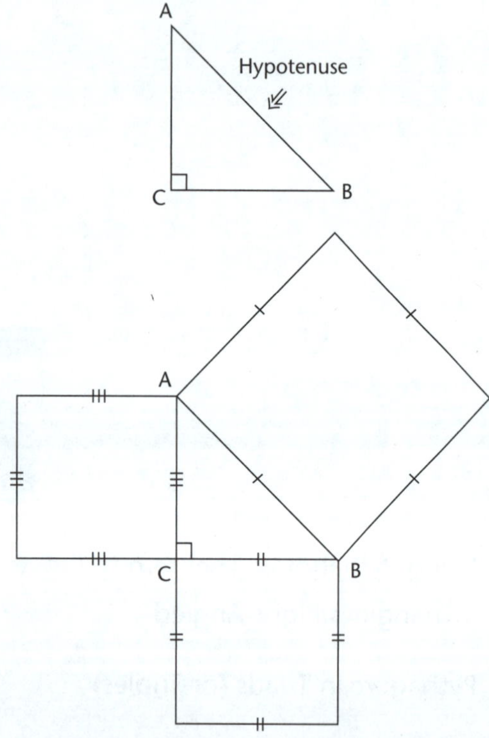


In any right-angled triangle the square of the hypotenuse is equal to the sum of the squares of the other two sides.

In simpler terms:

$c^2 = a^2 + b^2$

where $c$ is the length of the hypotenuse.

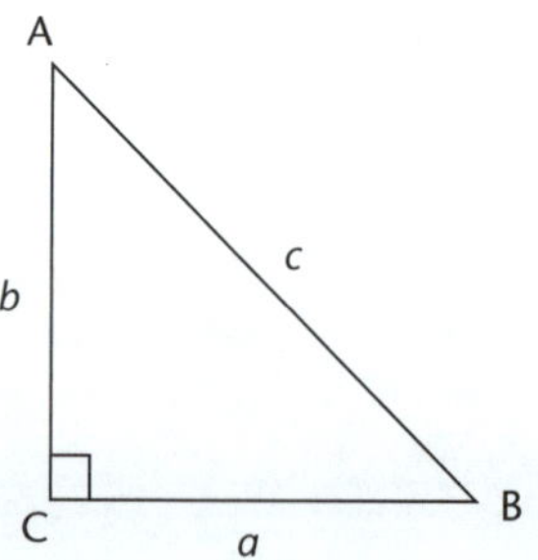

### For Example

In the following diagrams, state Pythagoras' Theorem using the markings on the diagrams:

**1**

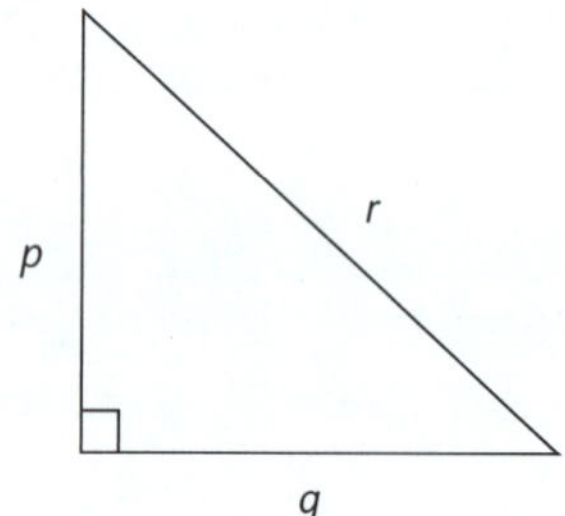


**2**

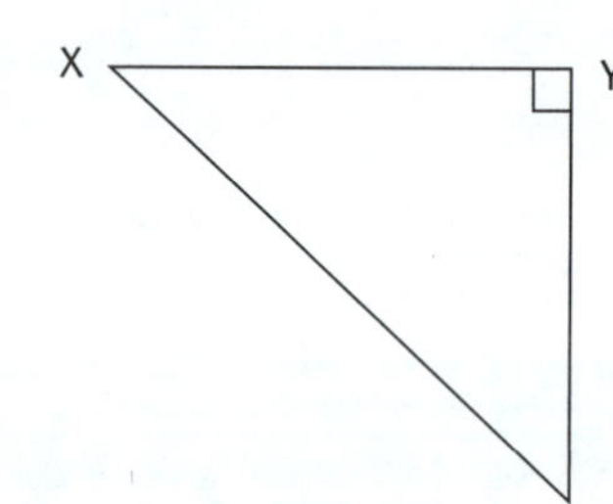


**3**

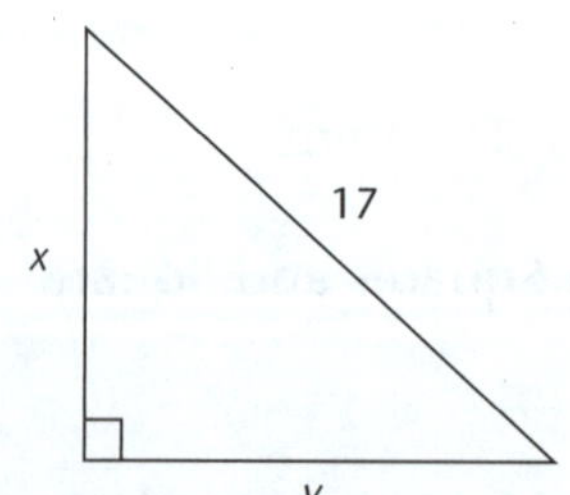


**4**

8

10

y

1

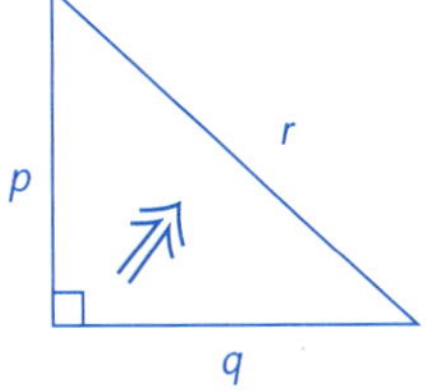


Locate the hypotenuse—draw an arrow directly across the triangle from the right-angle to the opposite side.

Hypotenuse = $r$

Then: $r^2 = p^2 + q^2$

[The hypotenuse squared = Sum of the squares of the other two sides.]

2

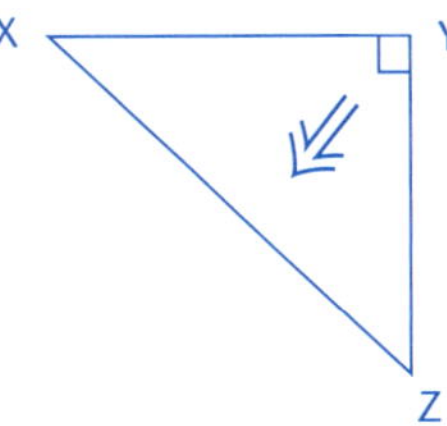


Sides are XZ, ZY, XY.

Hypotenuse is XZ (see arrow)

Then: $XZ^2 = XY^2 + YZ^2$

3

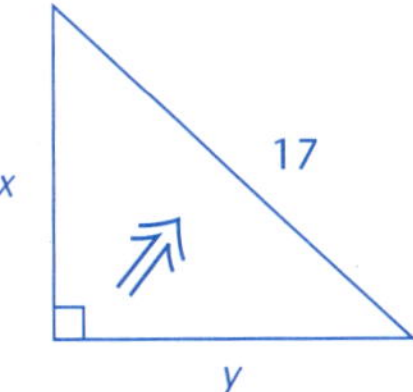


Hypotenuse = 17

Then: $17^2 = x^2 + y^2$

4

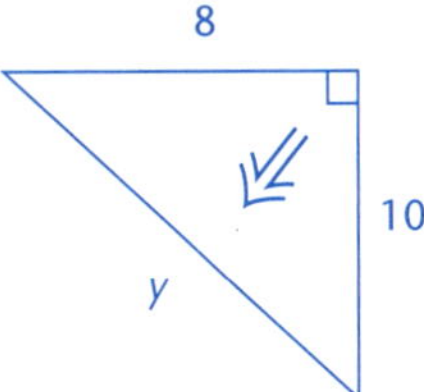


Hypotenuse = $y$

Then: $y^2 = 8^2 + 10^2$

# A Note on Squares and Square Roots

Pythagoras' Theorem requires familiarity with squares and square roots.

$$12^2 = 12 \times 12 = 144 \Rightarrow \sqrt{144} = 12$$

$$15^2 = 15 \times 15 = 225 \Rightarrow \sqrt{225} = 15$$

On your calculator, there will be a [$\sqrt{\ }$] button and a square [$x^2$] button. Or, they may be the same button, i.e. [$\sqrt{\ }$] (with $x^2$ above). In which case to square you must push: [INV] [$\sqrt{\ }$] (or [SHIFT] [$\sqrt{\ }$] or [2nd Function] [$\sqrt{\ }$]).

# Finding the Length of the Hypotenuse

Given the lengths of any two sides of a right-angled triangle, it is possible to calculate the length of the third side.

Consider this triangle to find the value of $t$:

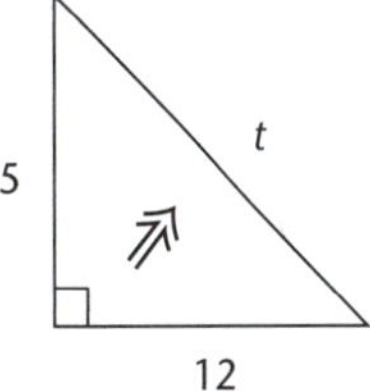


State Pythagoras' Theorem for the triangle.

Hypotenuse = $t$

$\therefore\ t^2 = 5^2 + 12^2$ [Evaluate $5^2 + 12^2$]

$= 25 + 144 = 169$

$\therefore\ t = \sqrt{169}$ [Calculate $\sqrt{169}$]

$= 13$

## For Example

1 Calculate the value of $y$ in this triangle:

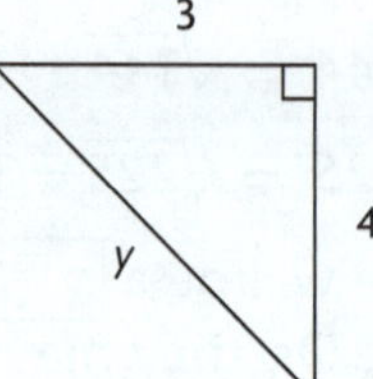


2 Find the length of the hypotenuse:

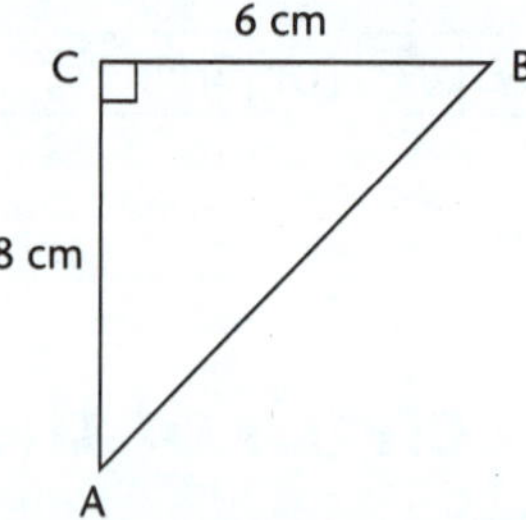


3 Calculate the length PQ in this triangle:

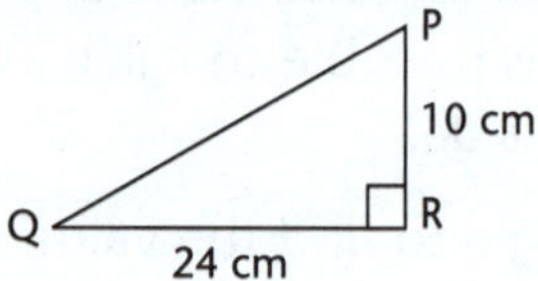


1

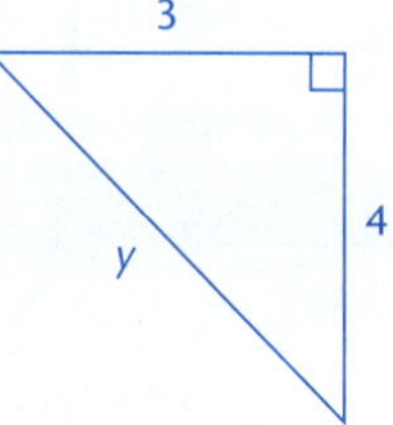


$y^2 = 3^2 + 4^2$

$= 9 + 16$

$= 25$

$\therefore y = \sqrt{25}$

$= 5$

2

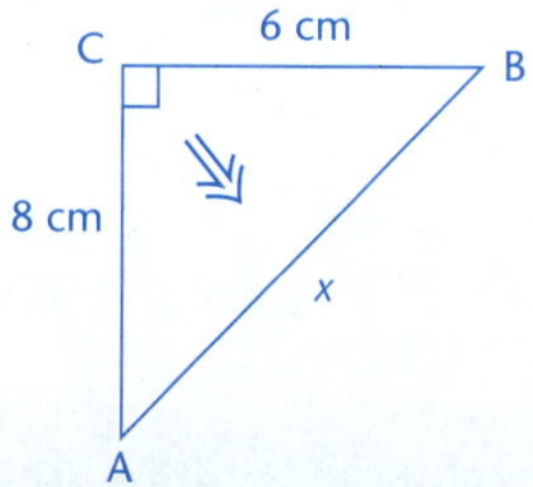


Hypotenuse = AB

Let length of AB be $x$ cm

$\therefore x^2 = 6^2 + 8^2$

$= 36 + 64$

$= 100$

$\therefore x = \sqrt{100}$

$= 10$

Then AB is 10 cm long.

3

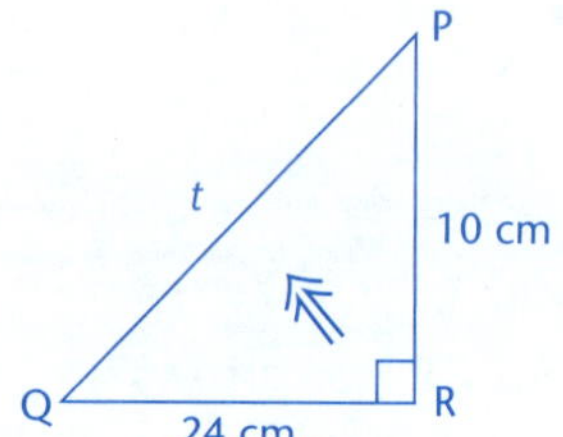


PQ is the hypotenuse

Let length of PQ be $t$ cm

$\therefore t^2 = 10^2 + 24^2$

$= 100 + 576$

$= 676$

$\therefore t = \sqrt{676}$

$= 26$

The length of PQ is 26 cm.

## Finding the Length of a Side Not the Hypotenuse

Two sides—one of which is the hypotenuse—are known and we need to find the length of the third side.

Consider this example:

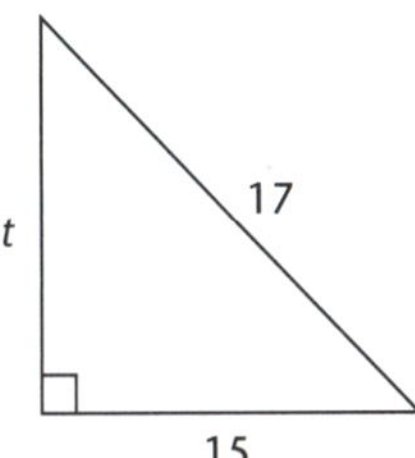


Note the hypotenuse is 17 units.

Using Pythagoras' Theorem:

$t^2 + 15^2 = 17^2$ — Keep the letter on the left side of the equation.

$\therefore t^2 = 17^2 - 15^2$ — Move $15^2$ to right side. It is then subtracted.

$= 289 - 225$

$= 64$

$\therefore t = \sqrt{64}$

$= 8$

## For Example

**1** Find the value of $t$ in this diagram:

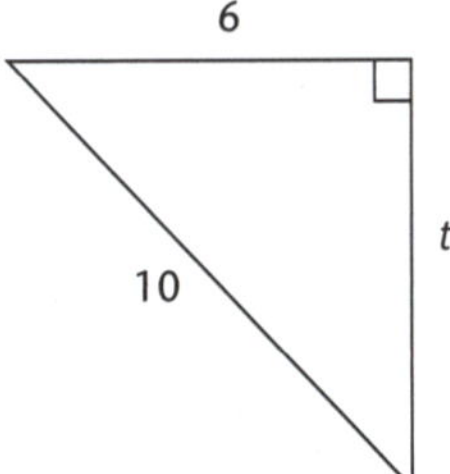


**2** Find the length of the side YZ given that XY = 15 cm and XZ = 12 cm.

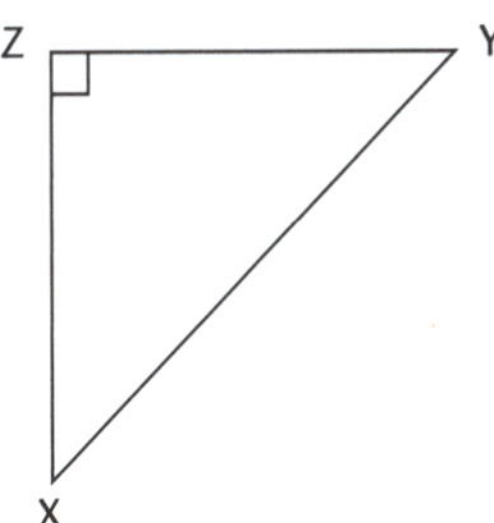


**1**

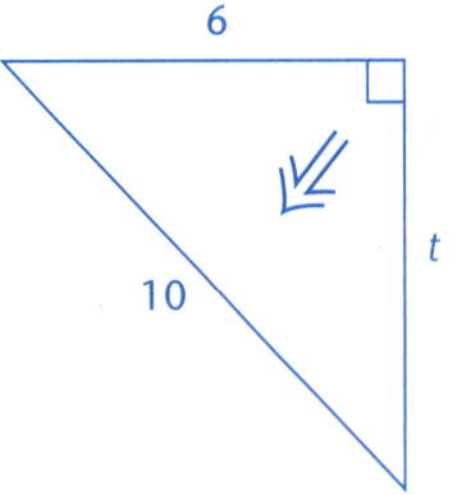


Hypotenuse = 10 units

$\therefore t^2 + 6^2 = 10^2$

$t^2 = 10^2 - 6^2$

$= 100 - 36$

$= 64$

$\therefore t = \sqrt{64}$

$= 8$

**2**

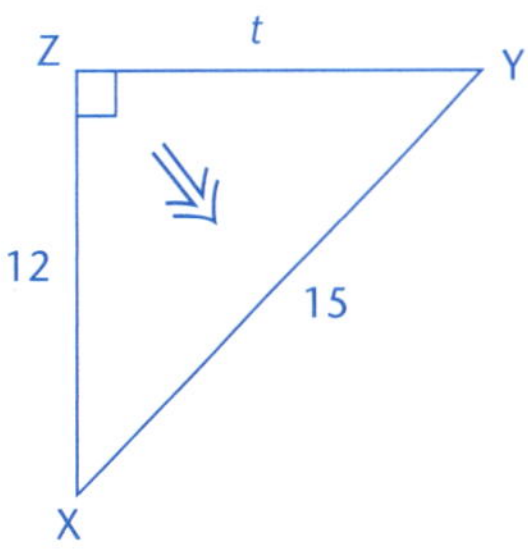


Let YZ = $t$

Hypotenuse = 15

$\therefore t^2 + 12^2 = 15^2$

$t^2 = 15^2 - 12^2$

$= 225 - 144$

$= 81$

$\therefore t = \sqrt{81}$

$= 9$

YZ is 9 cm long.

# Using Pythagoras' Theorem to Prove a Triangle is Right-Angled

For $\triangle ABC$ to be right-angled, we must be able to prove that $AB^2 = BC^2 + AC^2$.

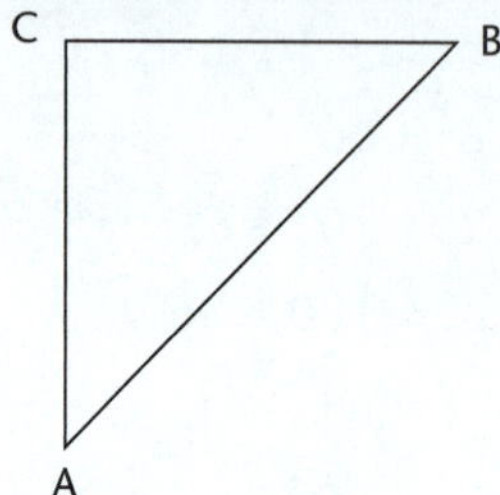

## For Example

**1**

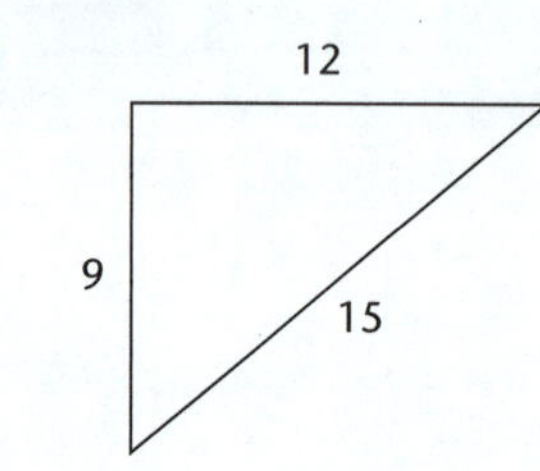


Show that the above triangle is right-angled.

**2**

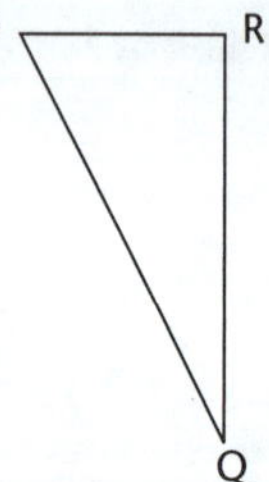


Prove $\triangle PQR$ is right-angled, given that PR = 5 cm, RQ = 12 cm and PQ = 13 cm.

**3** Prove $\triangle XYZ$ is right-angled, given that XY is 25 cm, ZY is 20 cm and XZ is 15 cm long.

**1**

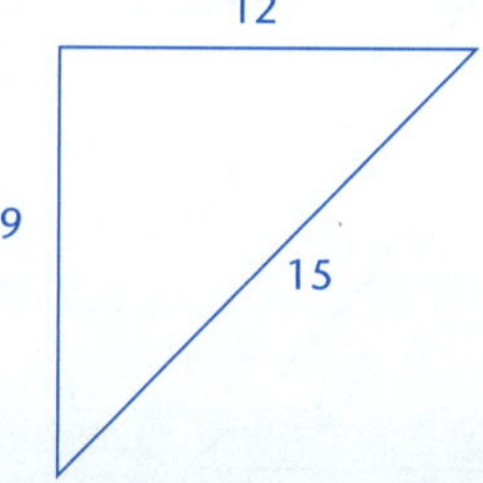


Hypotenuse is the longest side, i.e. 15 units.

We have to prove that:

$15^2 = 9^2 + 12^2$

Now $\text{LHS} = 15^2$

$= 225$

$\text{RHS} = 9^2 + 12^2$

$= 81 + 144$

$= 225$

$\therefore 15^2 = 9^2 + 12^2$

Then $\triangle$ is right-angled.

> LHS means 'left-hand side' of expression.
>
> RHS means 'right-hand side' of expression.

**2**

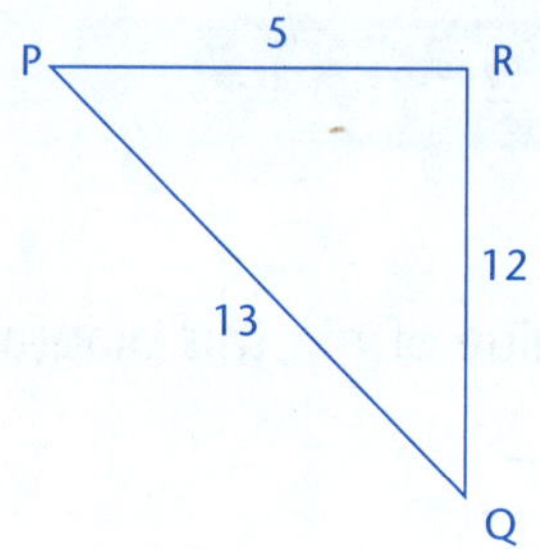


Hypotenuse = 13

For $\triangle PQR$ to be right-angled, we must prove that:

$13^2 = 5^2 + 12^2$

$\text{LHS} = 13^2$

$= 169$

$\text{RHS} = 5^2 + 12^2$

$= 25 + 144$

$= 169$

$\therefore 13^2 = 5^2 + 12^2$

i.e. $\triangle PQR$ is right-angled.

3

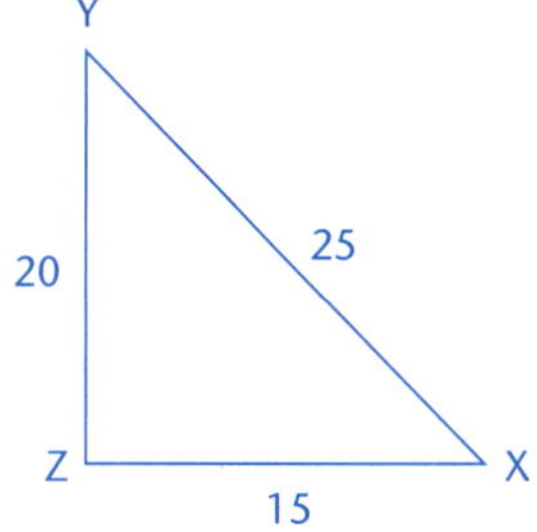


Hypotenuse is XY, i.e. 25 cm (the longest side).

For △XYZ to be right-angled, we must prove that:

$25^2 = 20^2 + 15^2$

$\text{LHS} = 25^2$
$= 625$

$\text{RHS} = 20^2 + 15^2$
$= 400 + 225$
$= 625$

$\therefore 25^2 = 20^2 + 15^2$

i.e. △XYZ is right-angled.

## Pythagorean Triads (or Triples)

1 The numbers $\{a, b, c\}$ form a Pythagorean triad if:

$c^2 = a^2 + b^2$

and

$a$, $b$, and $c$ are positive whole numbers.

i.e. {3, 4, 5} is a Pythagorean triad because $3^2 + 4^2 = 5^2$

| $\text{LHS} = 3^2 + 4^2$ | $\text{RHS} = 5^2$ |
|---|---|
| $= 9 + 16$ | $= 25$ |
| $= 25$ | $= \text{LHS}$ |

2 But {1.2, 1.6, 2} is *not* a Pythagorean triad, even though:

$2^2 = 1.2^2 + 1.6^2$

because 1.2 and 1.6 are not whole numbers.

### For Example

1 Find the value of $t$ such that:

**a** $\{6, 8, t\}$ is a Pythagorean triad and

**b** $\{t, 12, 13\}$ is a Pythagorean triad.

1 **a** $\{6, 8, t\}$ is a Pythagorean triad.

$\therefore t^2 = 6^2 + 8^2$
$= 36 + 64$
$= 100$
$\therefore t = \sqrt{100}$
$= 10$

**b** $\{t, 12, 13\}$ is a Pythagorean triad.

$\therefore t^2 + 12^2 = 13^2$
$t^2 = 13^2 - 12^2$
$= 169 - 144$
$= 25$
$\therefore t = \sqrt{25}$
$= 5$

## Using the Calculator for Pythagoras' Theorem

Sometimes a Pythagoras question results in an answer such as $\sqrt{10}$, $\sqrt{13}$, etc. These numbers do not have whole number square roots. However, these answers are the **exact** solution to the Pythagoras question.

The answer to these questions can either be left in **exact** form ($\sqrt{\ }$), or evaluated correct to a given number of decimal places (approximate answer) using the square root function on your calculator.

Exact form is often stated as **surd** form. That is, an answer left as $\sqrt{17}$ (for example) has been left in exact form or surd form.

## For Example

1 Calculate the value of $t$ from the following diagrams correct to one decimal place:

a

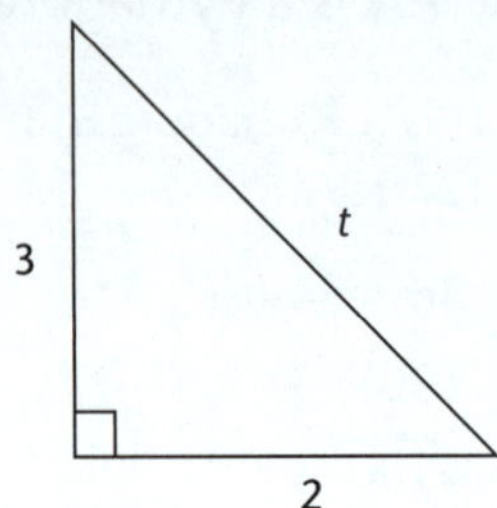


b

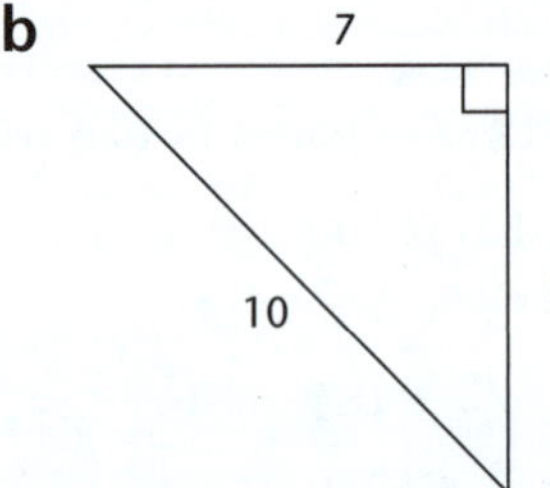


2

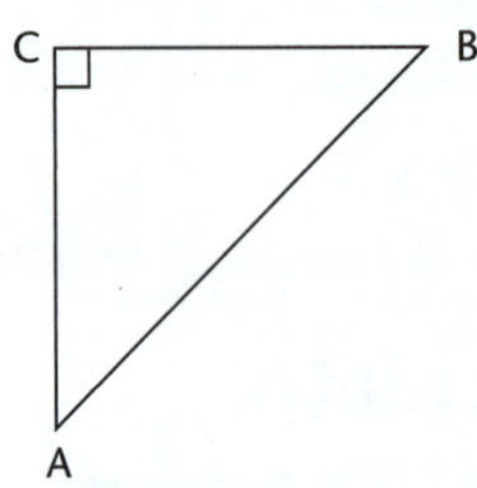


Given that AB = 7 cm, AC = 4 cm, calculate the length of BC correct to one decimal place.

3 In the $\triangle PQR$, PR = 5 cm, RQ = 4 cm and $\angle PRQ = 90°$. Calculate the length of PQ correct to three significant figures.

1 a

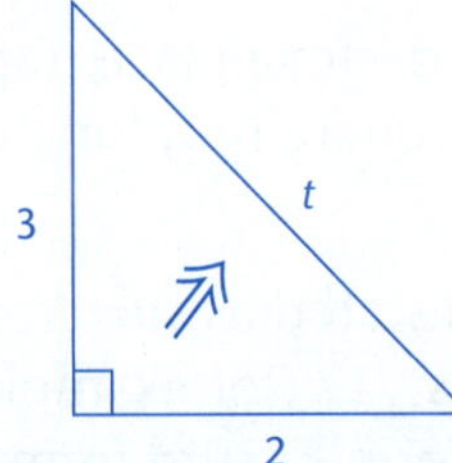


Using Pythagoras' Theorem:

$$t^2 = 3^2 + 2^2$$
$$= 9 + 4$$
$$= 13$$
$$\therefore t = \sqrt{13} \quad \text{[Exact solution]}$$
$$= 3.605\,551\,3$$
$$= 3.6 \text{ (to 1 decimal place)}$$

b

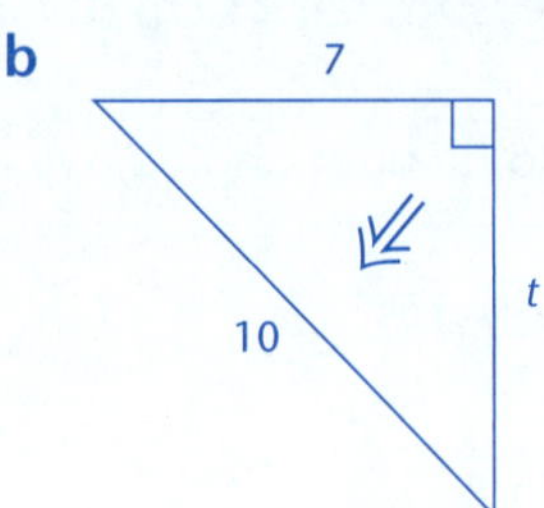


$$t^2 + 7^2 = 10^2$$
$$\therefore t^2 = 10^2 - 7^2$$
$$= 100 - 49$$
$$= 51$$
$$\therefore t = \sqrt{51} \quad \text{[Exact solution]}$$
$$= 7.141\,428\,4$$
$$= 7.1 \text{ (to 1 decimal place)}$$

2

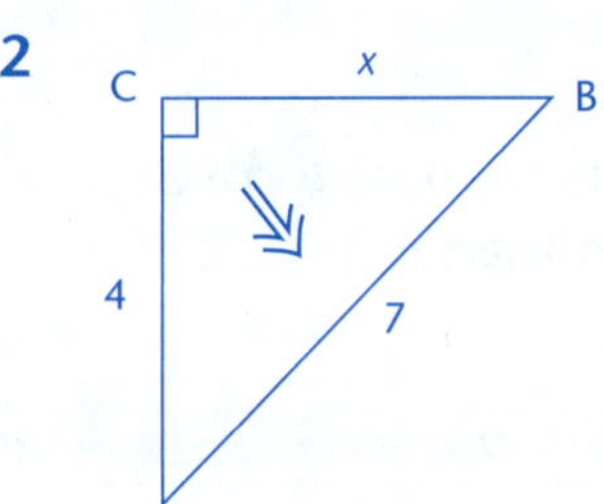


Let BC = $x$ cm (AB is hypotenuse).

$$\therefore x^2 + 4^2 = 7^2$$
$$x^2 = 7^2 - 4^2$$
$$= 49 - 16$$
$$= 33$$
$$\therefore x = \sqrt{33} \quad \text{[Exact solution]}$$
$$= 5.744\,562\,6$$
$$= 5.7 \text{ (to 1 decimal place)}$$

That is, BC is 5.7 cm long.

**3** Let PQ be $y$ cm (PQ is hypotenuse).

$\therefore y^2 = 5^2 + 4^2$

$= 25 + 16$

$= 41$

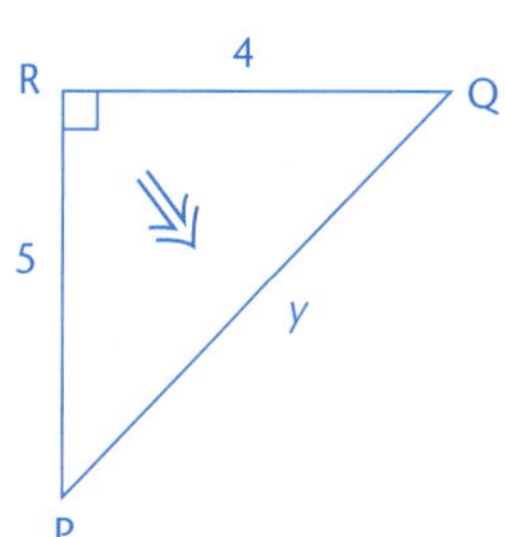


$\therefore y = \sqrt{41}$ [Exact solution]

$= 6.403\,124\,2$

$= 6.40$ (to 3 significant figures)

PQ is 6.40 cm long.

## Problem Solving Using Pythagoras' Theorem

To be able to use Pythagoras' Theorem to solve a problem a right-angle must be present.

**1**

A rectangle has sides of 9 cm and 12 cm. Calculate the length of the diagonal.

**2**

A 7-metre ladder must reach 5.5 metres up a brick wall. How far from the wall must the foot of the ladder be placed? (Answer correct to one decimal place.)

**3**

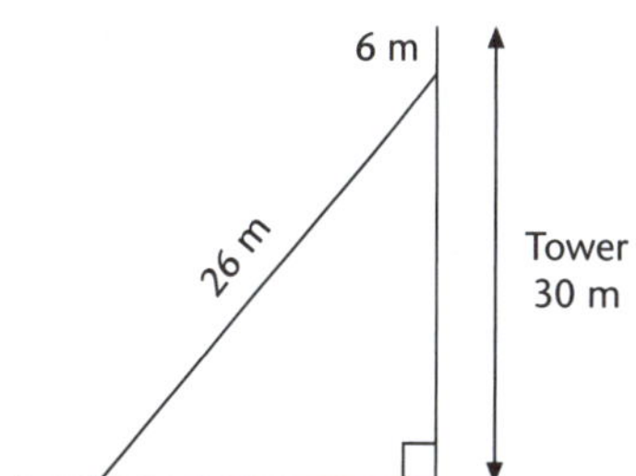


A support wire is to be attached 6 m from the top of a 30 m tower. If the length of wire used is 26 m long, how far from the base of the tower must the wire be attached?

**1**

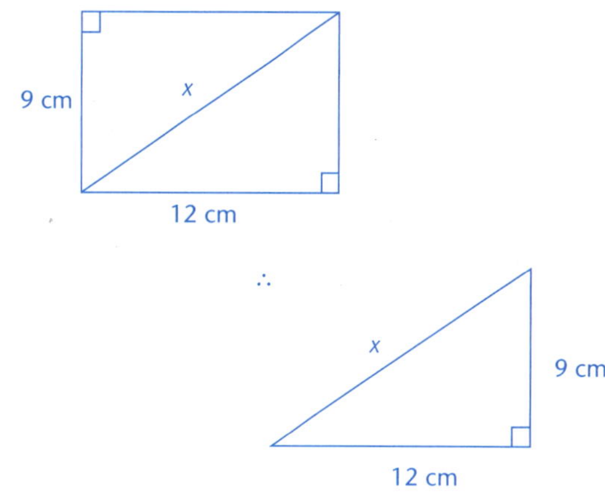

Let diagonal have length $x$ cm.

From the triangle, $x$ is the hypotenuse.

$\therefore x^2 = 9^2 + 12^2$

$= 81 + 144$

$= 225$

$\therefore x = \sqrt{225}$

$= 15$

The diagonal has length 15 cm.

**2** Consider only the triangle. Let the unknown side be $d$ m.

The hypotenuse = 7 m.

By Pythagoras' Theorem:

$d^2 + 5.5^2 = 7^2$

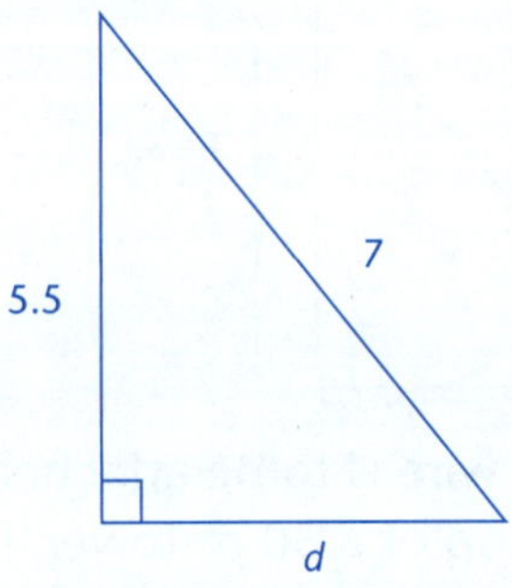


$\therefore d^2 = 7^2 - 5.5^2$

$= 49 - 30.25$

$= 18.75$

$\therefore d = \sqrt{18.75}$

$= 4.330127$

$= 4.3$ (1 decimal place)

The foot of the ladder must be placed 4.3 metres from the wall.

**3** $30 - 6 = 24$

$\therefore$ The wire is attached 24 m above the base.

Let the distance from the base be $y$ metres.

Hypotenuse = 26 metres.

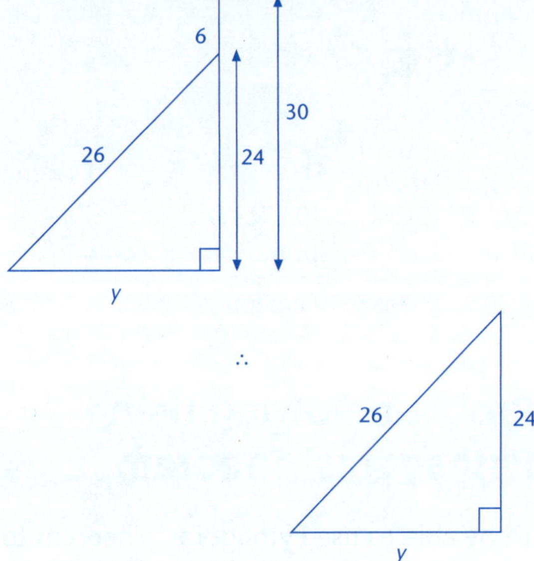


By Pythagoras' Theorem:

$y^2 + 24^2 = 26$

$\therefore y^2 = 26^2 - 24^2$

$= 676 - 576$

$= 100$

$\therefore y = \sqrt{100}$

$= 10$

$\therefore$ The wire is attached 10 metres from the base of the tower.

## PRACTISE, PRACTISE

Go to p. 286 for quick answers, or to pp. 332–340 for worked solutions.

**1** Write down Pythagoras' Theorem for these diagrams: pp. 110–111

**a**

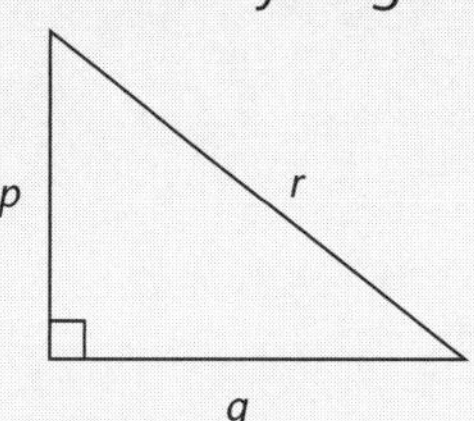


**b**

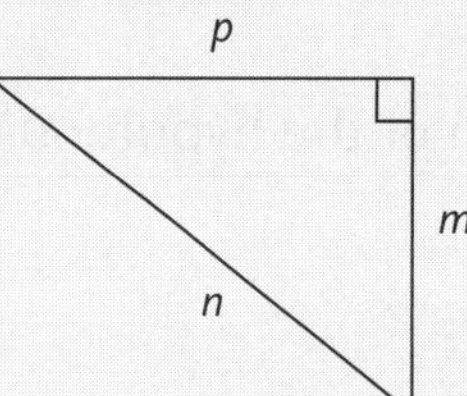


**c**

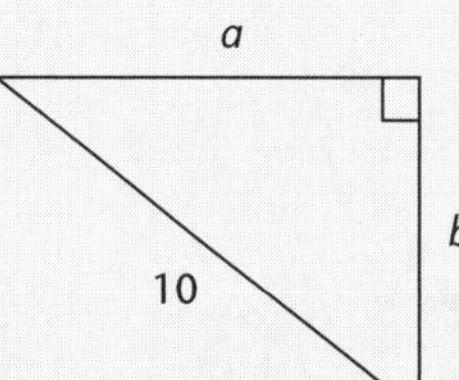


**d**

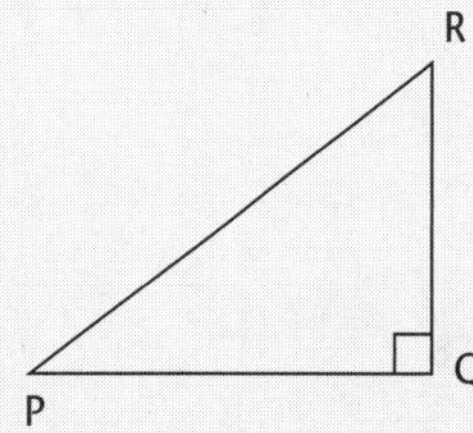


**e**

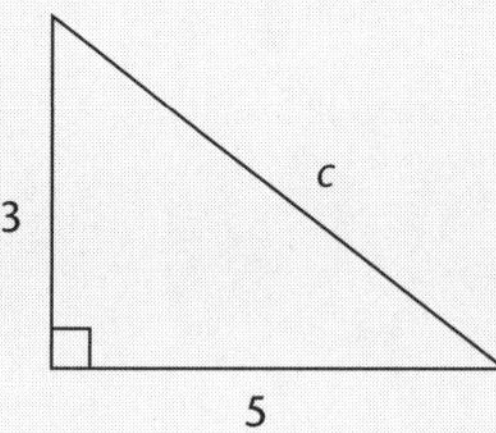


**f**

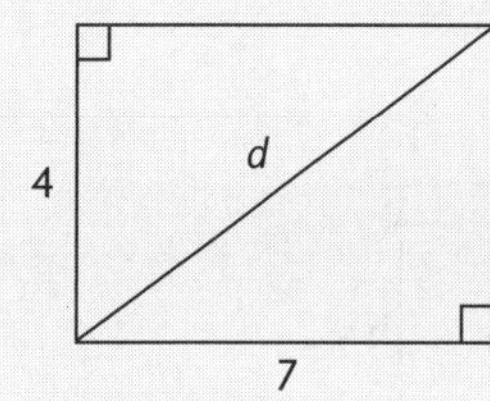


**2** Complete the statements under each diagram: pp. 110–111

**a**

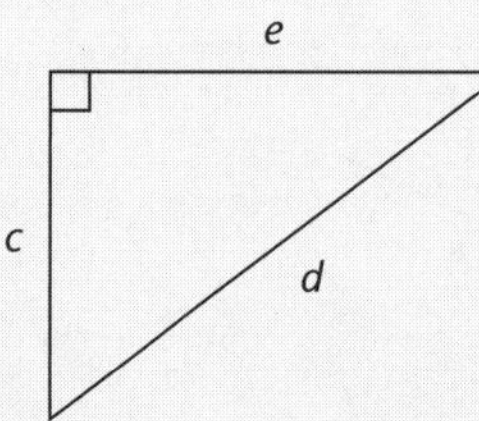


$d^2 =$

**b**

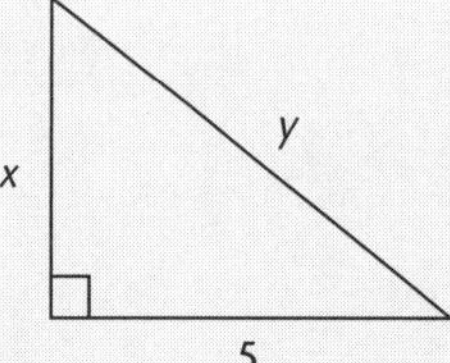


$y^2 =$

**c**

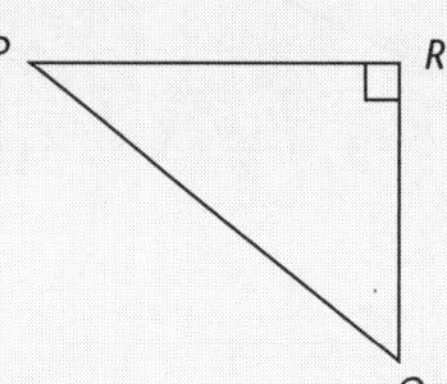


$PQ^2 =$

**d**

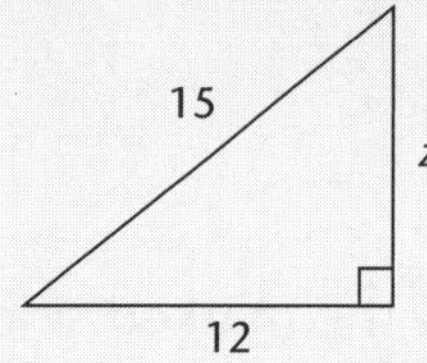


$z^2 =$

**e**

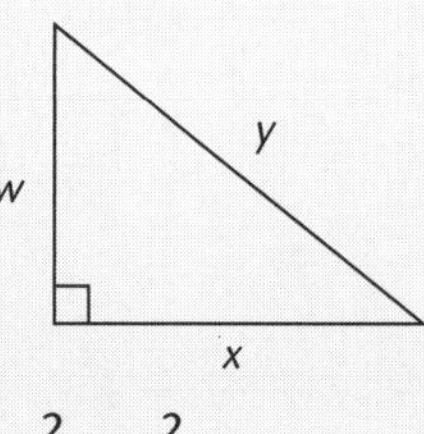
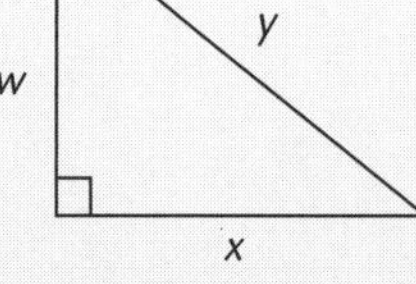


$w^2 + x^2 =$

**f**

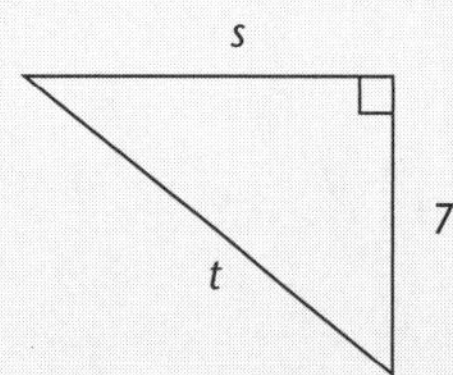


$s^2 + 7^2 =$

**3** Evaluate: p. 111

**a** $7^2 =$ **b** $11^2 =$

**c** $17^2 =$ **d** $29^2 =$

**e** $31^2 =$ **f** $\sqrt{144} =$

**g** $\sqrt{900} =$ **h** $\sqrt{256} =$

**i** $\sqrt{625} =$ **j** $\sqrt{16900} =$

**4** Calculate the length of the hypotenuse in each diagram: pp. 111–112

**a**

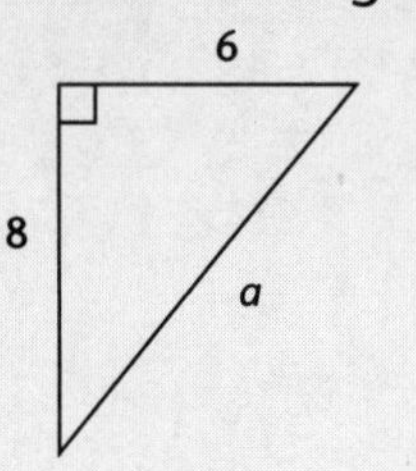

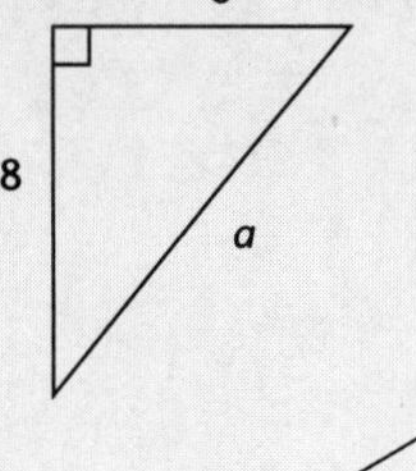

**b**

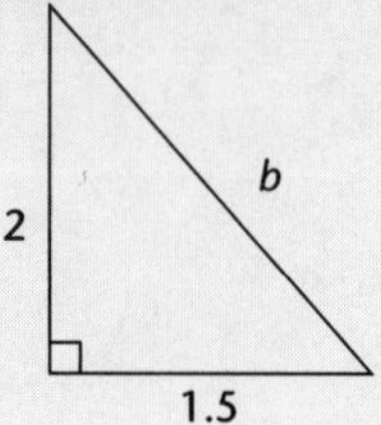


**c**

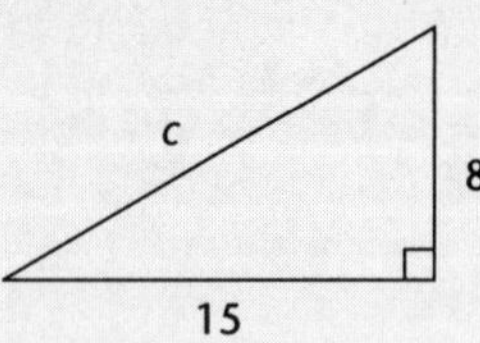


**d**

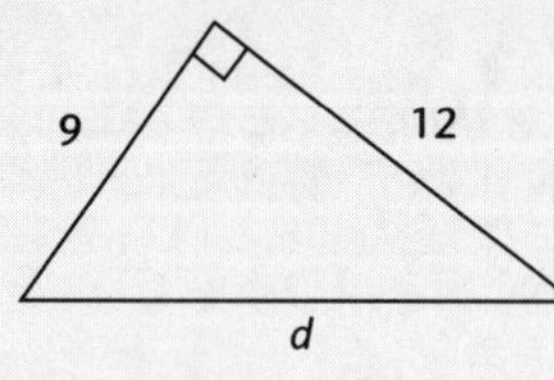


**e**

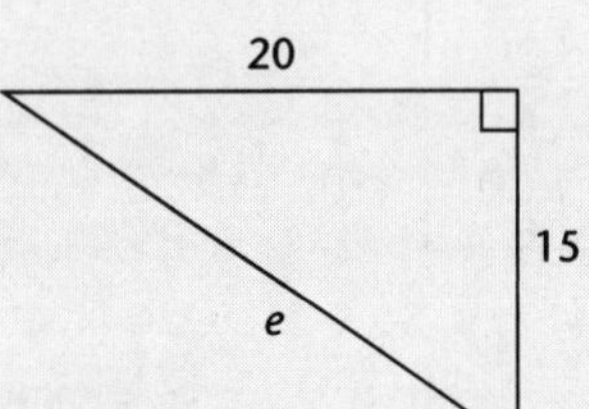


**f**

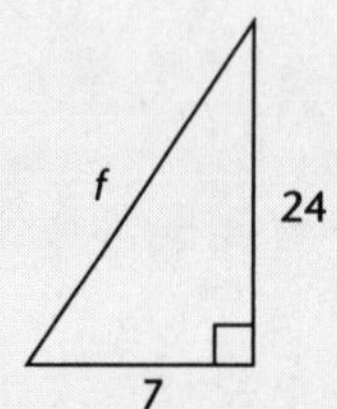


**g**

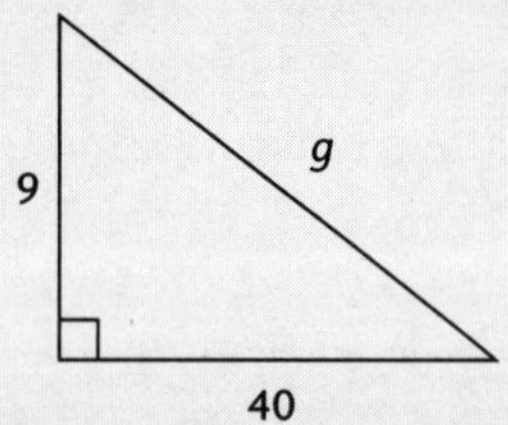


**h**

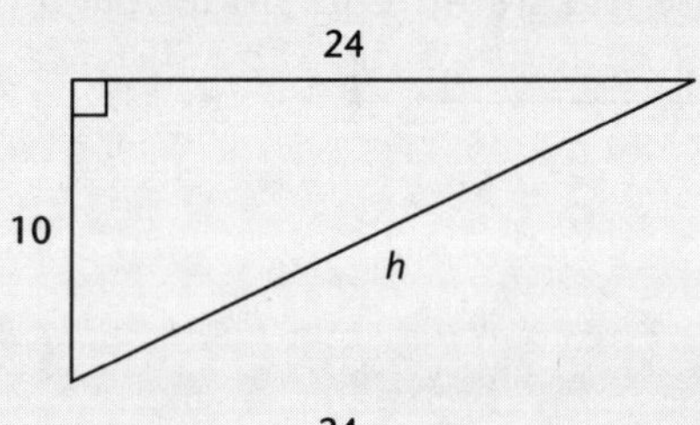


**i**

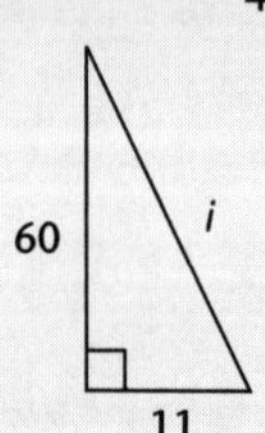


**j**

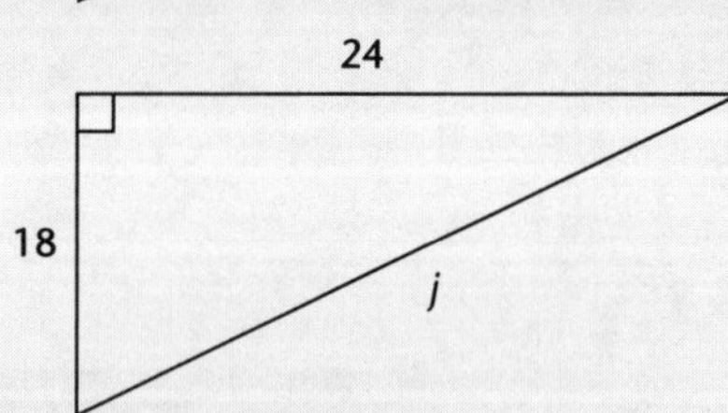


**5** Calculate the length of the marked side in these diagrams: pp. 112–113

**a**

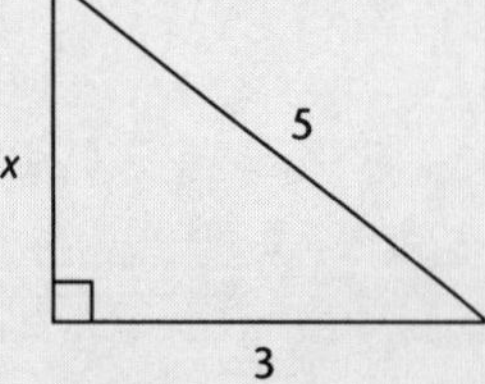


**b**

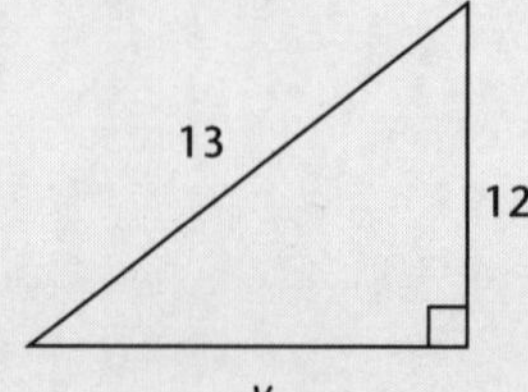

**c**

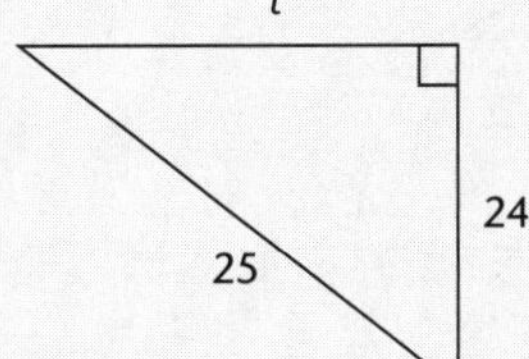


**d**

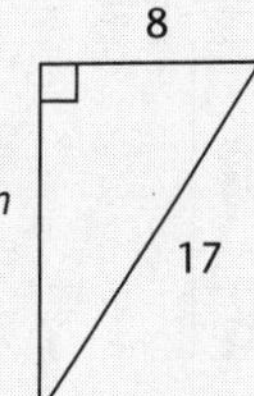


**e**

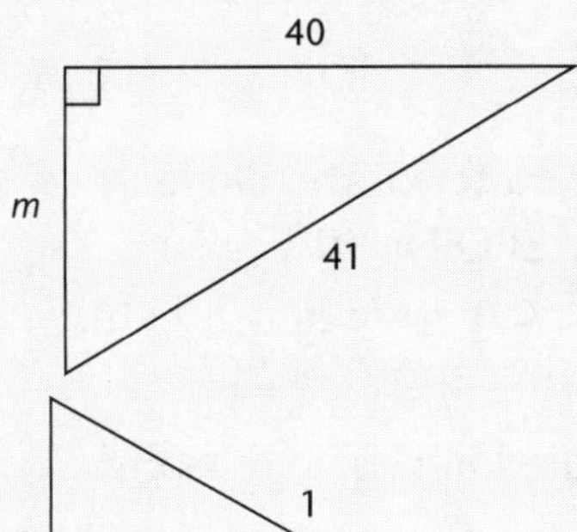


**f**

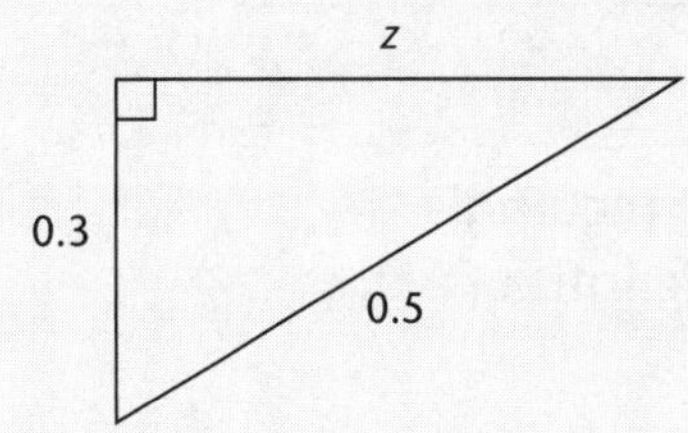


**g**

1
p
0.8

**h**

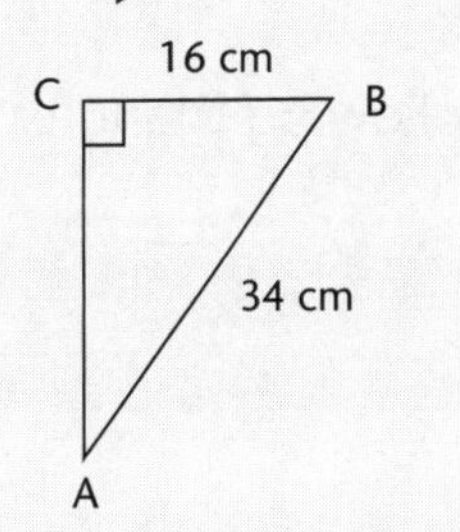


**6** Calculate the size of the marked side:

pp. 111–113

**a**

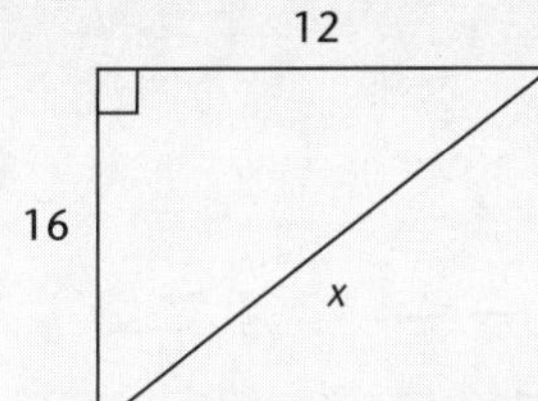


**b**

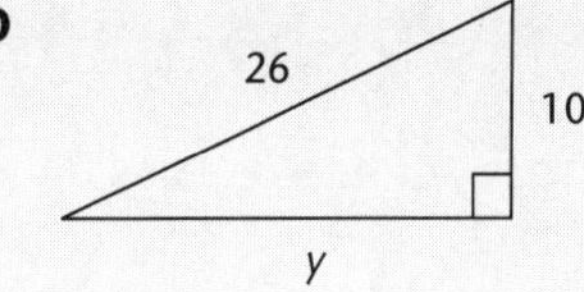


**c**

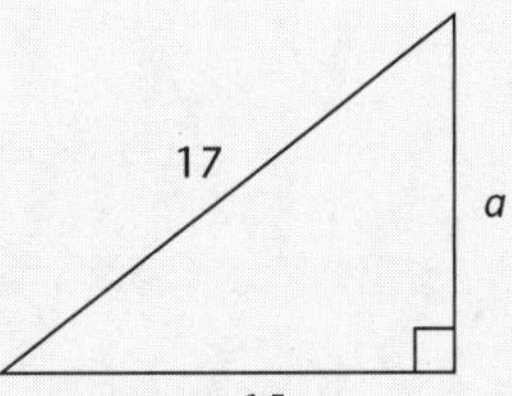


**d**

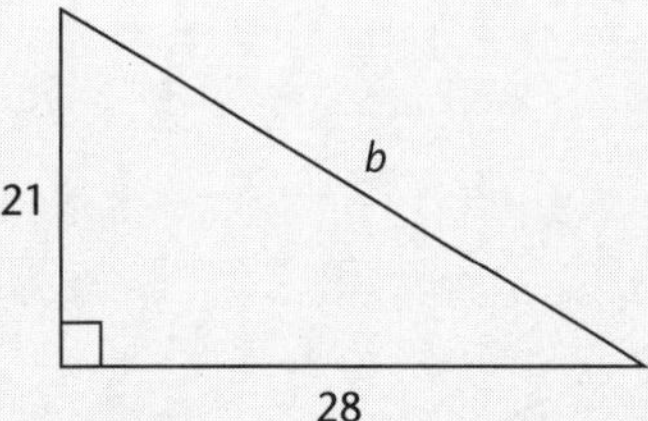


**e**

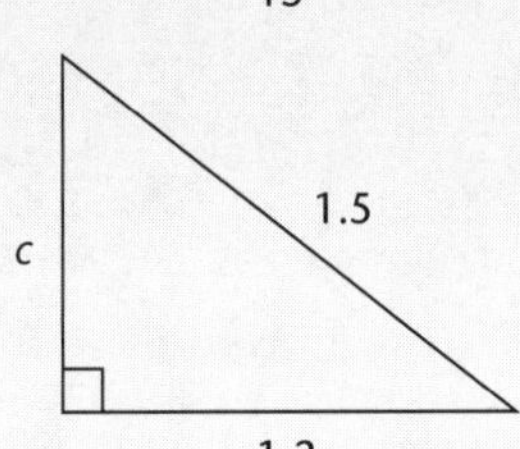


**f**

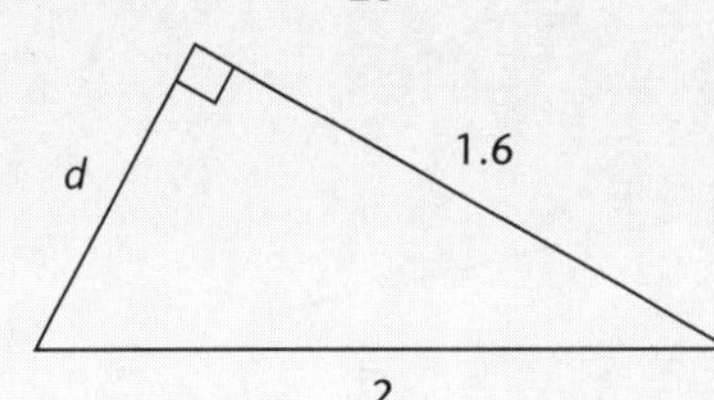


**g**

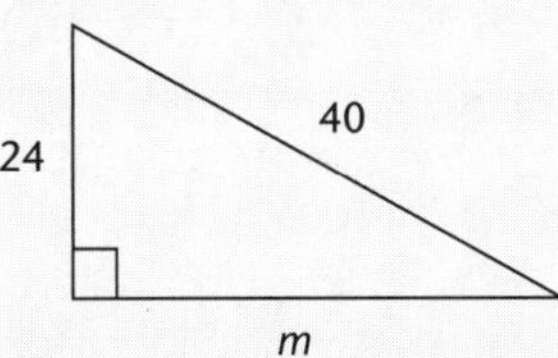


**h**

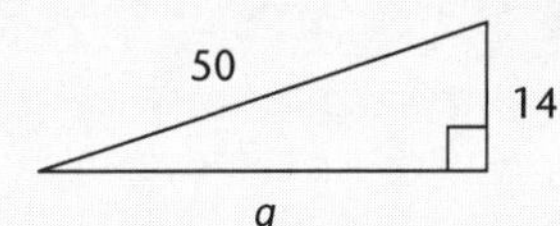

**i**

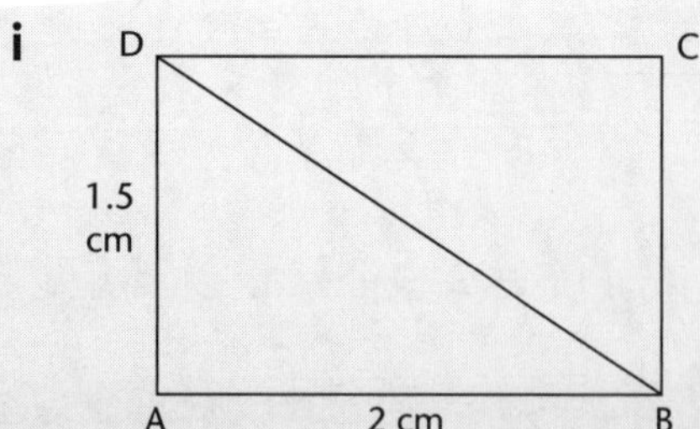


ABCD is a rectangle.
Find the length of BD.

**j**

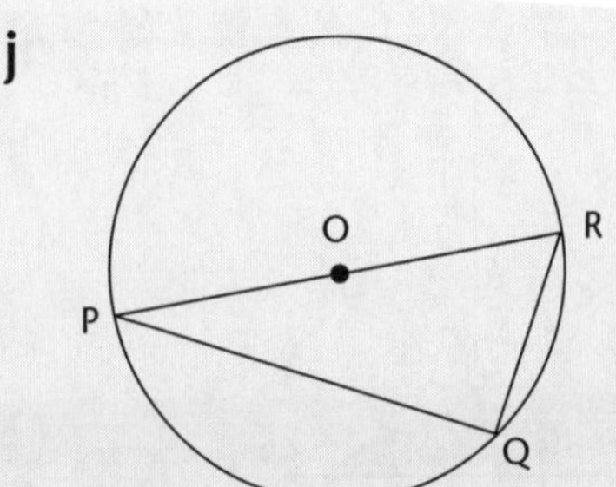


PR is a diameter of the circle centre O. $\angle PQR = 90°$.
If OP is 25 cm and QR is 30 cm, calculate the length of PQ.

**7** Decide which of the following triangles are right-angled and give reasons: pp. 114–115

**a**

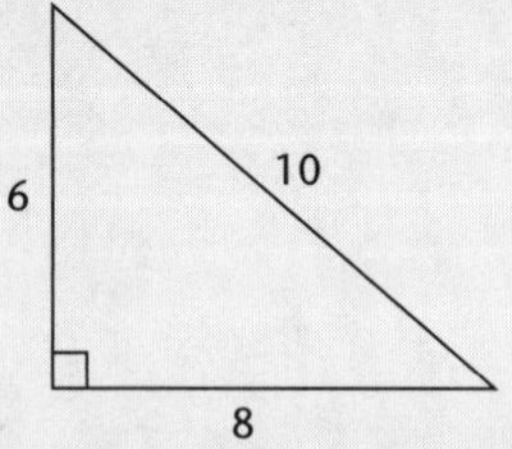
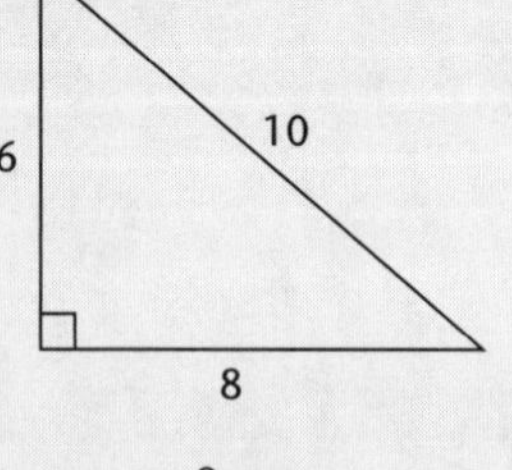


**b**

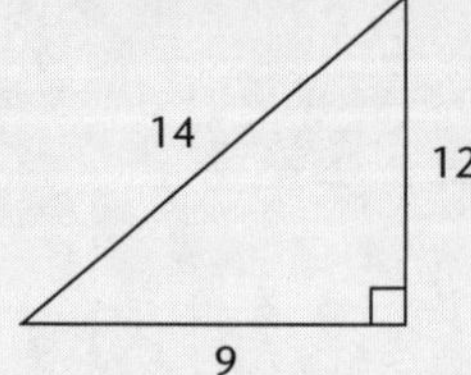


**c**

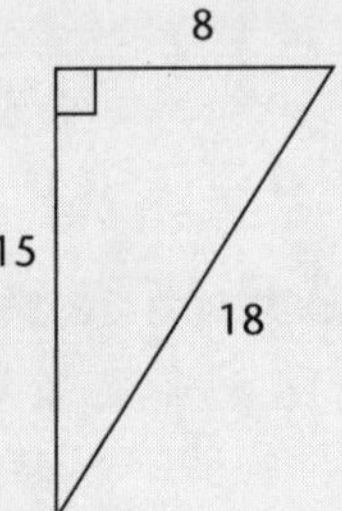


**d**

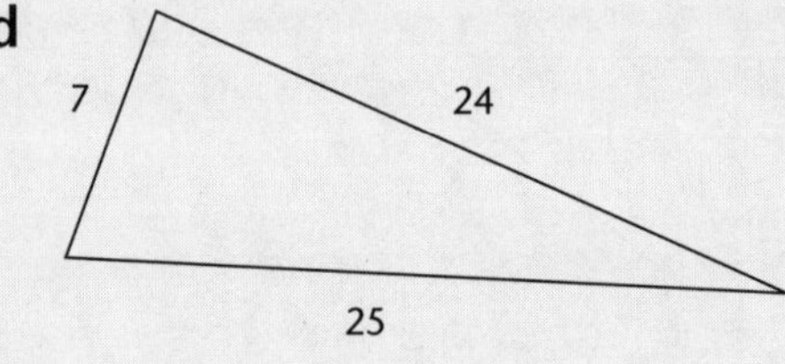


**e**

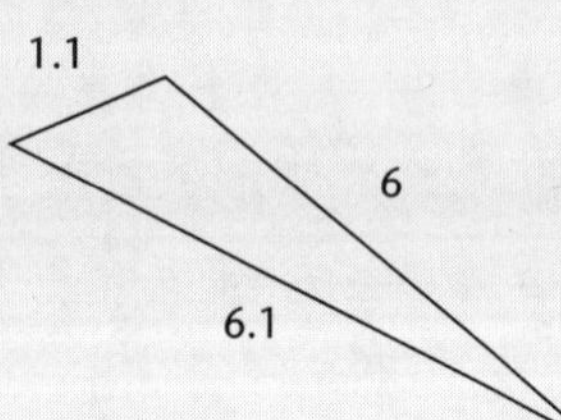


**f**

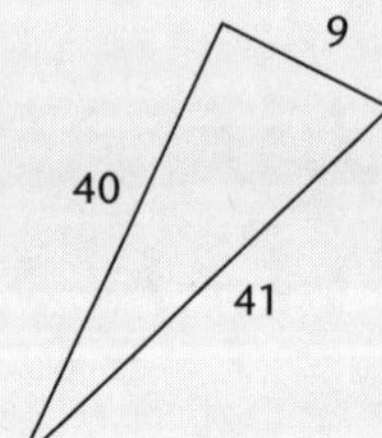


**g**

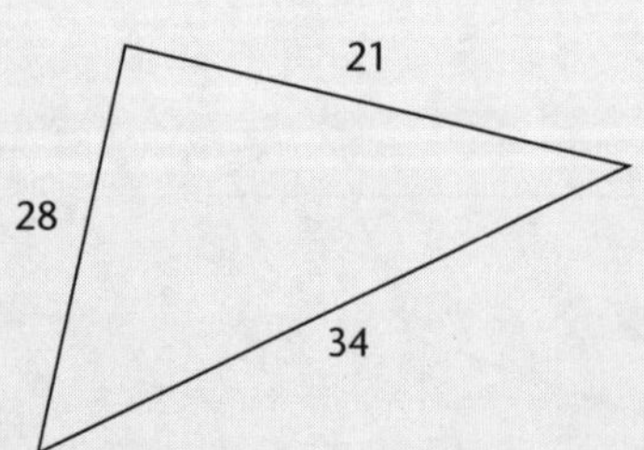


**h**

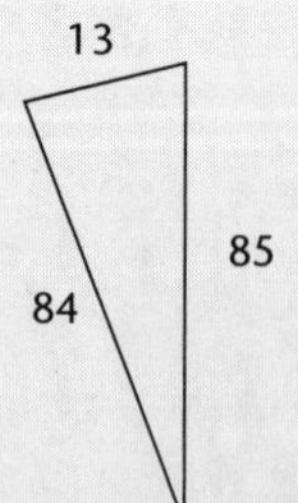


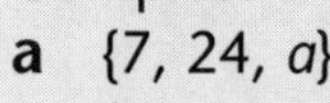

**8** Complete the Pythagorean triads: p. 115

**a** $\{7, 24, a\}$ **b** $\{9, 12, y\}$

**c** $\{18, n, 30\}$ **d** $\{9, t, 41\}$

**e** $\{13, n, 85\}$ **f** $\{21, 72, p\}$

**9** Calculate the length of the hypotenuse (correct to one decimal place): pp. 115–117

**a**

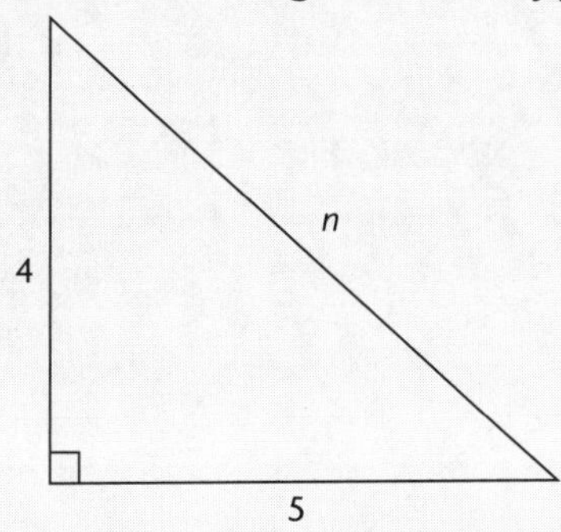


**b**

7

2

*w*

**c**

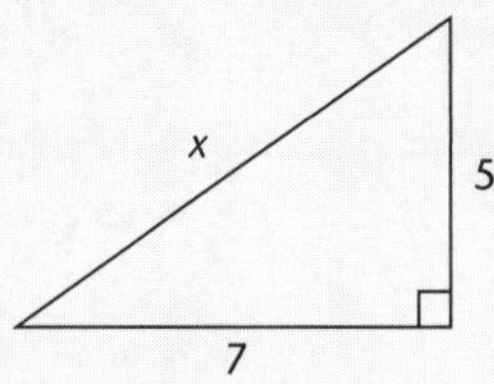


**d**

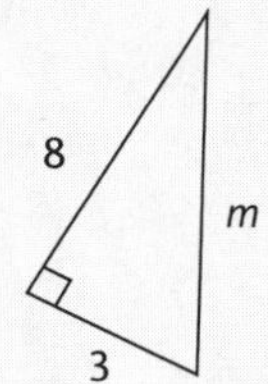


**e**

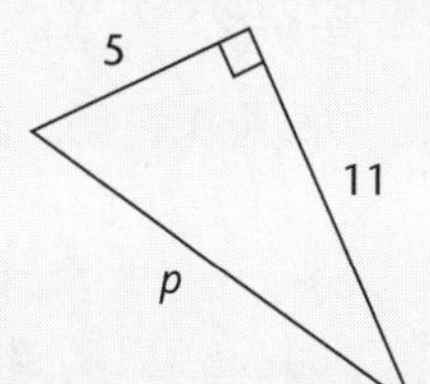


**f**

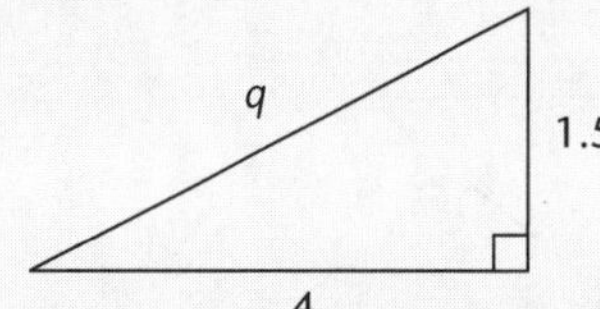


**g**

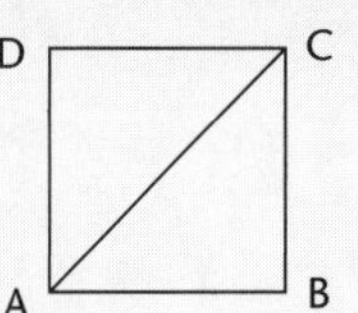


ABCD is a square with AD = 1 cm. Calculate the length of AC.

**h**

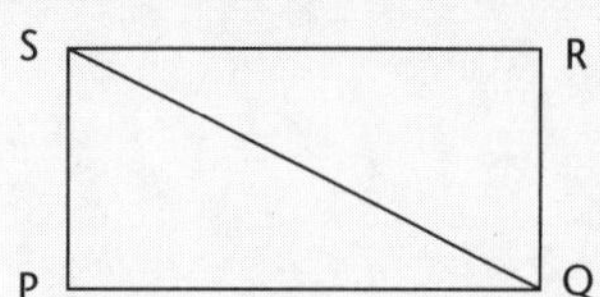


PQRS is a rectangle with PQ = 5 cm, RQ = 3 cm. Calculate the length of QS.

**10** Calculate the length of the marked side (correct to one decimal place): pp. 115–117

**a**

7

4

*a*

**b**

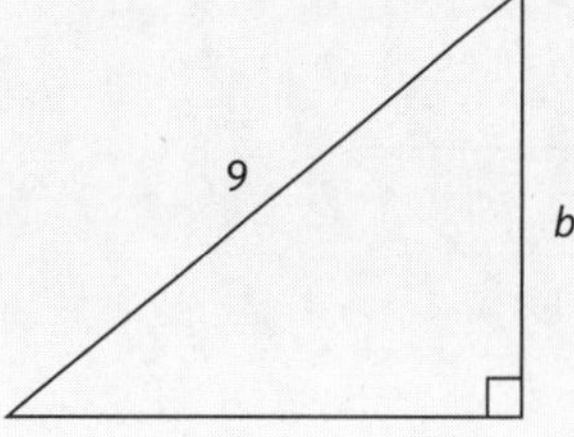


**c**

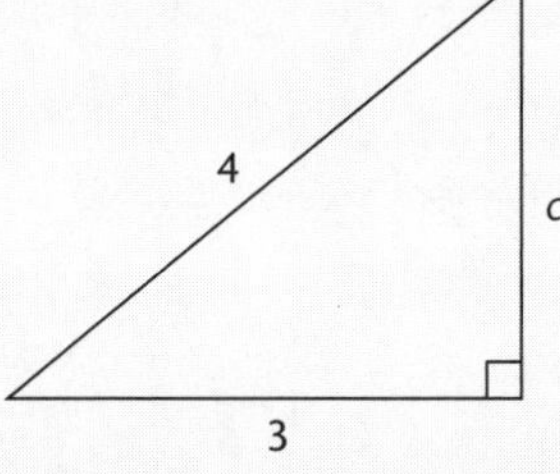


**d**

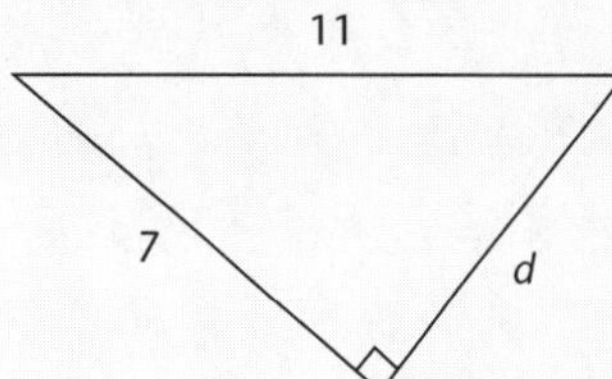

**e**

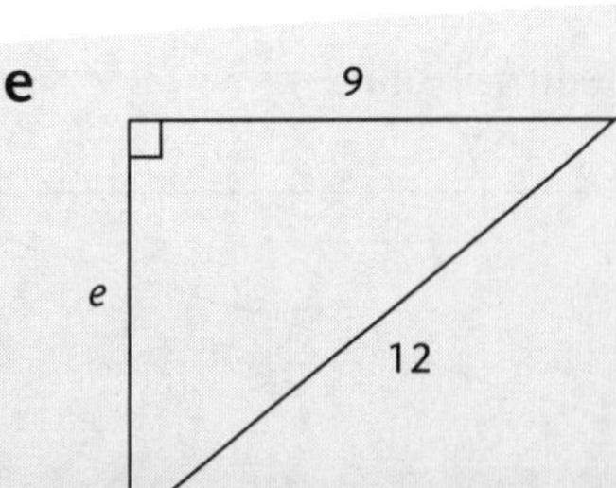


**f**

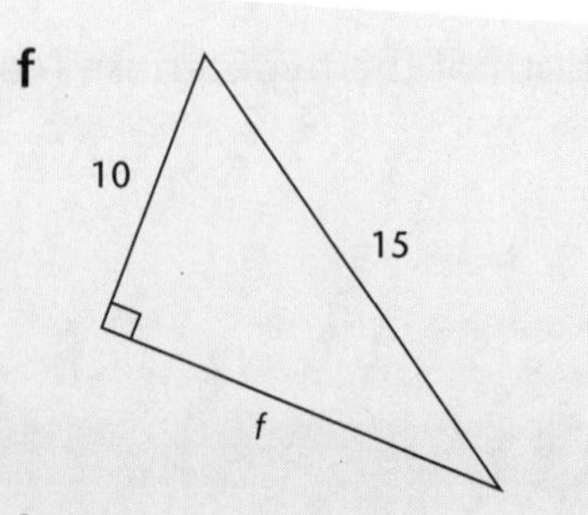


**g**

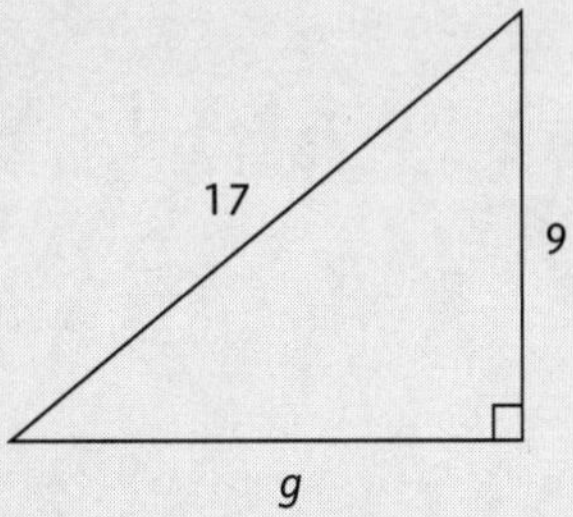

17
9
*g*

**h**

4
*h*
14

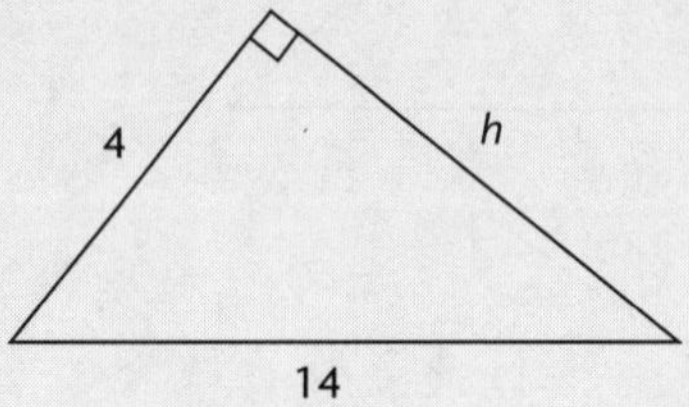

**11** Calculate the length of the marked side (correct to three significant figures): pp. 115–117

**a**

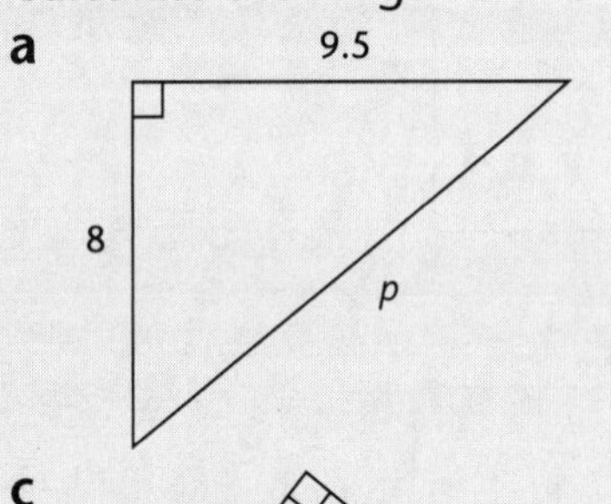


**b**

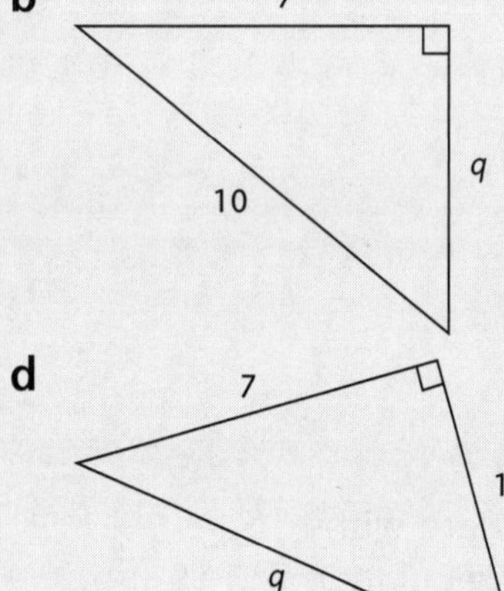


**c**

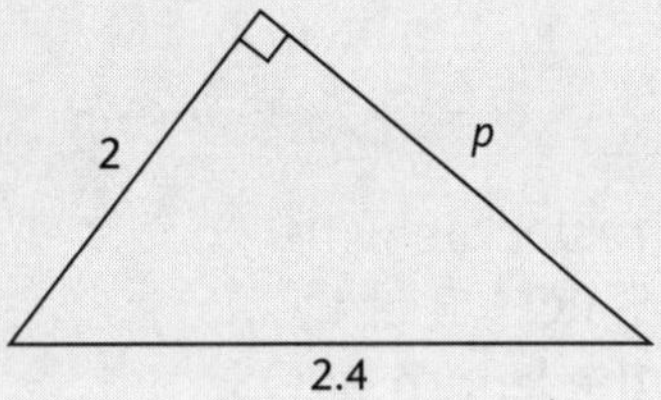

2
*p*
2.4

**d**

**12** Calculate the length of the marked side, leaving your answer in exact or surd form: pp. 115–117

**a**

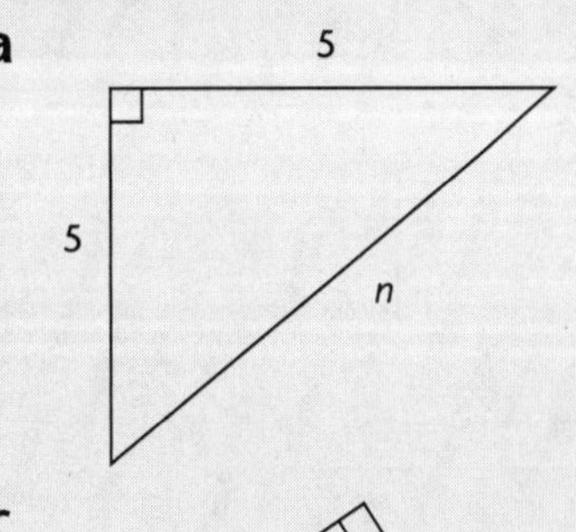


**b**

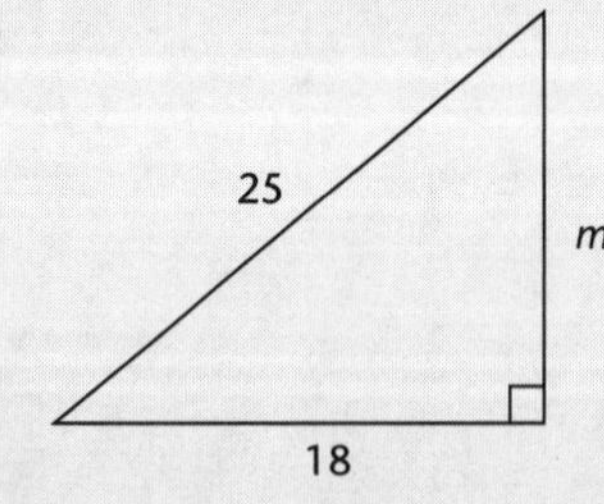


**c**

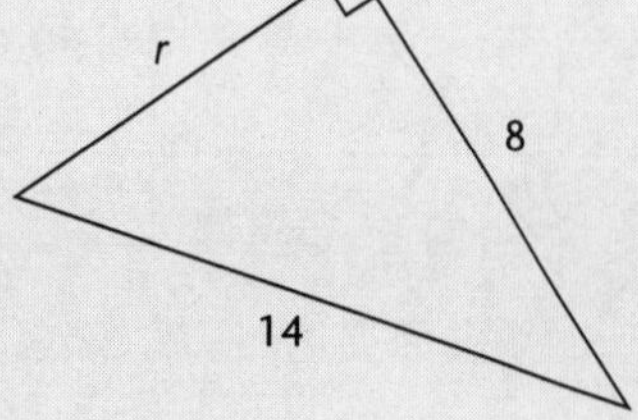

*r*
8
14

**d**

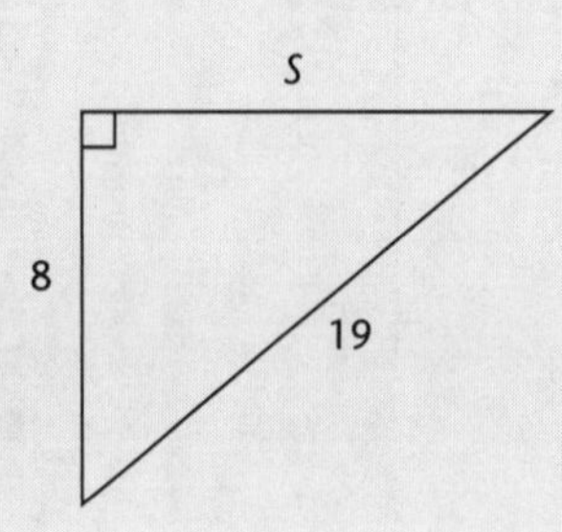

**13** A rectangle is 16 cm long and 12 cm wide. Calculate the length of the diagonal of the rectangle.

pp. 117–118

**14** Calculate the length of a diagonal of a square with sides of 3 cm. Answer correct to one decimal place.

pp. 117–118

**15** A metal brace 35 cm long is screwed into place so that one end is 21 cm from a wall. How far up the wall does the brace reach?

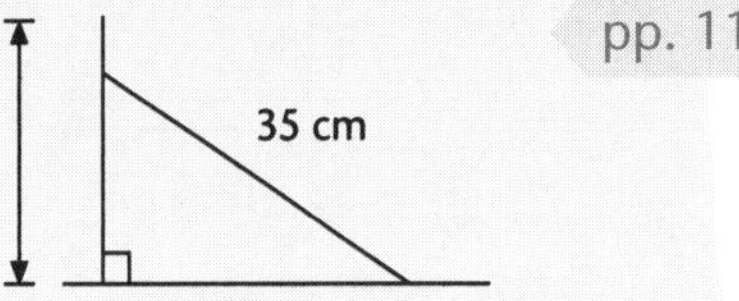


pp. 117–118

**16** An 8-metre ladder reaches 7.5 metres up a vertical brick wall. How far is the foot of the ladder from the wall? Answer correct to one decimal place.

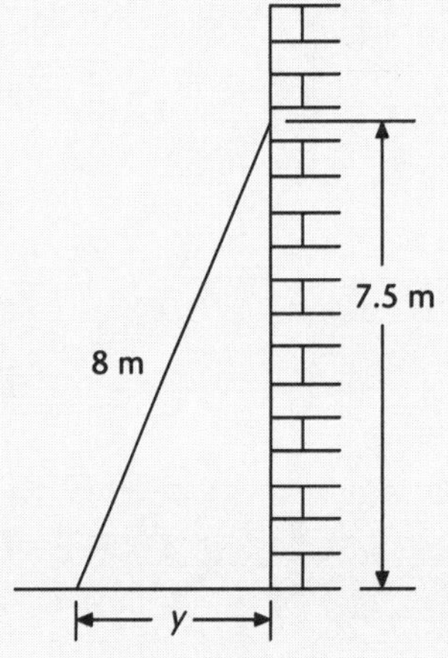


pp. 117–118

**17** A pollution tower 45 metres high is supported by wires 40 metres long. Each wire is attached at ground level 21 metres from the base of the tower. How far from the top of the tower are the wires attached? Answer correct to four significant figures.

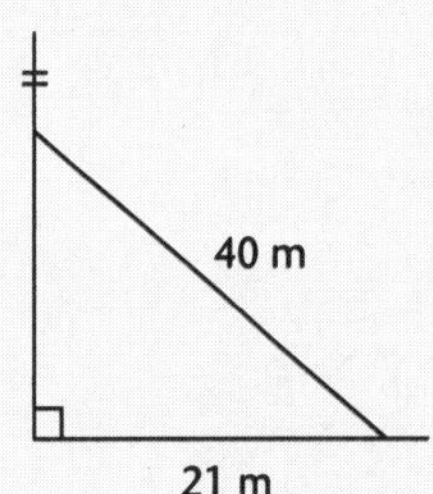


pp. 117–118

**18** The square MNOP has diagonal of length 18 cm. Calculate the length of a side of the square. Answer correct to one decimal place.

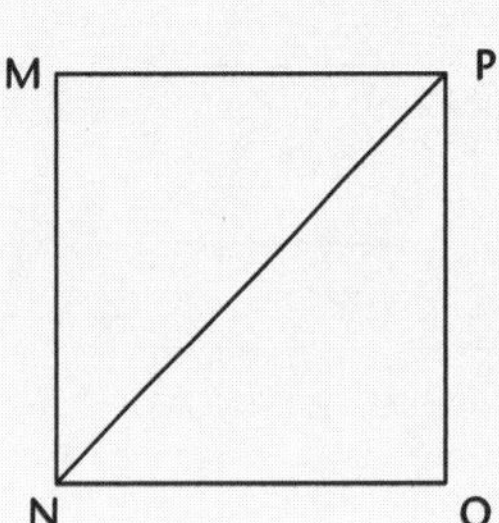


pp. 117–118

**19**

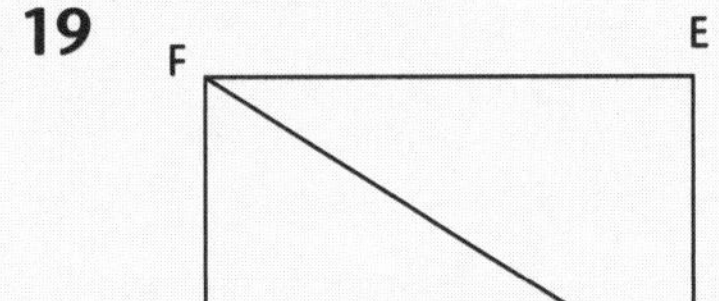


A rectangle CDEF has length 20 cm and diagonal 25 cm. Calculate the width of the rectangle.

pp. 117–118

**20** 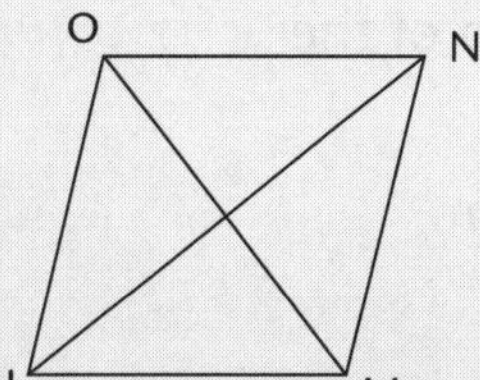


LMNO is a rhombus with diagonals 24 cm and 10 cm. Calculate the length of each side of the rhombus.

[ Note: the diagonals of a rhombus bisect each other at right-angles.
All sides of a rhombus are of equal length. ]

pp. 117–118

**21** 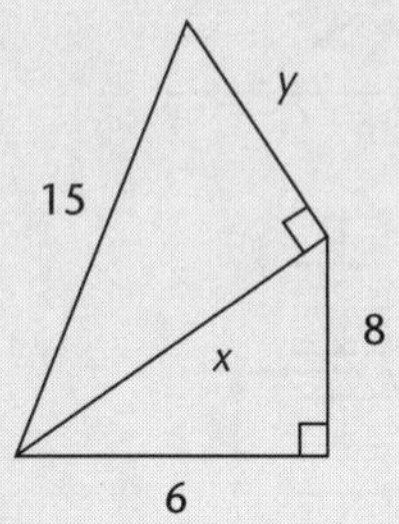


Calculate the value of $x$ and hence calculate the length $y$. Answer correct to three significant figures.

pp. 117–118

**22** 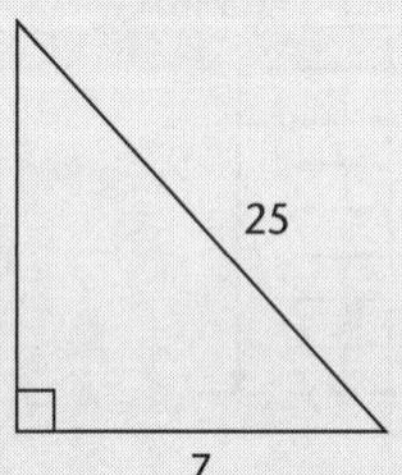


A sail is to be cut in the shape of a right-angled triangle. If the longest side is 25 m and the shortest is 7 m, calculate the length of the third side.

pp. 117–118

**23** 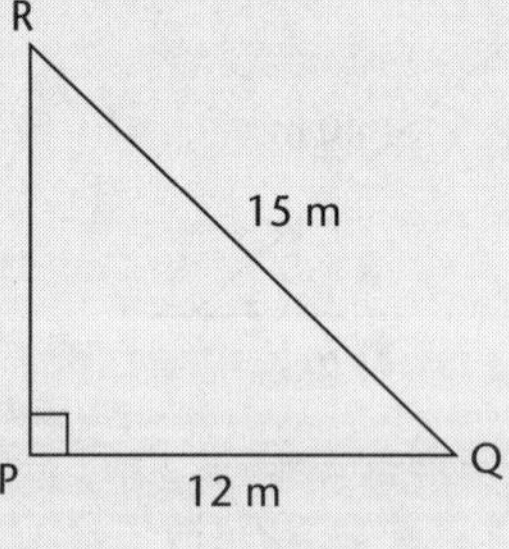


Calculate the perimeter of the triangle PQR.

pp. 117–118

**24** 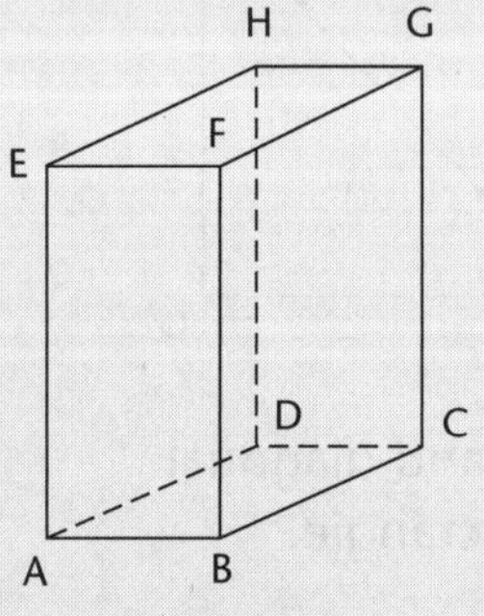


ABCDEFGH is a rectangular prism with AB = 9 cm, BC = 12 cm and CG = 20 cm. By first calculating the length of the diagonal AC, calculate the shortest distance that Florrie the Fly must travel between A and G.

pp. 117–118

**25** Consider the same diagram as for Question 24. Harry the Ant sets off from A to meet Florrie at G. If Harry travels by the most direct route, how much further (if any) must Harry travel than Florrie?

pp. 117–118

**26**

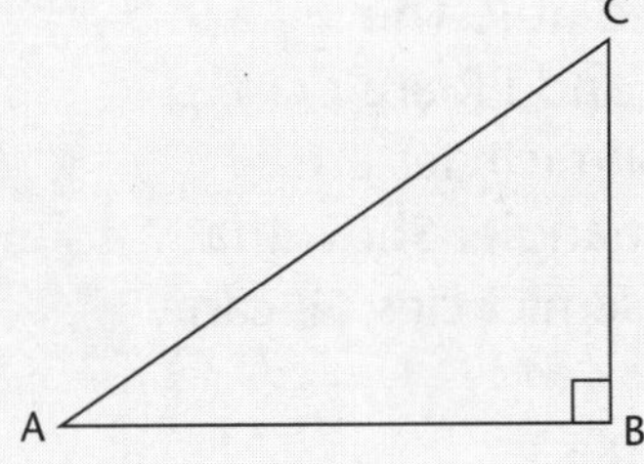


A section of railway line AC, 61 metres long, rises by 11 metres between A and C. Calculate the gradient of the line as a simple fraction.

$$\text{Gradient} = \frac{\text{Rise}}{\text{Run}}$$
$$= \frac{BC}{AB}$$

pp. 117–118

**27**

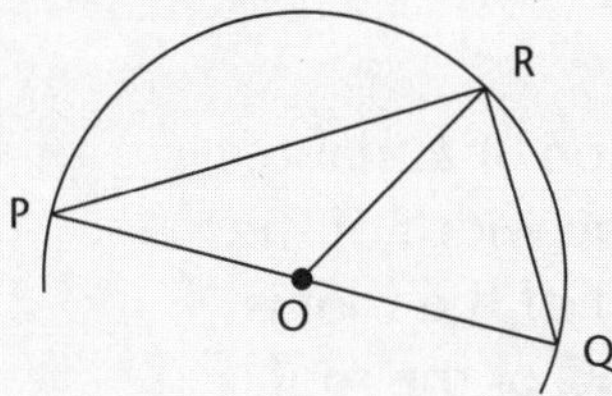


PQR is a triangle on the diameter of the circle. The angle PRQ = 90°. The radius of the circle, OR, is 7.5 cm long. If QR = 9 cm, calculate the length of PR.

pp. 117–118

**28**

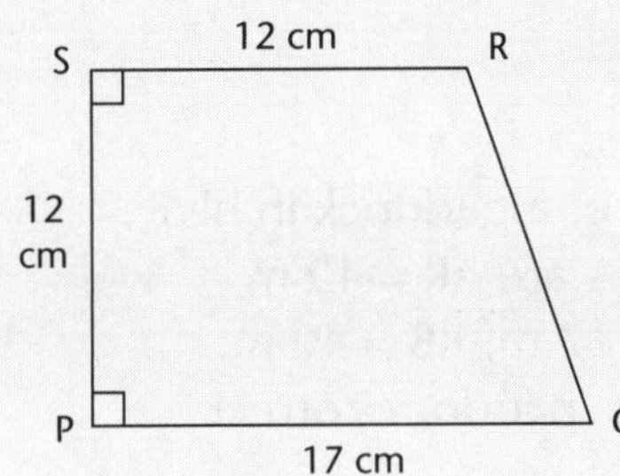


Calculate the perimeter of the trapezium PQRS.

pp. 117–118

**29**

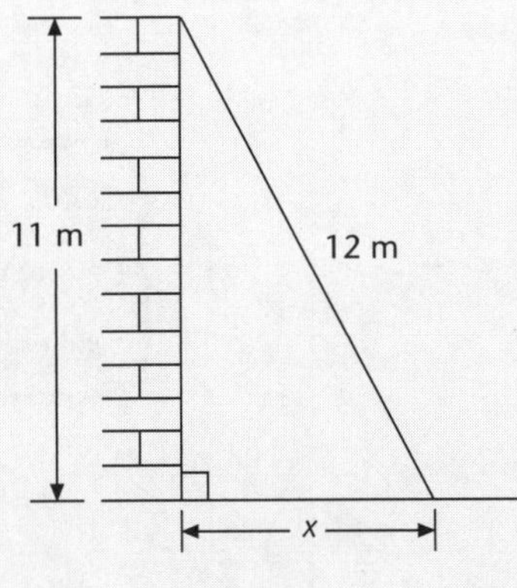


A 12-metre ladder is to be placed against a wall to reach a window 11 metres above the ground. How far from the wall must the foot of the ladder be placed (correct to one decimal place)?

pp. 117–118

**30**

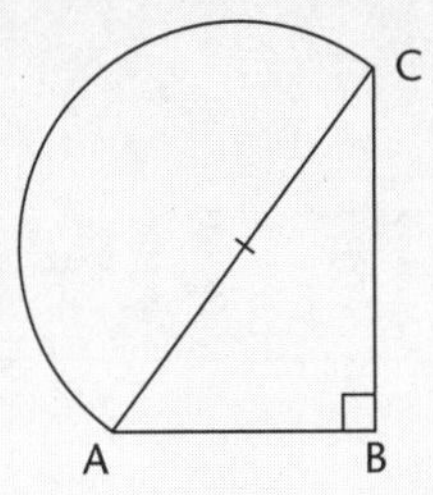


ABC is a triangle right-angled at B.
A semicircle is drawn with AC as the diameter.
If AB = 12 cm and BC = 16 cm, calculate the area of the semicircle to the nearest $cm^2$.
(See Chapter 8 for the formula.)

pp. 117–118

**31**

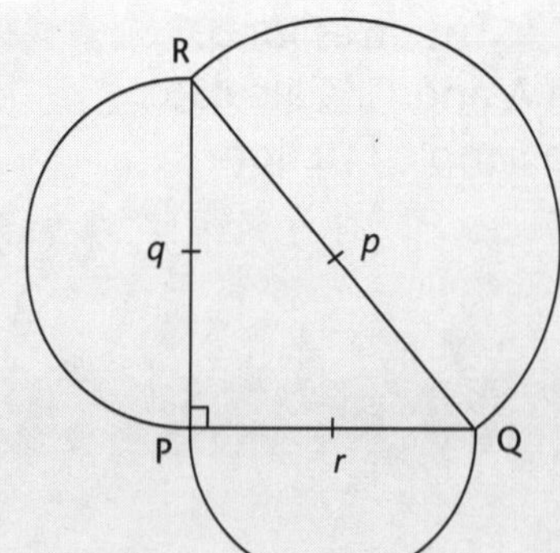


PQR is a triangle right-angled at P. The lengths of the sides PQ, QR and PR are $r$ cm, $p$ cm and $q$ cm respectively. Semicircles are drawn on each side of the triangle. Show that the sum of the areas of the semicircles, $A$, can be expressed as $A = \frac{1}{4}\pi p^2$.
(See Chapter 8 for the formula.)

pp. 117–118

**32**

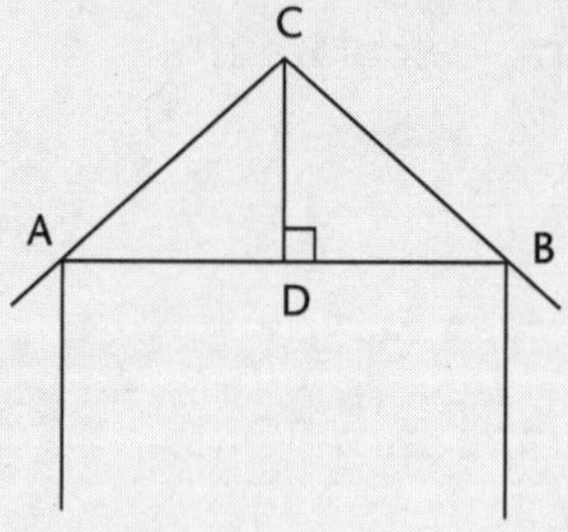


The roof of a house is in the shape of a triangle when viewed from the side. If the width of the house AB is 24 m and the height of the roof is 8 m, calculate the slant height BC of the roof. It can be assumed that the perpendicular height CD is in the centre of AB.

pp. 117–118

**33**

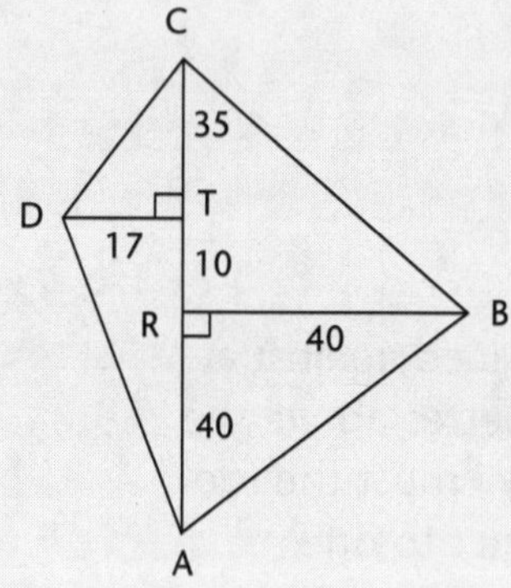


A surveyor makes this sketch of a paddock in his note book. The measurements are AR = 40 m, RT = 10 m, TC = 35 m, DT = 17 m, RB = 40 m. Calculate the perimeter of the paddock correct to the nearest metre.

pp. 117–118

Go to p. 286 for quick answers, or to pp. 332–340 for worked solutions.

## YOUR CHECKLIST

**For a complete understanding of this topic you must be able to:**

| ✓ | Skill | | Page |
|---|---|---|---|
| ✓ | Locate the hypotenuse in a right-angled triangle | | p. 110 |
| ✓ | State Pythagoras' Theorem for any right-angled triangle | | pp. 110–111 |
| ✓ | Use Pythagoras' Theorem to calculate the length of the hypotenuse of any right-angled triangle | | pp. 111–112 |
| ✓ | Use Pythagoras' Theorem to calculate the length of a shorter side of any right-angled triangle | | pp. 112–113 |
| ✓ | Use Pythagoras' Theorem to prove that a triangle is right-angled | | pp. 114–115 |
| ✓ | Use Pythagoras' Theorem to show that a set of three numbers is a Pythagorean triad | | p. 115 |
| ✓ | Understand the difference between an exact answer and an approximation | | pp. 115–117 |
| ✓ | Round answers to a given number of decimal places | | pp. 115–117 |
| ✓ | Apply Pythagoras' Theorem to problem solving involving right-angled triangles. | | pp. 117–118 |

**Now you are ready to do the tests!**

**(25 marks)**

1 The hypotenuse of this triangle is:

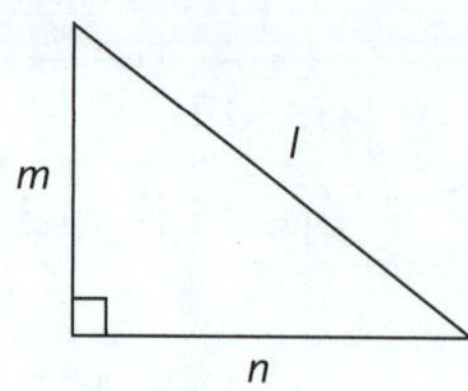


(1 mark)

2 State Pythagoras' Theorem for the triangle:

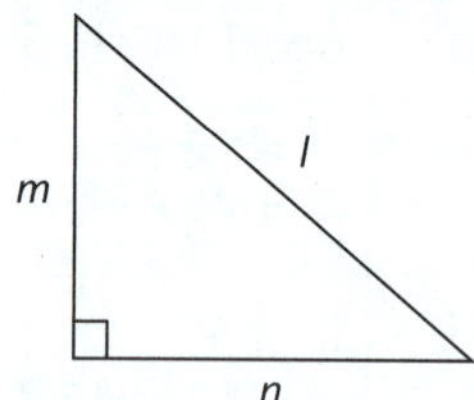


(1 mark)

3 Calculate the value of:

**a** $19^2$ **b** $21^2$ **c** $\sqrt{289}$ (3 marks)

4 Write correct to two decimal places: $\sqrt{50}$ (1 mark)

5 Find the length of the hypotenuse:

**a**

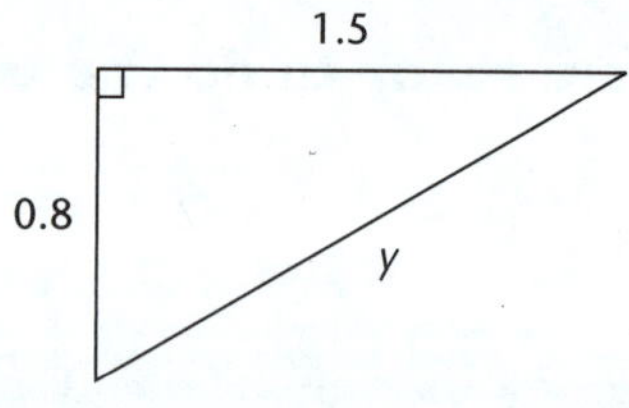


**b**

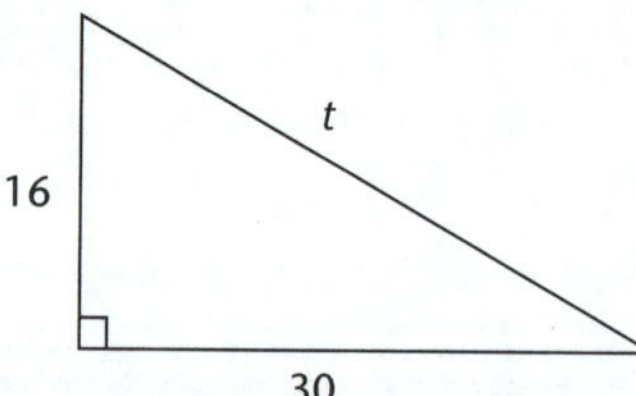


(4 marks)

6 Find the value of the marked side:

**a**

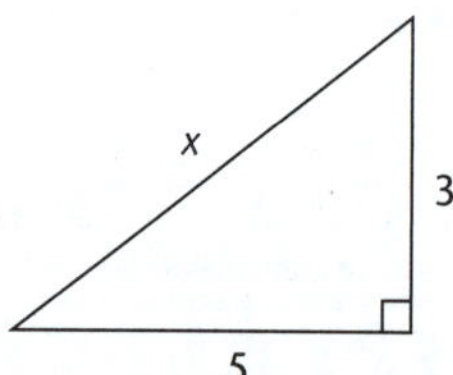


(To 1 decimal place)

**b**

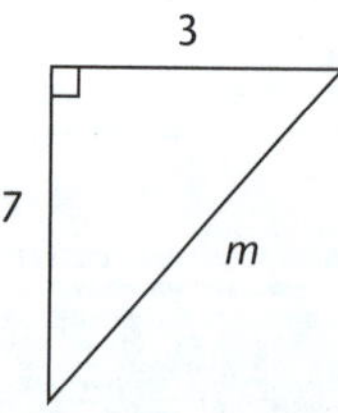


(Exact value)

**c**

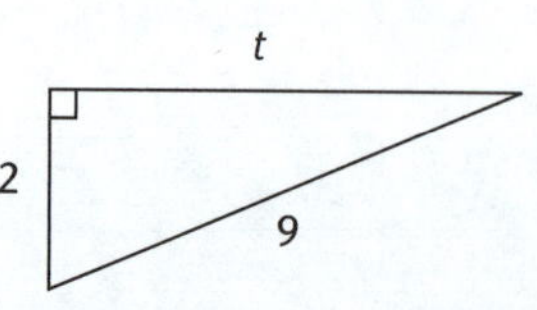


(To 2 decimal places)

(6 marks)

7 Find the length of the side BC:

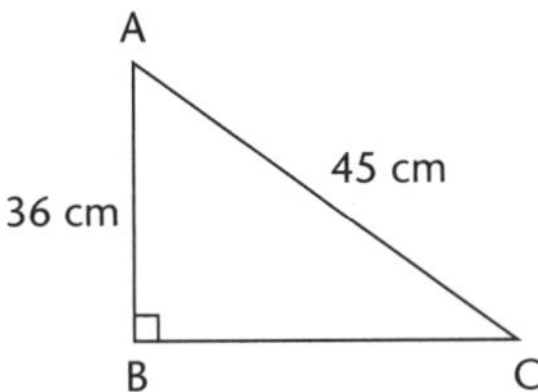


(2 marks)

8 Determine whether these triangles are right-angled:

**a**

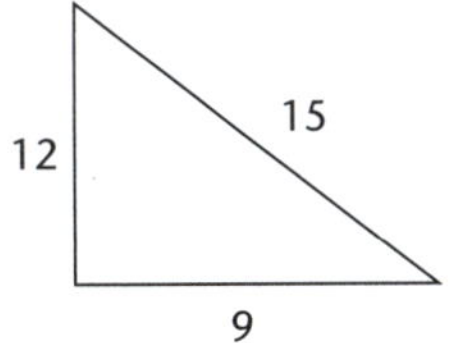


**b**

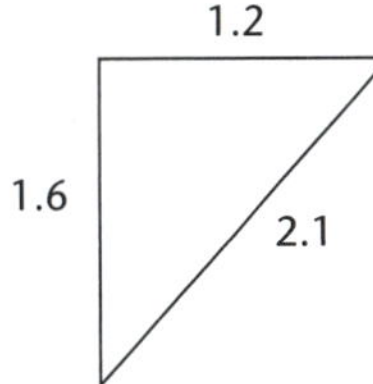


(4 marks)

9 Determine whether {15, 20, 25} is a Pythagorean triad. (1 mark)

10

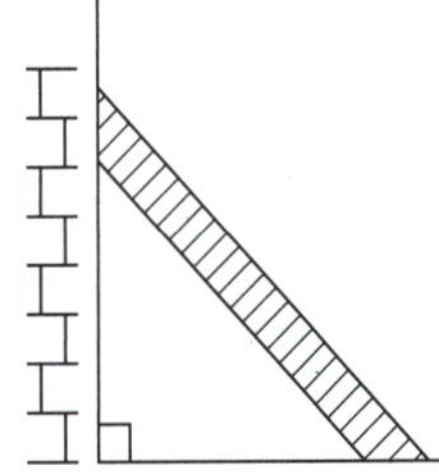

A 3.5-metre ladder reaches 3.1 metres up a wall. Mark this information on your diagram and calculate the distance that the foot of the ladder is from the base of the wall. Answer correct to one decimal place. (2 marks)

☞ Quick answers on page 292
☞ Worked solutions on page 385

**Your Feedback** $\frac{\square}{25} \times 100\% = \square\%$

**(25 marks)**

1 State Pythagoras' Theorem for the triangle PQR:

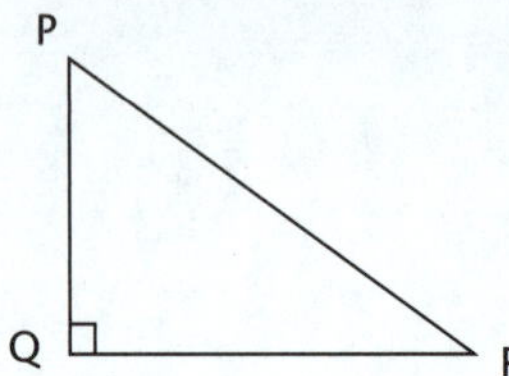


(1 mark)

2 Calculate the value of:

a $37^2$ b $\sqrt{1681}$ (2 marks)

3 Write correct to one decimal place:

$\sqrt{95}$ (1 mark)

4 Find the length of the hypotenuse:

a

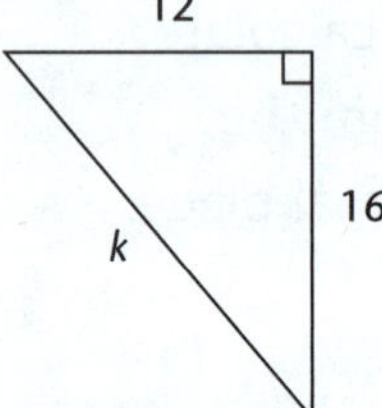


b

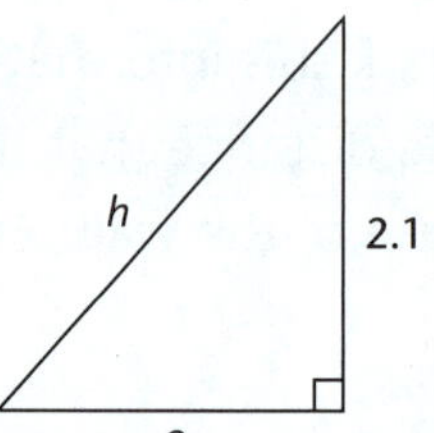


(4 marks)

5 Find the value of the marked side:

a

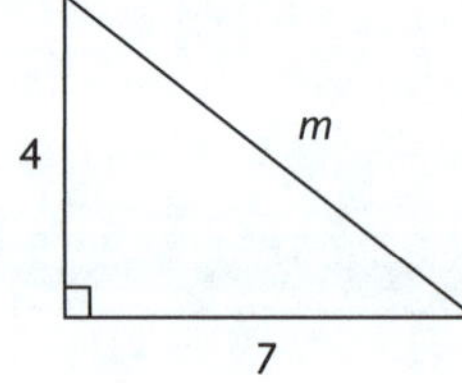


b

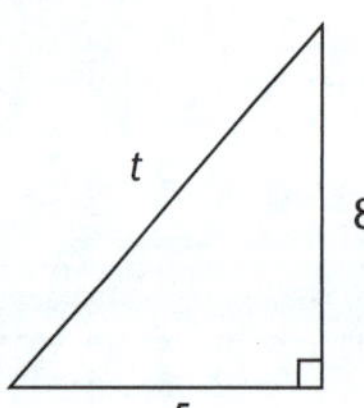


c

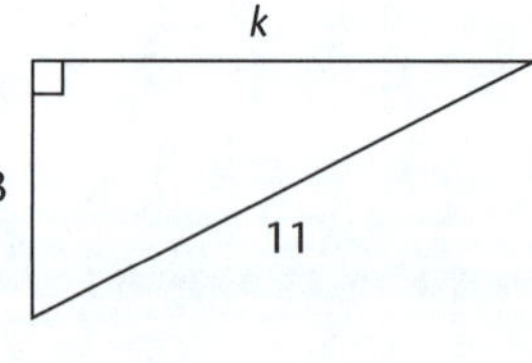


(To 2 decimal places) (To 3 significant figures) (To 1 decimal place) (6 marks)

6 Find the lengths of the marked sides as surds:

a

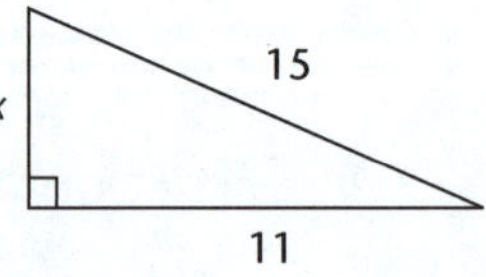


b

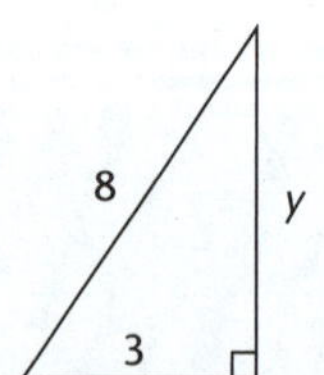


(4 marks)

7 Determine whether these triangles are right-angled:

a

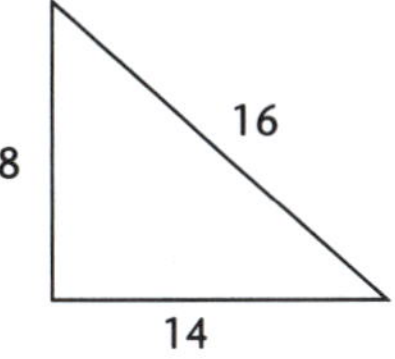


b

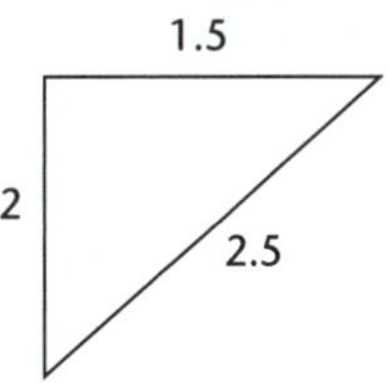


(2 marks)

8 If {21, 28, $x$} is a Pythagorean triad, calculate the value of $x$. (2 marks)

9 Calculate the length of the diagonal of a square of length 6 cm.
Answer correct to one decimal place. (3 marks)

**Your Feedback** $\frac{\square}{25} \times 100\% = \square\%$

☞ Quick answers on page 292
☞ Worked solutions on page 385

**(30 marks)**

1 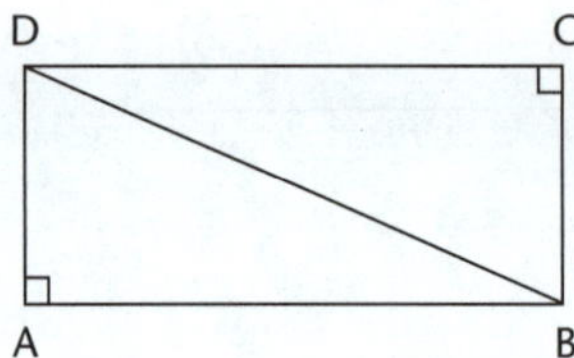


Write down Pythagoras' Theorem for the triangle BCD. (1 mark)

2 Calculate the value of:

a $18^2 + 22^2$ b $\sqrt{36^2 + 27^2}$ (4 marks)

3 Answer correct to two decimal places: $\sqrt{19^2 - 18^2}$ (2 marks)

4 Find the length of the marked side in each diagram:

a 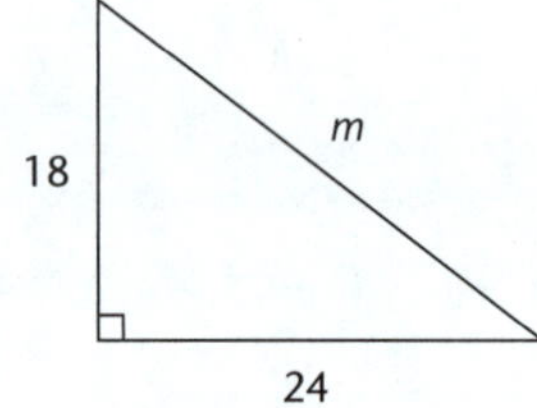


b 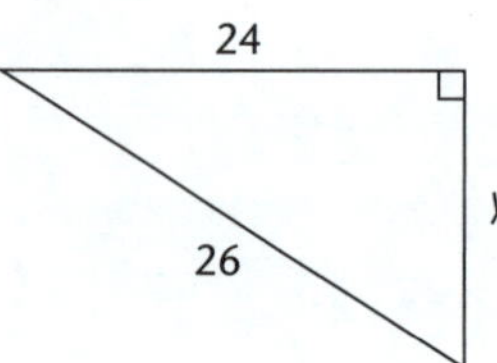


c 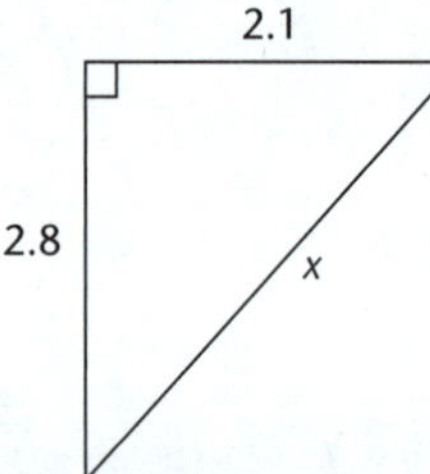


d 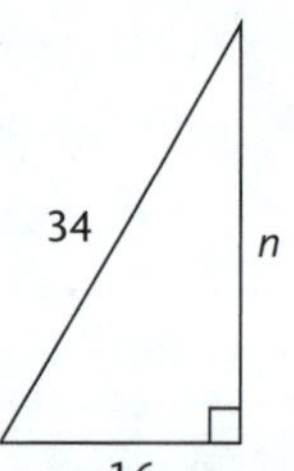


(8 marks)

5 Find the length of the marked sides, correct to one decimal place where necessary:

a 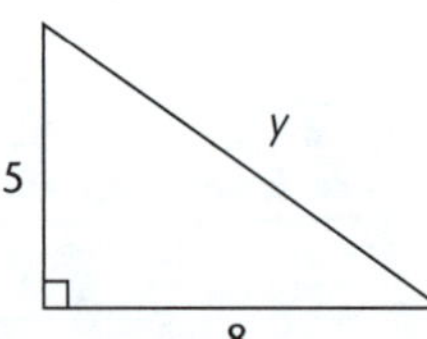


b 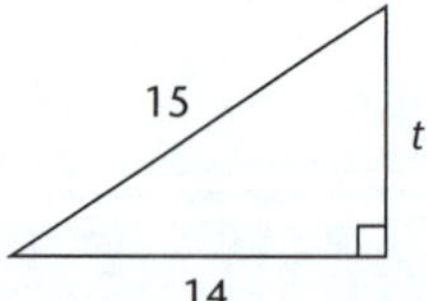


(4 marks)

c 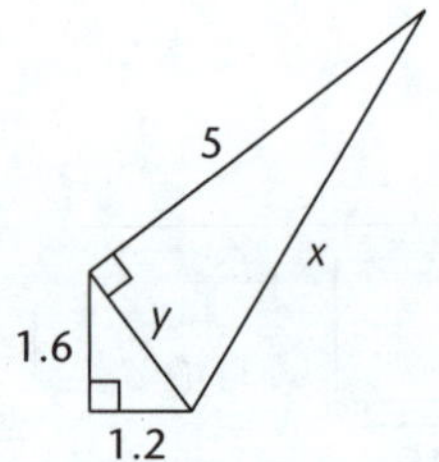


(4 marks)

6 Given that AC = 24 cm, AE = 7 cm, CD = 11 cm, calculate the length of DE:

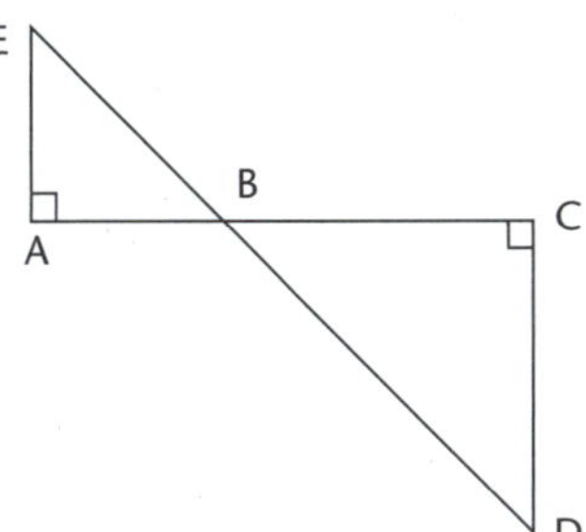


(3 marks)

7 Calculate the perimeter (correct to one decimal place):

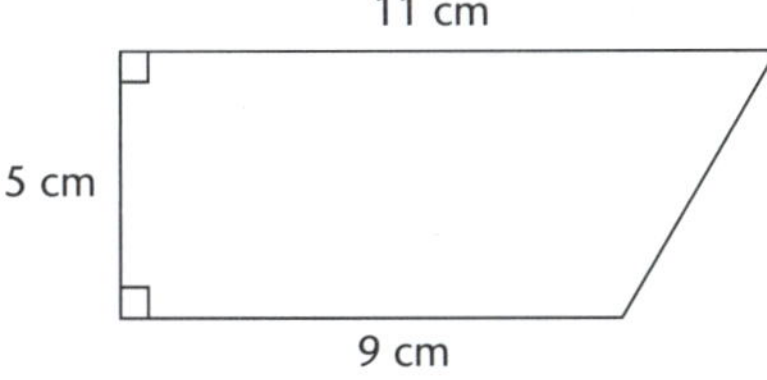


(2 marks)

8 A TV tower 32 m high is to be supported by four wires, each fastened 6 m from the top of the mast. Each wire connects to hooks on the ground 15 m from the tower. Calculate the length of wire required. Answer correct to one decimal place.

(2 marks)

☞ **Quick answers on page 292**
☞ **Worked solutions on page 386**

**Your Feedback** $\frac{\square}{30} \times 100\% = \square\%$

# MEASUREMENT 8

- Area of Simple Shapes
- Parts of the Circle
- Diameter and Radius
- A Note on π
- A Note on the Angle Sum of a Circle and Other Matters
- Circumference of a Circle
- Area of a Circle
- Calculating the Radius or Diameter given Circumference or Area
- Volumes of Prisms
- Volume of Cylinders
- Capacity
- Time

# KEYWORDS

| | |
|---|---|
| Arc | Quadrant |
| Area | Quadrilateral |
| Capacity | Radius |
| Circle | Rhombus |
| Circumference | Sector |
| Composite | Semicircle |
| Cylinder | Segment |
| Diameter | Surface area |
| Kite | Time difference |
| Local time | Time zone |
| Parallelogram | Trapezium |
| Pi | Volume |

# Area

## Area of Simple Shapes

| Name | Figure | Area |
|---|---|---|
| Rectangle | 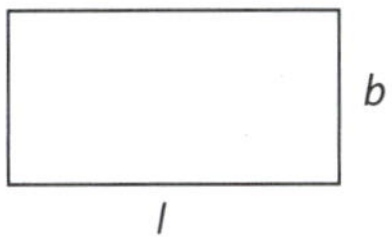  | Area = Length × Breadth<br>$\therefore A = lb$ |
| Square | 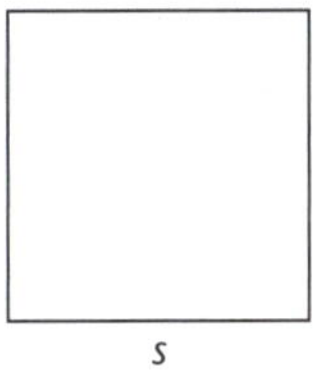  | Area = $\text{Side}^2$<br>$\therefore A = s^2$ |
| Triangle | 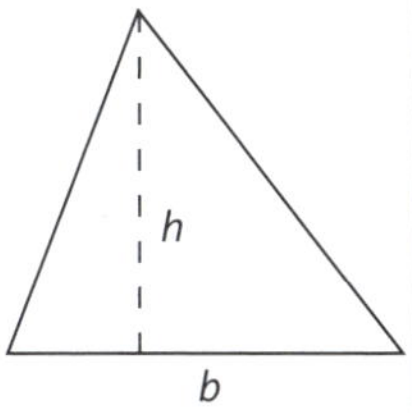  | Area<br>$= \frac{1}{2}$ Base × Height<br>$\therefore A = \frac{1}{2}bh$ |

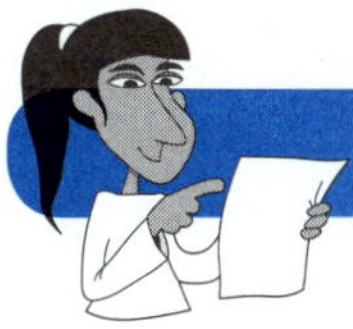

## For Example

1 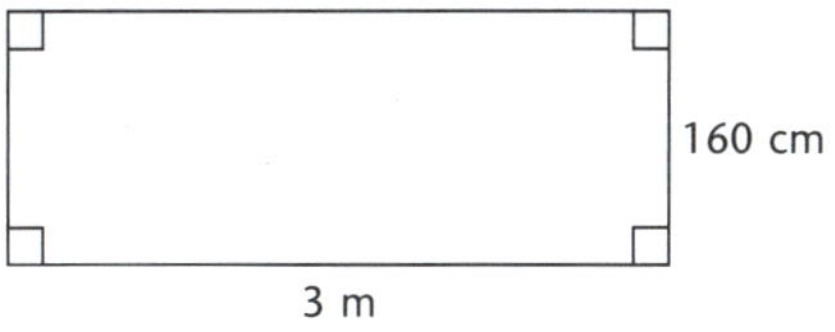


2 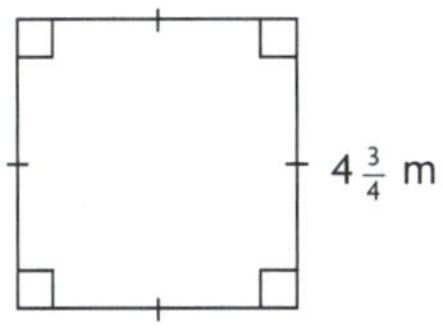


3 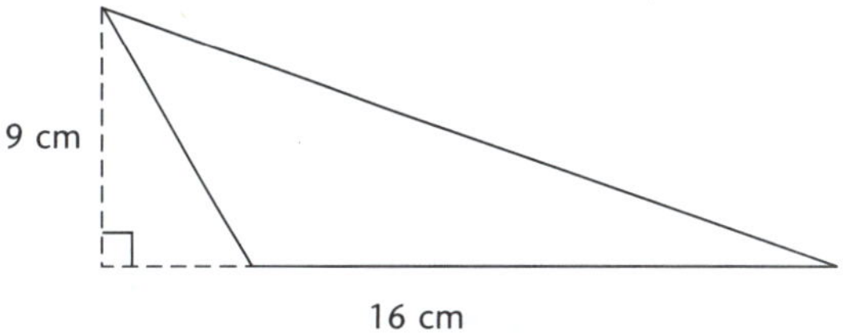


1 Now, 160 cm = 1.6 m:

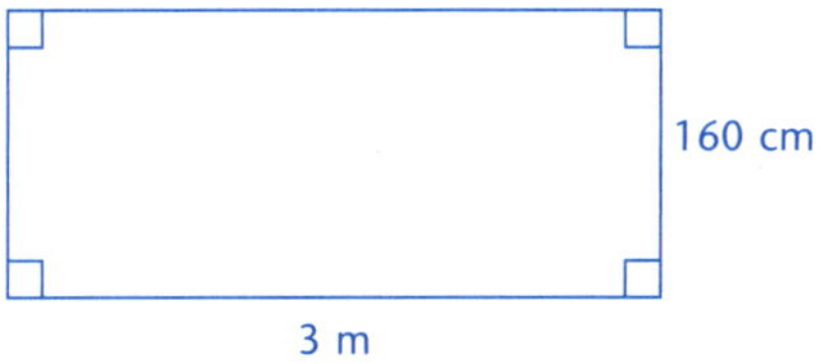


$\therefore A = lb$

$= 3 \times 1.6$

$= 4.8$

$\therefore$ The area is 4.8 $m^2$.

2 $A = s^2$

$= 4\frac{3}{4} \times 4\frac{3}{4}$

$= 22\frac{9}{16}$

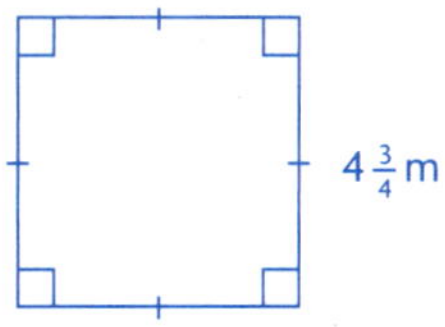


$\therefore$ The area is $22\frac{9}{16}$ $m^2$.

**3**

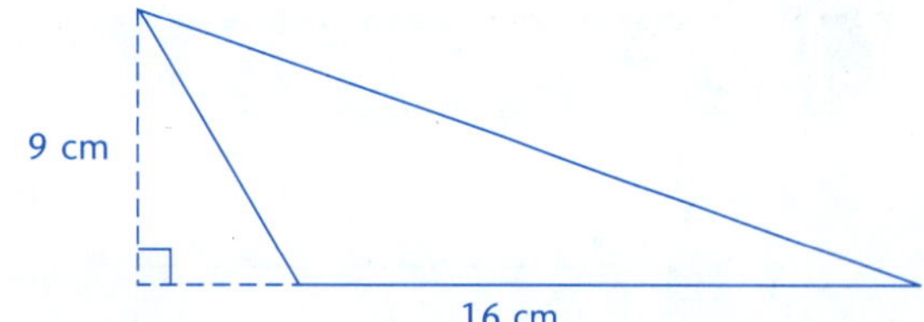


$$A = \tfrac{1}{2}bh$$
$$= \tfrac{1}{2} \times 16 \times 9$$
$$= 72$$

$\therefore$ The area is 72 cm$^2$.

## Area of Composite Figures

Two simple shapes can join together to form a composite figure. To find the area of this figure we split it back into simple shapes.

**For Example**

Find the area:

**1**

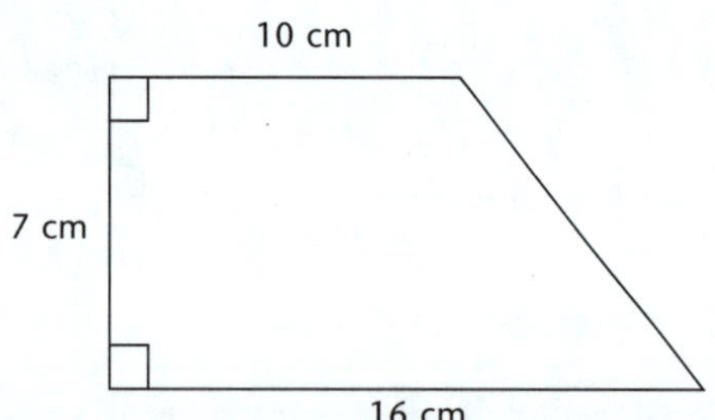


**2**

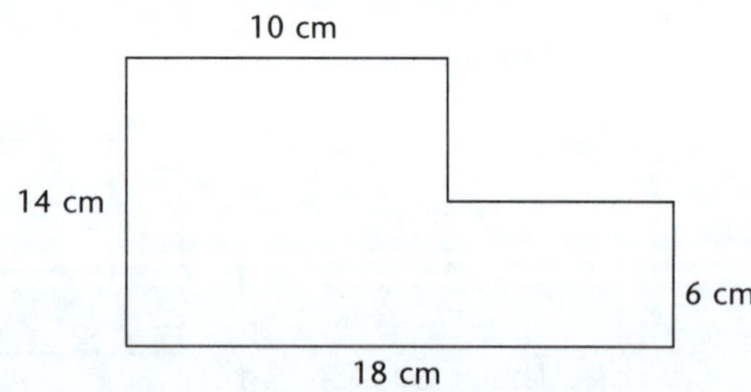


**3**

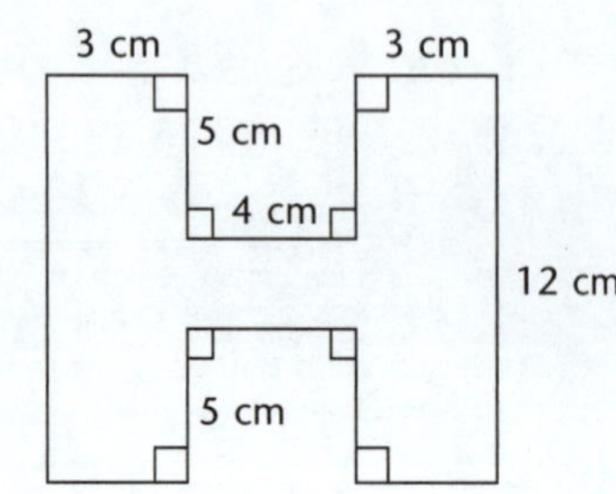


**4**

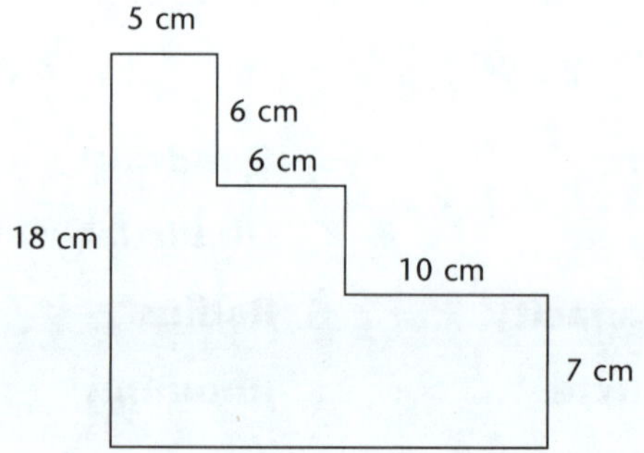


**1** $A = A_1 + A_2$

Shape is a rectangle and a triangle:

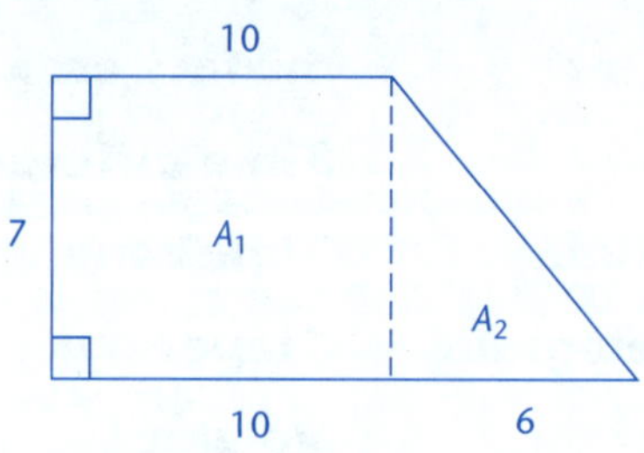


$$= 7 \times 10 + \tfrac{1}{2} \times 6 \times 7$$
$$= 70 + 21$$
$$= 91$$

$\therefore$ The area is 91 cm$^2$.

**2**

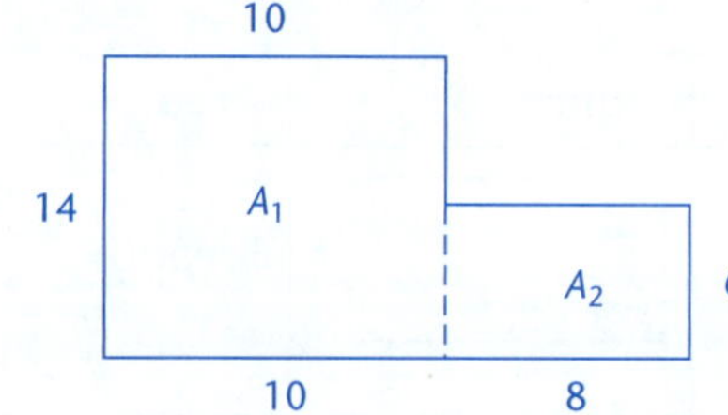


Shape is two rectangles:

$\therefore A = A_1 + A_2$
$$= 14 \times 10 + 8 \times 6$$
$$= 140 + 48$$
$$= 188$$

$\therefore$ The area is 188 cm$^2$.

**3**

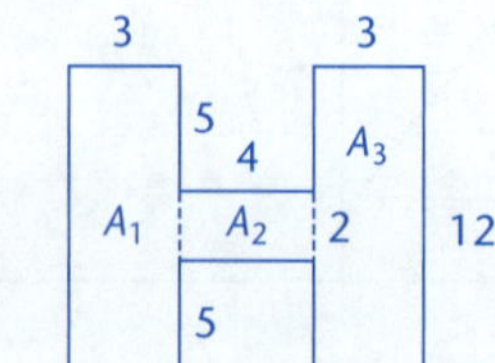

Shape is three rectangles:

$\therefore A = A_1 + A_2 + A_3$

$= 12 \times 3 + 4 \times 2 + 12 \times 3$

$= 36 + 8 + 36$

$= 80$

$\therefore$ The area is 80 cm$^2$.

**4**

5, 6, 6, 18, $A_1$, $A_2$, 12, 10, $A_3$, 7

Shape is three rectangles:

$\therefore A = A_1 + A_2 + A_3$

$= 18 \times 5 + 12 \times 6 + 10 \times 7$

$= 90 + 72 + 70$

$= 232$

$\therefore$ The area is 232 cm$^2$.

## Areas of Shaded Regions

With these questions a subtraction of two areas will give the required region.

Find the area of the shaded region.

**1**

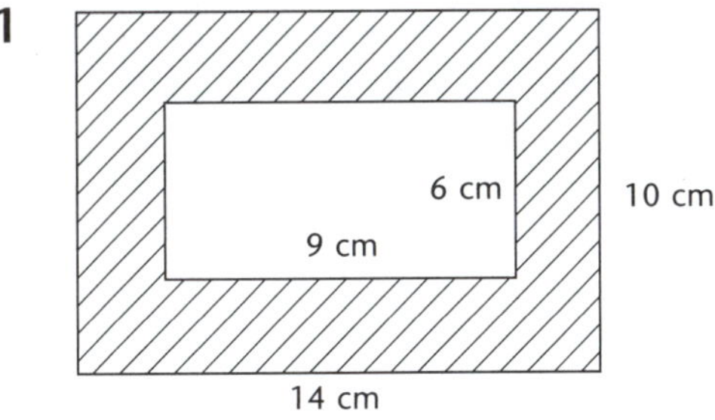


**2**

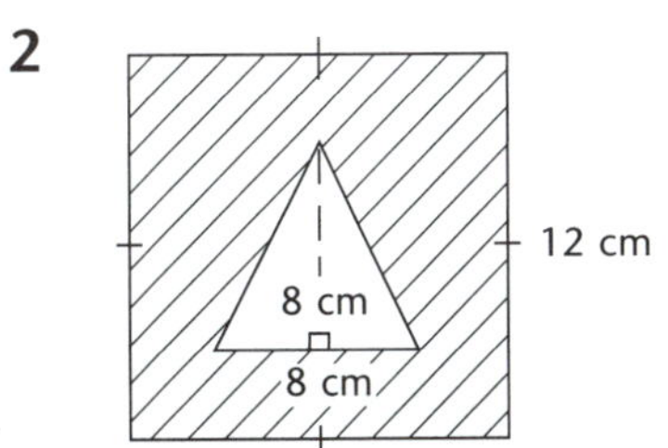


**3**

16 m, 2 m border, 20 m

**1** Area $= 14 \times 10 - 9 \times 6$

[Outer rectangle – Inner rectangle]

$= 140 - 54$

$= 86$

$\therefore$ The area is 86 cm$^2$.

**2** Area $= 12 \times 12 - \frac{1}{2} \times 8 \times 8$

[Square – Triangle]

$= 144 - 32$

$= 112$

$\therefore$ The area is 112 cm$^2$.

**3** The 2-metre border will mean the dimensions of an inside rectangle are 16 m by 12 m.

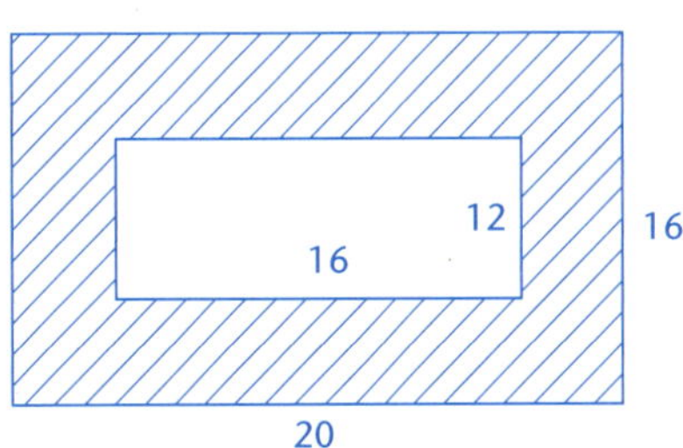


$\therefore$ Area $= 20 \times 16 - 16 \times 12$

$= 320 - 192$

$= 128$

$\therefore$ The area is 128 m$^2$.

## Areas of Special Quadrilaterals

| Name | Figure | Area ($A$) |
|---|---|---|
| Parallelogram | $h$, $b$ | Base × Height<br>$A = bh$ |
| Trapezium | $a$, $h$, $b$ | $A = \frac{1}{2}h(a + b)$ |
| Rhombus or kite | A, B, C, D, $h$<br>AC = $x$<br>BD = $y$ | $\frac{1}{2}$ Product of diagonals<br>$A = \frac{1}{2}xy$ |

### For Example

Find the area of:

**1**

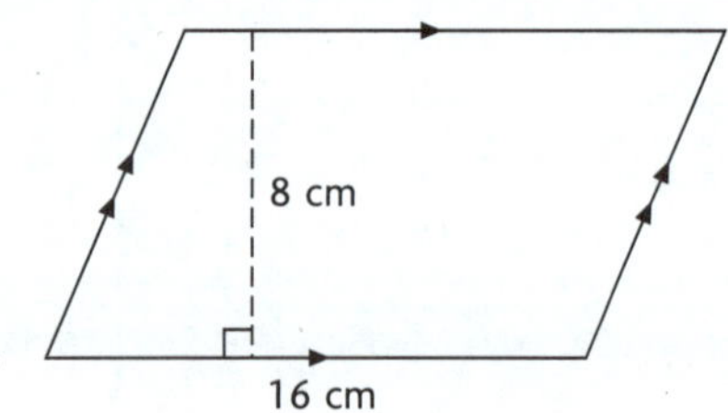


**2**

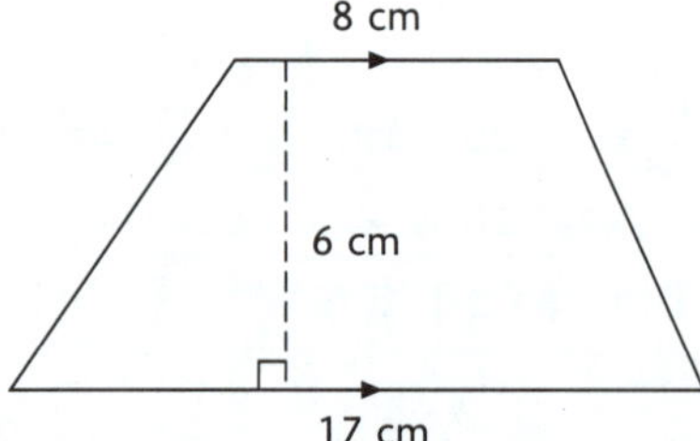


**3**

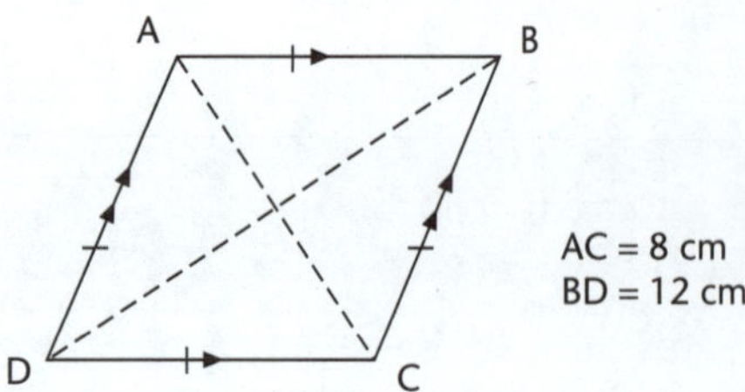


**4**

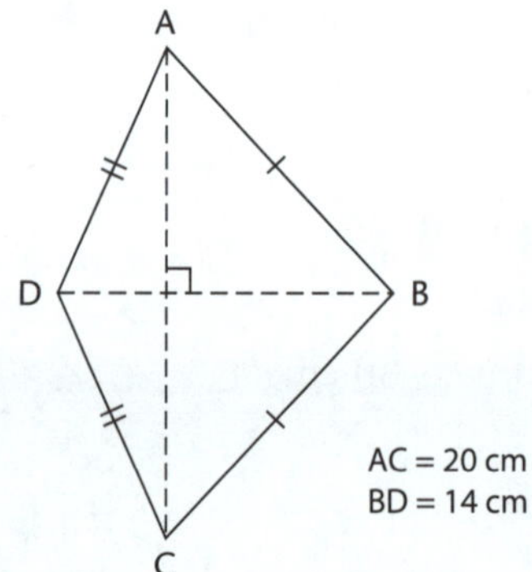


**1** Area $= bh$
$= 16 \times 8$
$= 128$
$\therefore$ Area is 128 cm$^2$.

**2** Area $= \frac{1}{2}h(a + b)$
$= \frac{1}{2} \times 6(8 + 17)$
$= 3(25)$
$= 75$
$\therefore$ Area is 75 cm$^2$.

**3** Area $= \frac{1}{2}xy$
$= \frac{1}{2} \times 8 \times 12$
$= 48$
$\therefore$ Area is 48 cm$^2$.

**4** Area $= \frac{1}{2}xy$
$= \frac{1}{2} \times 20 \times 14$
$= 140$
$\therefore$ Area is 140 cm$^2$.

## Some Important Area Results

Here are some conversions that should be learned.

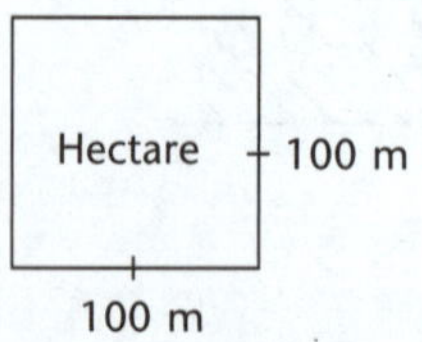


Area $= 100 \times 100$
$= 10\,000$
$\therefore$ Area is $10\,000$ m$^2$
**$\therefore$ $10\,000$ m$^2$ = 1 hectare (ha).**

Also, two identical squares:

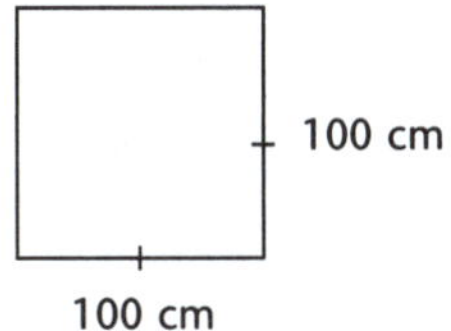


Area = 100 × 100
= 10 000
∴ The area is 10 000 $cm^2$.

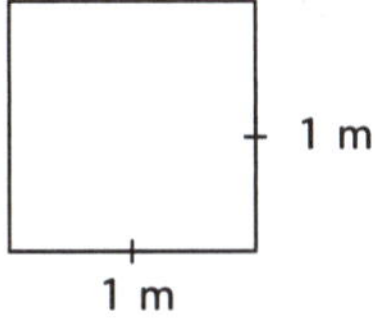


Area = 1 × 1
= 1
∴ The area is 1 $m^2$
∴ **10 000 $cm^2$ = 1 $m^2$.**

## For Example

**1** Find the area, in hectares:

**a**

130 m

462 m

**b**

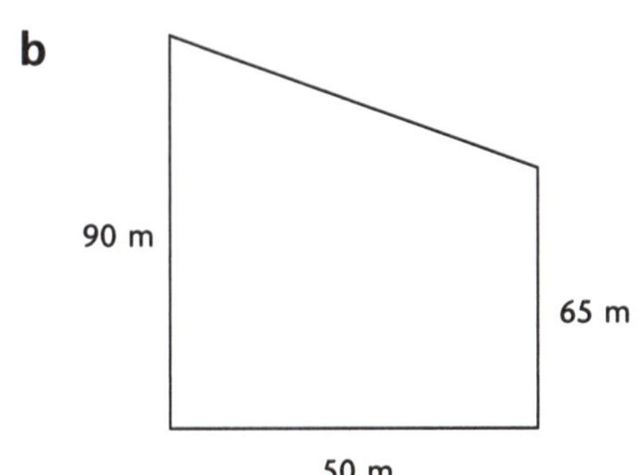


**2** Find the area, in $m^2$:

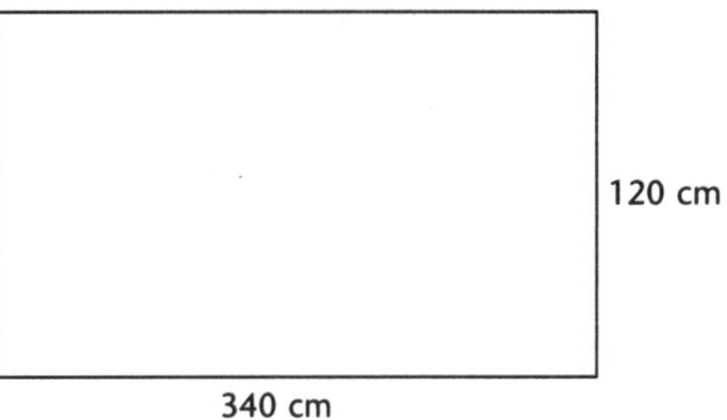


**1 a** Area = 462 × 130
= 60 060
∴ area is 60 060 $m^2$

∴ area (hectares) = 60 060 ÷ 10 000
= 6.006
∴ The area is 6.006 hectares.

**b**

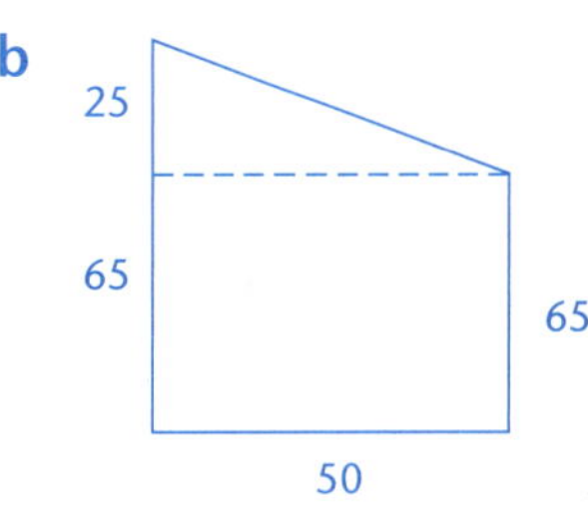


Area = $65 \times 50 + \frac{1}{2} \times 50 \times 25$
= 3250 + 625
= 3875
∴ area is 3875 $m^2$
∴ area (hectares) = 3875 ÷ 10 000
= 0.3875
∴ The area is 0.3875 hectares.

**2** Two alternatives:

**i** Area = 3.4 × 1.2
= 4.08

1.2 m

3.4 m

∴ The area is 4.08 $m^2$.

ii

Area = $340 \times 120$

$= 40\,800$

$\therefore$ area is $40\,800$ cm$^2$

$\therefore$ area (m$^2$) $= 40\,800 \div 10\,000$

$= 4.08$

$\therefore$ The area is 4.08 m$^2$.

## Parts of the Circle

1

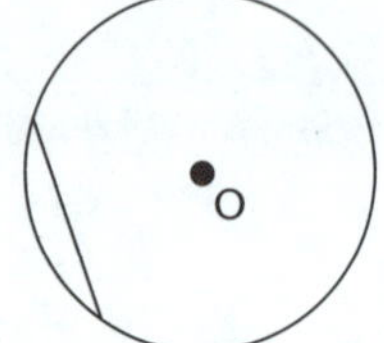


2

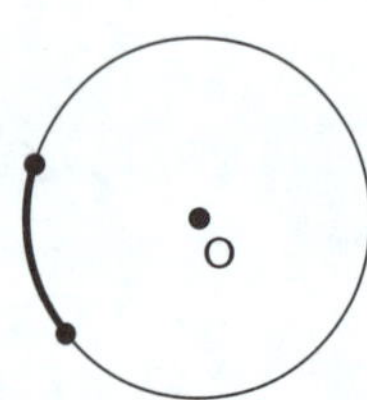


3

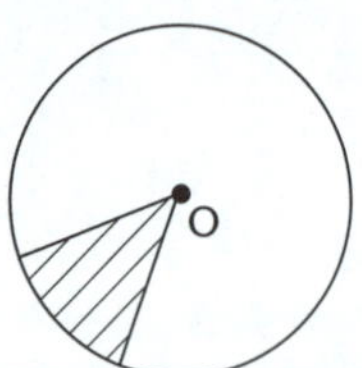


4

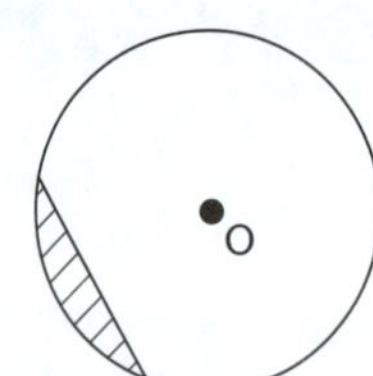


5

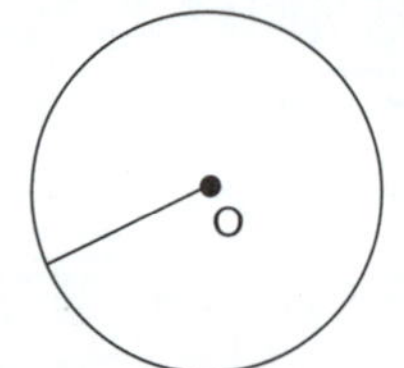


6

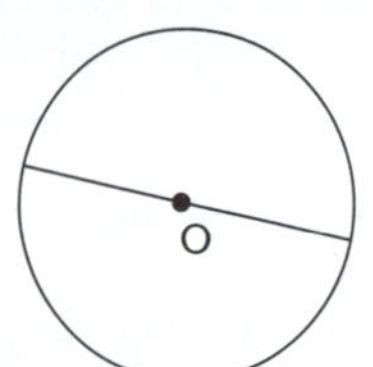


7

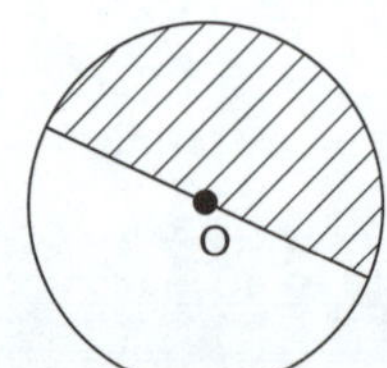


8

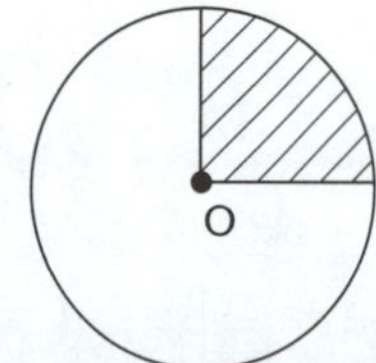


1 Chord  2 Arc

3 Sector  4 Segment

5 Radius  6 Diameter

7 Semicircle  8 Quadrant

The circumference of a circle is the distance around the outside of the circle.

## Diameter and Radius

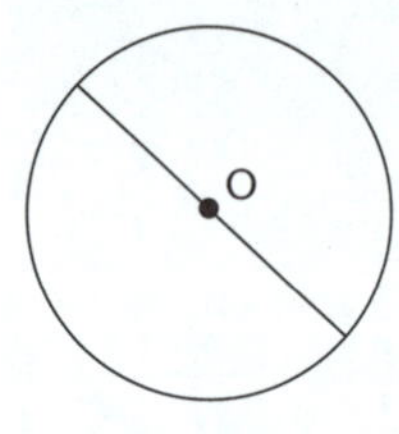


Diameter

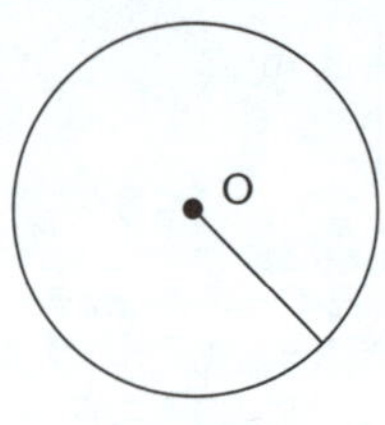


Radius

[O is the centre.]

Length of diameter = 2 × Length of the radius

$d = 2r$

*or* Radius = $\frac{1}{2}$ of diameter

$r = \frac{1}{2}d$

### For Example

1 If the radius of a circle is 15 cm, calculate the diameter.

2 A bike wheel has 36 cm diameter. Calculate the radius of the wheel.

1 $r = 15$

Now $d = 2r$

$= 2 \times 15$

$= 30$

The diameter of the circle is 30 cm.

2 $d = 36$

Now $r = \frac{1}{2}d$

$= \frac{1}{2} \times 36$

$= 18$

The radius of the wheel is 18 cm.

## A Note on π

The constant π is used in many circle questions. There are a number of commonly used approximations for π ($3\frac{1}{7}$, 3.14, etc). If a particular value of π is required it will be included in the question. Where a value is *not* given, the π button on the calculator should be used.

## A Note on the Angle Sum of a Circle and Other Matters

A semicircle is half a circle.

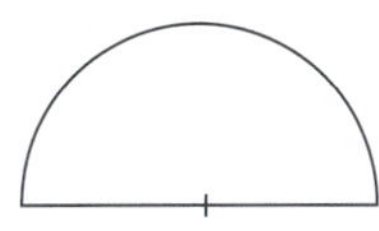

A quadrant is one-quarter of a circle.

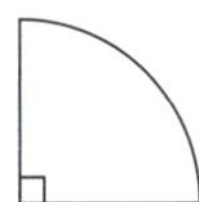

Other fractions of a circle can be calculated using the fact that the angle sum of a circle is 360°.

i.e. Fraction of circle $= \frac{60°}{360°}$

$= \frac{1}{6}$

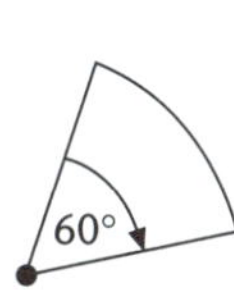


This sector then has $\frac{1}{6}$ of the area of the whole circle.

The length of the arc, PQ, is $\frac{1}{6}$ of the circumference of the full circle.

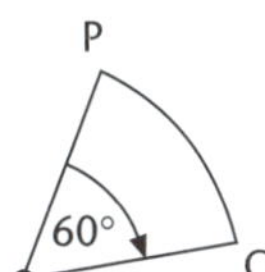


**For Example**

Calculate the fraction of a circle represented in these diagrams:

**1**

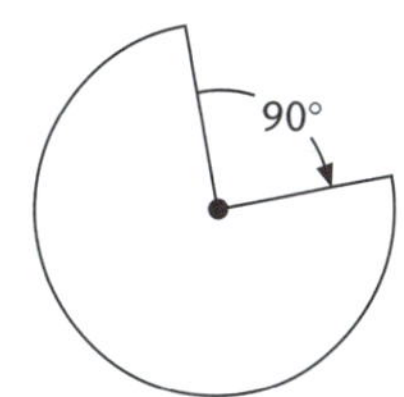


**2**

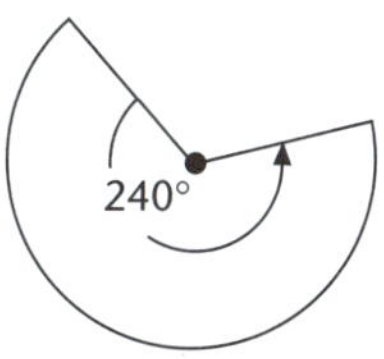


**1** Angle at centre = 360° − 90°

= 270°

Fraction of circle $= \frac{270°}{360°}$

$= \frac{3}{4}$

**2** Fraction of a circle $= \frac{240°}{360°}$

$= \frac{2}{3}$

## The Circumference of a Circle

The circumference of a circle is the perimeter of the circle—the distance around the outside of the circle.

Circumference = 2 × π × Radius of circle

$C = 2\pi r$

*or*

Circumference = π × Diameter of circle

$C = \pi d$

$\pi$ = 3.141 592 653 589 793 238 462 643
383 279 502 884 197 169 399 375
105 820 974 944 592 307 816 406
286 208 998 628 034 825 342 117
067 982 148 086 513 282 306 647
093 844 609 550 582 231 725 359
408 128

Correct to 150 decimal places.

Remember, $\pi$ = Circumference ÷ Diameter, or there are $\pi$ number of diameters around the circumference.

## For Example

1 Calculate the circumference of these circles. Answer correct to one decimal place:

a 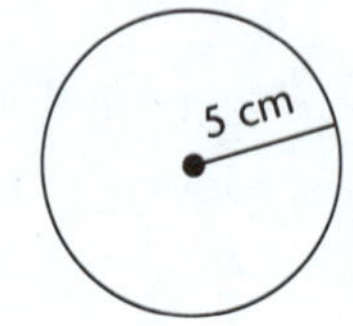


b 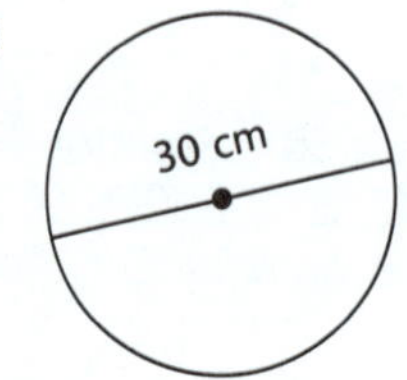


2 Find the circumference of a circle with radius 8 cm. Leave your answer in terms of $\pi$.

3 Calculate the circumference of a circle with radius 0.12 m. Use $\pi = 3.142$. Answer correct to three decimal places.

4 Calculate the circumference of a circle with diameter 21 cm. Use $\pi = 3\frac{1}{7}$.

5 Calculate the perimeter of these figures. Answer correct to two decimal places:

a 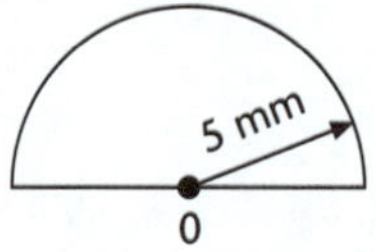


b 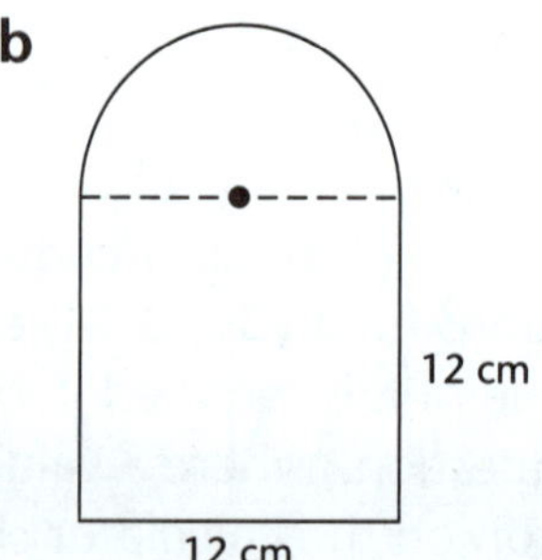


Note: for this chapter, the formula used for all questions will be $C = \pi d$.

1 a $r = 5$

$d = 2r$

$\therefore d = 10$

$C = \pi d$

$= \pi \times 10$

$= 31.415927$

$= 31.4$ (to 1 decimal place)

$\therefore$ The circumference is 31.4 cm.

b $d = 30$

$C = \pi d$

$= \pi \times 30$

$= 94.24778$

$= 94.2$ (to 1 decimal place)

$\therefore$ The circumference is 94.2 mm.

2 $C = \pi d$

$= \pi \times 16$

$= 16\pi$

$\therefore$ The circumference is $16\pi$ cm.

3 $r = 0.12$

$d = 2 \times r$

$= 0.24$

$C = \pi d$ [Given $\pi = 3.142$]

$= 3.142 \times 0.24$

$= 0.75408$

$= 0.754$ (to 3 decimal places)

$\therefore$ The circumference is 0.754 m.

4 $d = 21$

$C = \pi d$

$= 3\frac{1}{7} \times 21$ [Given $\pi = 3\frac{1}{7}$]

$= \frac{22}{\not{7}_1} \times \frac{\not{21}^3}{1}$

$= 66$

$\therefore$ The circumference is 66 cm.

5 a Perimeter = Circumference of semicircle + Length of diameter

Diameter = 2 × Radius

$= 2 \times 5$

$= 10$

Now $C$ of semicircle s $= \frac{1}{2}(\pi d)$

[$\frac{1}{2}$ circumference of circle]

$\therefore C_s = \frac{1}{2} \times \pi \times 10$

$= 15.707963$

$= 15.71$ (to 2 decimal places)

$\therefore P = C_s + d$

$= 15.71 + 10$

$= 25.71$

$\therefore$ The perimeter is 25.71 mm.

b Perimeter = 12 + 12 + 12 + Circumference of semicircle

$= 36 + C_s$

Now $C_s = \frac{1}{2}\pi d$ [$d = 12$]

$= \frac{1}{2} \times \pi \times 12$

$= 18.849556$

$= 18.85$ (to 2 decimal places)

Perimeter = 36 + 18.85

$= 54.85$

$\therefore$ The perimeter is 54.85 cm.

## For Example

1 A circular fish pond has a radius of 4 metres. Calculate the length of edging required to surround the fountain.

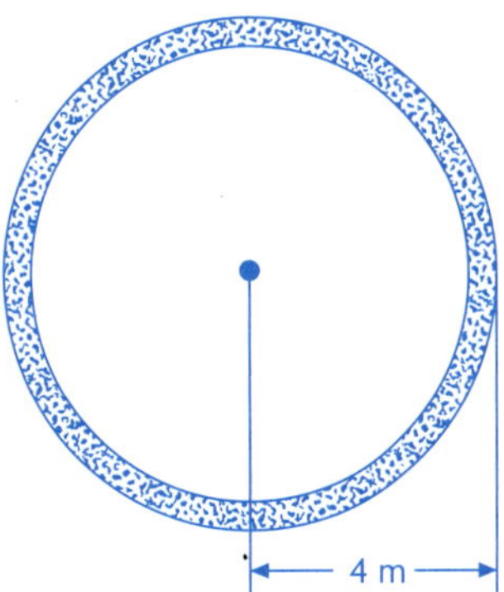


1 $r = 4$

$\therefore d = 2 \times r$

$= 2 \times 4$

$= 8$

$C = \pi d$

$= \pi \times 8$

$= 25.132741$

$= 25.1$ (to 1 decimal place)

Circumference = 25.1 metres.
The length of edging required is 25.1 metres.

## Area of a Circle

Area = $\pi \times$ (radius of circle)$^2$

$A = \pi r^2$ $\Rightarrow$ This means $A = \pi \times r \times r$

## For Example

1 Find the area of these circles. Answer correct to one decimal place:

a 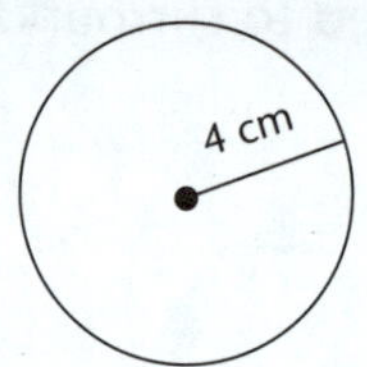


b 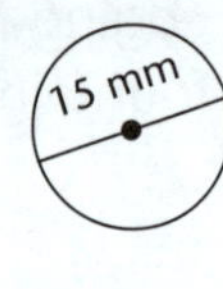


2 Using $\pi = 3.14$, find the area of a circle with a:

a radius 3 m

b diameter 20 cm.

3 Find the area of a circle radius 14 mm. Use $\pi = \frac{22}{7}$.

4 Calculate the area of these figures. Answer correct to one decimal place:

a 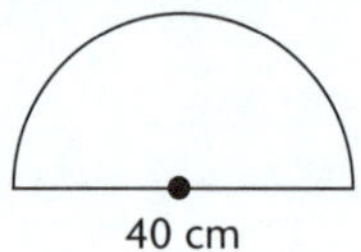


b 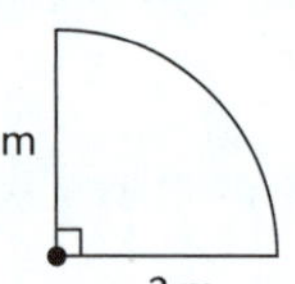


c 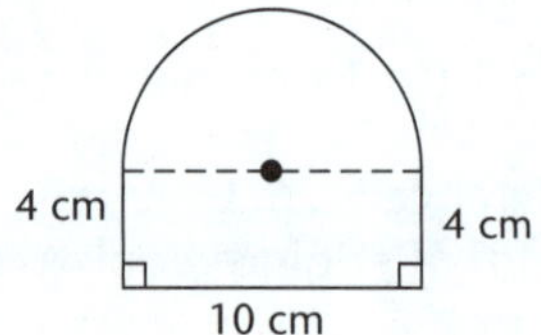


d 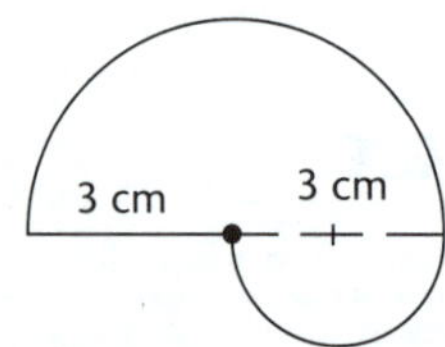


1 a $r = 4$

$A = \pi r^2$

$= \pi \times 4 \times 4$

$= 50.265\,482$

$= 50.3$ (to 1 decimal place)

The area is 50.3 $cm^2$.

b $d = 15$

$r = \frac{1}{2} \times 15$

$= 7.5$

$A = \pi r^2$

$= \pi \times 7.5 \times 7.5$

$= 176.714\,59$

$= 176.7$ (to 1 decimal place)

The area is 176.7 $mm^2$.

2 a $r = 3$

$A = \pi r^2$

$= 3.14 \times 3 \times 3$

$= 28.26$

The area is 28.26 $m^2$.

b $d = 20$

$r = \frac{1}{2} \times 20$

$= 10$

$A = \pi r^2$

$= 3.14 \times 10 \times 10$

$= 314$

The area is 314 $cm^2$.

3 $r = 14$

$A = \pi r^2$

$= \frac{22}{\cancel{7}_1} \times \cancel{14}^2 \times 14$

$= 616$

The area is 616 $mm^2$.

4 a Figure is a semicircle (half of a circle).

$A = \frac{1}{2}\pi r^2$

$= \frac{1}{2} \times \pi \times 20 \times 20$

$= 628.318\,53$

$= 628.3$ (to 1 decimal place)

[$d = 40$, $\therefore r = \frac{1}{2} \times 40 = 20$]

The area of the semicircle is 628.3 $cm^2$.

b The area of the quadrant

$= \frac{1}{4}\pi r^2$ [Quarter of a circle.]

$= \frac{1}{4} \times \pi \times 2 \times 2$ [$r = 2$]

$= 3.141\,592\,7$

$= 3.1$ (to 1 decimal place)

The area of the quadrant is 3.1 $m^2$.

**c** Total area = Area of semicircle + Area of rectangle.

Area of semicircle $\left[\begin{array}{l} d = 10 \\ \therefore r = 5 \end{array}\right]$

$= \frac{1}{2}\pi r^2$

$= \frac{1}{2} \times \pi \times 5 \times 5$

$= 39.269\,908$

$= 39.3$ (to 1 decimal place)

Area rectangle $= lb$

$= 10 \times 4$

$= 40$

Total area $= 39.3 + 40$

$= 79.3$

The area of the figure is 79.3 $cm^2$.

**d** Total area = Area of semicircle 1 + Area of semicircle 2

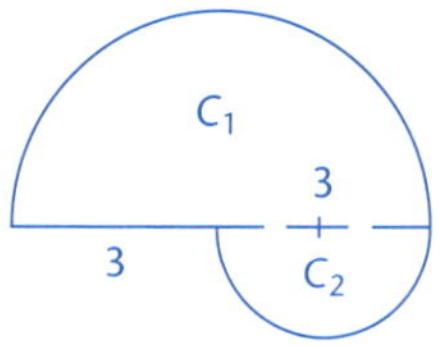


$A$ of $C_1 = \frac{1}{2}\pi r^2$ $[r = 3]$

$= \frac{1}{2} \times \pi \times 3 \times 3$

$= 14.137\,167$

$A$ of $C_2 = \frac{1}{2}\pi r^2$ $\left[\begin{array}{l} d = 3 \\ \therefore r = 15 \end{array}\right]$

$= \frac{1}{2} \times \pi \times 1.5 \times 1.5$

$= 3.534\,291\,7$

Total area $= 14.137 + 3.534$

$= 17.671$

$= 17.7$ (to 1 decimal place)

The total area of the figure is 17.7 $cm^2$.

## For Example

1 A circular section is cut from a square of metal 4 cm wide such that the diameter of the circle is exactly the width of the square. Calculate the amount of metal remaining.

1 Area of square $= lb$

$= 4 \times 4$

$= 16$

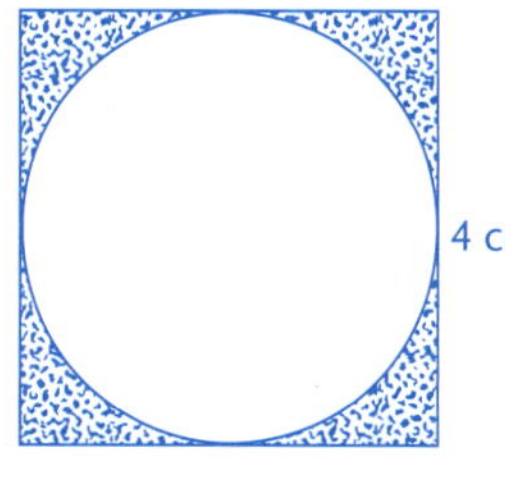


$\left[\begin{array}{l} d = 4 \\ \therefore r = 2 \end{array}\right]$

Area of circle $= \pi r^2$

$= \pi \times 2 \times 2$

$= 12.566\,371$

Area remaining = Area of square − Area of circle

$= 16 - 12.566\,371$

$= 3.433\,629\,4$

$= 3.43$ (to 2 decimal places)

The area of metal remaining = 3.43 $cm^2$.

## Calculating the Radius or Diameter Given the Circumference or Area

From the formula $C = \pi d$, we get:

$$d = \frac{C}{\pi} \quad or \quad r = \frac{C}{2\pi}$$

From the formula $A = \pi r^2$ we get:

$$r = \sqrt{\frac{A}{\pi}}$$

## For Example

1 Calculate the diameter of a circle with circumference 40 m. Answer correct to three decimal places.

2 A wheel has a circumference of 190 cm. Calculate the radius of the wheel. Leave the answer in terms of $\pi$.

3 The circumference of a circle is 60 cm. Find the radius of the circle. Answer correct to one decimal place.

4 The area of a circle is 240 cm$^2$. Find the radius of the circle. Answer correct to two decimal places.

**1** $d = \dfrac{C}{\pi}$

$= \dfrac{40}{\pi}$

$= 12.732\,395$

$= 12.732$ (3 decimal places)

The diameter is 12.732 m.

**2** $r = \dfrac{C}{2\pi}$

$= \dfrac{190}{2\pi}$

$= \dfrac{95}{\pi}$

The radius is $\dfrac{95}{\pi}$ cm.

**3** $r = \dfrac{C}{2\pi}$

$= \dfrac{60}{2\pi}$

$= 9.549\,297$

$= 9.5$ (1 decimal place)

The radius is 9.5 cm.

**4** $r = \sqrt{\dfrac{A}{\pi}}$

$= \sqrt{\dfrac{240}{\pi}}$

$= 8.740\,387$

$= 8.74$ (2 decimal places)

The radius is 8.74 cm.

## Volumes of Prisms

A prism is a solid shape with a uniform cross-section (i.e. the area of the cross-section parallel to the face is the same no matter where the solid is cut).

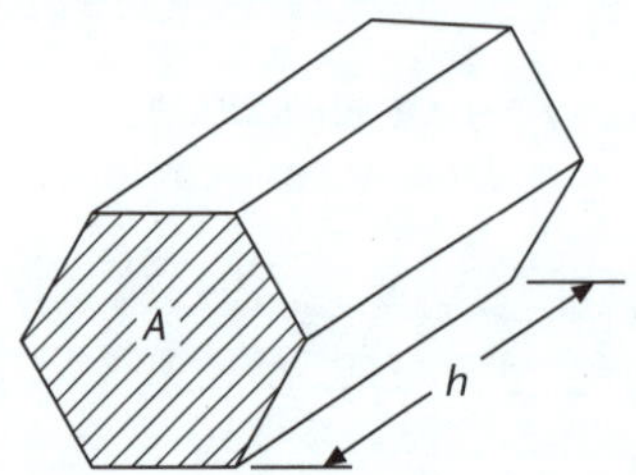


Volume = Area of base × Height

$\therefore V = Ah$

(The base will be identical to any cross-section.)

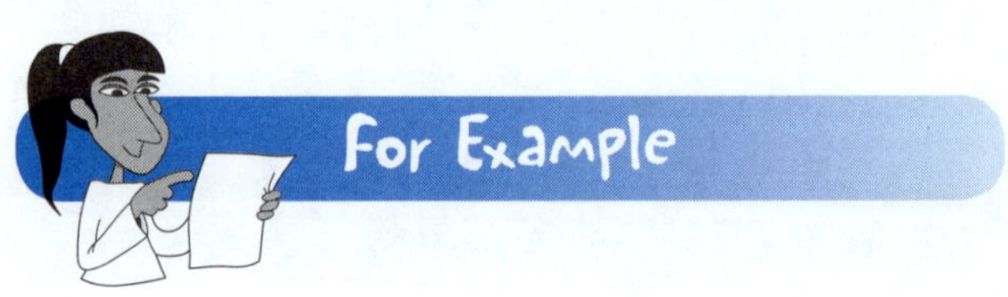


Find the volume of:

1

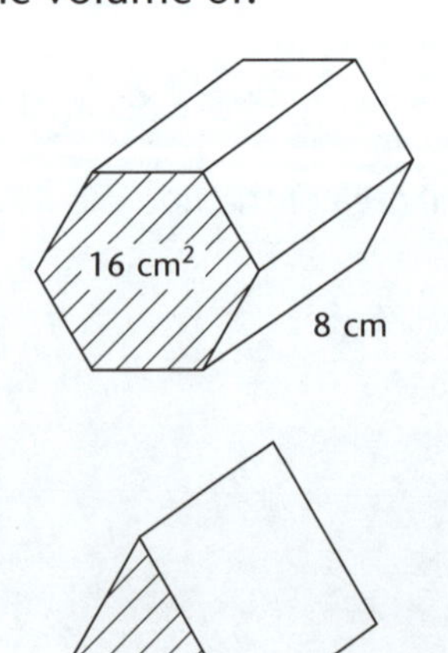


2

**3**

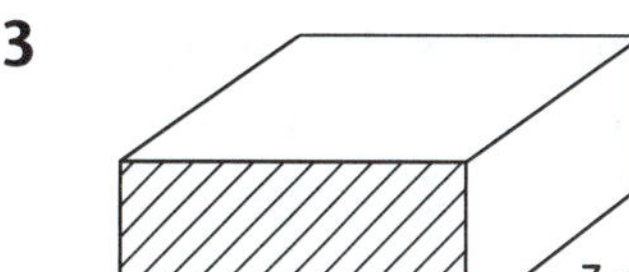


**4**

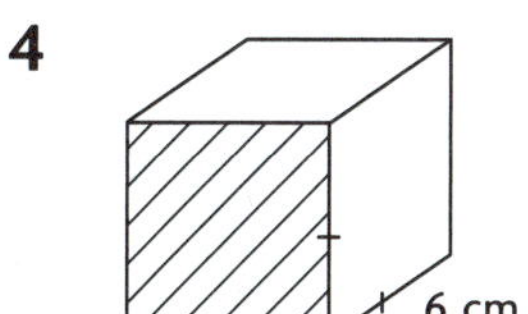


**5**

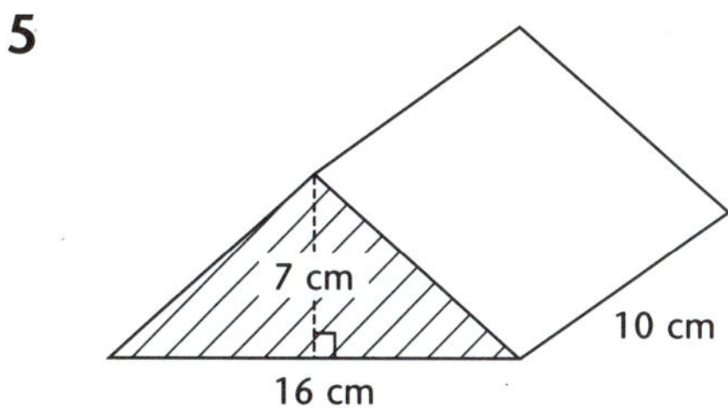


**6**

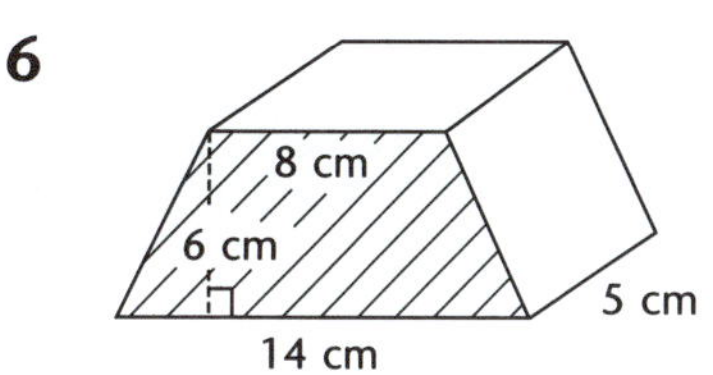


**7**

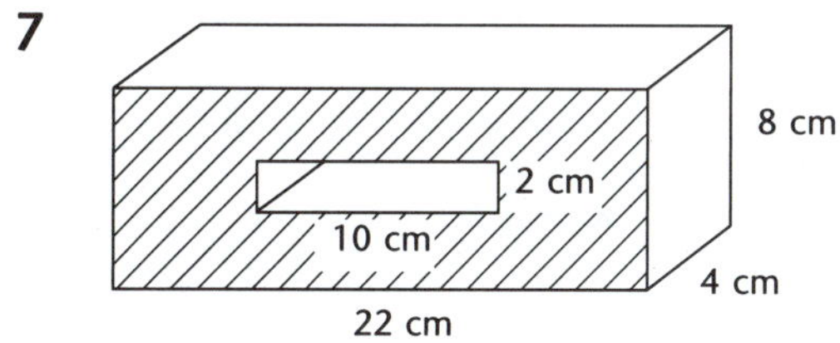


**8**

20 cm
12 cm
6 cm
14 cm

**1** Volume $= Ah$

$= 16 \times 8$

$= 128$

$\therefore$ The volume is 128 cm$^3$.

**2** Volume $= Ah$

$= 25 \times 6.2$

$= 155$

$\therefore$ The volume is 155 mm$^3$.

**3** Volume $= Ah$

$=$ Length $\times$ Breadth $\times$ Height

$= 12 \times 7 \times 5$

$= 420$

$\therefore$ The area is 420 cm$^3$.

**4** Volume $= Ah$

$=$ side$^3$

$= 6^3$

$= 216$

$\therefore$ The volume is 216 cm$^3$.

**5** Volume $= Ah$

$= (\frac{1}{2} \times 16 \times 7) \times 10$

$= 56 \times 10$

$= 560$

$\therefore$ The volume is 560 cm$^3$.

**6** Volume $= Ah$

$= \frac{1}{2} \times 6 \times (8 + 14) \times 5$

$= 3(22) \times 5$

$= 66 \times 5$

$= 330$

$\therefore$ The volume is 330 cm$^3$.

**7** Area of cross-section $= 22 \times 8 - 10 \times 2$

$= 176 - 20$

$= 156$

$\therefore$ Volume $= Ah$

$= 156 \times 4$

$= 624$

$\therefore$ The volume is 624 cm$^3$.

**8** Area of cross-section $= 14 \times 12 + \frac{1}{2} \times 14 \times 8$

$= 168 + 56$

$= 224$

$\therefore$ Volume $= Ah$

$= 224 \times 6$

$= 1344$

$\therefore$ The volume is 1344 cm$^3$.

## Volumes of Cylinders

A cylinder has a circular cross-section.

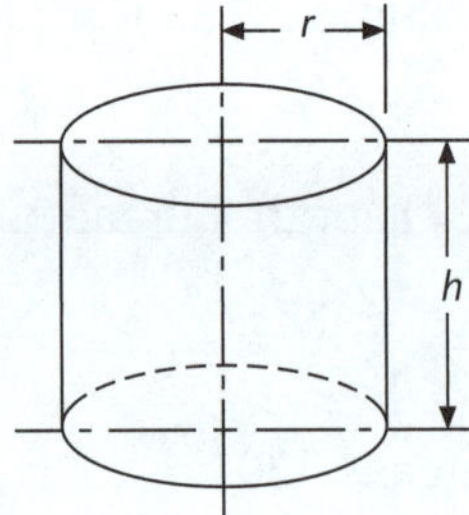


Volume = $A \times h$
where $A$ = cross-sectional area
$h$ = height of prism
$V = \pi r^2 \times h$
$\therefore V = \pi r^2 h$

**For Example**

1 Calculate the volume of these cylinders. Answer correct to one decimal place:

**a**

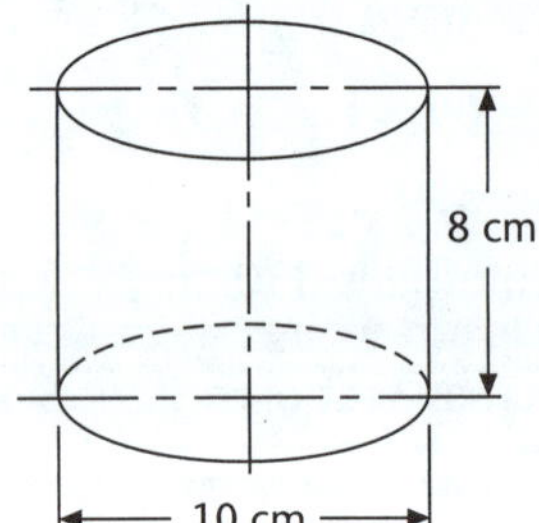


**b**

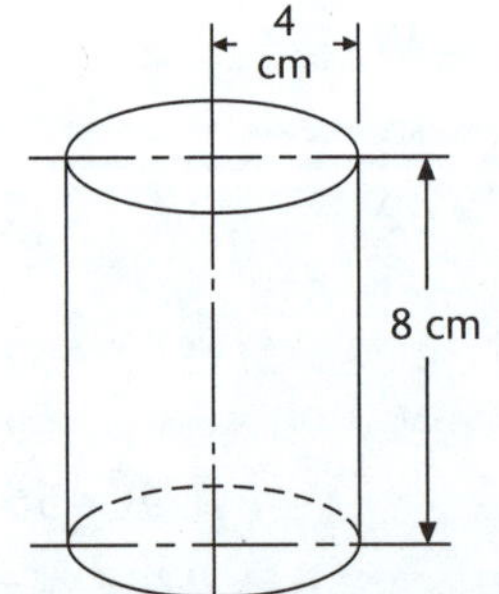


2 Calculate the volume of a cylinder with base radius 5 cm and height 14 cm. Use $\pi = \frac{22}{7}$.

3 A cylindrical hole is drilled through a 1 m cube of concrete. At its narrowest points the hole is 10 cm from each edge of the block.

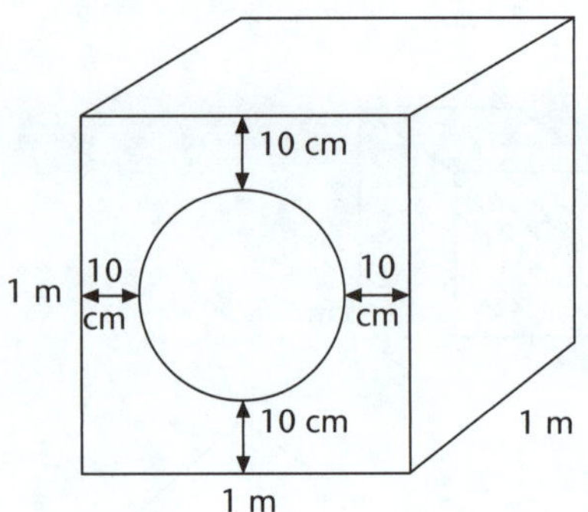


Calculate the:

**a** volume of concrete removed

**b** volume of concrete remaining.

1 **a** $d = 10$

$\therefore r = 5$ $\left[\begin{array}{r} r = 5 \\ \therefore h = 8 \end{array}\right]$

$V = \pi r^2 h$

$= \pi \times 5 \times 5 \times 8$

$= 628.318\,53$

$= 628.3$ (to 1 decimal place)

The volume of the cylinder is 628.3 $cm^3$.

**b** $r = 4,\ h = 8$

$V = \pi r^2 h$

$= \pi \times 4 \times 4 \times 8$

$= 402.123\,86$

$= 402.1$ (to 1 decimal place)

The volume of the cylinder is 402.1 $cm^3$.

2 $V = \pi r^2 h$

$= \frac{22}{\cancel{7}_1} \times 5 \times 5 \times \cancel{14}^2$

$= 1100$

$\left[\begin{array}{l} r = 5 \\ h = 14 \\ \pi = \frac{22}{7} \end{array}\right]$

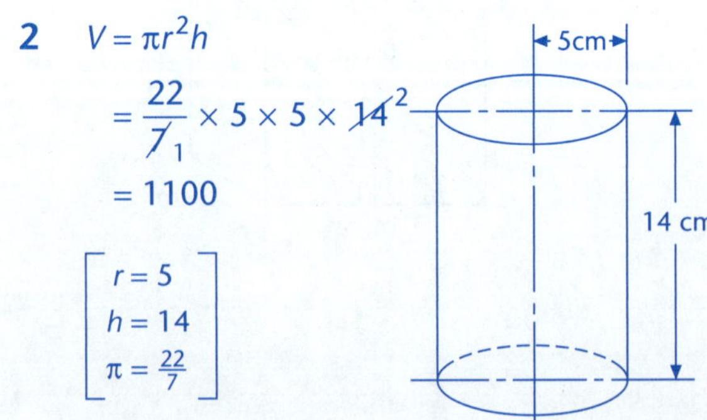


The volume of the cylinder is 1100 $cm^3$.

**3** **a** Diameter of circle = 1 metre – 20 cm

$= 80$ cm

$= 0.8$ m

Radius = 0.4 m

$V = \pi r^2 h$

$= \pi \times 0.4 \times 0.4 \times 1$

$= 0.5026548$

$= 0.503$ (to 3 decimal places)

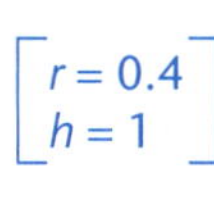


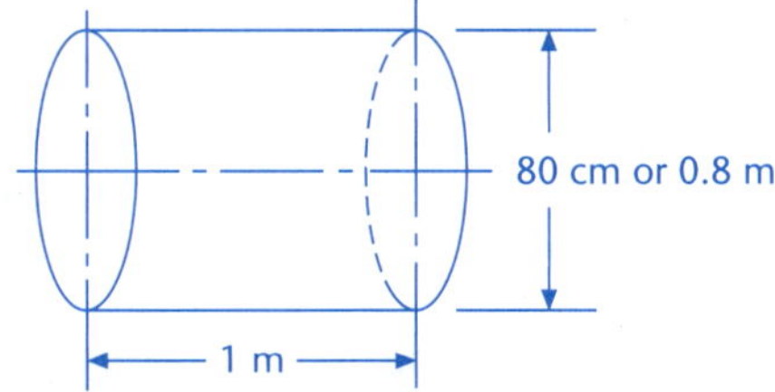


The volume of concrete removed is 0.503 m$^3$.

**b** Volume of cube = $lbh$

$= 1 \times 1 \times 1 = 1$ m$^3$

Volume remaining

= Volume of cube – Volume of cylinder

$= 1 - 0.503$

$= 0.497$

Volume remaining is 0.497 m$^3$.

## Capacity

These results are important:

- 1 mL = 1 cm$^3$ $\therefore$ 1 L = 1000 cm$^3$
- Also 1 kL = 1 m$^3$

For Example

**1** Convert:

**a** 3500 cm$^3$ = ________ L

**b** 2.3 kL = ________ m$^3$

**2** Find the capacity in litres:

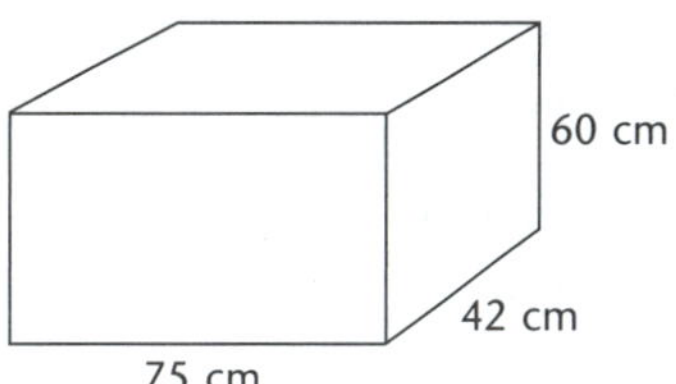


**1** **a** $3500 \div 1000 = 3.5$ $\therefore$ 3.5 L

**b** 2.3 kL = 2.3 m$^3$

**2** Volume = $75 \times 60 \times 42$

$= 189\,000$

$\therefore$ volume is 189 000 cm$^3$

$\therefore$ capacity = $189\,000 \div 1000$

$= 189$

The capacity is 189 L.

For Example

A cylindrical water tank has a diameter of 4 metres and a height of 3 metres. Calculate the kilolitres the tank can hold. Answer correct to one decimal place.

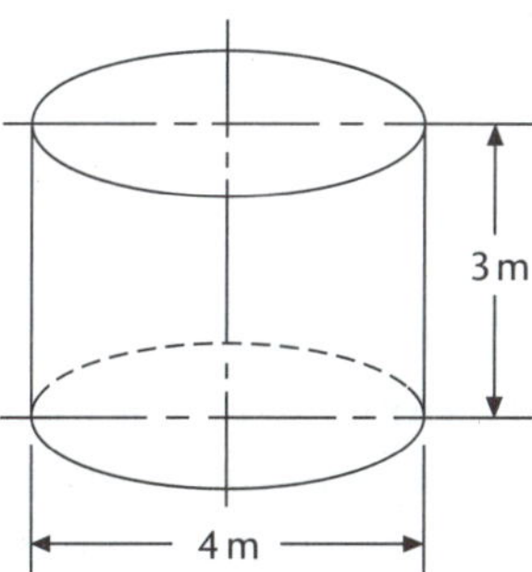


Diameter = 4 metres

$\therefore r = 2$

$V = \pi r^2 h$

$= \pi \times 2 \times 2 \times 3$

$= 37.69912$

$= 37.7$ (to 1 decimal place)

[r = 2, h = 3]

Volume is 37.7 m$^3$

The capacity of the tank is 37.7 kL.

# Time

| Important Units |
|---|
| 60 seconds = 1 minute |
| 60 minutes = 1 hour |
| 24 hours = 1 day |
| 7 days = 1 week |
| 52 weeks = 1 year |
| 12 months = 1 year |
| 365 days = 1 year |
| 366 days = 1 leap year |
| 10 years = 1 decade |
| 100 years = 1 century |
| 1000 years = 1 millennium |

| Days in each Month | | | |
|---|---|---|---|
| January | 31 | July | 31 |
| February | 28* | August | 31 |
| March | 31 | September | 30 |
| April | 30 | October | 31 |
| May | 31 | November | 30 |
| June | 30 | December | 31 |
| *February has 29 days in a leap year | | | |

## Notation

**12-hour notation:** we normally number the 12 hours from midnight to midday and the 12 hours from midday to midnight the same, but use 'am' or 'pm' to show which is meant. Thurs, 7:15 am might be breakfast time, while 7:15 pm might be homework time.

**24-hour notation:** the hours are numbered from 00 to 23. Thus, breakfast time would be 0715, while homework time would be 1915.

Converting between the two forms may require adding or subtracting 12 hours.

Like this:

3:25 pm = 1525 hours
4:04 am = 0404 hours
2048 hours = 8:48 pm

## For Example

**1** **a** Convert:

- **i** 3 days to hours
- **ii** 180 seconds to minutes
- **iii** 75 minutes to hours
- **iv** 3 hours 20 minutes to minutes.

**b** A bus departs at 11:42 am and arrives at 12:31 pm. How long was the trip?

**c** How many days from May 20th to August 7th?

**d** **i** Convert to 12-hour notation:
(A) 2016
(B) 2337

**ii** Convert to 24-hour notation:
(A) 9:47 am
(B) 7:58 pm

**e** A repairman charges $42 per hour. If he begins a job at 9:40 am and finishes at 11:20 am, how much will he be paid?

**2** Daniel consulted his bus timetable:

| | | | | | |
|---|---|---|---|---|---|
| Maryland | 8:45 | 10:21 | 11:57 | 14:32 | 16:48 |
| Wallsend | 8:49 | 10:25 | 12:01 | 14:36 | 16:52 |
| Jesmond | 8:53 | 10:29 | 12:05 | 14:40 | 16:56 |
| Lambton | 8:56 | 10:32 | 12:08 | 14:43 | 16:59 |
| Broadmeadow | 9:00 | 10:36 | 12:12 | 14:47 | 17:03 |
| Newcastle | 9:12 | 10:48 | 12:24 | 14:59 | 17:15 |

**a** How long does the trip take from Maryland to Newcastle?

**b** The 14:32 from Maryland arrived in Newcastle 4 minutes late. At what time did it reach its destination?

**c** Daniel wanted to catch the bus to Newcastle from Wallsend. If he got to the bus stop at 11:20, how long did he wait to catch the bus?

**d** If Daniel lives 2 minutes from the Maryland bus stop and leaves home at quarter past ten, what time will he get to Lambton?

**3** Different time zones operate across the world. The table shows the local times at four cities when it is 9 pm in Melbourne:

| Melbourne | 9 pm | Tuesday |
|---|---|---|
| Perth | 7 pm | Tuesday |
| Calcutta | 4:30 pm | Tuesday |
| Washington | 6 am | Tuesday |

Using the table:

**a** What is the time difference between Washington and Melbourne?

**b** If it is midnight in Perth, what is the local time in Calcutta?

**1 a i** $3 \times 24 = 72$

$\therefore$ 72 hours

**ii** $180 \div 60 = 3$

$\therefore$ 3 minutes

**iii** $75 \div 60 = 1\frac{1}{4}$ or 1.25

$\therefore 1\frac{1}{4}$ hours or 1.25 hours

**iv** $3 \times 60 + 20 = 180 + 20$

$= 200$

$\therefore$ 200 minutes

**b** 11:42 } 18 min
12:00 } 31 min
12:31

Now $18 + 31 = 49$

$\therefore$ Time taken is 49 minutes.

**c** Days to end of May + June + July + 7 days

$= 11 + 30 + 31 + 7 = 79$

$\therefore$ 79 days

**d i** (A) 2016 = 8:16 pm

(B) 2337 = 11:37 pm

**ii** (A) 9:47 am = 0947

(B) 7:58 pm = 1958

**e** Time: 9:40 } 20 min
10:00 } 1 h
11:00 } 20 min
11:20

$\therefore$ 1 h 40 min $= 1\frac{2}{3}$ hours

$\therefore$ Pay $= \$42 \times 1\frac{2}{3}$

$= \$70$

$\therefore$ The repairman is paid \$70.

**2 a** 8:45 } 15 min
9:00 } 12 min
9:12

$\therefore 15 + 12 = 27$

$\therefore$ Trip from Maryland takes 27 minutes.

**b** Normal time of arrival = 14:59

New (late) time = 15:03

**c** 11:20 } 40
12:00 } 1
12:01

$\therefore 40 + 1 = 41$

$\therefore$ 41 minutes

Daniel waited 41 minutes.

**d** Catch the bus at 10:21 $\therefore$ arrive at 10:32.

**3** **a** Time difference = 9 pm from 6 am

= 9 + 6

= 15

∴ Melbourne is 15 hours ahead of Washington.

**b** Time difference:

7:00 pm from 4:30 pm = 2 h 30 min

Time in Calcutta:

Midnight minus 2 h 30 min = 9:30 pm

∴ In Calcutta it is 9:30 pm.

## PRACTISE, PRACTISE

Go to p. 287 for quick answers, or to pp. 341–353 for worked solutions.

**1** Find the area of: pp. 137–139

**a**

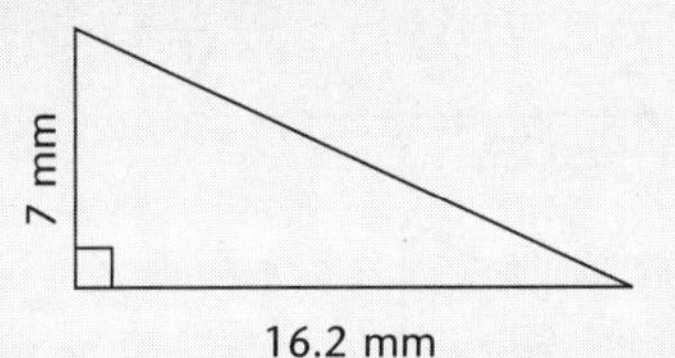


**b**

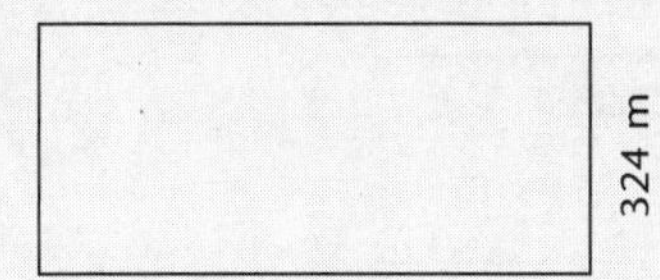


**c**

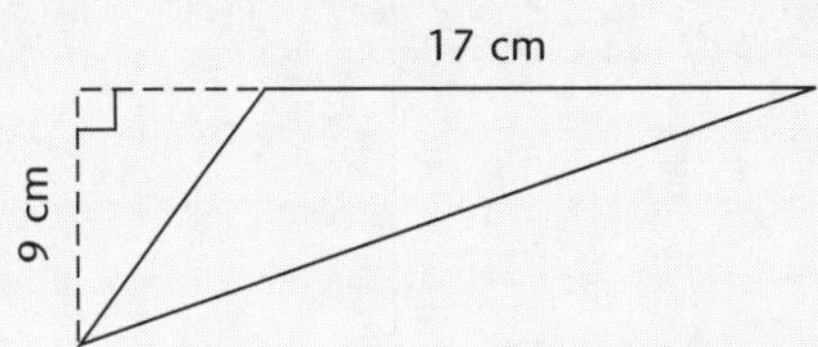


**d**

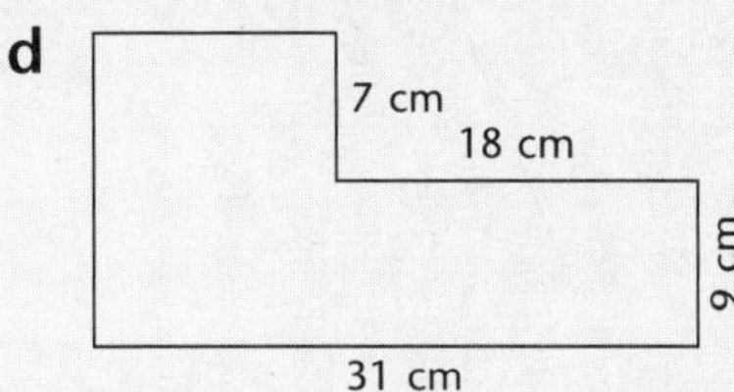


**e**

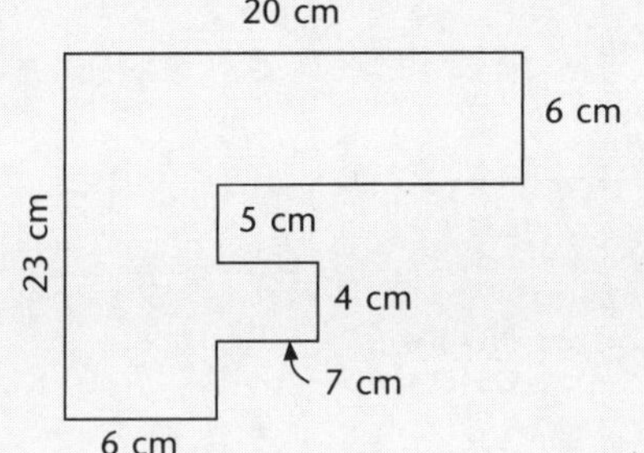


**f**

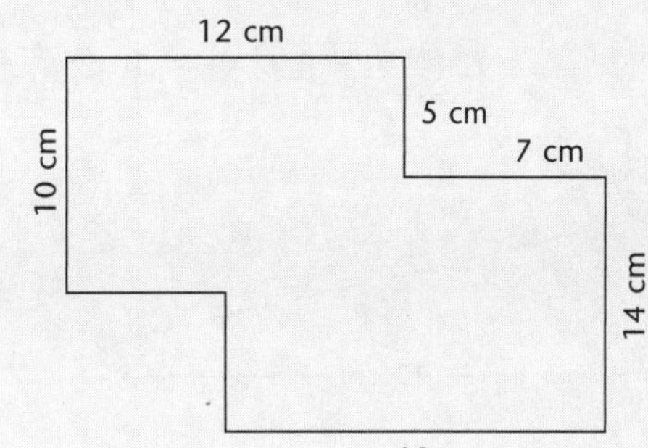


**2** Find the area of the shaded region: p. 139

**a**

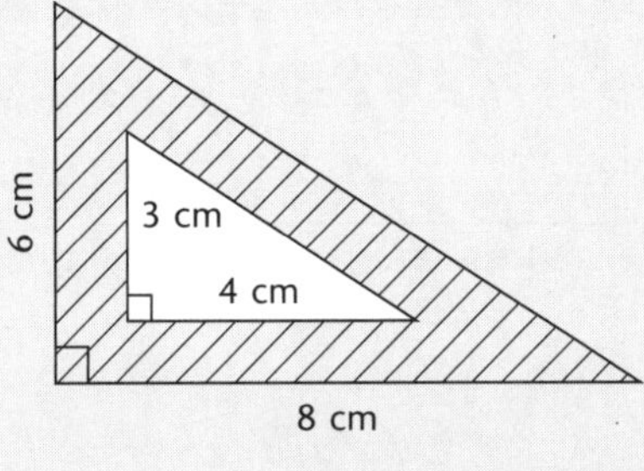


**b**

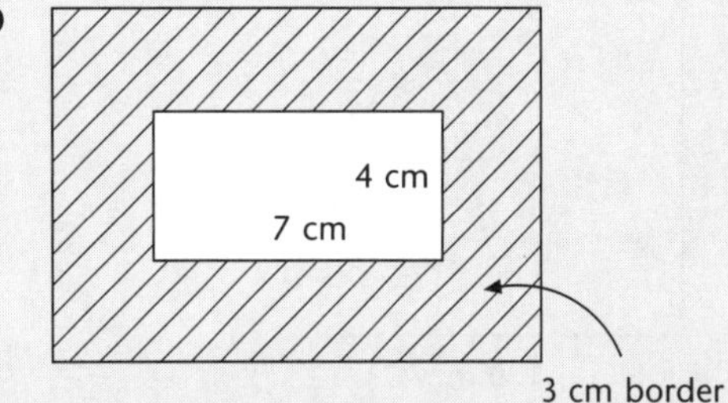


**c**

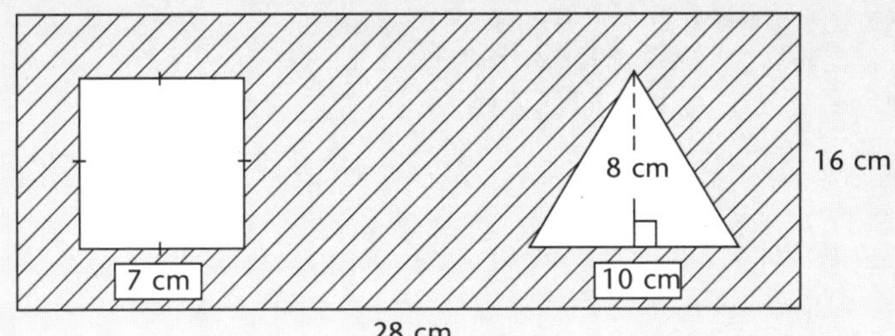

**3** Find the area: p. 140

**a**

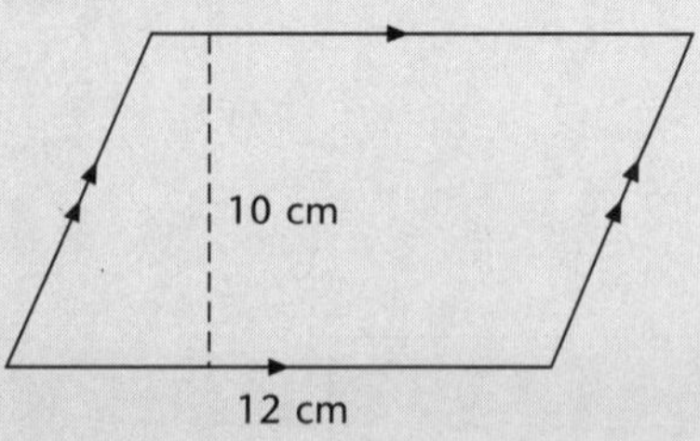


**b**

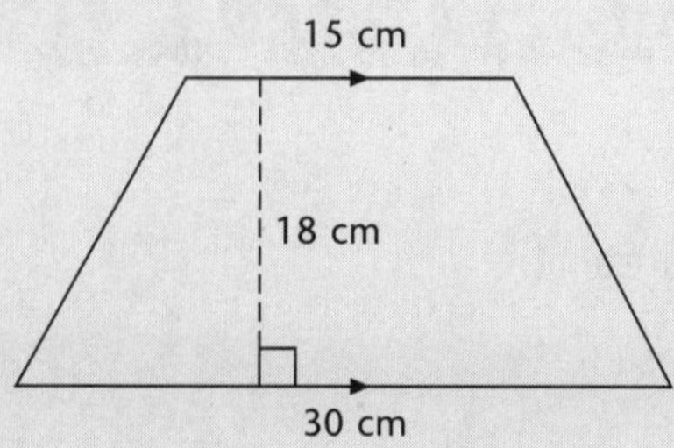


**c**

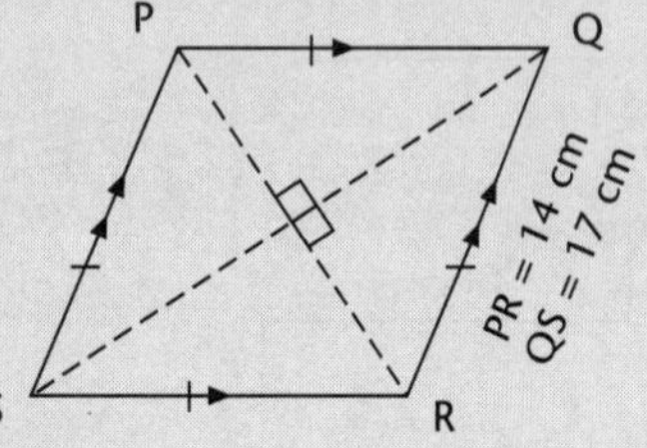


**d**

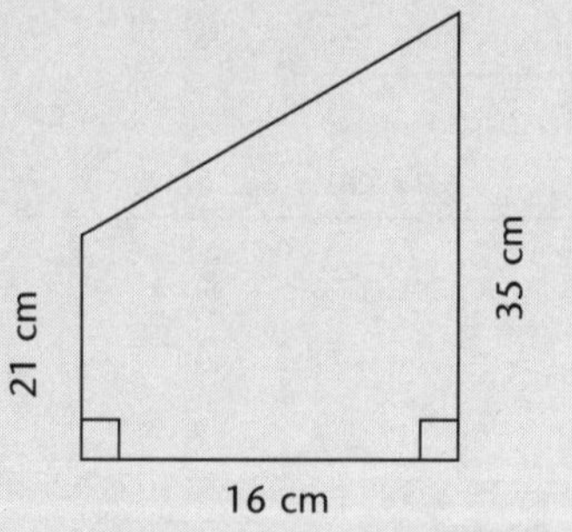


**e**

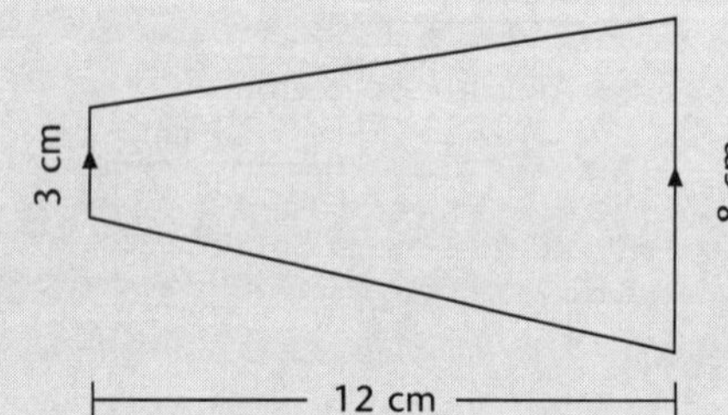


**f**

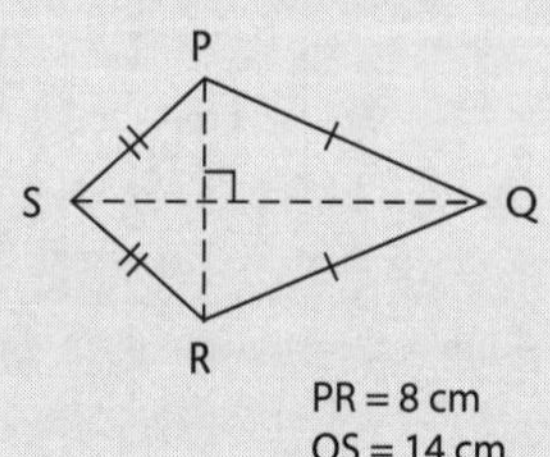


**4** Find the area, in hectares: pp. 140–142

**a**

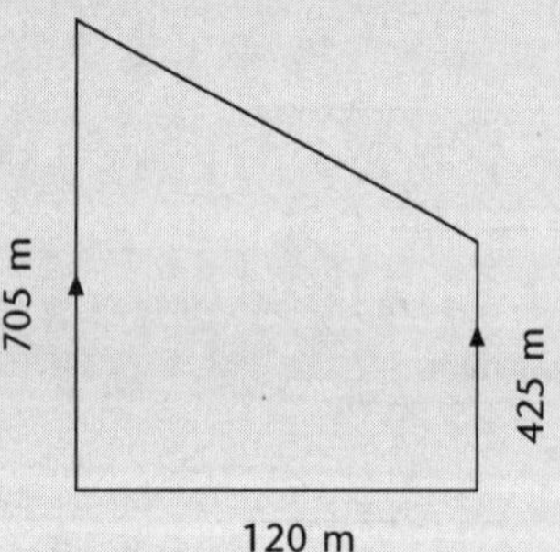


**b**

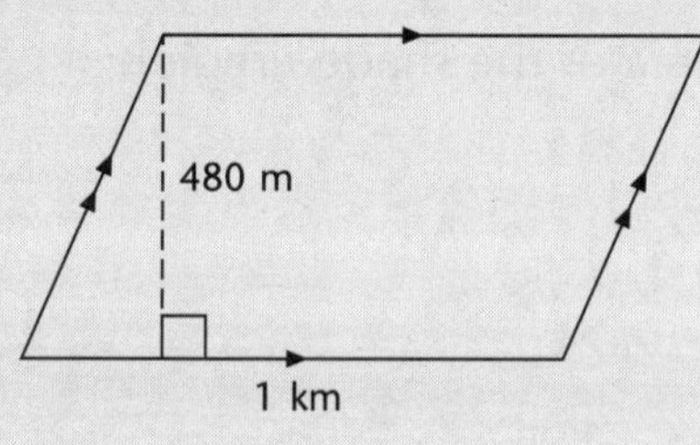


**5** Convert: pp. 140–142

**a** $30\,000\text{ m}^2 =$ ______ ha **b** $4.72\text{ ha} =$ ______ $\text{m}^2$

**c** $65\,400\text{ cm}^2 =$ ______ $\text{m}^2$ **d** $11.57\text{ m}^2 =$ ______ $\text{cm}^2$

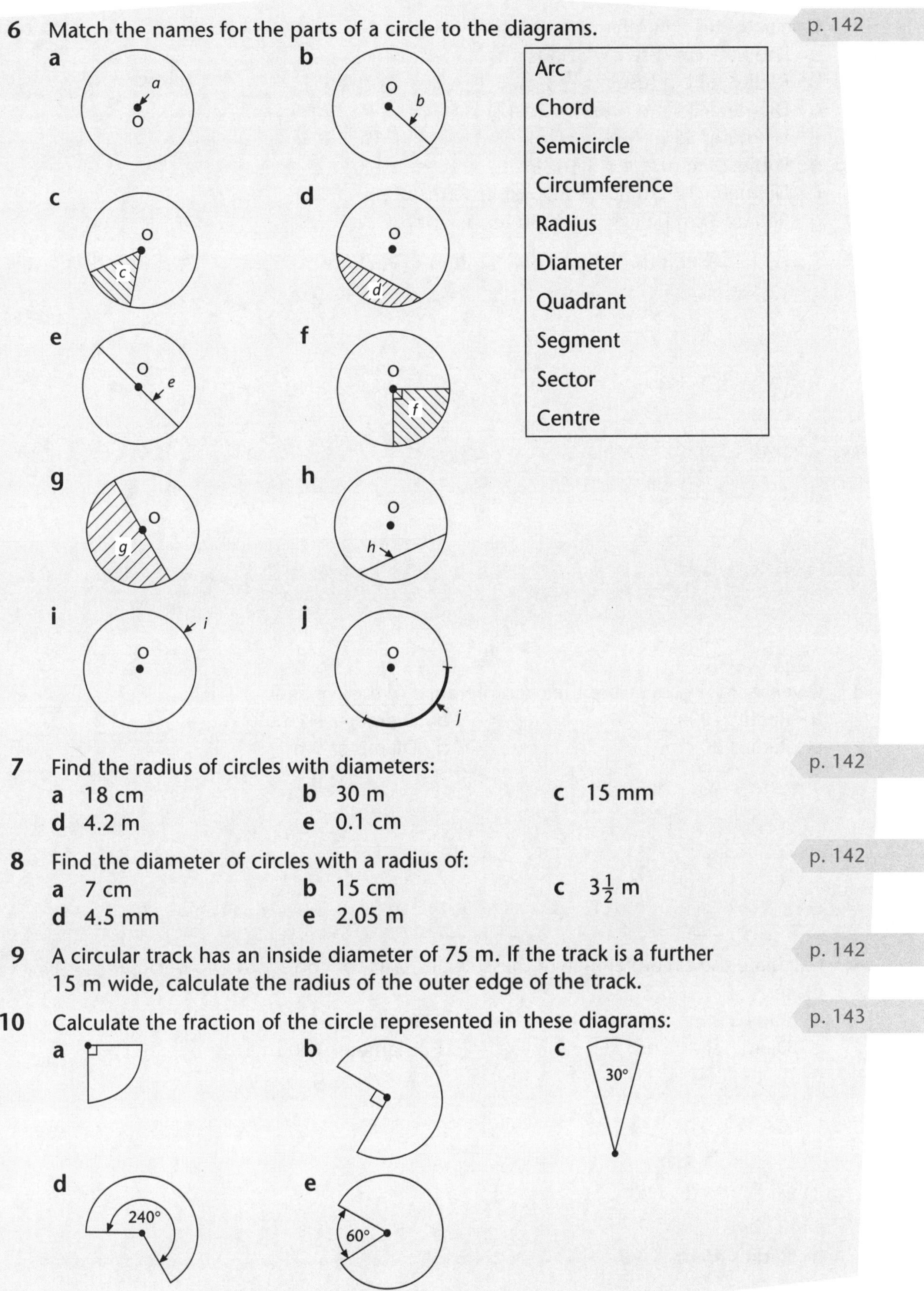

**6** Match the names for the parts of a circle to the diagrams. p. 142



| |
|---|
| Arc |
| Chord |
| Semicircle |
| Circumference |
| Radius |
| Diameter |
| Quadrant |
| Segment |
| Sector |
| Centre |

**7** Find the radius of circles with diameters: p. 142

**a** 18 cm  **b** 30 m  **c** 15 mm
**d** 4.2 m  **e** 0.1 cm

**8** Find the diameter of circles with a radius of: p. 142

**a** 7 cm  **b** 15 cm  **c** $3\frac{1}{2}$ m
**d** 4.5 mm  **e** 2.05 m

**9** A circular track has an inside diameter of 75 m. If the track is a further 15 m wide, calculate the radius of the outer edge of the track. p. 142

**10** Calculate the fraction of the circle represented in these diagrams: p. 143

**11** Calculate the circumference of a circle with a: pp. 143–145

- **a** Radius 5 cm (use $\pi = 3.14$)
- **b** Radius 28 cm (use $\pi = \frac{22}{7}$)
- **c** Diameter 10 cm (use $\pi = 3.142$)
- **d** Diameter 35 m (use $\pi = 3\frac{1}{7}$)
- **e** Radius 2 m (use $\pi = 3.142$)
- **f** Diameter 12 cm (leave answer in terms of $\pi$)
- **g** Radius 8 cm (leave answer in terms of $\pi$)

**12** Given $\pi = \frac{22}{7}$ find the circumference of these circles: pp. 143–145

**a**

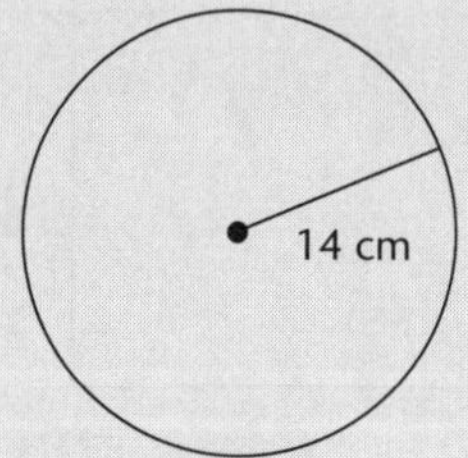


**b**

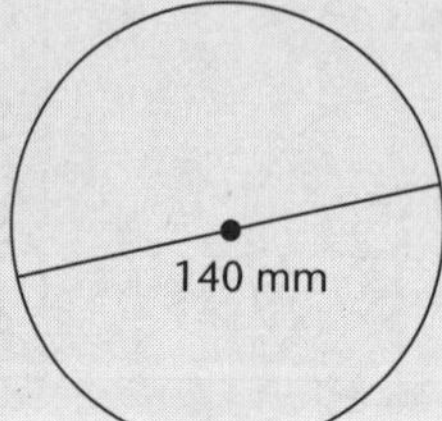


**c**

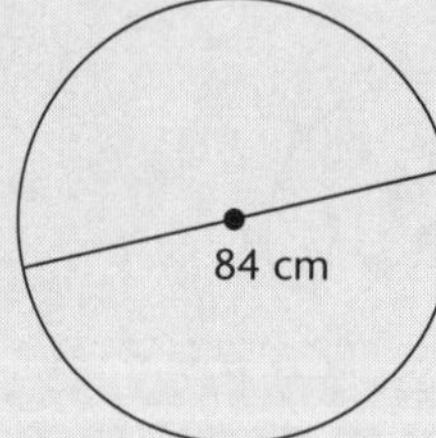


**d**

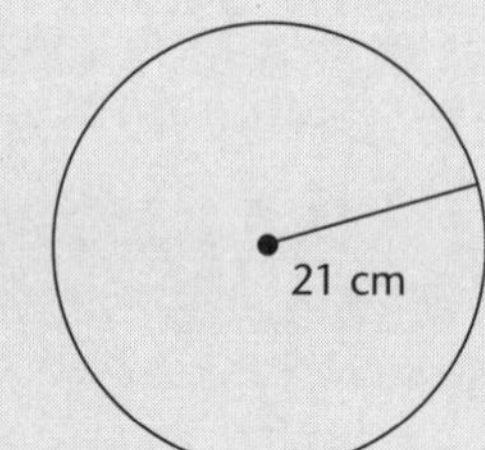


**e**

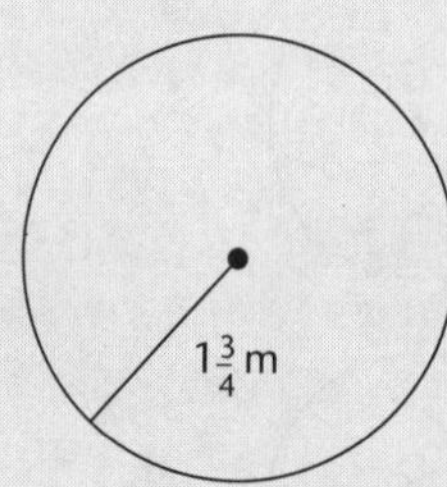


**f**

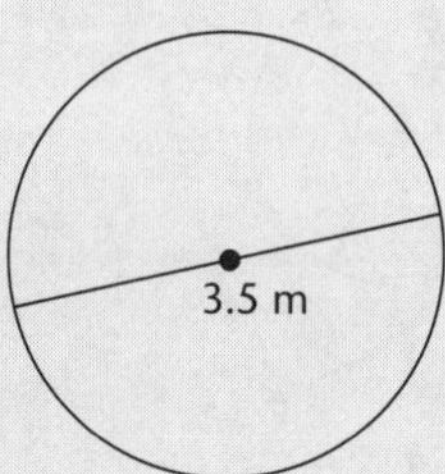


**13** Given $\pi = 3.14$, calculate the circumference of these circles: pp. 143–145

- **a** Radius 100 mm
- **b** Diameter 4 cm
- **c** Radius 20 cm
- **d** Diameter 9 m

**e**

15 m

**f**

30 cm

**g**

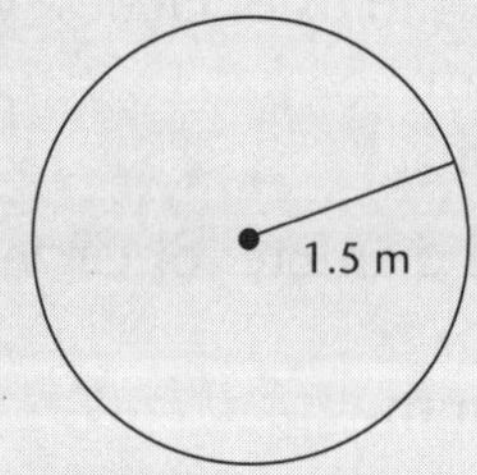


**14** Calculate the circumference of these circles. Answer correct to two decimal places: pp. 143–145

- **a** Radius 1 m
- **b** Radius 0.5 cm
- **c** Diameter 40 cm
- **d** Diameter 100 m

**e**

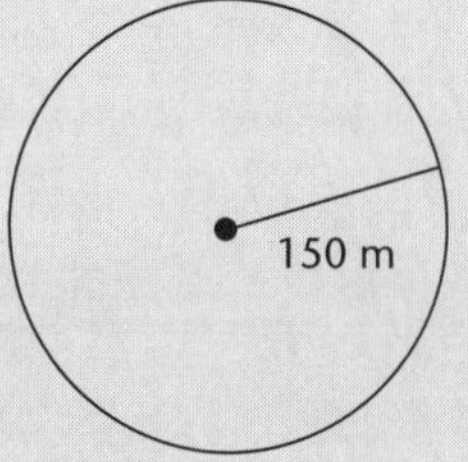


**f**

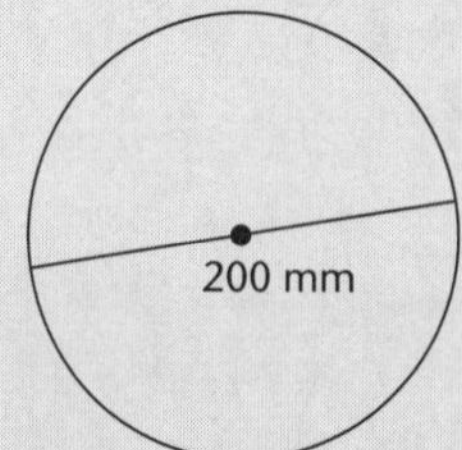


**g**

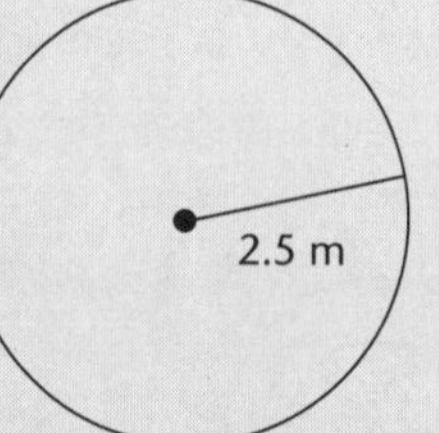


- **h** Radius 45 m

**15** A circular dish has a diameter of 24 cm. Calculate the distance around the outside of the dish. Answer correct to one decimal place. pp. 143–145

**16** A circular track has an inside radius of 70 m. Calculate the distance around the track. Leave the answer in terms of $\pi$. pp. 143–145

**17** Calculate the perimeter of these shapes. Answer correct to one decimal place: pp. 143–145

**a**

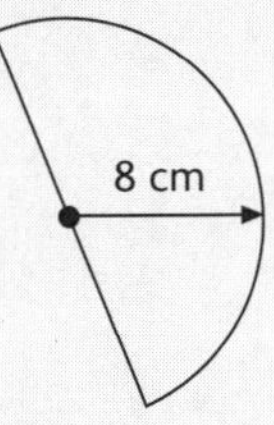


**b**

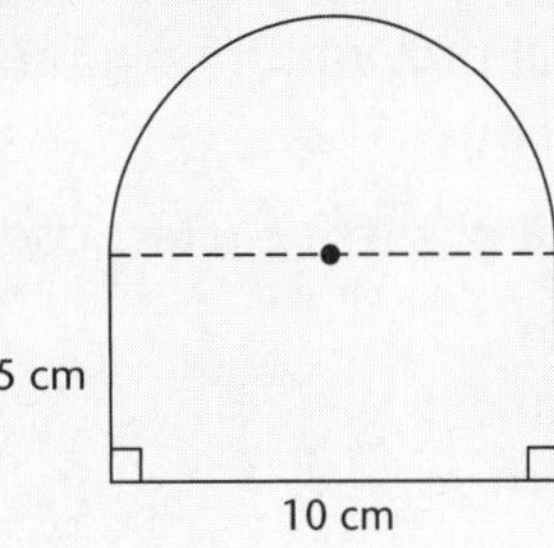


**c**

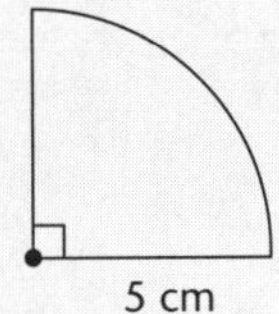


**d**

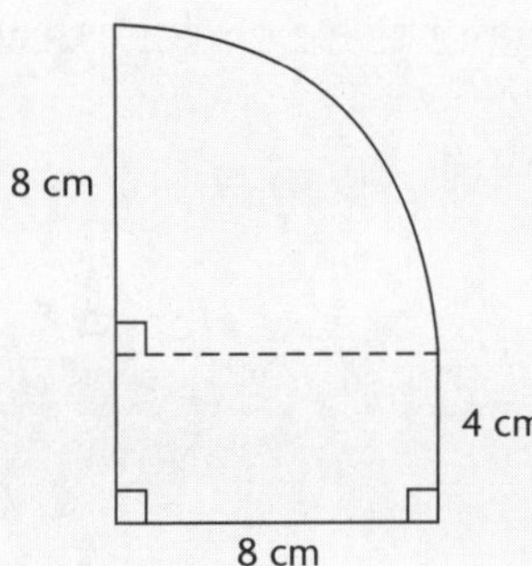


**e**

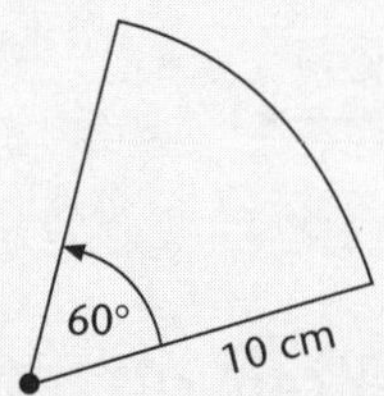


**f**

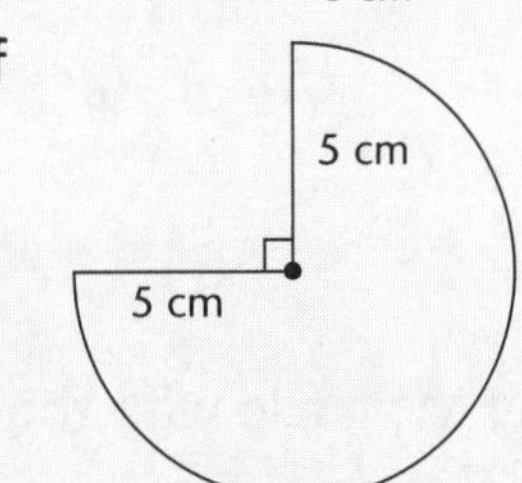


**18** Calculate the diameter of a circle with the following circumferences. Answer correct to one decimal place: pp. 147–148

**a** $C = 40$ cm **b** $C = 250$ cm **c** $C = 4.5$ cm
**d** $C = 1000$ m **e** $C = 6400$ km

**19** Calculate the radius of the circle with circumference (answer correct to one decimal place): pp. 147–148

**a** $C = 50$ cm **b** $C = 2.5$ m **c** $C = 0.4$ cm
**d** $C = 100$ m **e** $C = 314$ m

**20** A circular bike track is to be built such that the inner circumference is 200 m. Calculate the radius of the proposed track to the nearest metre. pp. 147–148

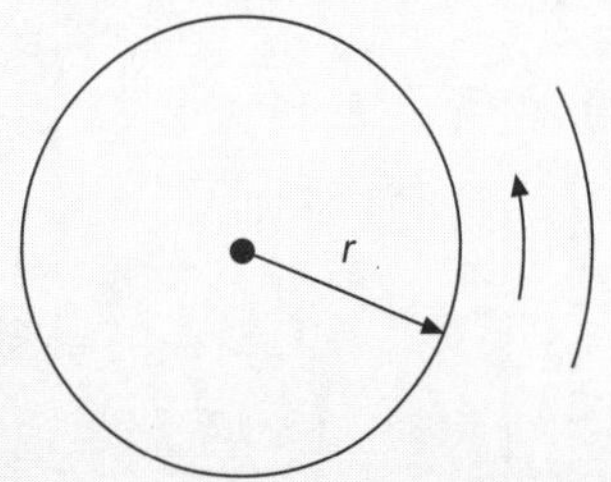

21 An ice-skating arena is to be built so that it has two straight sides and a semicircle at each end. The two straight sides are 60 m in length while the semicircles have a radius of 15 m. Calculate the distance around the arena correct to the nearest metre. If Sven tendered for fencing the arena at $17.50 a metre (all material provided), how much in total will Sven charge?

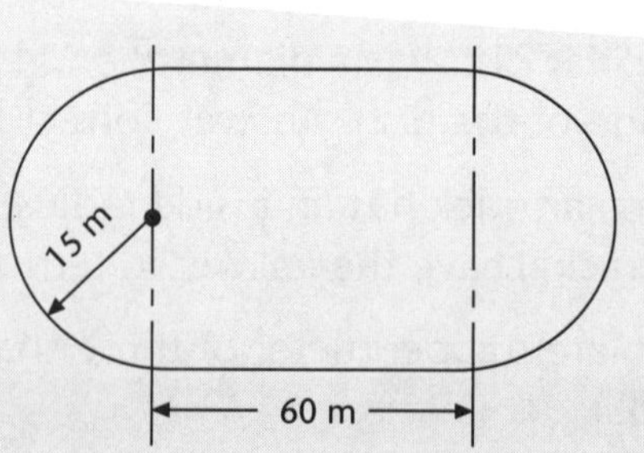


pp. 143–145

22 Calculate the area of these circles. Use $\pi = 3\frac{1}{7}$ (or $\frac{22}{7}$):

pp. 145–147

a Radius = 14 cm
b Radius = 21 cm
c Radius = $3\frac{1}{2}$ cm
d Diameter = 14 m
e Diameter = 7 m
f Radius = 1 cm

23 Calculate the area of these circles. Answers correct to one decimal place:

pp. 145–147

a Radius = 10 m
b Radius = 5 cm
c Radius = 100 cm
d Diameter = 20 cm

e
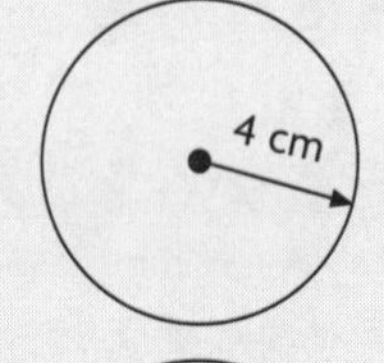


f
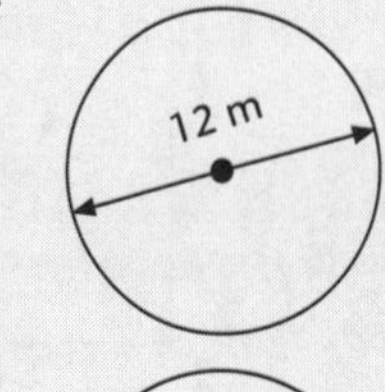


g
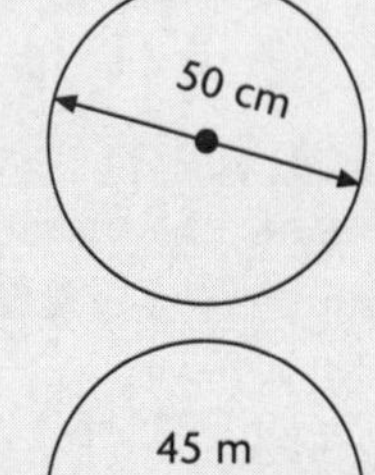


h
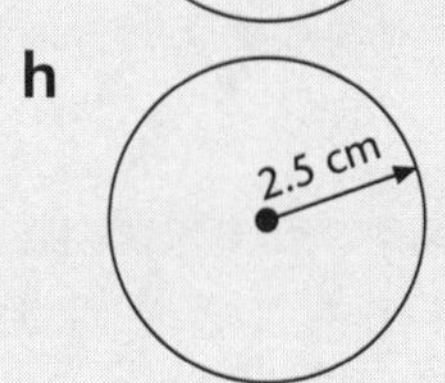


i
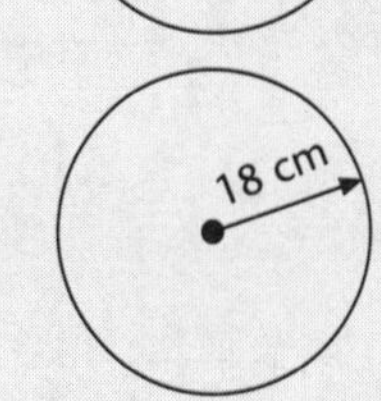


j


24 Calculate the radius of the circle with the following areas. Answer correct to two decimal places:

pp. 145–147

a Area = 35 cm$^2$
b Area = 100 cm$^2$
c Area = 155.6 mm$^2$
d Area = 59.4 cm$^2$

25 Find the diameter of a circle with the following areas. Answer correct to two decimal places:

pp. 145–147

a Area = 100 mm$^2$
b Area = 40 cm$^2$

26 Calculate the area of these shapes. Answer correct to one decimal place:

pp. 145–147

a
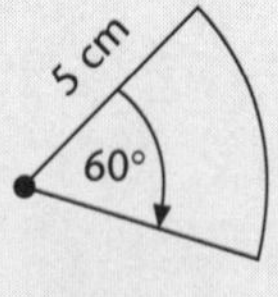


b
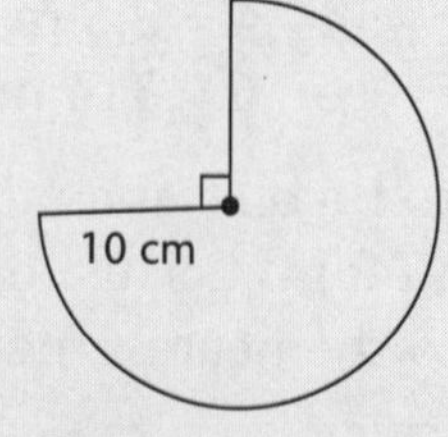


c
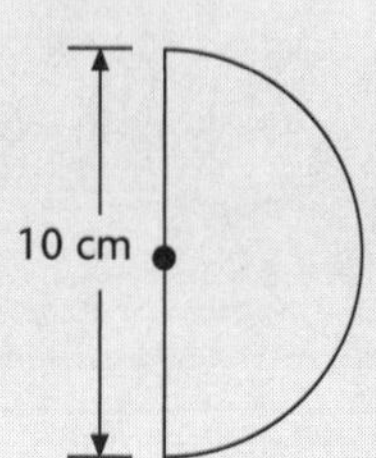


d

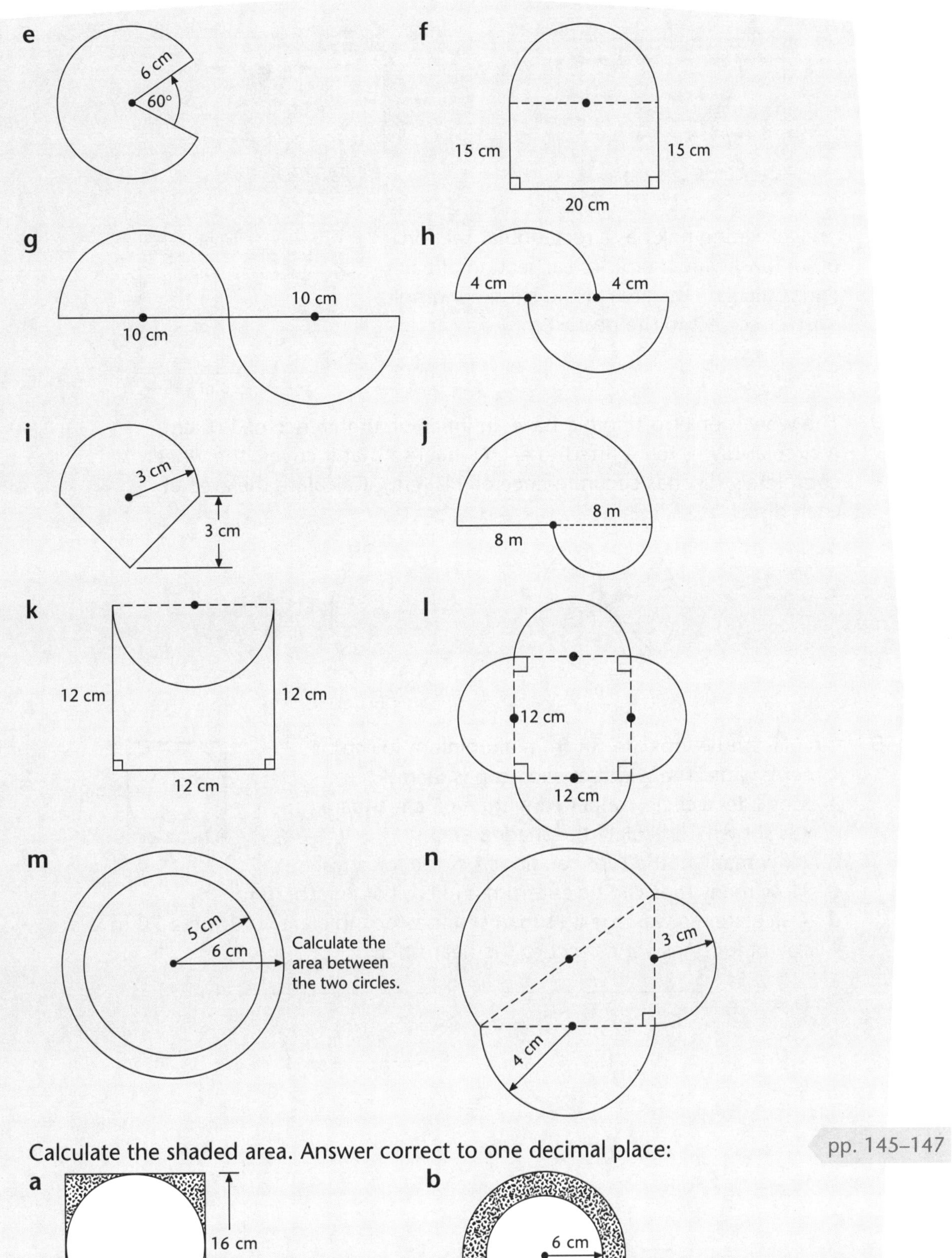


**27** Calculate the shaded area. Answer correct to one decimal place:

pp. 145–147

**a**

16 cm

16 cm

**b**

6 cm

20 cm

**c**

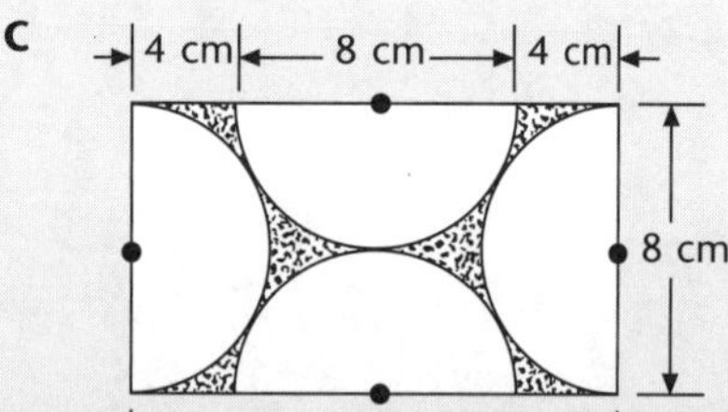


**d**



**28** An ice-skating rink has a rectangular section 60 m long with a semicircular section at each end. Calculate the area of the ice-skating rink. Answer correct to the nearest m$^2$.

pp. 145–147

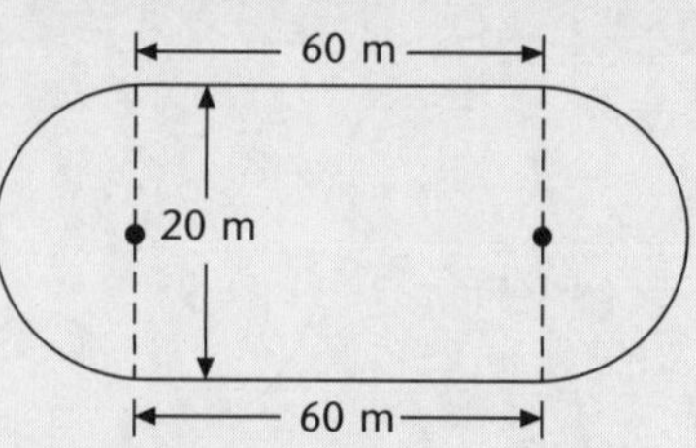


**29** The wheels of a trotting gig have an inner circumference of 110 cm. A circular disc is to be fitted to each wheel so that it covers the wheel completely (i.e. has circumference of 110 cm). Calculate the area of each disc.

pp. 145–147

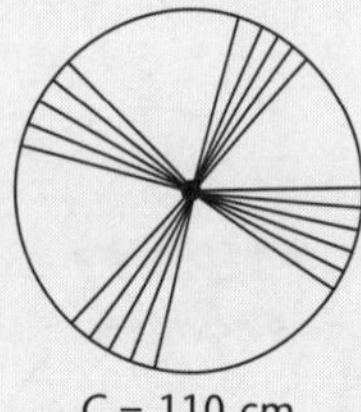


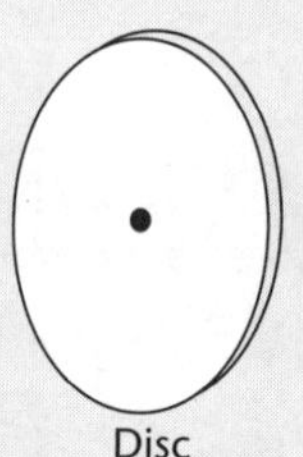


**30** Circular bottle tops are cut from aluminium foil strips 1 metre wide. The width of each cap is 5 cm.

pp. 145–147

**a** Consider a circle contained within a 5 cm square (as shown). Calculate the shaded area.



**b** How many bottle tops can fit across the foil strip?

**c** How many tops can be cut from a foil roll of length 20 m?

**d** Calculate the wastage if as many caps as possible are cut from a 20 m roll of foil? Answer correct to the nearest cm$^2$.

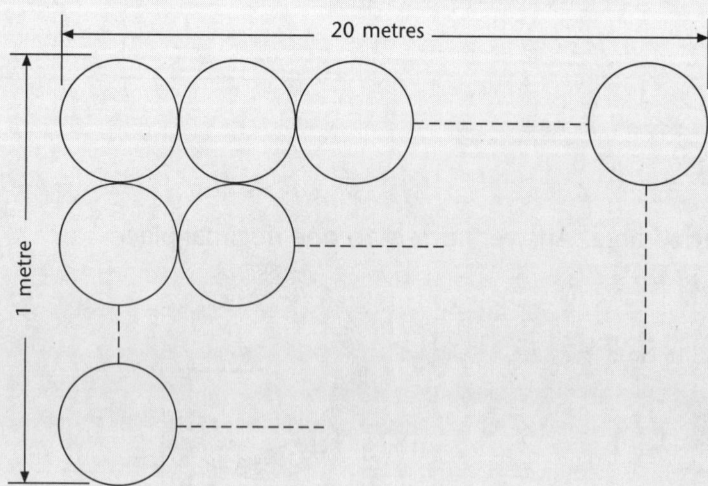

**31** Find the volume: pp. 148–149

a

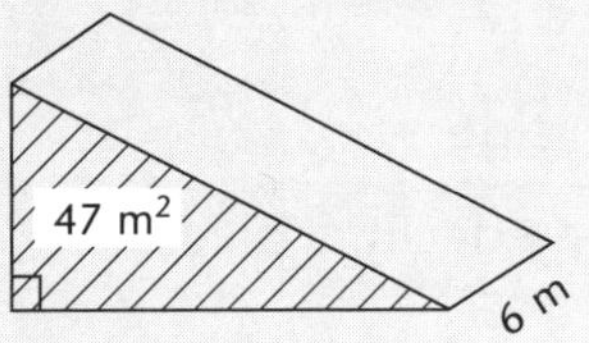


b

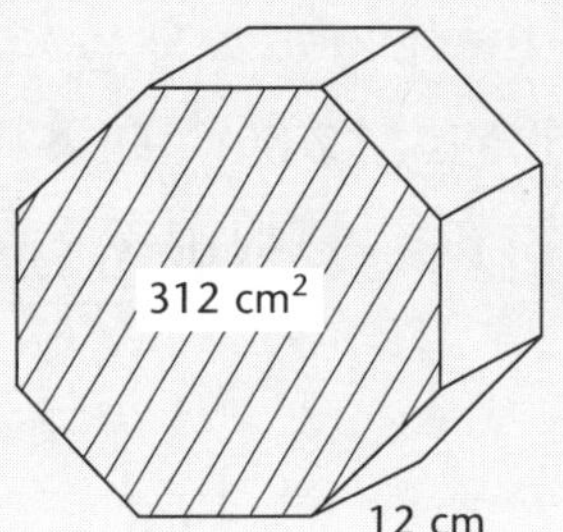


c

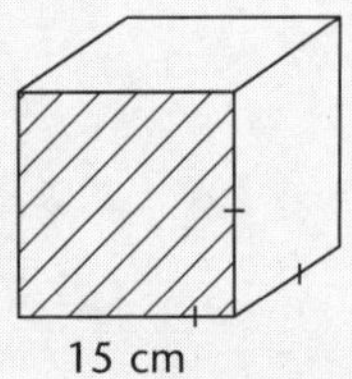


d

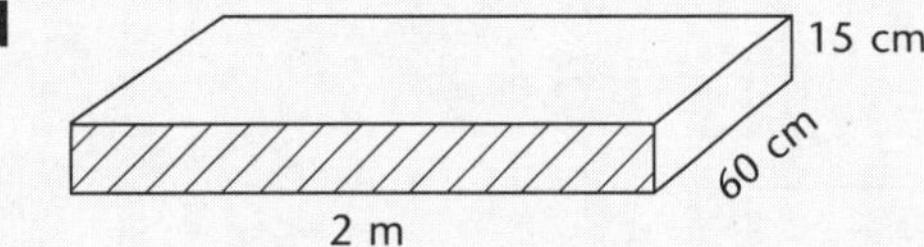


e

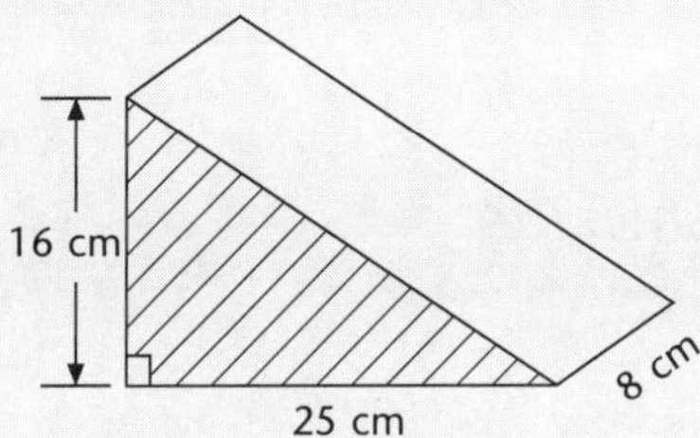


f

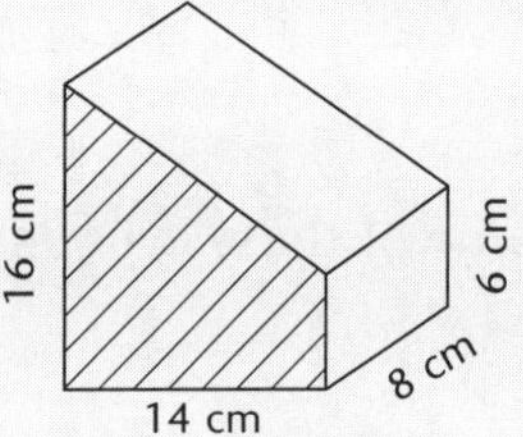


g

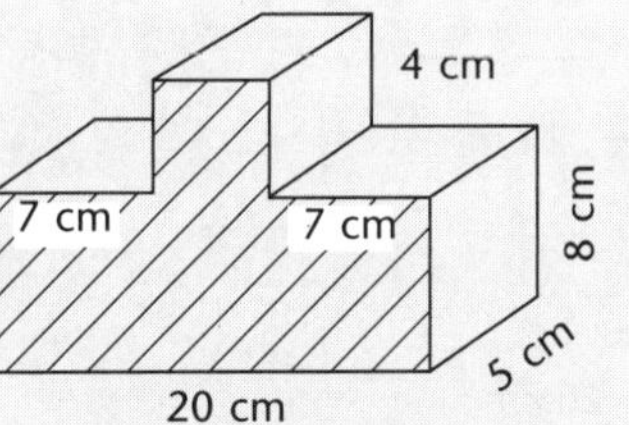


h

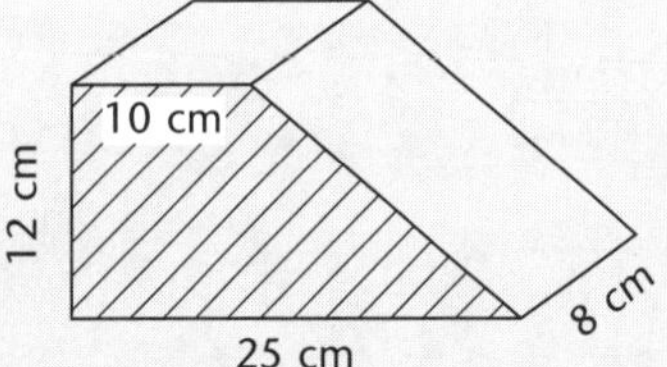


i

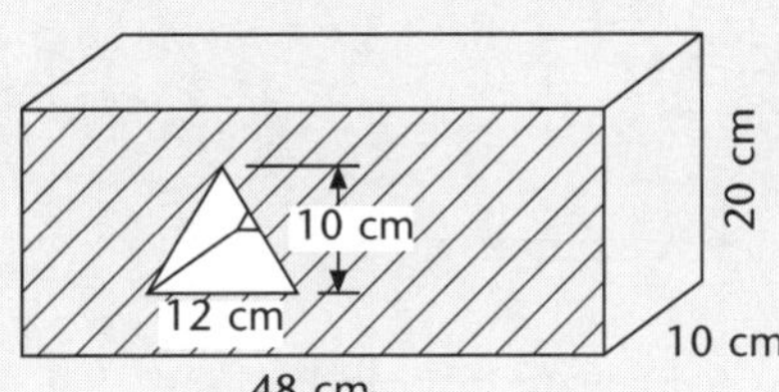


j

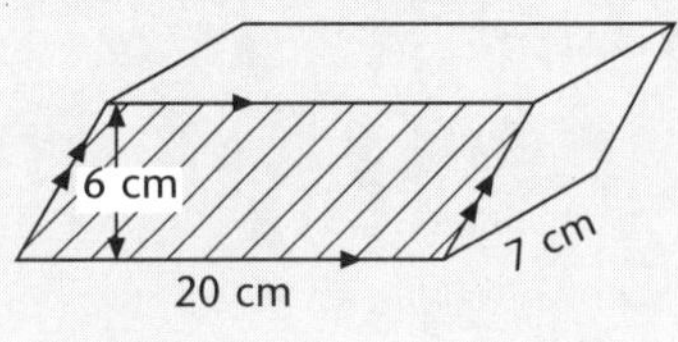


**32** Complete the table for the volumes of prisms: pp. 148–149

| Area of base | Height | Volume of prism |
|---|---|---|
| $36\ m^2$ | 11 m | |
| $76\ mm^2$ | | $760\ mm^3$ |
| | 24 mm | $960\ mm^3$ |

**33** A rectangular prism has a length of 12 cm, a breadth of 7 cm and a volume of 420 cm$^3$. Find the height. pp. 148–149

**34** The volume of cube is 512 cm$^3$. Find the length of each side. pp. 148–149

**35** Find the volume of these cylinders. Answer correct to one decimal place: pp. 150–151

**a**

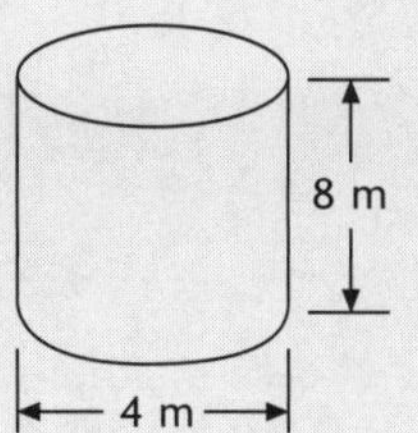


**b**

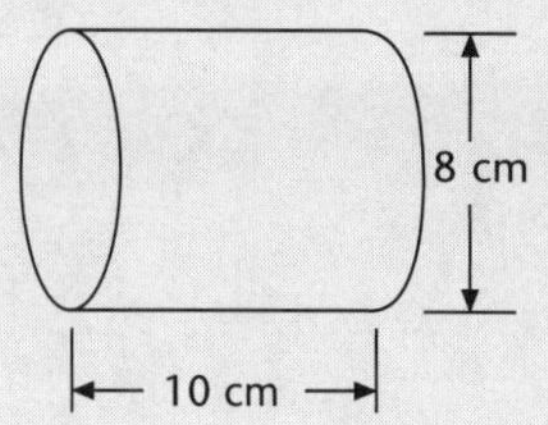


**c**

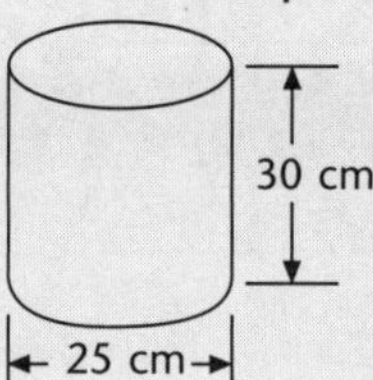


**d**

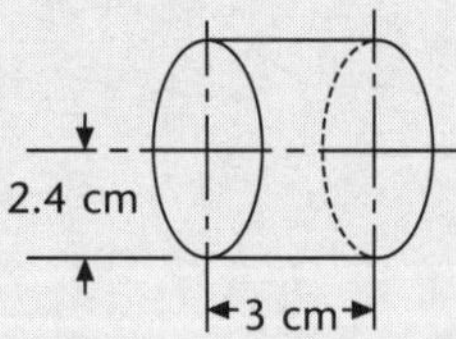


**e**

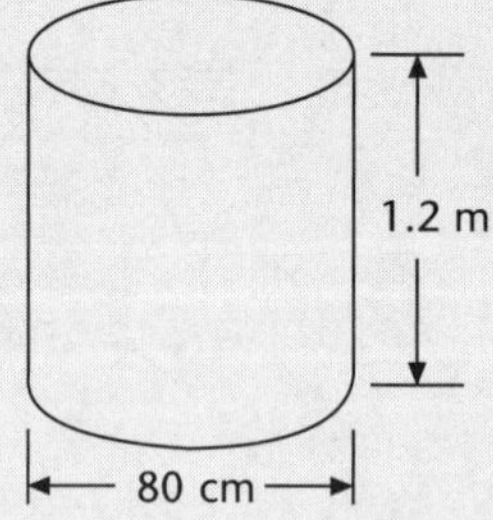


**36** Calculate the volume of these cylinders. Use $\pi = 3\frac{1}{7}$: pp. 150–151

**a**

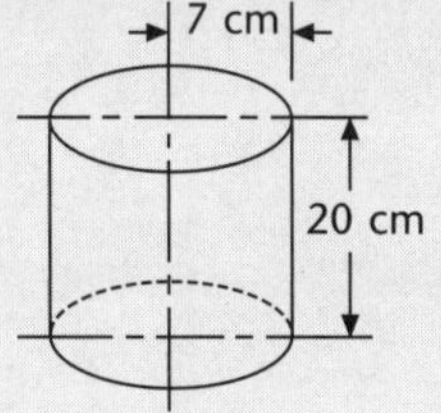


**b**

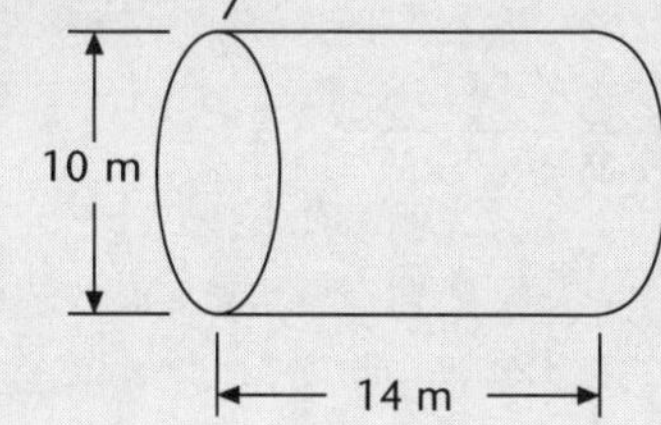


**c**

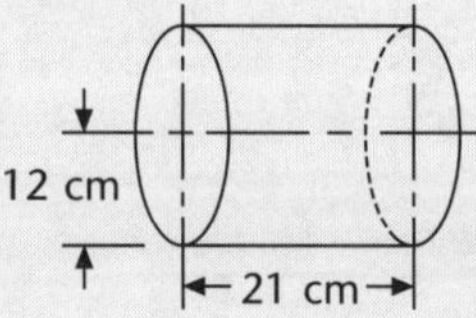


**d**

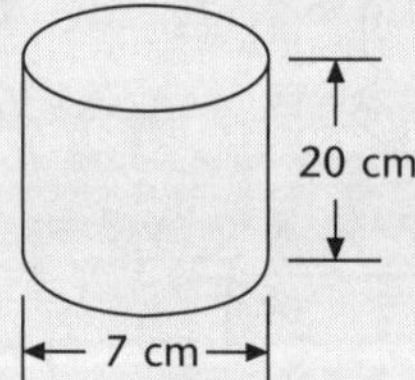


**e**

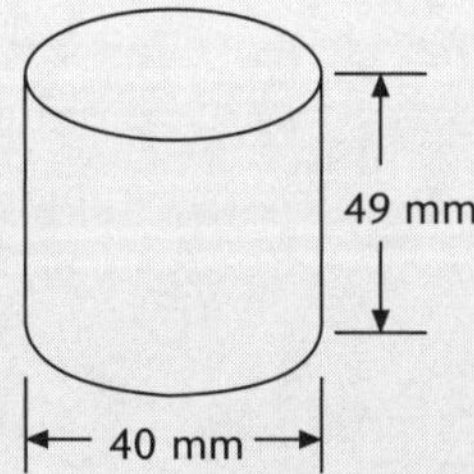


**f**

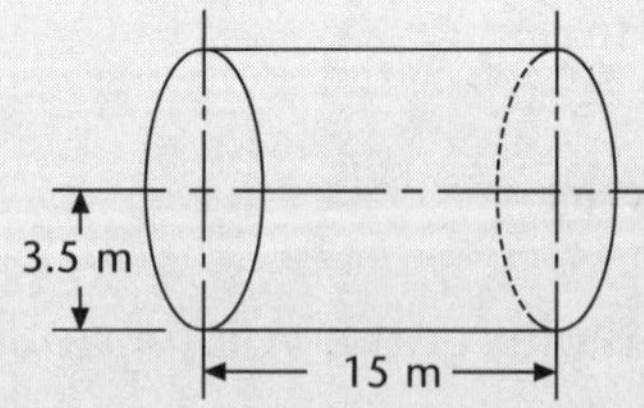

**37** A watering trough is made by slicing a cylinder down the centre, giving a solid with semicircles at each end. Calculate the volume of the trough correct to the nearest 0.1 $m^3$. If the trough is filled to the top, calculate the capacity of the trough in kilolitres. If during the day the cows drink 35% of the water, calculate the amount of water needed to refill the trough. p. 151

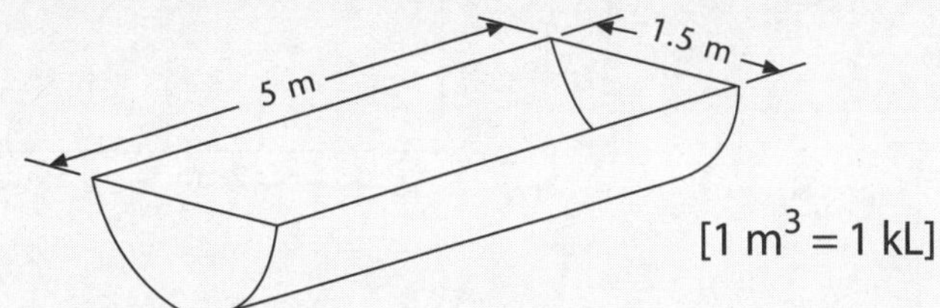


[1 $m^3$ = 1 kL]

**38** A cylindrical concrete water tank with a base diameter of 12 m and a height of 6 m is filled with water. Calculate the volume of water required to fill the tank. Answer correct to one decimal place. pp. 150–151

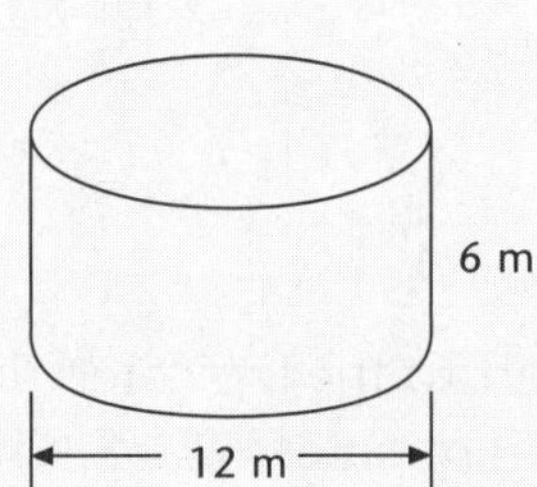


**39** A letter box is made using a half cylinder on a square prism. If the width of the rectangle is 24 cm and the length is 35 cm, calculate the volume of the letter box. Answer correct to the nearest $cm^3$. pp. 150–151

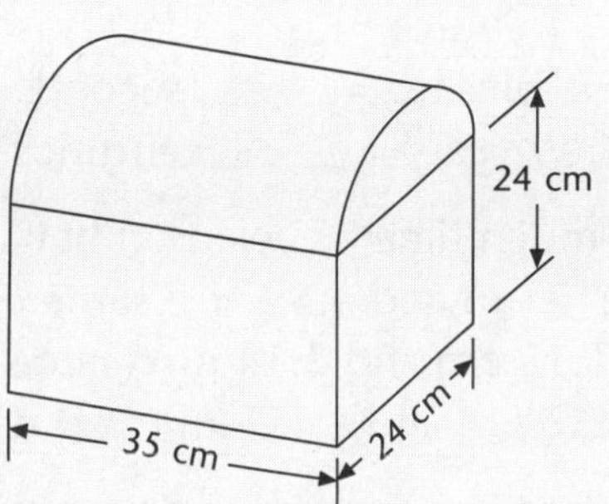


**40** A cubic concrete block of side 50 cm has four circular holes of diameter 20 cm drilled through the block. Calculate the volume of concrete remaining. Answer correct to the nearest centimetre$^3$. pp. 150–151

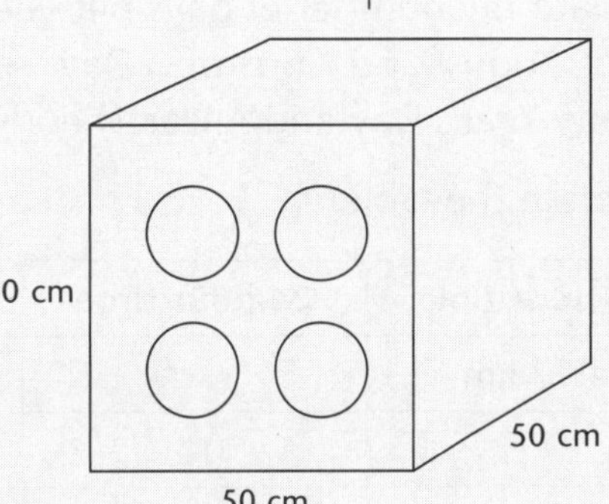


**41** The volume of a prism is 1420 $cm^3$. What will be its capacity in litres? p. 151

**42** Convert: p. 151

**a** 3500 $cm^3$ ______ L  **b** 28 $cm^3$ = ______ mL
**c** 760 mL = ______ $cm^3$  **d** 7850 mL = ______ L
**e** 58 L = ______ $cm^3$  **f** 1000 L = ______ $cm^3$

**43** Find the volume of a cube with side each 100 cm. Give your answer in $cm^3$. Hence, complete: p. 151

1 $m^3$ = ______ $cm^3$
But 1 L = 1000 $cm^3$
∴ 1 $m^3$ = ______ L
i.e. 1 $m^3$ = ______ kL

**44** Find the capacity of the following (in L): p. 151

a

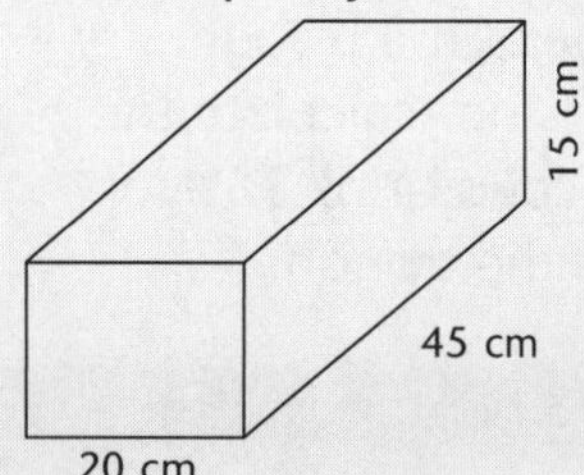

b

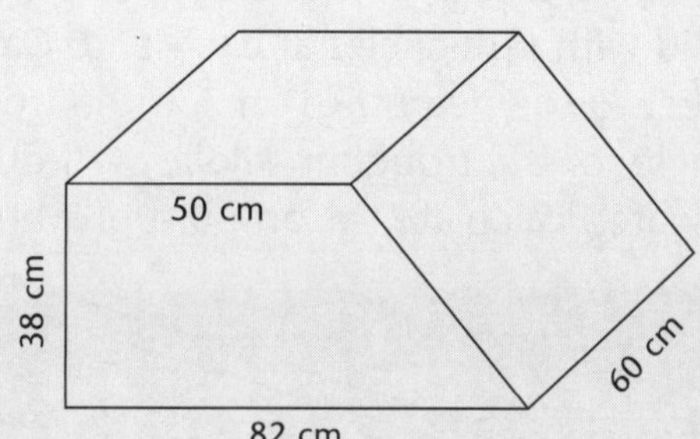


c

2 m

**45** Convert to the indicated units: pp. 152–154

a 300 minutes = ______ hours

b 4 days = ______ hours

c $6\frac{1}{2}$ minutes = ______ seconds

d $\frac{3}{4}$ hour = ______ minutes

e 0.2 minutes = ______ seconds

f 60 hours = ______ days

g 2000 years = ______ centuries

h 60 years = ______ decades

**46** How much time has elapsed between: pp. 152–154

a 4:17 am and 4:52 am (same day)?

b 11:19 am and 3:14 pm (same day)?

c 3:37 pm and 4:12 am (next day)?

**47** Calculate the number of days between: pp. 152–154

a 4th August and Christmas Day

b New Year's Day and Anzac Day (in a leap year)

**48** Complete the table: pp. 152–154

| 12-hour time | 24-hour time |
|---|---|
| 4:17 am | |
| | 1142 |
| 3:47 pm | |
| | 2148 |

**49** The digital clock below is 6 minutes fast: pp. 152–154

4:02

What is the correct time?

**50** Mitchell went to his bed at 8:45 pm and 15 minutes later fell asleep. He then slept for 9 hours 20 minutes. What time did Mitchell: pp. 152–154

a fall asleep?

b wake up next morning?

**51** Laura begins work at 10:45 am and finishes at 7:15 pm. She is paid at the rate of $12 per hour for this work. pp. 152–154

**a** How many hours did she work?

**b** What is her pay for the day's work?

**52** In winter, when it is noon in Sydney it is 11:30 am in Adelaide and 10 am in Perth. What is the local time in: pp. 152–154

**a** Adelaide when it is 3:40 pm in Sydney?

**b** Sydney when it is 4:52 pm in Perth?

**c** Adelaide when it is 11:43 pm in Perth?

**53** Express in 12-hour digital time: pp. 152–154

**a**

**b**

**54** Here is a local bus timetable. pp. 152–154

| | |
|---|---|
| Macquarie St | 4:39 |
| John St | 4:48 |
| Grace St | 5:02 |
| Coolah Av | 5:17 |
| Parallel St | 5:22 |
| River Rd | 5:29 |
| City Centre | 5:36 |

**a** How long is the bus ride from:

**i** Grace St to Parallel St?

**ii** John St to City Centre?

**b** If the bus is five minutes late, what time will it arrive at City Centre?

**c** Fiona catches a later bus that travels the same route at the same speed. If it leaves Grace St at 7:42, what time will it reach City Centre?

**55** This table shows the time difference between major cities around the world (daylight saving alterations have been ignored). pp. 152–154

| | | | |
|---|---|---|---|
| Los Angeles: | 4 am | Calcutta: | 5:30 pm |
| New York: | 7 am | Tokyo: | 9 pm |
| London: | Noon | Sydney: | 10 pm |
| Jerusalem: | 2 pm | Auckland: | Midnight |

**a** How many hours is Auckland ahead of Sydney?

**b** How many hours is New York behind Jerusalem?

**c** If it is 4 pm in London, what time is it in Los Angeles?

**d** A golf tournament finishes in Los Angeles at 5:30 pm on Sunday. What time is it in Sydney?

Go to p. 287 for quick answers, or to pp. 341–353 for worked solutions.

## YOUR CHECKLIST

**For a complete understanding of this topic you must be able to:**

| | | | |
|---|---|---|---|
| ✓ | Find areas of squares, rectangles, triangles | | pp. 137–138 |
| ✓ | Find area of composite figures | | pp. 138–139 |
| ✓ | Find area of shaded regions | | p. 139 |
| ✓ | Find area of special quadrilaterals | | p. 140 |
| ✓ | Convert between units such as $m^2$ and ha | | pp. 140–142 |
| ✓ | Name the parts of a circle | | p. 142 |
| ✓ | Find the diameter and radius of circles | | p. 142 |
| ✓ | Find the circumference of circles | | pp. 143–145 |
| ✓ | Find the area of circles | | pp. 145–147 |
| ✓ | Calculate the radius or diameter of a circle, given circumference or area | | pp. 147–148 |
| ✓ | Find volume of prisms using $V = Ah$ | | pp. 148–149 |
| ✓ | Calculate the volume of cylinders | | pp. 150–151 |
| ✓ | Find the capacity of solids | | p. 151 |
| ✓ | Perform operations involving time units | | p. 152 |
| ✓ | Interpret and use tables relating to time | | pp. 152–154 |
| ✓ | Compare times and calculate time differences between major cities of the world. | | pp. 152–154 |

**Now you are ready to do the tests!**

**(30 marks)**

1 Find the area.

a
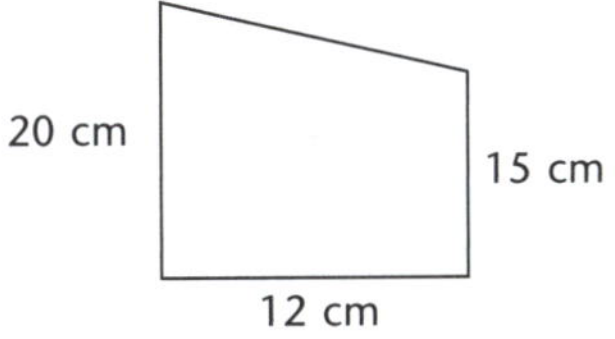


b
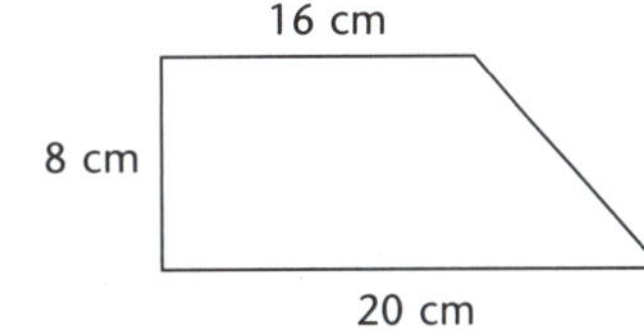


(4 marks)

2 Find the volume.

a
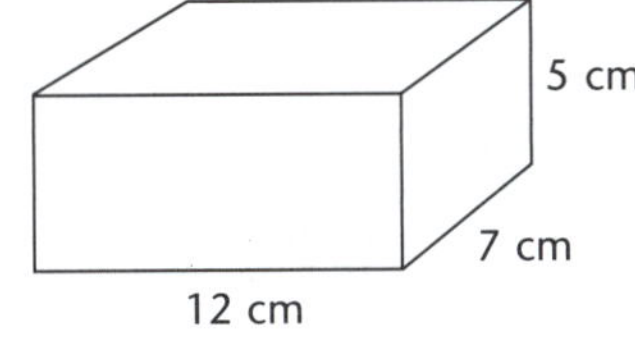


b
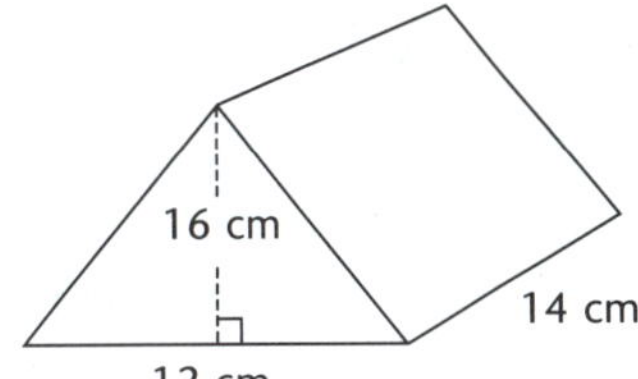


(4 marks)

3 Find the capacity in L.

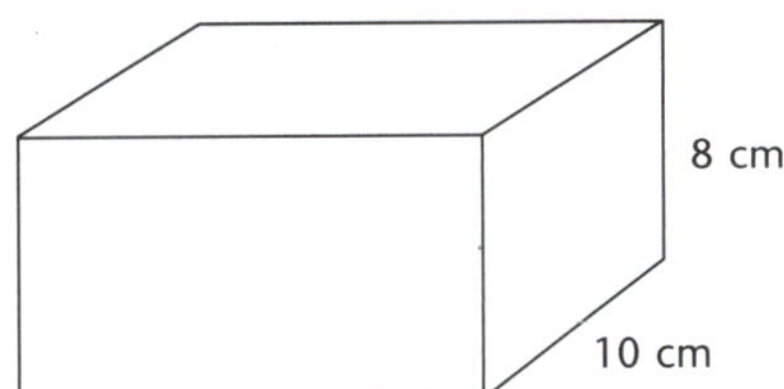


(3 marks)

4 Find the circumference of the circles. Answer correct to two decimal places.

a
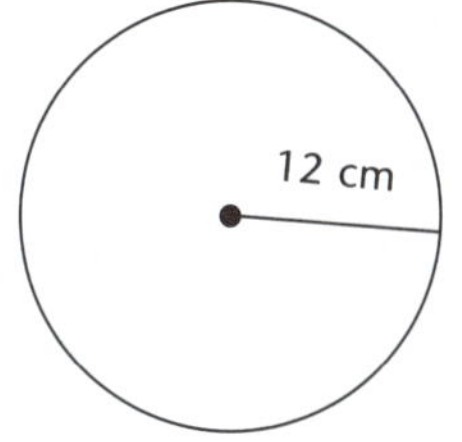


b
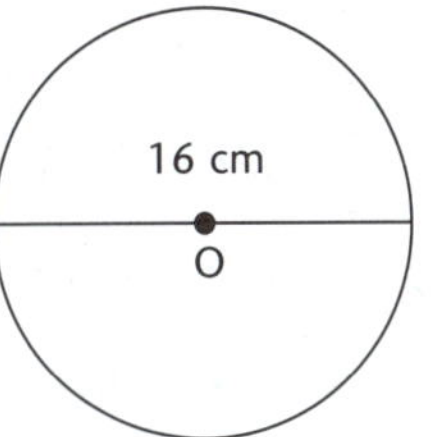


(4 marks)

5 Find the area of the circles. Answer correct to three decimal places.

a
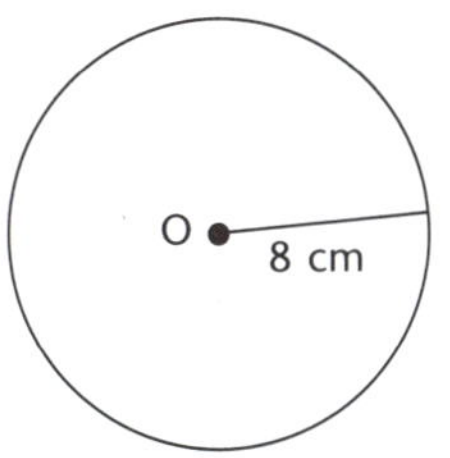


b
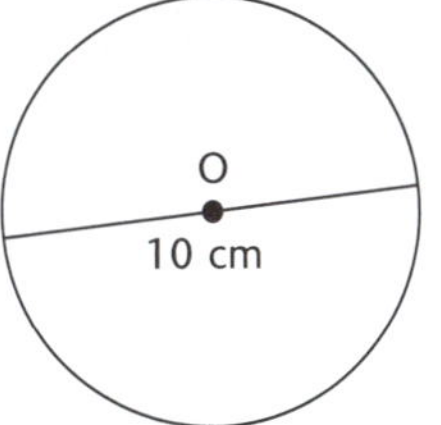


(4 marks)

6 Find the area and circumference of a circle with radius 2 cm, leaving your answer in terms of $\pi$. (2 marks)

*(cont.)*

7 Find the volume of the cylinder. Answer correct to two decimal places.

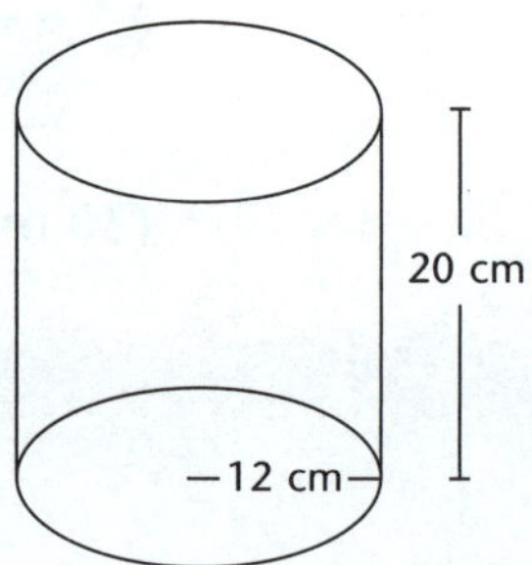


(2 marks)

8 The table shows the local times across the world when it is 9 pm in Sydney.

a What is the time difference between London and Los Angeles?

b When it is 3 am in London, what time is it in Perth?

c If it is 10 pm in Sydney, find the local time in New York.

| City | Local time |
|---|---|
| Sydney | 2100 |
| Perth | 1900 |
| London | 1100 |
| New York | 0600 |
| Los Angeles | 0300 |

(3 marks)

9 The chart details the tides in Sydney.

| Day | High | Low |
|---|---|---|
| 6 July | 0025 (1.62 m) | 0709 (0.37 m) |
| | 1326 (1.32 m) | 1908 (0.61 m) |
| 7 July | 0125 (1.53 m) | 0759 (0.38 m) |
| | 1424 (1.42 m) | 2018 (0.61 m) |
| 8 July | 0225 (1.45 m) | 0849 (0.39 m) |
| | 1522 (1:51 m) | 2134 (0.57 m) |
| 9 July | 0333 (1.38 m) | 0943 (0.41 m) |
| | 1620 (1.62 m) | 2250 (0.49 m) |

a Six July is a Sunday. When is the first high tide on the following Tuesday?

b What is the lowest tide across the four days?

c Find the time difference between consecutive low tides on 9 July.

d At Swansea, the tides occur 30 minutes later than at Sydney. Bruce wants to fish on the first low tide on 7 July. What time will the low tide occur at Swansea?

(4 marks)

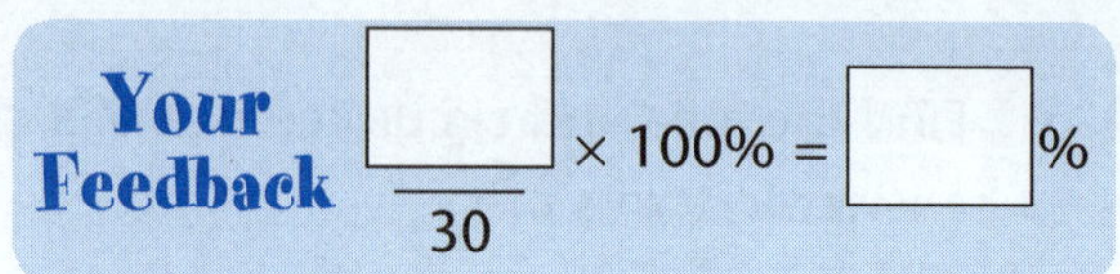



☞ **Quick answers on page 293**

☞ **Worked solutions on page 387**

**(35 marks)**

**1** Find the area:

**a**

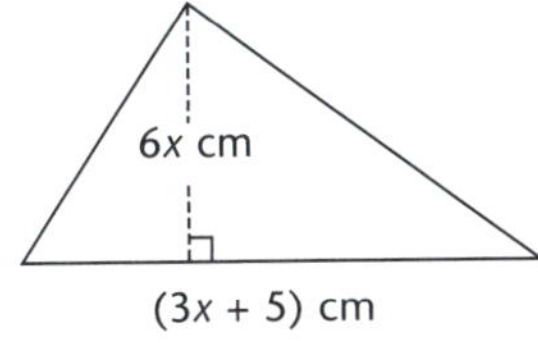


**b**

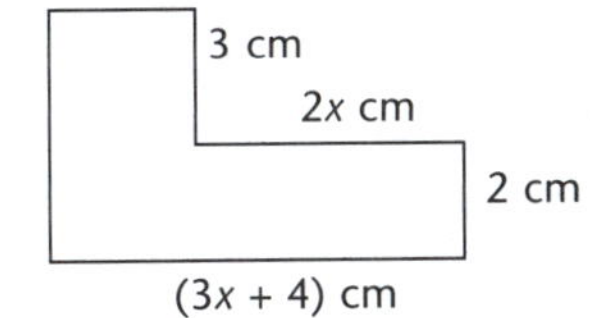


(4 marks)

**2** Find the volume:

**a**

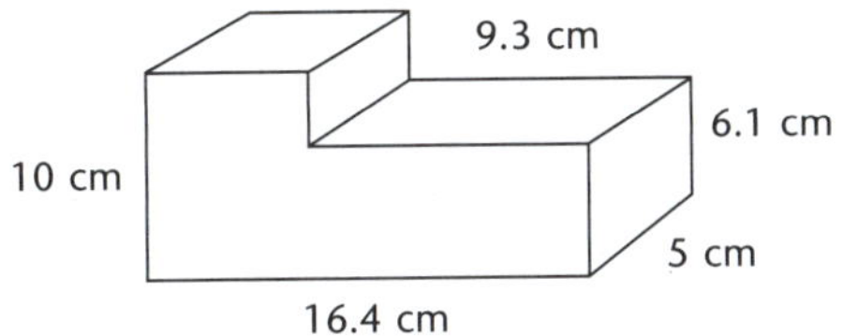


**b**

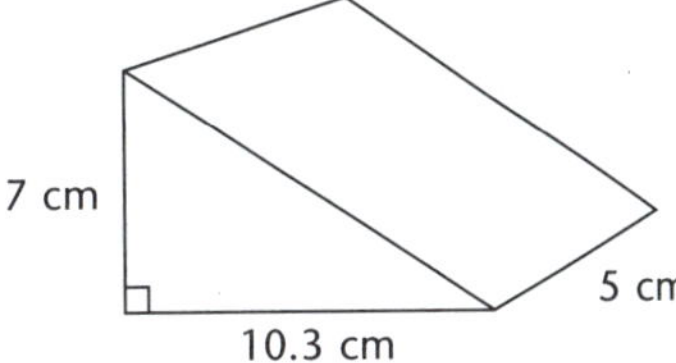


(4 marks)

**3** Find the capacity of a:

**a** rectangular prism with sides 18 cm, 12 cm, 9 cm

**b** triangular prism with area of triangular face 26 $m^2$ and perpendicular height 16 m. (4 marks)

**4** Find the perimeter of the following figures. Answer correct to three decimal places:

**a**

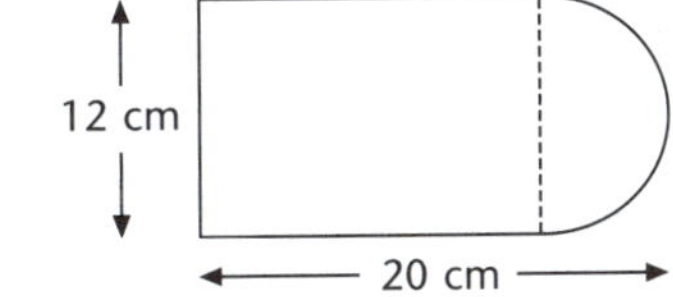


**b**

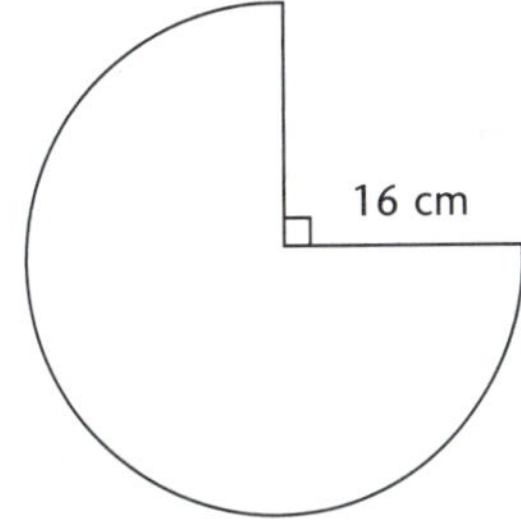


**c**

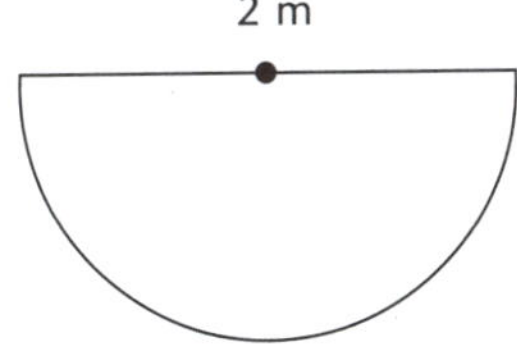


**d**

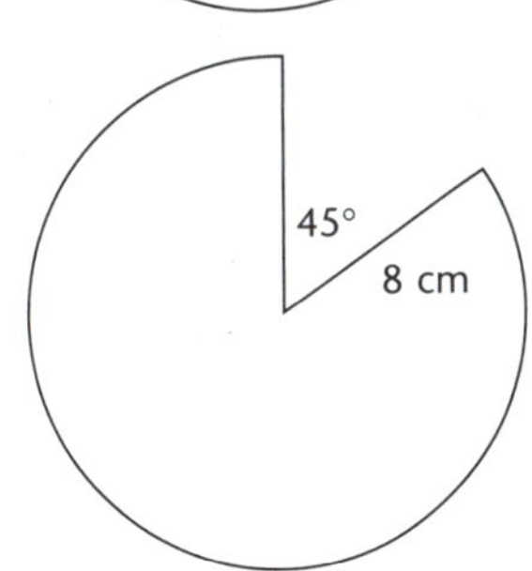


(8 marks)

**5** Find the radius of a circle with circumference 240 cm. Answer correct to one decimal place. (2 marks)

*(cont.)*

**6** Find the area (correct to 2 decimal places):

**a**

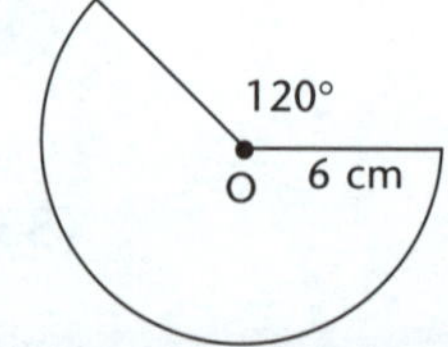


**b**

28 cm

40 cm

(4 marks)

**7** Find the volume, to nearest cm$^3$:

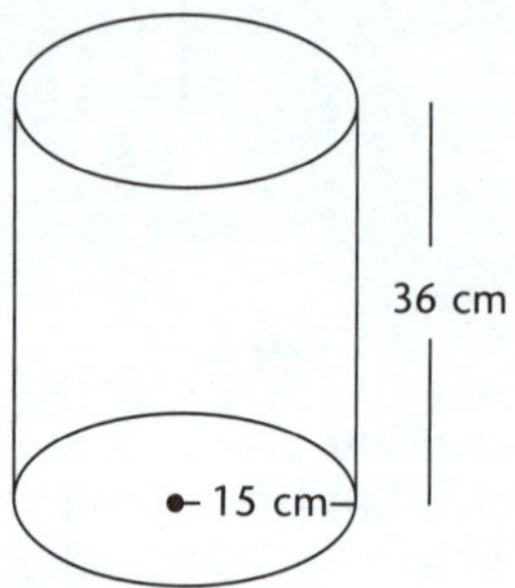


(2 marks)

**8** Find the capacity of a cylinder with a diameter of 48 cm and a height of 60 cm. Answer to the nearest litre. (3 marks)

**9** The table shows a local bus timetable:

| | | | |
|---|---|---|---|
| Silsoe St | 1131 | 1157 | 1218 |
| Park Ave | 1146 | 1212 | 1233 |
| Charlestown Rd | 1158 | – | 1245 |
| Newcastle Rd | 1207 | 1229 | 1254 |
| Thomas St | 1214 | – | 1301 |
| Nelson St | 1218 | – | 1305 |
| Maryland Dr | 1223 | 1240 | 1310 |

**a** Pete catches the 1131 bus at Silsoe St to travel to Thomas St. How long does the trip take?

**b** The express bus leaves Silsoe St at 11:57 am. How much quicker is this bus compared with the normal service to Maryland Drive?

**c** Sandy arrived at Park Ave and missed the 1146 bus by 5 minutes. How long will she wait for the next bus to take her to Newcastle Rd?

**d** Scott gets to the bus stop on Newcastle Rd at 12:20 pm to catch a bus to Nelson St. What time will he arrive at his destination? (4 marks)

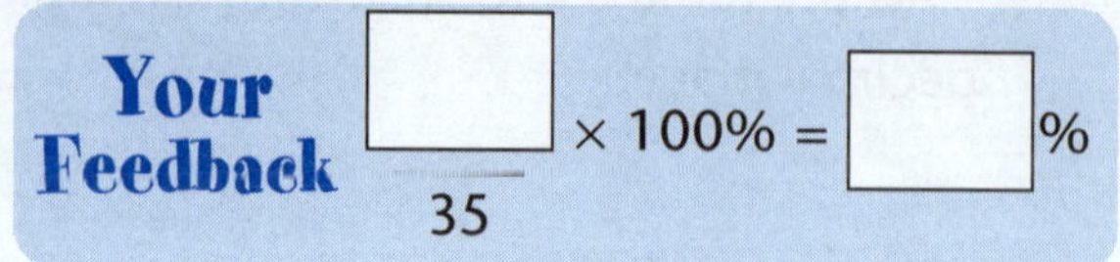


☞ **Quick answers on page 293**
☞ **Worked solutions on page 388**

**(40 marks)**

1 Find the area:

a

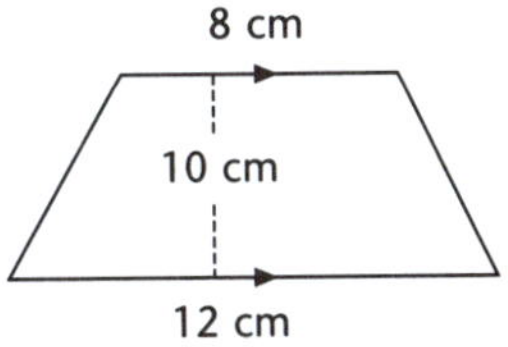


b

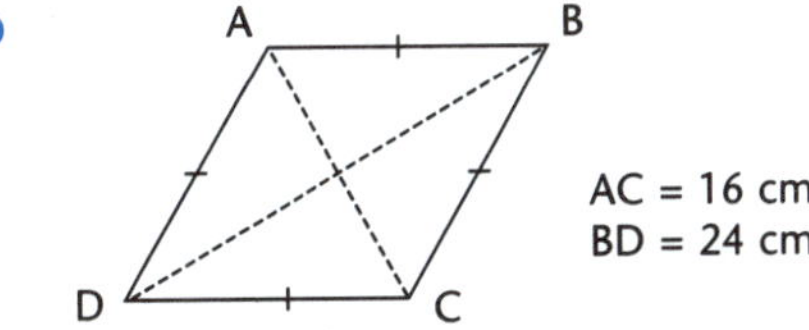


(4 marks)

2 Find the area of the shaded region:

a

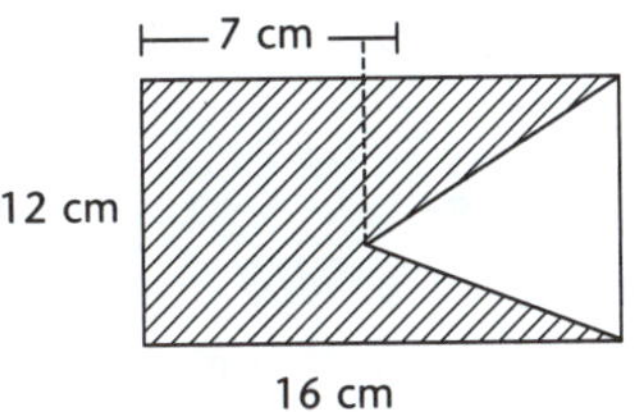


b

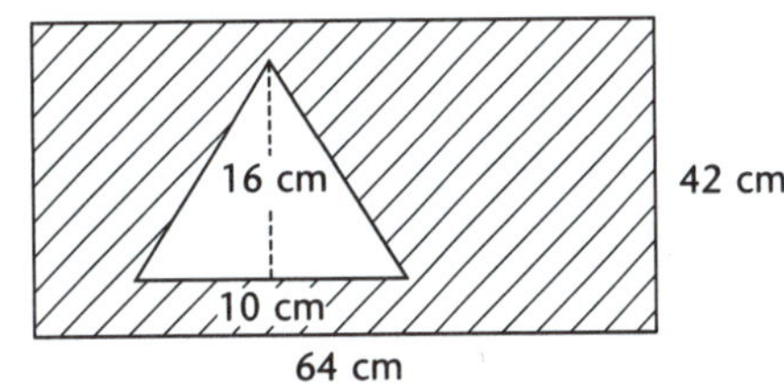


(4 marks)

3 A swimming pool is 8 m long, 4 m wide and is filled to a depth of 1.8 m.

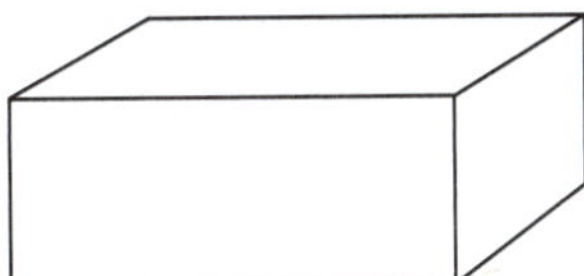

Calculate the cost of filling the pool if water costs $1.25/kilolitre. (3 marks)

4 Find the area of the kite ABCD if AC = 12 cm and BD = 7 cm:

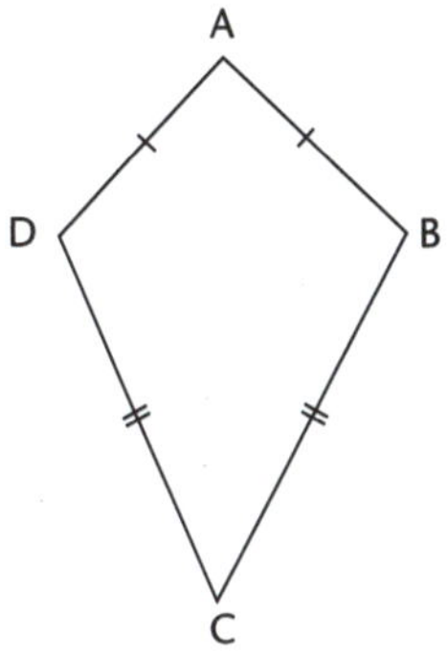


(2 marks)

5 The circle has an area of $36\pi$ cm$^2$.

a Find the radius.

b Find the area of the shaded sector, AOB, correct to two decimal places.

c Find the length of the arc AB, correct to two decimal places.

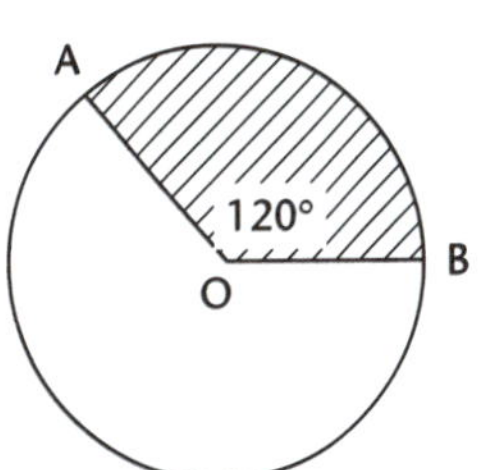


(5 marks)

*(cont.)*

6 A wheel has a radius of 36 cm. How many complete rotations of the wheel are required to travel 2 km? (2 marks)

7 Find the perimeter, correct to four decimal places:

**a**

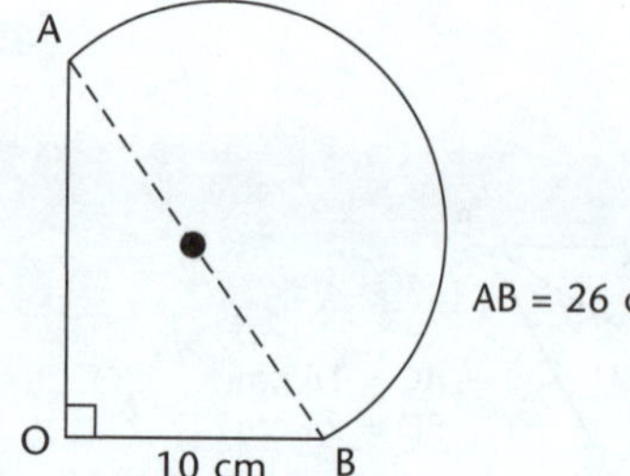


**b**

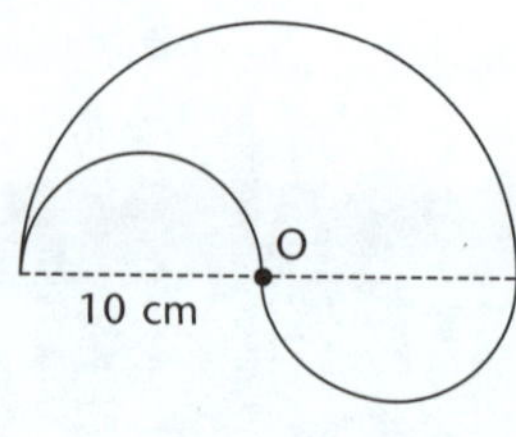


(4 marks)

8 Find the area of the shaded region, correct to two decimal places:

**a**

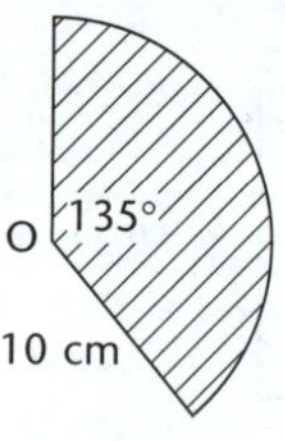


**b**

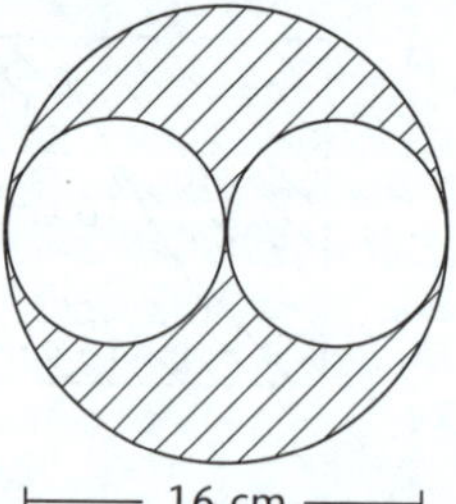


**c**

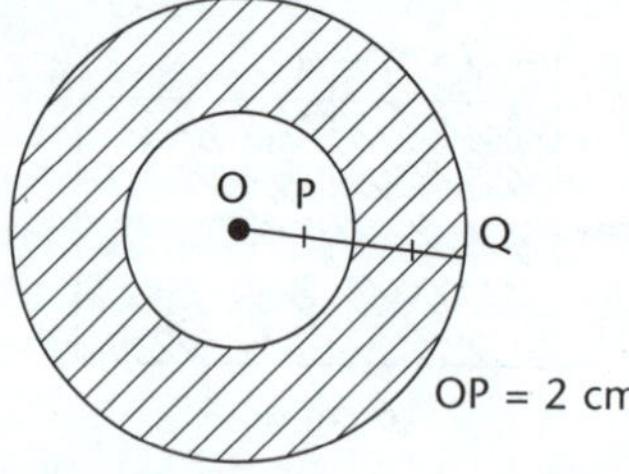


**d**

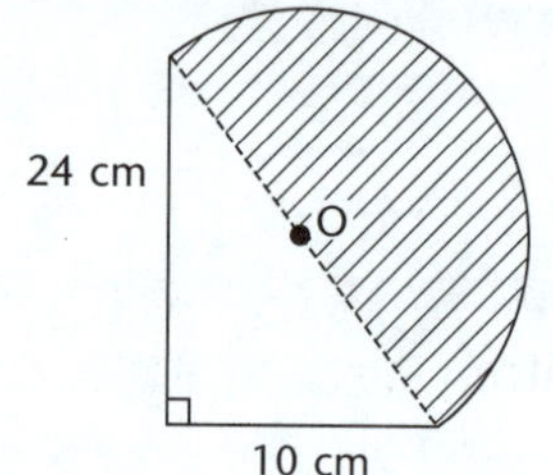


**e**

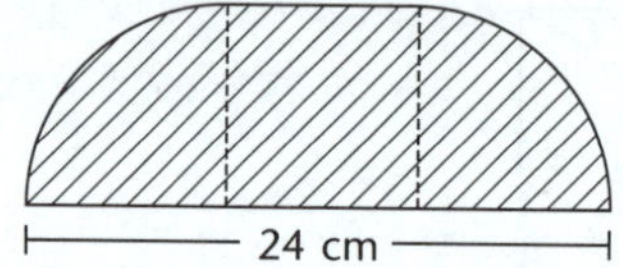


**f**

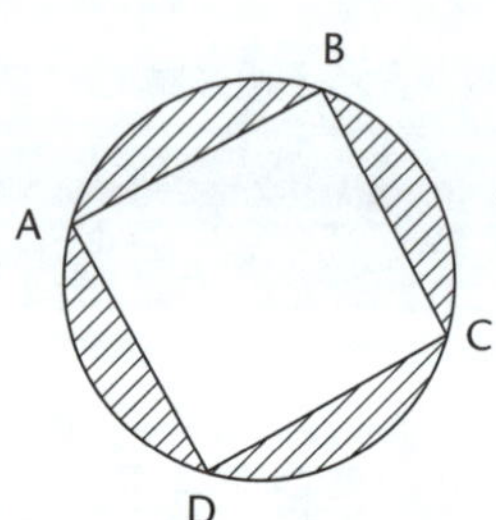


ABCD is a square where AB = 4 cm. (12 marks)

9 Find the capacity of a cylinder with a radius of 0.6 m and a height of 80 cm. Answer to the nearest 0.1 L. (2 marks)

10 Kristie sets her watch at the correct time of 3:00 pm. However, the watch is losing 2 minutes every hour. When it is 6:00 pm, what time does Kristie's watch show? (2 marks)

**Your Feedback** $\frac{\square}{40} \times 100\% = \square\%$


☞ **Quick answers on page 293**

☞ **Worked solutions on page 390**

# 9 RATIO AND RATES

- Ratio
- Using Ratio to Solve Problems
- Rates
- Scale Drawing

**KEYWORDS**

**Distance** **Scale drawing**

**Rate** **Speed**

**Ratio** **Time**

**Scale**

## Ratio

Ratio is the comparison of two or more quantities expressed in the same units. The ratio of *a* to *b* is written as $a:b$ $\left(\text{or sometimes } \frac{a}{b}\right)$.

Order is important so that $a:b$ is different to $b:a$. The value of the ratio $a:b$ is $\frac{a}{b}$, of $b:a$ is $\frac{b}{a}$.

Examples of ratios are 2:5, 5:2, 3:4, 4:3, 7:2. The value of the ratio 2:5 is $\frac{2}{5}$ while the value of 5:2 is $\frac{5}{2}$.

### Equivalent Ratios

Ratios are simplified in the same way as fractions—we divide (or multiply) both components of the ratio by the same number (except 0).

**For Example**

Simplify:

| | | | |
|---|---|---|---|
| **1** | 6:4 | **2** | 4:6 |
| **3** | 30:50 | **4** | 16:64 |
| **5** | 120:10 | **6** | 40:20:10 |
| **7** | $\frac{3}{4}:\frac{1}{4}$ | **8** | $\frac{2}{3}:1$ |
| **9** | $\frac{1}{2}:\frac{1}{4}$ | **10** | $1\frac{1}{2}:2$ |
| **11** | $\frac{4}{5}:1\frac{1}{2}$ | **12** | 0.2:0.8 |
| **13** | 0.5:1.5:4 | | |

**1** $6:4 = \frac{6}{2}:\frac{4}{2}$ (dividing both by 2)
$= 3:2$

**2** $4:6 = \frac{4}{2}:\frac{6}{2}$
$= 2:3$

**3** 30:50 = 3:5 (divide both parts by 10)

**4** 16:64 = 1:4 (divide by 16)

**5** $12\cancel{0}:1\cancel{0} = 12:1$
(cross off zeros—divide by 10)

**6** $4\cancel{0}:2\cancel{0}:1\cancel{0} = 4:2:1$ (divide all by 10)

**7** $\frac{3}{4}:\frac{1}{4} = \frac{3}{4} \times 4:\frac{1}{4} \times 4$ (multiplying both by 4)
$= 3:1$

**8** $\frac{2}{3}:1 = \frac{2}{3} \times 3:1 \times 3$
$= 2:3$

**9** $\frac{1}{2}:\frac{1}{4} = \frac{1}{2} \times 4:\frac{1}{4} \times 4$
$= 2:1$
or $\frac{1}{2}:\frac{1}{4} = \frac{2}{4}:\frac{1}{4}$
$= 2:1$ (multiply both by 4)

**10** $1\frac{1}{2}:2 = 1\frac{1}{2} \times 2:2 \times 2$
$= 3:4$
or $1\frac{1}{2}:2 = \frac{3}{2}:2$
$= \frac{3}{2}:\frac{4}{2}$
$= 3:4$ (multiplying by 2)

**11** $\frac{4}{5}:1\frac{1}{2} = \frac{4}{5} \times 10:1\frac{1}{2} \times 10$
$= \frac{8}{10} \times 10:\frac{15}{10} \times 10$
$= 8:15$

[Multiply both parts by 10 as is LCD of 5 and 2.]

**12** $0.2:0.8 = 0.2 \times 10:0.8 \times 10$
$= 2:8$
$= 1:4$ (dividing by 2)

**13** $0.5:1.5:4 = 5:15:40$ (multiplying by 10)
$= 1:3:8$ (dividing by 5)

Sometimes we find the value of the pronumeral to complete the simplification.

## For Example

Find the value of pronumerals:

1 $x:4 = 15:20$

2 $4:28 = 2:x$

1 $15:20 = 3:4$

$\therefore x = 3$

2 $4:28 = 2:14$

$\therefore x = 14$

Often we have to change one part of the ratio so that both parts have the same units. We then simplify the resulting ratio. For example, we might have to change m to cm.

## For Example

Simplify:

| | | | |
|---|---|---|---|
| **1** | 20 cents:80 cents | **2** | 40 cents:\$2 |
| **3** | \$1.20:\$6 | **4** | 6 days:3 weeks |
| **5** | 4 cm:1 metre | **6** | 2 L:150 mL |
| **7** | 8 sec:2 min | **8** | 4 h:1 day |
| **9** | $1\frac{1}{2}$ mL:1 L | **10** | 400 kg:2 t |
| **11** | \$20:\$40:\$80 | **12** | 7.2 L:21.6 L |
| **13** | $3x:12x$ | **14** | $21ab:14ac$ |
| **15** | $\frac{1}{p}:p$ | **16** | $16ab:4ab^2$ |

1 20 cents:80 cents = 20:80
= 1:4 (divide by 20)

2 40 cents:\$2 = 40 cents:200 cents
= 40:200
= 1:5 (divide by 40)

3 \$1.20:\$6 = 1.2:6
= 6:30 (multiply by 5)
[Or 120c:600c = 1:5]
= 1:5 (divide by 6)

4 6 days:3 weeks = 6 days:21 days
= 6:21
= 2:7

5 4 cm:1 m = 4 cm:100 cm
= 4:100
= 1:25

6 2 L:150 mL = 2000 mL:150 mL
= 2000:150 (zeros cancelled)
= 200:15
= 40:3 (dividing by 5)

7 8 sec:2 min = 8 sec:120 sec
= 8:120
= 1:15

8 4 h:1 day = 4 h:24 h
= 4:24
= 1:6

9 $1\frac{1}{2}$ mL:1 L = $1\frac{1}{2}$ mL:1000 mL
= $1\frac{1}{2}$:1000
= 3:2000 (multiply by 2)

10 400 kg:2 t = 400 kg:2000 kg (1 t = 1000 kg)
= 400:2000 (zeros cancelled)
= 4:20
= 1:5

11 \$20:\$40:\$80 = 20:40:80
= 1:2:4

12 7.2 L:21.6 L = 7.2:21.6
= 72:216 (multiply by 10)
= 1:3 (divide by 72)

13 $3x:12x = \frac{3x}{x}:\frac{12x}{x}$ (divide both parts by $x$)

$= 3:12$

$= 1:4$

14 $21ab:14ac = \frac{21ab}{a}:\frac{14ac}{a}$ (divide by $a$)

$= 21b:14c$

$= 3b:2c$ (÷ by 7)

15 $\frac{1}{p}:p = \frac{1}{p} \times p:p \times p$

$= 1:p^2$

16 $16ab:4ab^2 = \frac{\overset{4}{\cancel{16}}\,\overset{1}{\cancel{a}}\,\overset{1}{\cancel{b}}}{\underset{1}{\cancel{4}}\,\underset{1}{\cancel{a}}\,\underset{1}{\cancel{b}}}:\frac{\overset{1}{\cancel{4}}\,\overset{1}{\cancel{a}}\,\overset{1}{\cancel{b}}\,b}{\underset{1}{\cancel{4}}\,\underset{1}{\cancel{a}}\,\underset{1}{\cancel{b}}}$ (divide by $4ab$)

$= 4:b$

## Finding a Ratio of Two or More Quantities

From the information in a problem a ratio is formed and simplified if necessary.

**For Example**

1 A netball team won eight games in a season and lost four games. What is the ratio of games won to games lost?

2 Sylvio mixes six buckets of gravel, four buckets of sand and two buckets of cement into a batch of concrete. Determine the ratio of gravel to sand to cement that was used.

3 A street in a new subdivision has 40 blocks of which 25 have been built on. What is the ratio of occupied to vacant blocks?

4 The results of an election were announced:

| Candidate | No. of votes |
|---|---|
| Christine Howard | 16 000 |
| Noel Horton | 12 000 |
| Peter Chapman | 20 000 |

What is the ratio of the votes for:

a Christine Howard to Noel Horton?

b Peter Chapman to the total votes?

5 In a class survey the hair colour of boys and girls was recorded:

| Colour | Boys | Girls |
|---|---|---|
| Blonde | 3 | 4 |
| Brown | 5 | 7 |
| Black | 2 | 4 |
| Red | 2 | 1 |

a Write down the number of students:

i who are boys

ii who are girls

iii in the class.

b What is the ratio of:

i blonde boys to blonde girls?

ii blonde girls to black-haired girls?

iii brown-haired to black-haired to red-haired students?

iv non-blonde students to the total number of students?

6 Find the ratio of:

a shaded squares to unshaded squares

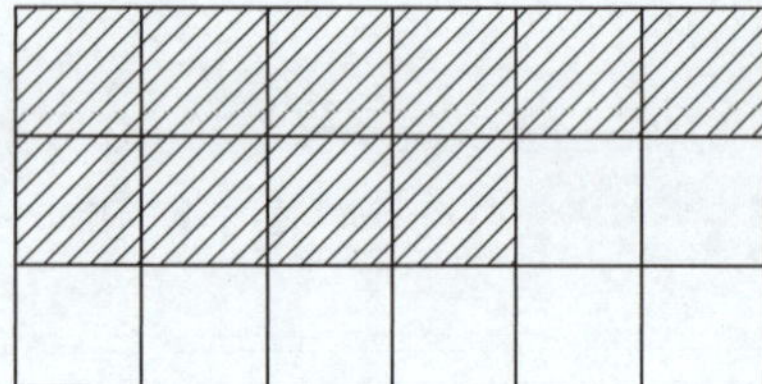

b shaded squares to total number of squares.

7 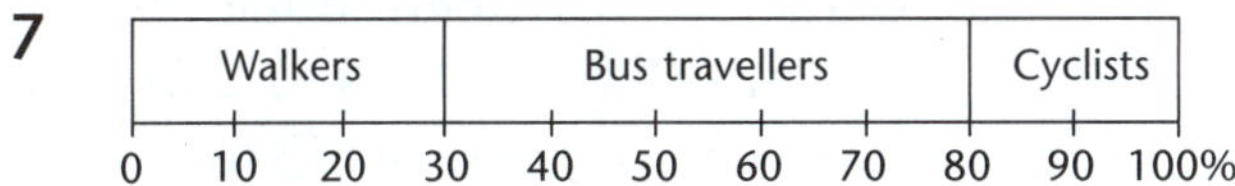


Students in Year 8 were surveyed to determine their method of travelling to school each day and the results recorded in the above graph. Find the ratio of:

a bus travellers to walkers

b bus travellers to total number of students surveyed.

8 

4 cm

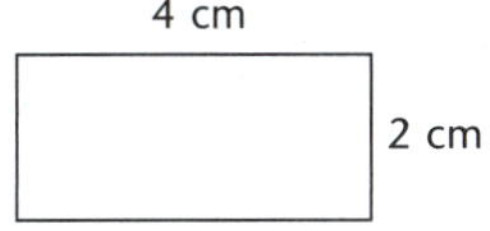


8 cm

What is the ratio of the:

a lengths of the two rectangles?

b perimeters of the two rectangles?

c areas of the two rectangles?

1 8:4 = 2:1

2 6:4:2 = 3:2:1

3 Vacant = 40 − 25
= 15
∴ 25:15 = 5:3

4 a $16\,\cancel{000}:12\,\cancel{000} = 16:12$
$= 4:3$

b Total = 16 000 + 12 000 + 20 000
= 48 000
∴ Ratio = $20\,\cancel{000}:48\,\cancel{000}$
= 20:48
= 5:12

5 a i Boys = 3 + 5 + 2 + 2
= 12
∴ There were 12 boys.

ii Girls = 4 + 7 + 4 + 1
= 16
∴ There were 16 girls.

iii Total = 12 + 16
= 28
∴ There were 28 students.

b i 3:4

ii 4:4 = 1:1

iii 12:6:3 = 4:2:1

iv (28 − 7):28 = 21:28
= 3:4

6 a Shaded = 10, unshaded = 8
∴ 10:8 = 5:4

b Shaded = 10, total = 18
∴ 10:18 = 5:9

7 a Bus travellers = 50, walkers = 30
Ratio = 50%:30%
= 50:30
= 5:3

b Bus travellers = 50, total students = 100
Ratio = 50%:100%
= 50:100
= 1:2

8 a 4 cm:8 cm = 4:8 = 1:2

b Perimeter of small rectangle = 2(4 + 2)
= 2(6) = 12
∴ The perimeter is 12 cm.

Perimeter of large rectangle = 2(8 + 4)
= 2(12)
= 24
∴ The perimeter is 24 cm.

$\therefore$ Ratio of perimeters = 12 cm:24 cm
= 12:24
= 1:2

c Area of small rectangle = $4 \times 2$
= 8

$\therefore$ The area is 8 cm$^2$.

Area of large rectangle = $8 \times 4$
= 32

$\therefore$ The area is 32 cm$^2$.

$\therefore$ Ratio of areas = 8 cm$^2$:32 cm$^2$
= 8:32
= 1:4

# Using Ratios to Solve Problems

## Using the Unitary Method

The **unitary method** can be used in ratio problems. It involves finding the value of one item (or one *unit*) and then multiplying this value to find the required answer.

1 The ratio of cats to dogs is 4:5. If there are 30 dogs, how many cats?

2 On a bus trip the ratio of males to females is 4:7. If there are 28 females on the bus, how many males?

3 Three sums of money are in the ratio of 2:3:5. If the smallest amount is \$4.80, find the largest amount.

4 It is known that an alloy is made up of copper, tin and zinc in the ratio 14:17:10. In a certain batch, 85 kg of tin is used. How much copper and zinc will be required to make the alloy?

5 Joe's will stated that his estate should be divided such that for each \$2 that Carol receives, Keith should receive \$5. If Keith gained \$74 000 from Joe's estate, how much will Carol receive?

6 The ratio of the ages of three brothers is 10:8:5. If the youngest is 15 years, how old is the eldest?

1 Cats to dogs = 4:5 [Order is important.]
Now, 30 dogs:
$\therefore$ 5 parts = 30 [We can use 'parts' or 'shares'.]
1 part = 6
4 parts = $(4 \times 6)$
= 24
$\therefore$ There are 24 cats. [Check: 24:30 = 4:5]

2 Males to females = 4:7
Now, 28 females:
$\therefore$ 7 parts = 28
1 part = 4
4 parts = 16 [Check: 16:28 = 4:7]
$\therefore$ There are 16 males.

3 Sums of money = 2:3:5
$\therefore$ smallest amount = 2 parts
$\therefore$ 2 parts = \$4.80
1 part = \$2.40
$\therefore$ 5 parts = \$12
$\therefore$ \$12 is largest amount.

4 Copper, tin, zinc = 14:17:10
$\therefore$ 17 parts = 85
1 part = 5
$\therefore$ 14 parts = 70
and 10 parts = 50
$\therefore$ 70 kg copper and 50 kg zinc are required.

5 Carol:Keith = \$2:\$5

= 2:5

∴ Keith's share is 5 parts (Carol's share is 2 parts)

∴ 5 parts = \$74 000

1 part = \$14 800 (74 000 ÷ 5)

2 parts = \$29 600 (14 800 × 2)

∴ Carol's share of Joe's estate is \$29 600.

6 Brothers' ages = 10:8:5 (5 is the youngest)

∴ 5 parts = 15

1 part = 3

10 parts = 30

∴ The eldest brother is 30 years old.

## Dividing Quantities in a Given Ratio

To divide quantities in a given ratio, find the total of the parts and then multiply by fractions.

**For Example**

1 Divide \$250 in the ratio of 7:3.

2 Share \$72 in the ratio of 4:3:2.

3 If \$200 is to be split in the ratio of 3:2, find the smaller amount.

4 A 2-litre container is used to make up an orange fruit juice drink. 500 mL of concentrate is poured into the container and then water is used to fill the remainder of the container.

a What is the ratio of concentrate to water in the fruit juice drink?

b Two-hundred and forty millilitres of the fruit juice drink is poured into a glass. How much of the drink is water?

5 

The ratio of two supplementary angles is 4:5. What is the size of each angle?

6 Two sisters, Margaret and Robyn, contribute \$8000 and \$4000 respectively to buy a car. Three years later they decide to sell the car for \$7200 and split the money in the same ratio as their investment. What was Margaret's share?

7 The ratio of a father's age to that of his son is 9:2. If the sum of their ages is 44, how old is the father?

8 In an orchard the ratio of orange trees to lemon trees is 3:1, while the ratio of lemon trees to mandarin trees is 2:5.

a What is the ratio of orange trees to lemon trees to mandarin trees?

b If there are 390 citrus trees, how many of each variety are in the orchard?

1 Total parts = 7 + 3

= 10

∴ 10 parts = \$250

∴ $\frac{7}{10} \times \$250 = \$175$

$\frac{3}{10} \times \$250 = \$75$

∴ \$175 and \$75 are the parts.

2 Total parts = 4 + 3 + 2

= 9

∴ 9 parts = \$72

∴ $\frac{4}{9} \times \$72 = \$32$

$\frac{3}{9} \times \$72 = \$24$

$\frac{2}{9} \times \$72 = \$16$

∴ \$32, \$24, \$16 are the parts.

**3** 5 parts = \$200

$\therefore \frac{2}{5} \times \$200 = \$80$

∴ The smaller amount is \$80.

**4** **a** Water = 2000 − 500

= 1500

∴ 1500 mL water

∴ Concentrate:water = 500 mL:1500 mL

= 1:3

∴ The ratio of concentrate to water is 1:3.

**b** Total parts = 1 + 3

= 4

∴ 4 parts = 240 mL

$\therefore \frac{3}{4} \times 240 = 180$

∴ 180 mL of the drink is water.

**5** Supplementary angles add to 180°.

As 4 + 5 = 9, there are 9 parts:

∴ 9 parts = 180°

$\frac{4}{9} \times 180 = 80$

$\frac{5}{9} \times 180 = 100$

∴ The angles are 80° and 100°.

**6** Ratio of investments = \$8000:\$4000

= 2:1

∴ total parts = 2 + 1

= 3

$\therefore \frac{2}{3} \times \$7200 = \$4800$

∴ Margaret's share of the car is \$4800.

**7** Total parts = 9 + 2

= 11

$\therefore \frac{9}{11} \times 44 = 36$

∴ The father is 36 years old.

**8** **a** Orange:lemon = 3:1 = 6:2,

Lemon:mandarin = 2:5

∴ 6:2:5 is ratio of orange to lemon to mandarin.

**b** Total parts = 6 + 2 + 5

= 13

∴ 13 parts = 390

$\frac{6}{13} \times 390 = 180$

$\frac{2}{13} \times 390 = 60$

$\frac{5}{13} \times 390 = 150$

∴ There are 180 orange trees, 60 lemon trees and 150 mandarin trees in the orchard.

## Rates

In ratio we compared two or more quantities with the same units. Rates are a comparison of quantities expressed in different units.

### Calculating Rates

When we calculate a rate we express it as for a single unit (e.g. km/h, runs/over, \$/week).

**For Example**

**1** Jenny made a 12-minute phone call and was charged \$6. Find the cost of the phone call per minute.

**2** A motorist was charged \$85.95 for 50 litres of petrol. Find the cost of the petrol in cents per litre.

**3** Which is the better buy: a 300 g packet of muesli for \$3.90 or a 500 g packet of muesli for \$6.45?

**4** Complete the table and determine the best buy.

| Volume | Price | Cost/10 mL |
|---|---|---|
| 300 mL | \$1.20 | |
| 750 mL | \$2.10 | |
| 1 L | \$2.95 | |

5 Land rates are paid at the rate of 1.65 cents per dollar of the land's valuation.

a Find the cost of the land rates on a property valued at $345 000.

b What is the value of the property if land rates are set at $3382.50?

6 Kellie's car used 28 litres of petrol, costing $49.28, on a trip of 400 kilometres.

a Find the car's petrol consumption expressed as L/100 km.

b What was the cost of Kellie's journey in cents/km?

1 Cost/minute = $6 ÷ 12 = 50c

∴ Jenny's phone call cost 50c/minute.

2 Cost per litre = $85.95 ÷ 50

= $1.719

∴ The petrol costs 171.9c per litre.

3 300 g: Cost per 100 g = $3.90 ÷ 3

= $1.30

∴ $1.30c/100 g

500 g: Cost per 100 g = $6.45 ÷ 5

= $1.29

∴ $1.29c/100 g

∴ The better buy (cheaper) is the 500 g packet.

4

| Volume | Price | Cost/10 mL |
|---|---|---|
| 300 mL | $1.20 (120c) | 120 ÷ 30 = 4 ∴ 4c/10 mL |
| 750 mL | $2.10 (210c) | 210 ÷ 75 = 2.8 ∴ 2.8c/10 mL |
| 1 L (1000 mL) | $2.95 (295c) | 295 ÷ 100 = 2.95 ∴ 2.95c/10 mL |

∴ The best buy is the 750 mL container.

5 Rate = 1.65 cents/$ value

= $0.0165

a ∴ Cost = 345 000 × 0.0165 ($)

= $5692.50

∴ The land rates cost $5692.50.

b ∴ As cost = value × rate

∴ $3382.50 = value × $0.0165

i.e. value = $\frac{3382.5}{0.0165}$

= 205 000

∴ The value of the land is $205 000.

6 a Consumption = 28 L/400 km

= 7 L/100 km

(dividing both by 4)

∴ Kellie's car uses 7 L/100 km.

b Cost/km = $49.28 ÷ 400

= $0.1232

∴ The cost of the trip was 12.32 cents/km.

## Using Given Rates

Here a rate is maintained for a longer period of time, or for a larger quantity.

**For Example**

1 Ken Short, a finalist in the World Athletics Championship, can run 100 metres in 10 seconds. Convert this to km/h.

2 My heart beats 72 times each minute. How many times did it beat in the year 2000?

3 A tap is leaking at the rate of 15 mL every 10 seconds. Find the amount of water wasted each day.

4 A pump can pump from a dam at the rate of 8 litres/second.

a Express this rate in kL/h.

b If Simon started up the pump on Tuesday at 6 am, at what time will one megalitre of water have been pumped?

1 Speed = 100 m/10 sec
= 600 m/minute (multiply both by 6)
= 36000 m/hour (multiply both by 60)
= 36 km/h (36 000 m = 36 km)

2 Beats = 72/minute
= 4320/h (multiply both by 60)
= 103 680/day (multiply both by 24)
= 37 946 880/year (multiply both by 366—a leap year)

∴ My heart beat 37 946 880 times in the year 2000.

3 Amount of water = 15 mL/10 sec
= 90 mL/minute (multiply both by 6)
= 5400 mL/h (multiply both by 60)
= 5.4 L/h
= 129.6 L/day (multiply both by 24)

∴ 129.6 litres of water is wasted each day.

4 a Rate = 8 L/second
= 480 L/minute (multiply both by 60)
= 28 800 L/h (multiply both by 60)
= 28.8 kL/h

∴ The pump pumps at 28.8 kL/h.

b 1 megalitre = 1000 kilolitres

∴ time to pump 1 ML
$= \frac{1000}{28.8}$ h
= 34.722 222 (hours)
= 34 hours 43 minutes 20 seconds

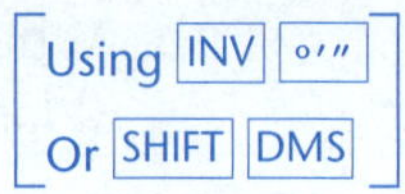


∴ Tuesday 6 am plus 1 day 10 hours 43 min 20 sec
= Wednesday 4:43:20 pm

∴ The pump will have pumped 1 ML on Wednesday afternoon at 4:43:20 p.m.

## Distance, Speed, Time

$$\text{Speed} = \frac{\text{Distance}}{\text{Time}} \qquad S = \frac{D}{T}$$

$$\text{Time} = \frac{\text{Distance}}{\text{Speed}} \qquad T = \frac{D}{S}$$

$$\text{Distance} = \text{Speed} \times \text{Time} \qquad D = S \times T$$

For Example

1 A car travels 360 km in 5 hours. What is its average speed in km/h?

2 I can walk at a speed of 8 km/h. How far will I walk in $3\frac{1}{2}$ hours?

3 A train travels 400 kilometres and averages 75 km/h. How long will the journey take?

4 A car left Cessnock at 2:30 pm and averaged 80 km/h to arrive at its destination at 10 pm the same day. How far did the car travel?

5 George left his home at 12:20 pm and travelled 175 km to Pokolbin arriving there at 2:40 pm. What was his average speed?

1 Average speed $= \frac{\text{Distance}}{\text{Time}}$

$= \frac{360}{5}$

$= 72$

$\therefore$ The car's average speed is 72 km/h.

2 Distance = Speed × Time

$= 8 \times 3\frac{1}{2}$

$= 8 \times \frac{7}{2}$

$= 28$

$\therefore$ I can walk 28 km in that time.

3 Time $= \frac{\text{Distance}}{\text{Speed}}$

$= \frac{400}{75}$

$= 5.\dot{3}\ (5\frac{1}{3})$

$\therefore$ The journey will take $5\frac{1}{3}$ hours, or 5 hours 20 minutes.

4 Time taken = 2.30 pm to 10 pm

$= 7\frac{1}{2}$ hrs

$\therefore$ Distance $= 80 \times 7\frac{1}{2}$

$= 600$

$\therefore$ The car travelled 600 km.

5 Time taken = 12:20 pm to 2:40 pm

= 2 h 20 min

$= 2\frac{1}{3}$ h

$\therefore$ Speed $= \frac{175}{2\frac{1}{3}}$ $\quad \left[175 \div 2\frac{1}{3} = 175 \times \frac{3}{7}\right]$

$= 75$

$\therefore$ George's average speed was 75 km/h.

## Scale Drawing

All scale drawings have the same shape as the original but are either reduced or enlarged. The **ratio** of the scale drawing to the original is called the **scale**.

This scale can be expressed in a variety of ways:

1 cm to 10 metres, $\frac{1}{100}$, 1:100 000, etc.

## For Example

1 Write each of the following in ratio form:

a 1 cm to 1 m  b 2 cm to 1 m

c 1 cm to 10 m  d 5 cm to 1 km

2 Copy and complete:

a 1:100 = 1 cm:_______ m

b 1:100 000 = 1 cm:_______ km

c 1:1 000 000 = 1 cm:_______ km

3 Using a scale of 1:100, what is the real length represented by the following:

a 1 cm

b 5.7 cm

c 6 mm

4 The plan of the house has been drawn using a scale of 1 cm to 2 m.

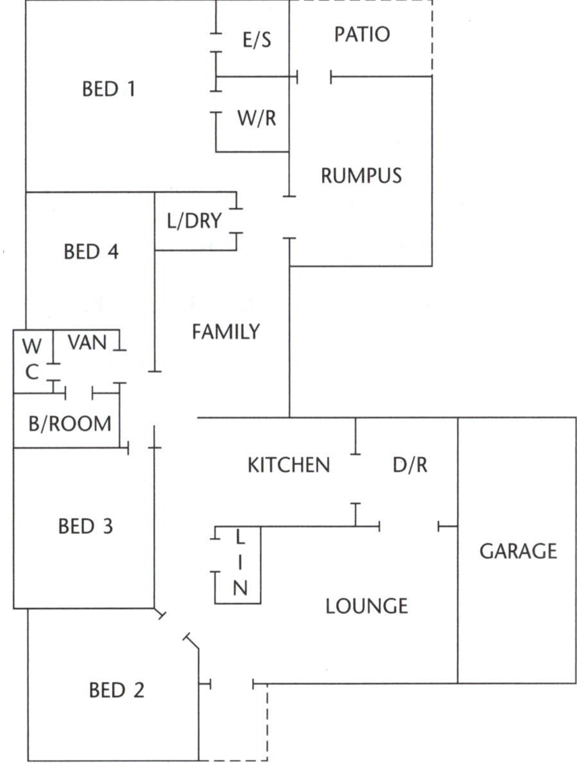

a Express the scale in ratio form.

b What are dimensions of the garage?

c The rumpus room is to be carpeted. Find the area to be carpeted.

d If the back patio is to be tiled, find the total cost, if the tiles cost $\$42/m^2$.

5 The rumpus room in the house above is to be furnished. (Each square in the diagram is 5 mm by 5 mm.)

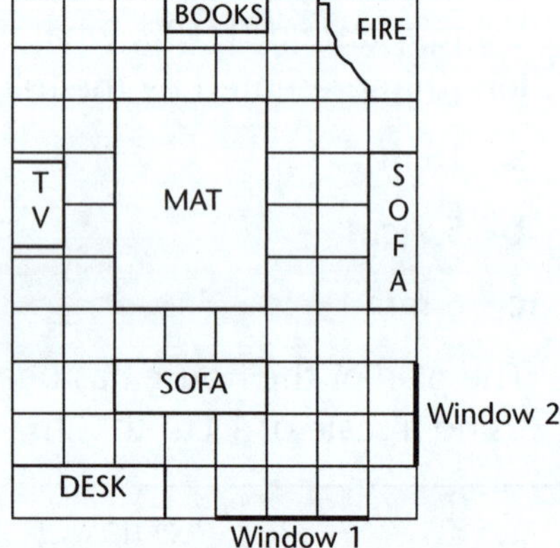


The scale used is 1 cm = 1 m.

a Express this scale as a ratio.

b What is the width of window 1?

c Find the area of the mat.

d Find the length of the desk.

1 a 1 cm:1 m = 1 cm:100 cm
= 1:100

b 2 cm:1 m = 2 cm:100 cm
= 2:100
= 1:50

c 1 cm:10 m = 1 cm:1000 cm
= 1:1000

d 5 cm:1 km = 5 cm:1000 m
= 5 cm:100 000 cm
= 5:100 000
= 1:20 000

2 a 1:100 = 1 cm:100 cm
= 1 cm:1 m

b 1:100 000 = 1 cm:100 000 cm
= 1 cm:1000 m
= 1 cm:1 km

c 1:1 000 000 = 1 cm:1 000 000 cm
= 1 cm:10 000 m
= 1 cm:10 km

3 a Length = $1 \times 100$
Now 100 cm = 1 m
∴ It would represent 1 metre.

b $5.7 \times 100 = 570$
∴ 570 cm = 5.7 m
It would represent 5.7 m.

c $6 \times 100 = 600$
∴ 600 mm = 60 cm
= 0.6 m
∴ It would represent 0.6 m.

4 a 1 cm to 2 m = 1 cm:2 m
= 1 cm:200 cm
= 1:200
∴ The scale in ratio form is 1:200.

b Garage measures 3.5 cm by 1.6 cm.
∴ Length = $3.5 \times 200$
= 700 cm
= 7 m
Width = $1.6 \times 200$
= 320 cm
= 3.2 m
∴ The garage is 7 m × 3.2 m.

**c** Rumpus room measurements are 2.5 cm by 2 cm.

$\therefore$ Length $= 2.5 \times 200$
$= 500$ cm
$= 5$ m

Width $= 2 \times 200$
$= 400$ cm
$= 4$ m

$\therefore$ Area $= 5 \times 4$
$= 20\ \text{m}^2$

$\therefore$ The area to be carpeted is $20\ \text{m}^2$.

**d** Back patio measurements are 2 cm by 1 cm.

$\therefore$ Length $= 2 \times 200$
$= 400$ cm
$= 4$ m

Width $= 1 \times 200$
$= 200$ cm
$= 2$ m

$\therefore$ Area $= 4 \times 2$
$= 8\ \text{m}^2$

$\therefore$ The area to be tiled is $8\ \text{m}^2$.

Now cost = \$42 × 8 = \$336

$\therefore$ The total cost of tiles will be \$336.

**5** **a** 1 cm to 1 m = 1 cm:100 cm
= 1:100

$\therefore$ The scale in ratio form is 1:100.

**b** Window 1 = 1.5 cm
i.e. 1.5 metres

$\therefore$ The width of window 1 is 1.5 metres.

**c** Mat is 2 cm by 1.5 cm
i.e. 2 m by 1.5 m

$\therefore$ Area $= 2 \times 1.5$
$= 3$

$\therefore$ The area of the mat is $3\ \text{m}^2$.

**d** Desk is 1.5 cm
i.e. 1.5 metres

$\therefore$ The desk is 1.5 metres long.

## PRACTISE, PRACTISE

Go to p. 288 for quick answers, or to pp. 354–359 for worked solutions.

**1** Express the following in simplest form: pp. 176–177

**a** 25:15 **b** 30:70 **c** 9:18:27
**d** 14:49 **e** 100:100 000 **f** 75:125

**2** Find the value of the pronumeral: p. 177

**a** 6:2 = $x$:1 **b** 15:45 = 1:$x$ **c** 12:8 = 3:$x$
**d** 7:5 = $x$:35 **e** 13:3 = $x$:9

**3** Simplify the following: pp. 176–178

**a** 3.5:1.5 **b** 2:1.4 **c** 8.4:4.2
**d** 3.25:4.25 **e** $\frac{3}{4}:\frac{1}{2}$ **f** $\frac{2}{3}:\frac{5}{6}$
**g** $\frac{3}{5}:1\frac{1}{2}$ **h** $2\frac{1}{4}:2\frac{1}{2}$ **i** $\frac{4}{5}$:2.2

**4** Simplify the following ratios by first changing to the same units: pp. 176–178

**a** \$1.50:40c **b** \$2.40:\$1.60
**c** 8 hours:2 days **d** 1 mm:1 cm:1 m
**e** 3 weeks:6 days **f** $1\frac{1}{2}$ min:45 sec
**g** 1 mL:1 kL **h** 3 decades:2 centuries

**5** Express in simplest form: pp. 176–178

**a** $12ab:6a$ **b** $3xy:15xy$
**c** $2a^2:(2a)^2$ **d** $\frac{3}{p}:3p$

**6** A motorbike was bought for \$700 and after 2 years was sold for \$900. Find the: pp. 178–180

**a** profit
**b** ratio of the:
  **i** cost price to selling price
  **ii** profit to cost price.

**7** A school has a teaching staff of 56. If 31 are male, find the: pp. 178–180

**a** number of female teachers
**b** ratio of the:
  **i** male teachers to female teachers
  **ii** total teaching staff to female teachers.

**8** On a particular day, there are 10 hours of sunlight. What is the ratio of the number of hours of sunlight to hours in the day? pp. 178–180

**9** A bag contains six black, four white and ten red marbles. What is the ratio of: pp. 178–180

**a** black to red marbles?
**b** red to non-red marbles?

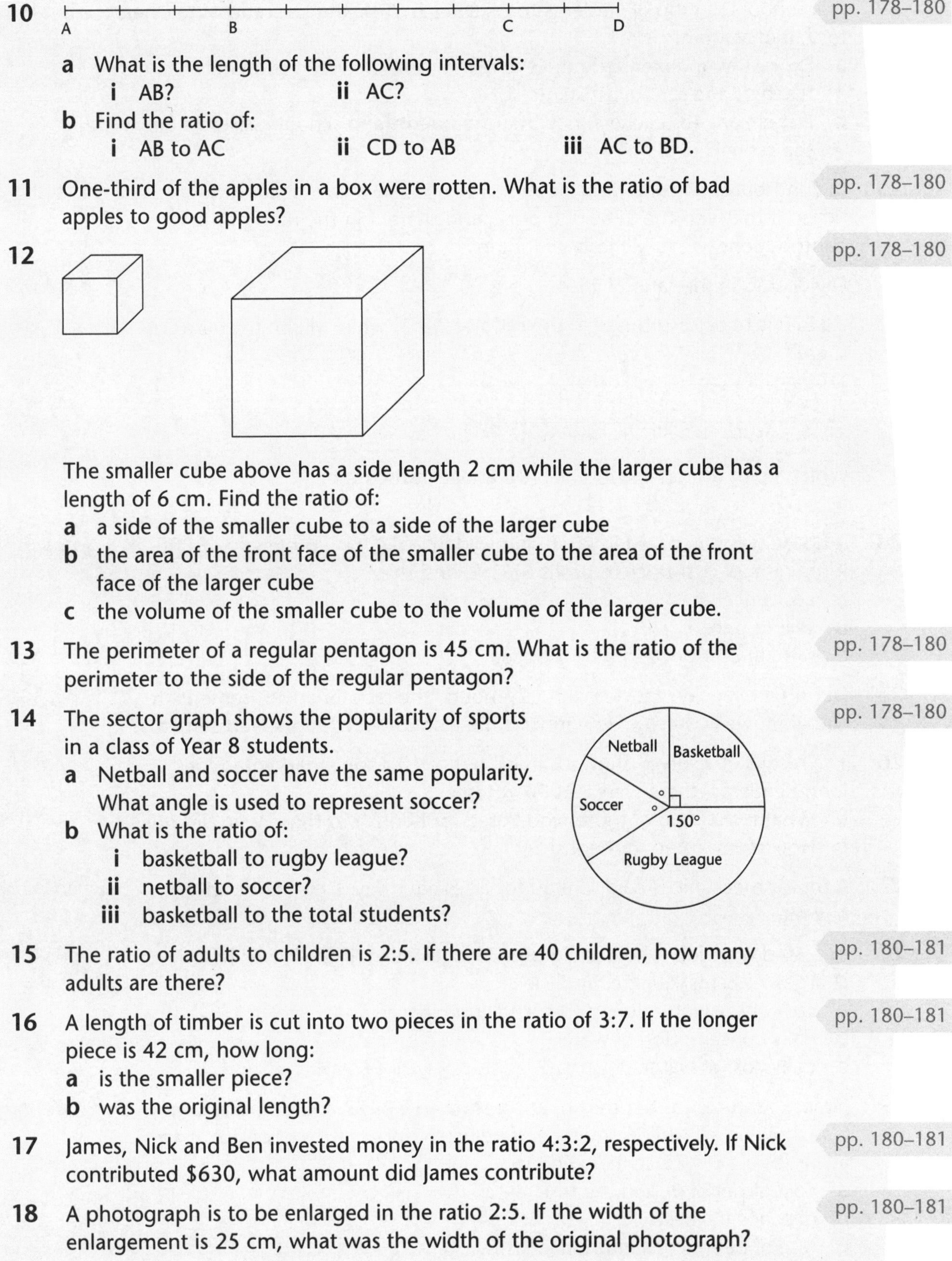

**10** pp. 178–180



**a** What is the length of the following intervals:
  **i** AB? **ii** AC?

**b** Find the ratio of:
  **i** AB to AC **ii** CD to AB **iii** AC to BD.

**11** One-third of the apples in a box were rotten. What is the ratio of bad apples to good apples? pp. 178–180

**12** pp. 178–180

The smaller cube above has a side length 2 cm while the larger cube has a length of 6 cm. Find the ratio of:

**a** a side of the smaller cube to a side of the larger cube

**b** the area of the front face of the smaller cube to the area of the front face of the larger cube

**c** the volume of the smaller cube to the volume of the larger cube.

**13** The perimeter of a regular pentagon is 45 cm. What is the ratio of the perimeter to the side of the regular pentagon? pp. 178–180

**14** The sector graph shows the popularity of sports in a class of Year 8 students. pp. 178–180



**a** Netball and soccer have the same popularity. What angle is used to represent soccer?

**b** What is the ratio of:
  **i** basketball to rugby league?
  **ii** netball to soccer?
  **iii** basketball to the total students?

**15** The ratio of adults to children is 2:5. If there are 40 children, how many adults are there? pp. 180–181

**16** A length of timber is cut into two pieces in the ratio of 3:7. If the longer piece is 42 cm, how long: pp. 180–181

**a** is the smaller piece?

**b** was the original length?

**17** James, Nick and Ben invested money in the ratio 4:3:2, respectively. If Nick contributed $630, what amount did James contribute? pp. 180–181

**18** A photograph is to be enlarged in the ratio 2:5. If the width of the enlargement is 25 cm, what was the width of the original photograph? pp. 180–181

**19** In a school the ratio of students to teachers is 20:1 and the ratio of teachers to school assistants is 4:1. pp. 180–181

**a** Express as an extended ratio the relationship between students, teachers and school assistants.

**b** If there are 48 teachers, how many students and school assistants are at the school?

**20** Jo contributes 80c and Josh the remainder to buy a two-dollar lottery ticket. If they win the $200 000 prize, and share it in the ratio of their contributions, what will Josh's winnings be? pp. 181–182

**21** Divide $75 in the ratio of 11:4. pp. 181–182

**22** If $270 is to be distributed in the ratio of 2:4:3, what will be the largest share? pp. 181–182

**23** pp. 181–182

A B C D E F G H I J K L M

Which point divides the interval AM above in the ratio:

**a** 5:7? **b** 1:11? **c** 1:1? **d** 3:1?

**24** A house was purchased in 2010 and sold in 2014 for a profit of $45 000. If the ratio of cost price to profit is 61:5, find the: pp. 181–182

**a** cost price

**b** selling price

**c** ratio of cost price to selling price.

**25** During a television movie lasting $2\frac{1}{2}$ hours, the ratio of advertisements to actual movie was 4:11. How much time is spent watching advertisements? pp. 181–182

**26** The results of a survey showed that 3 out of 10 people had voted for John Kentish. If there were 40 000 votes: pp. 181–182

**a** What is the ratio of those who voted for Kentish to those who did not?

**b** How many voted for Kentish?

**27** A motor mechanic is paid $99.60 for an 8-hour day. Find her hourly pay rate. pp. 182–183

**28** In a 30-day period a household uses 39.6 kilolitres. If the Water Board charges 75 cents/kilolitre, find the: pp. 182–183

**a** cost of the water during the 30-day period

**b** daily consumption of water

**c** daily cost of water.

**29** James' odometer at the start of a trip showed 89 472 and on his return was showing 91 386 (kilometres). If he used 239.25 litres of petrol at an average cost of 172 cents/L, find the: pp. 182–183

**a** cost of petrol during the trip

**b** distance travelled

**c** rate of petrol consumption in km/L

**d** cost of travelling per kilometre.

**30** The council charges a land rate of 0.9275 cents per dollar of valuation. Find the rates charged on a property valued at $190 000. pp. 182–183

**31** A bricklayer can lay 480 bricks in 6 hours. How many bricks would he lay, working at a similar rate, over a 10-hour period? pp. 182–183

**32** Three men take 4 hours to build a fence 20 metres long. Working at a similar speed, how long would it take: pp. 182–183

**a** one man to build the 20-metre fence?
**b** four men to build a 10-metre fence?

**33** Determine the better way to buy honey? pp. 182–183

**A** 250 g jar for $1.80 **B** 375 g jar for $2.55

**34** Which is the best way to purchase washing powder? pp. 182–183

**A** 1 kg for $2.31 **B** 1.5 kg for $3.60
**C** 2 kg for $4.80

**35** Convert: pp. 183–184

**a** 20 m/sec to km/h **b** 16 L/h to kL/day
**c** 3.4 cm/min to m/week.

**36** Convert: pp. 183–184

**a** 84 km/h to m/sec **b** 10.08 kL/week to L/min.

**37** Driving at 100 km/h on the Sydney–Newcastle freeway, Jordie passes the following sign. pp. 184–185

**Gosford**
**Exit**
**2 km**

If he maintains his speed, how long will it be until he leaves the freeway and turns off to Gosford?

**38** An 8-minute shower uses 250 litres of water. If Alice has two 8-minute showers a day and water costs 75c/kilolitre, what will be the cost of her showers for a year (365 days)? pp. 183–184

**39** Complete the table: pp. 184–185

| Distance | Speed | Time |
|---|---|---|
| 300 km | 60 km/h | |
| | 52 km/h | 3 hour |
| 760 km | | 8 hour |
| 540 km | 216 km/h | |
| | 60 km/h | 20 min |
| 120 km | | 30 min |

**40** Francisco leaves Wangaratta at 2:00 pm and arrives in Swan Hill, 320 kilometres away, after averaging 80 km/h. At what time does he arrive in Swan Hill? pp. 184–185

**41** A motorist travels from town A to town B, a distance of 25 kilometres in 20 minutes. Find her speed in km/h. pp. 184–185

**42** Jenny leaves Port Augusta at 1:30 pm and travels 540 kilometres, averaging 80 km/h. At what time does Jenny arrive at her destination? pp. 184–185

**43** pp. 184–185

| Sydney–Brisbane | |
|---|---|
| | Outward total (km) |
| Sydney | 0 |
| Hexham | 166 |
| Taree | 318 |
| Kempsey | 437 |
| Coffs Harbour | 548 |
| Ballina | 760 |
| Brisbane | 975 |

Car A leaves Sydney at midnight and averages 75 km/h. Car B leaves Sydney later that morning at 2 am and averages 90 km/h. Which car will get to Brisbane first, and how much later will the second car arrive?

**44** Express in ratio form: pp. 185–187

**a** 4 cm to 1 m **b** 1 cm to 3 km
**c** 5 mm to 1 km **d** 3 cm to 2 m

**45** A house plan has been drawn using a scale of 1:100. Find the real size of a: pp. 185–187

**a** lounge room of length 42 mm
**b** bedroom with width 31 mm
**c** shed of length 6.2 cm.

**46** A map is drawn using a scale of 1 cm:25 km. pp. 185–187

**a** Find the actual distance, if on the map:
  **i** Adelaide and Murray Bridge are 3 cm apart
  **ii** Southport and Rockhampton are 9.6 cm apart.
**b** Find the distance on the map if, in reality:
  **i** Mandurah and Perth are 75 km apart
  **ii** Devonport and Stanley are 125 km apart.

**47** The diagram is a scale drawing of a school hall that has a length of 35 metres. pp. 185–187

**a** By measuring the length of the hall, express the scale used as a ratio.
**b** What is the actual width of the hall?
**c** What are the actual dimensions of the dance floor?

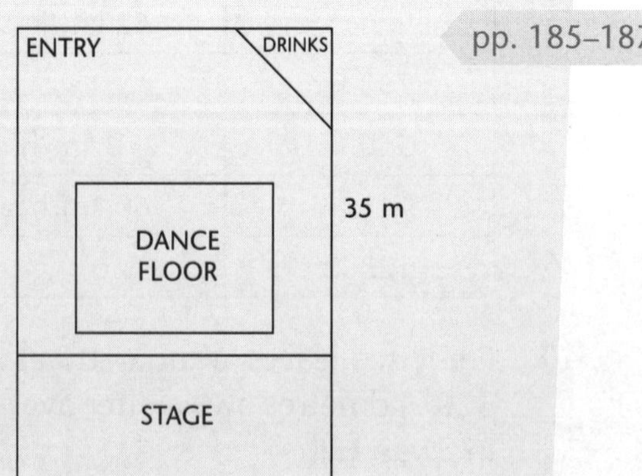

pp. 185–187

**48** The scale drawing shows the position of a house on a suburban block. The scale used is 1:500.

**a** Express the scale as 1 cm = ________ m.

**b** What is the real width of the front of the house?

**c** The owners of the house are considering installing a backyard pool that measures 25 metres by 12.5 metres. Will they be able to fit it into their backyard?

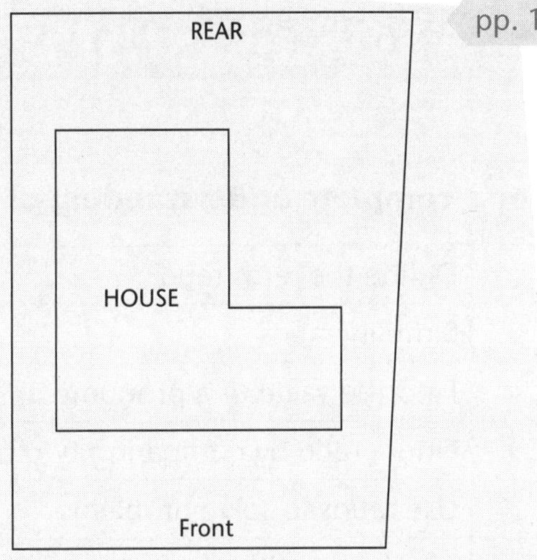


Go to p. 288 for quick answers, or to pp. 354–359 for worked solutions.

## YOUR CHECKLIST

**For a complete understanding of this topic you must be able to:**

| | | | |
|---|---|---|---|
| ✓ | Define the term ratio | | p. 176 |
| ✓ | Simplify ratios | | pp. 176–178 |
| ✓ | Find the value of a pronumeral in equivalent ratios | | p. 177 |
| ✓ | Find a ratio by comparing two or more quantities | | pp. 178–180 |
| ✓ | Use ratios to solve problems | | pp. 178–181 |
| ✓ | Divide quantities in a given ratio | | pp. 181–182 |
| ✓ | Define the term rate | | p. 182 |
| ✓ | Calculate a rate from given information | | pp. 182–184 |
| ✓ | Solve real-life problems involving distance, speed and time | | pp. 184–185 |
| ✓ | Determine the scale factor for a pair of similar figures. | | pp. 185–187 |

**Now you are ready to do the tests!**

**(25 marks)**

1 Simplify:

a 3:12 b 45:35 c $2:20c (3 marks)

2 Number line with points A, B, C, D, E, F, G, H, I

Find the ratio of:

a AB:EF b CE:AI c BG:AF. (3 marks)

3 A class contains 16 boys and 14 girls. Find the ratio of:

a boys to girls b girls to total number of students. (2 marks)

4 An orchard has 120 citrus trees, comprising orange and mandarin trees. If there are 80 orange trees, find the ratio of:

a orange trees to mandarin trees b total trees to orange trees. (2 marks)

5 In a cordial, the ratio of concentrate to water is 1:3. If there is 270 mL of water, how much concentrate is in the cordial? (3 marks)

6 A small town has a population of 464. If the ratio of adults to children in the town is 5:3, find the number of adults in the town. (3 marks)

7 A 30-metre cable is cut in the ratio 2:3. Find the longer length. (2 marks)

8 A car travels at a speed of 80 km/h for 3 hours. Find the distance travelled. (2 marks)

9 Fiona travels 560 km in 8 hours. Find her average speed. (2 marks)

10 A street directory has a scale of 1 cm = 400 m.

a Express the scale as a ratio in the form 1:________

b The post office and police station are 2 km apart. Find the distance on the street directory.

c The primary school and the bus depot are 6.4 cm apart on the street directory. Find the distance between the two places in kilometres. (3 marks)

☞ Quick answers on page 293
☞ Worked solutions on page 392

**Your Feedback** $\frac{\square}{25} \times 100\% = \square\%$

**(25 marks)**

1 Simplify:
 a $3.20:$4  b 15:12:6  c 1 metre:10 kilometre (3 marks)

2 Find the value of $x$:
 a $3:x = 12:16$  b $5:15 = x:30$ (2 marks)

3 Of 72 blocks of land in a new subdivision, 24 are already sold. Find the ratio of total blocks to unsold blocks. (2 marks)

4 At a school dance the ratio of teachers to students is 1:35. If there are 280 students, how many teachers are there? (2 marks)

5 Divide 540 g in the ratio of 7:2. (2 marks)

6 A piece of timber is cut in the ratio of 2:3:4. If the longest piece is 1.2 m, how long is the shortest piece? (2 marks)

7 A farmer planted 480 ha of land with wheat, oats and barley, in the ratio of 3:5:4. Find the area planted with oats. (2 marks)

8 A tank has a capacity of 2460 L. If it can be filled at the rate of 1.2 L/second, how long will it take to fill? (2 marks)

9 Sally travels from Coober Pedy to Renmark. She covers the 1050 km distance at an average speed of 90 km/h. How long did the trip take? (2 marks)

10 Brett left Albany at 7:30 am and drove 360 km to Bunbury. If he arrived at midday, what was his average speed? (2 marks)

11 A river is flowing at the rate of 6 km/h. How long would it take for a floating log to travel 15 km? (2 marks)

12 A map has a scale of 1:100 000. How far apart are two towns if the distance between them on the map is 4.5 cm? (2 marks)

☞ Quick answers on page 293
☞ Worked solutions on page 392

**Your Feedback** $\frac{\square}{25} \times 100\% = \square\%$

**(25 marks)**

1 Simplify:

a $\frac{3}{4} = 1$ b $5000 \text{ m}^2 = 2 \text{ ha}$ c $\frac{2}{3}:\frac{3}{4}$ (3 marks)

2 Find the value of $x$:

a $x:4 = 48:64$ b $25:35 = 10:x$ (2 marks)

3 A triangle has two angles of 60° and 40°. Find the ratio of the largest angle to the smallest angle in the triangle. (1 marks)

4 A DVD recorder is priced at $650, but is discounted by 15%. Express the discount as a ratio of the original price. (1 marks)

5 Divide $360 in the ratio of 2:3:5. (2 marks)

6 The perimeter of a rectangle is 80 cm and its sides are in the ratio of 2:3. Find the length of each side. (2 marks)

7 An alloy contains copper and iron in the ratio of 3:4. If the quantity of iron is 5.6 kg, find the amount of copper required to make the alloy. (2 marks)

8 A manufacturer produces aluminium cans at the rate of 240 cans/minute. How long would it take to produce 58 600 cans? (2 marks)

9 Mark walks at a speed of 5.2 km/h. How far will he walk in 25 minutes? Answer to nearest m. (2 marks)

10 John travels at 60 km/h for a distance of 500 km. Find the time taken for the journey. (2 marks)

11 An island is 3.6 km wide. How long would it be on a map using a scale of 1:40 000? (2 marks)

12 A survey is conducted with a group of students to find their mode of transport to school. The ratio of train travellers to bus travellers was 3:2, while the ratio of bus travellers to car travellers was 5:2.

a Find the ratio of train travellers to bus travellers to car travellers.

b If there were 48 car travellers, how many students went by train? (4 marks)

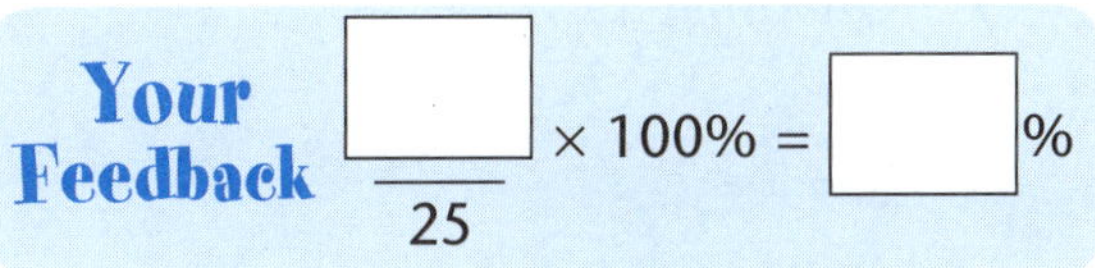


☞ Quick answers on page 293
☞ Worked solutions on page 393

# 10

# GEOMETRY AND CONGRUENCE

- Revision of Basic Geometrical Properties of Angles
- Parallel Lines
- Triangles
- Quadrilaterals
- Numerical Examples Based on Angle Properties
- Examples in Simple Deductive Geometry of Non-numerical Type of Questions
- Congruence
- Congruent Triangles

## KEYWORDS

| | |
|---|---|
| Alternate | Rectangle |
| Co-interior | Reflection |
| Complementary | Rhombus |
| Congruent | Rotation |
| Corresponding | Scale drawing |
| Deductive geometry | Similar |
| Enlargement | Square |
| Equilateral | Superimpose |
| Exterior angle | Supplementary |
| Geometry | Transformation |
| Isosceles | Translation |
| Kite | Transversal |
| Parallelogram | Trapezium |
| Proportion | Vertically opposite |
| Quadrilateral | |

## Revision of Basic Geometrical Properties of Angles

a **Complementary** angles are two angles that add up to 90°.

b **Supplementary** angles are two angles that add up to 180°.

c **Angles on a straight line:**

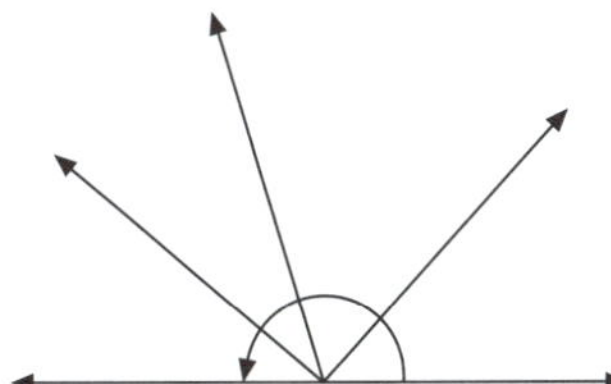

Angles on a straight line add up to 180°.

d **Vertically opposite angles:**

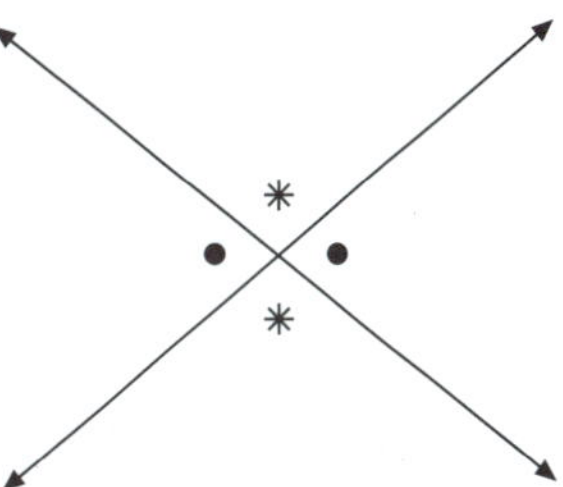

When two lines intersect, two pairs of **vertically opposite** angles are formed. Vertically opposite angles are equal.

e **Angles at a point:**

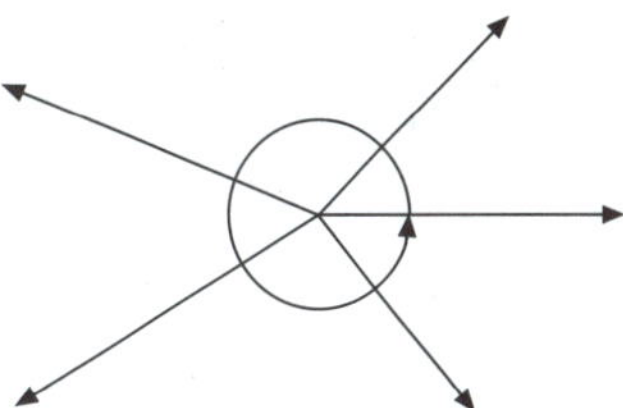

Angles at a point add up to 360°.

## Parallel Lines

When a pair of parallel lines is cut by a **transversal**, three types of special angles are formed. They are formed in pairs.

a

**Corresponding** angles are equal, (F angles).

b

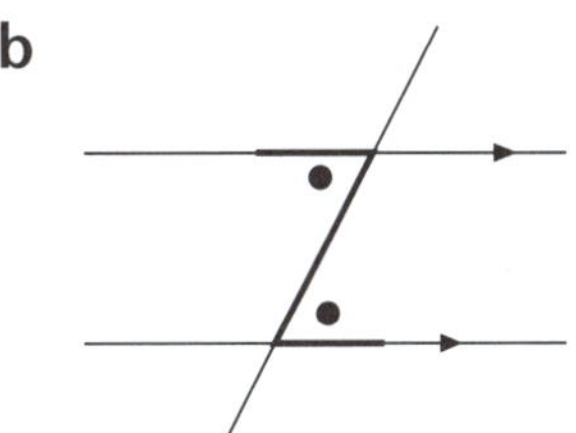

**Alternate** angles are equal, (Z angles).

c

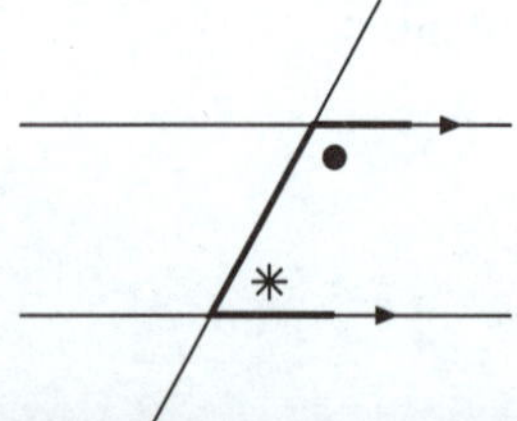

**Co-interior** angles are supplementary (that is, they add up to 180°), (C angles).

### Tests for Parallel Lines

Two lines are parallel if:

- a pair of corresponding angles are equal; or
- a pair of alternate angles are equal; or
- a pair of co-interior angles are supplementary.

## Triangles

**a** Angle sum of a triangle:

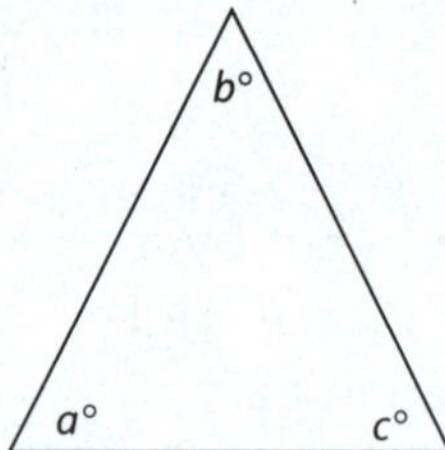


The angle sum of a triangle is 180°.

That is: $a + b + c = 180$

**b** Isosceles triangle:

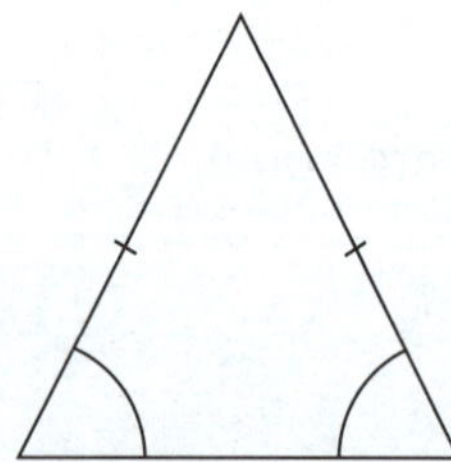

The base angles of an **isosceles triangle** are equal.

**c** Equilateral triangle:

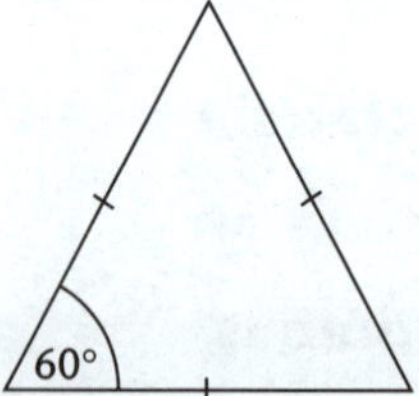


All sides are equal. Any angle in an **equilateral triangle** is 60°.

**d** Exterior angle of a triangle:

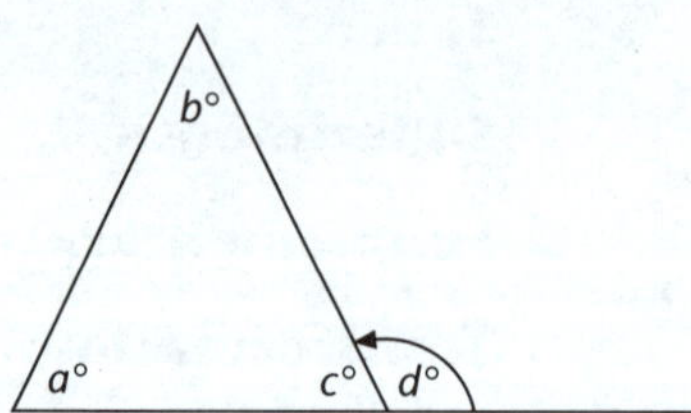


The **exterior angle** of a triangle is equal to the sum of the opposite interior angles.

That is: $d = a + b$

## Quadrilaterals

### Angle Sum of Quadrilaterals

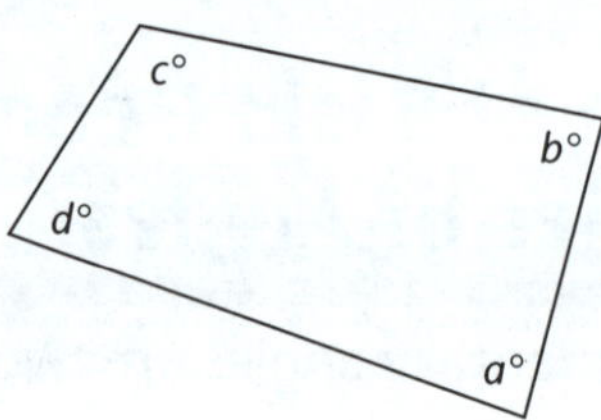


The angle sum of any quadrilateral is 360°.

That is: $a + b + c + d = 360$

### Properties of Special Quadrilaterals

A **trapezium** is a quadrilateral with one pair of opposite sides parallel:

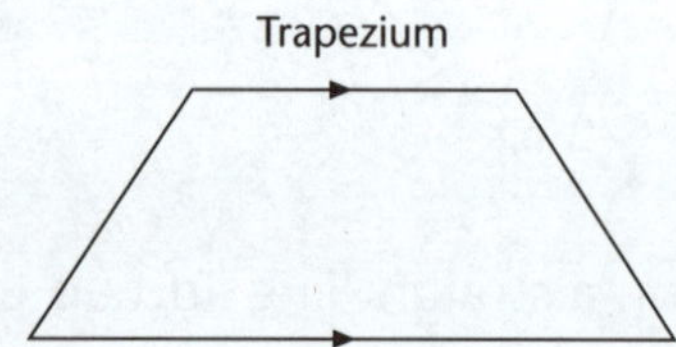

A **parallelogram** is a quadrilateral with both pairs of opposite sides parallel and equal:

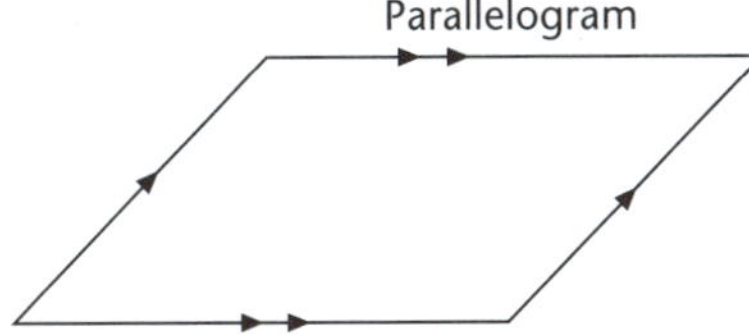


A **rhombus** is a quadrilateral with both pairs of opposite sides parallel and all sides equal:

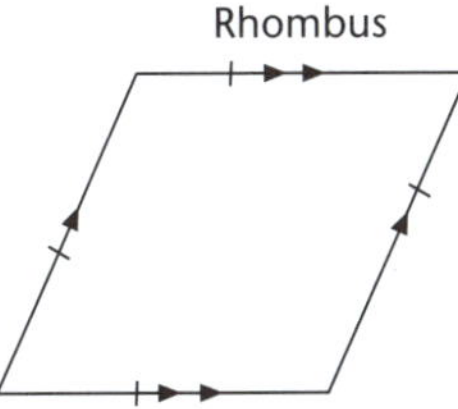


A **rectangle** is a quadrilateral with both pairs of opposite sides parallel and equal and all four angles right-angles:

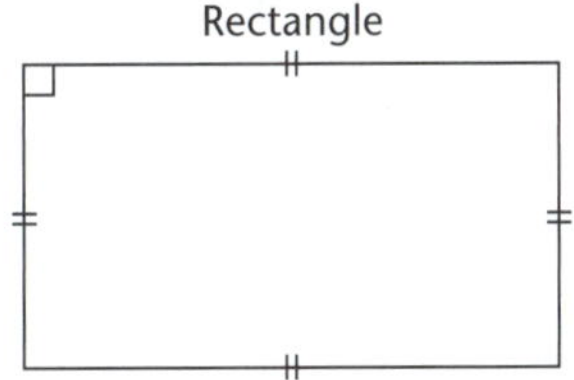


A **square** is a quadrilateral with all four sides equal and all four angles right-angles:

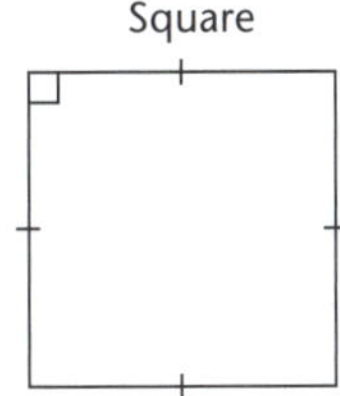


A **kite** is a quadrilateral with two pairs of adjacent sides equal:

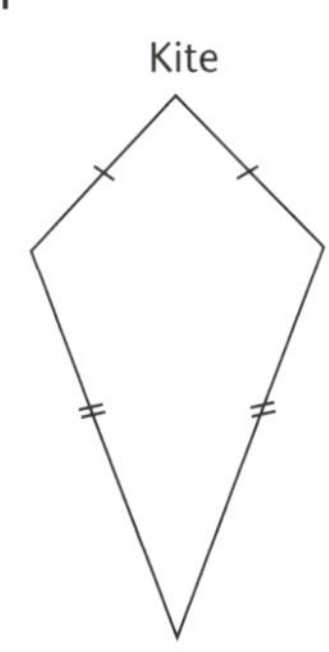


## Family of Quadrilaterals

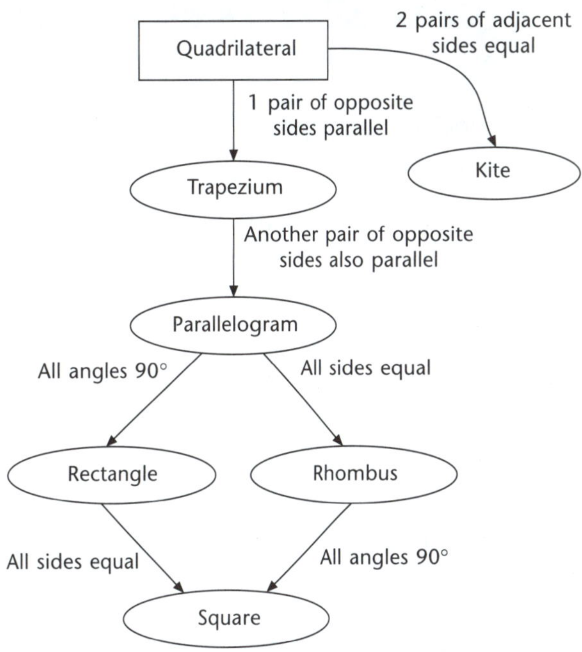


## Numerical Examples Based on Angle Properties

The value of each pronumeral is found using geometric rules.

**1** Find $x$, giving reasons for your answer.

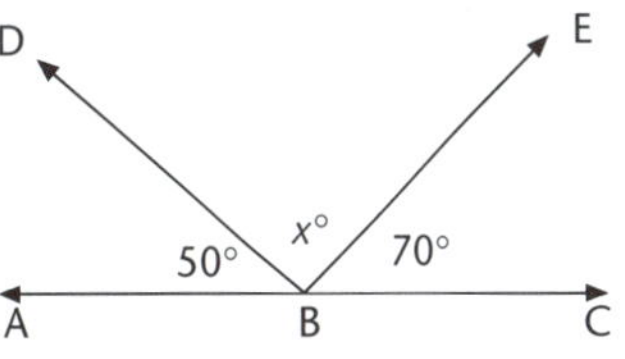


**1** $x + 50 + 70 = 180$

$\therefore x = 60$

[Angles on a straight line.]

## For Example

1 Find $x$ and $y$, giving reasons for your answer.

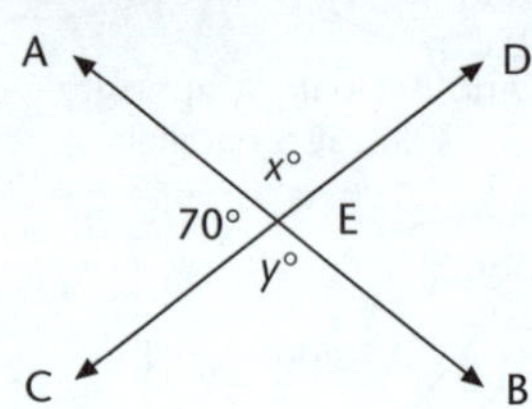


1 The lines AB and CD intersect at E.

$x = 110$ [∠AED is supplementary to ∠CEA.]

$y = 110$ [∠BEC is vertically opposite to ∠AED.]

## For Example

1 Find $x$, giving reasons for your answer.

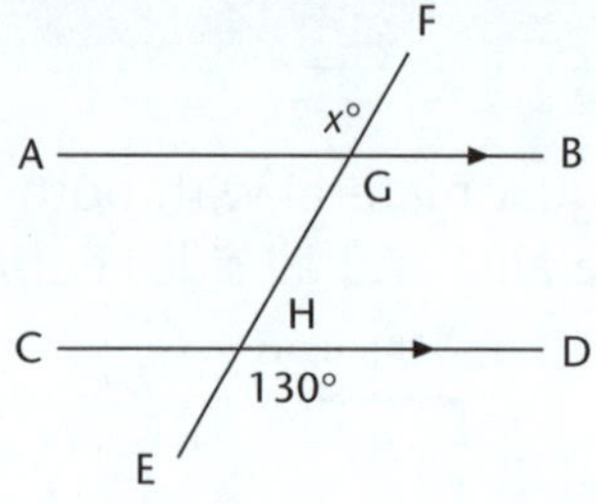


1 ∠CHG = 130° [Vertically opposite to ∠DHE.]

$x = 130$ [∠AGF is corresponding to ∠CHG and AB || CD.]

## For Example

1 Find $x$, giving reasons for your answer.

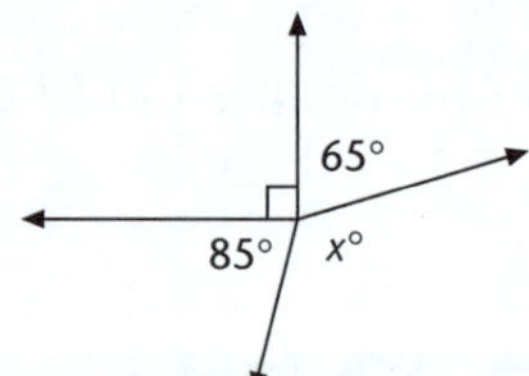


1 $x + 85 + 90 + 65 = 360$ [Angles at a point.]

$\therefore x + 240 = 360$

$x = 120$

## For Example

1 Find $x$, giving reasons for your answer.

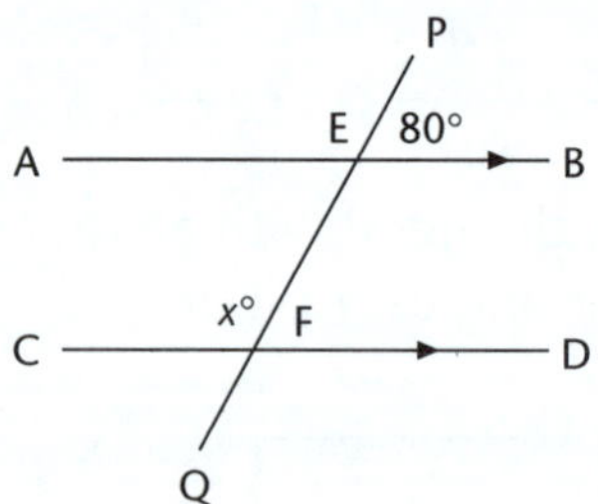


1 ∠FEA = 80° [Vertically opposite to ∠PEB.]

$x = 100$ [∠CFE is co-interior to ∠FEA and AB || CD.]

## For Example

**1** Find the size of ∠CEA, giving reasons for your answer.

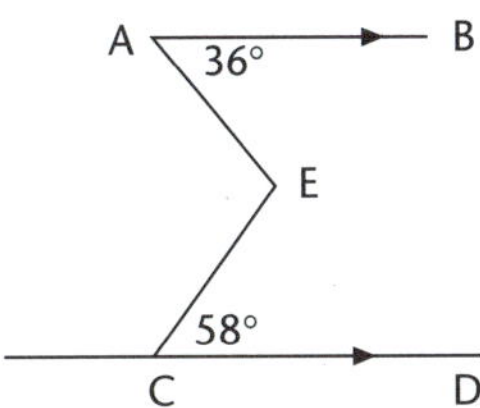


**1** Construct a line PQ through E which is parallel to AB*:

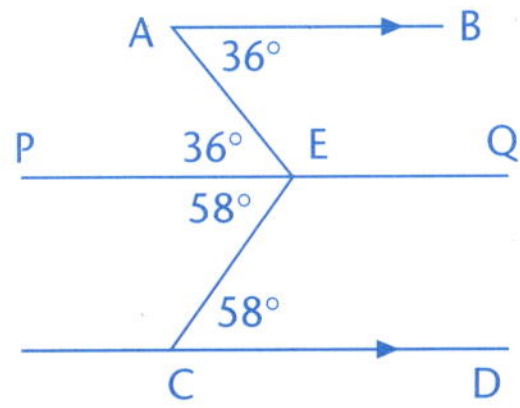


∠PEA = 36°

[Alternate to ∠BAE and AB || PQ.]

∠CEP = 58°

[Alternate to ∠ECD and CD || PQ.]

∴ ∠CEA = (36 + 58)° = 94°.

*Note: if two lines are parallel to a third line, they are parallel to one another.

## For Example

**1** Is line AB parallel to line CD? Give reasons for your answer.

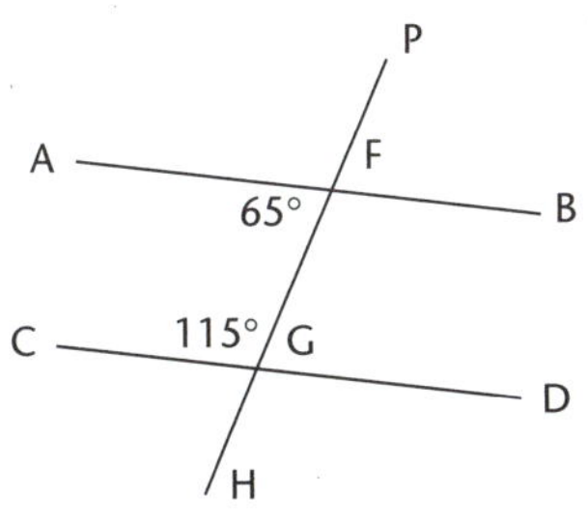


**1** AB || CD because a pair of co-interior angles are supplementary.

That is, ∠CGF + ∠GFA = (115 + 65)° = 180°.

## For Example

**1** Find $x$, giving reasons for your answer.

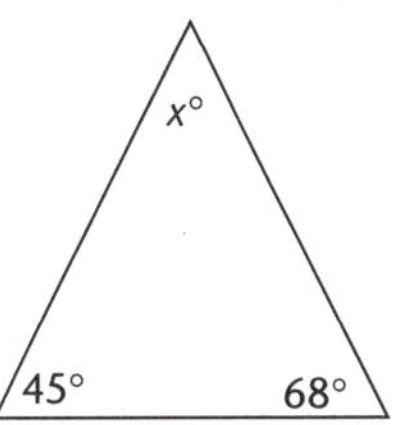


**1** $x + 68 + 45 = 180$ [Angle sum of triangle.]

$\therefore x + 113 = 180$

$x = 67$

## For Example

**1** In △ABC, AB = AC and ∠ABC = 63°. Find $x$, giving reasons for your answer.

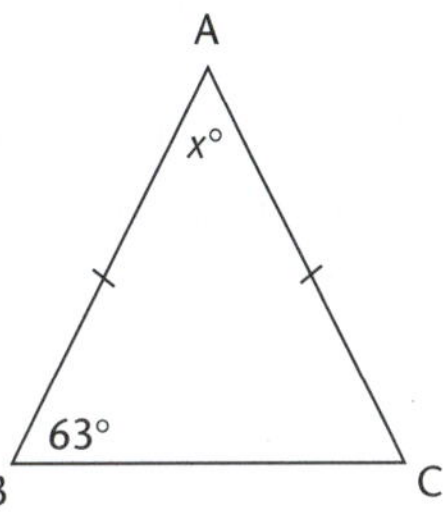


**1** ∠BCA = 63°

[Base angles of isosceles triangle ABC.]

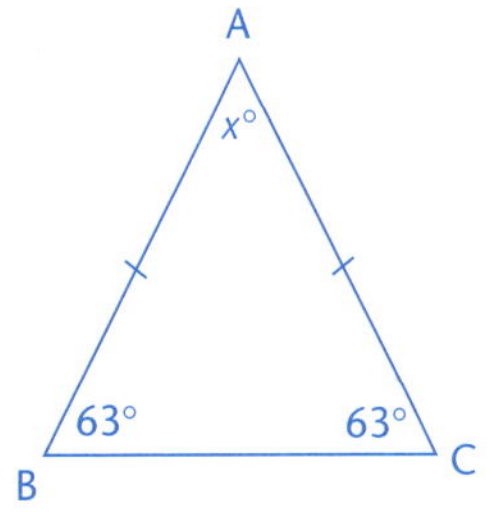

$\therefore x + 63 + 63 = 180$ [Angle sum of $\triangle$ABC.]

$\therefore x + 126 = 180$

$\therefore x = 54$

## For Example

**1** Find $x$, giving reasons for your answer.

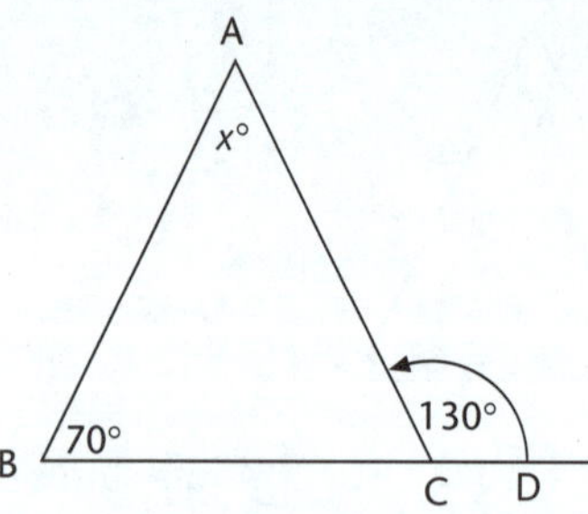


**1** $x + 70 = 130$

$\therefore x = 60$

[Exterior angle of a triangle is equal to the sum of the opposite interior angles.]

## For Example

**1** ABC is an equilateral triangle and BC || FG. Find $x$, giving reasons for your answer.

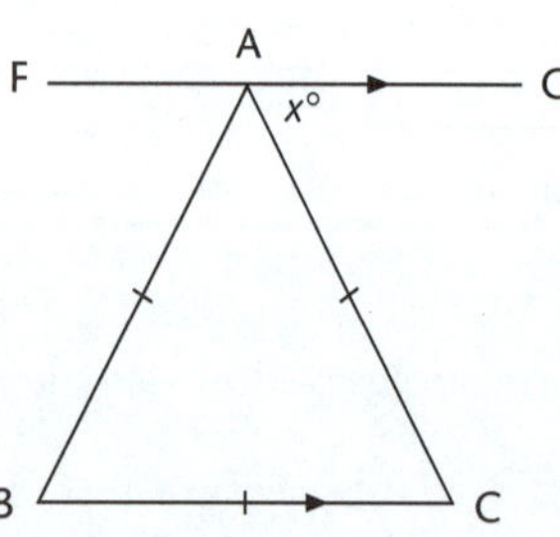


**1** $\angle BCA = 60°$

[Angle in an equilateral triangle.]

$\therefore x = 60$

[$\angle$GAC is alternate to $\angle$BCA and BC || FG.]

## For Example

**1** Find $x$, giving reasons for your answer.

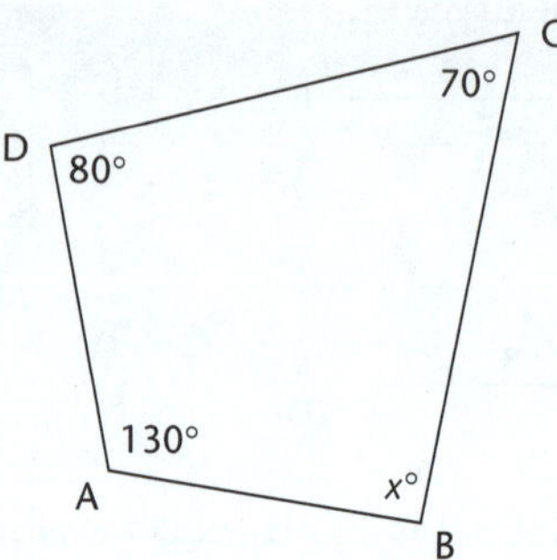


**1** $x + 130 + 80 + 70 = 360$

[Angle sum of quadrilateral ABCD.]

$\therefore x + 280 = 360$

$x = 80$

## For Example

**1** Find $x$, giving reasons for your answer.

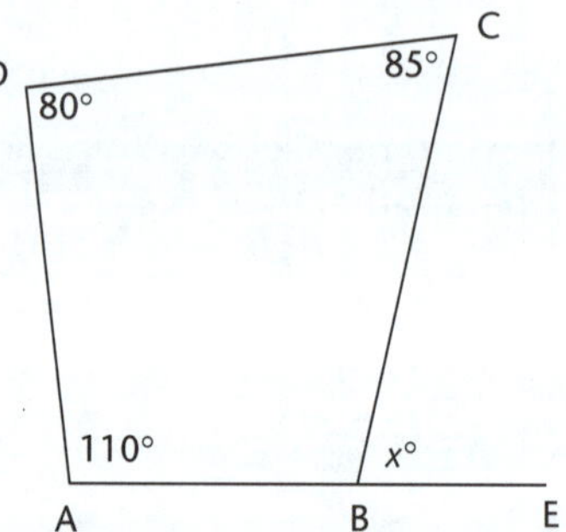


**1** $\angle ABC + 85° + 80° + 110° = 360°$

[Angle sum of quadrilateral ABCD.]

$\angle ABC + 275° = 360°$

$\angle ABC = 85°$

$\therefore x + 85 = 180$

[$\angle$CBE is supplementary to $\angle$ABC.]

$\therefore x = 95$

# Examples in Simple Deductive Geometry of Non-numerical Type of Questions

Pronumerals are used to represent angles to establish geometric properties.

## For Example

1 Prove that $\angle DBF = 90°$

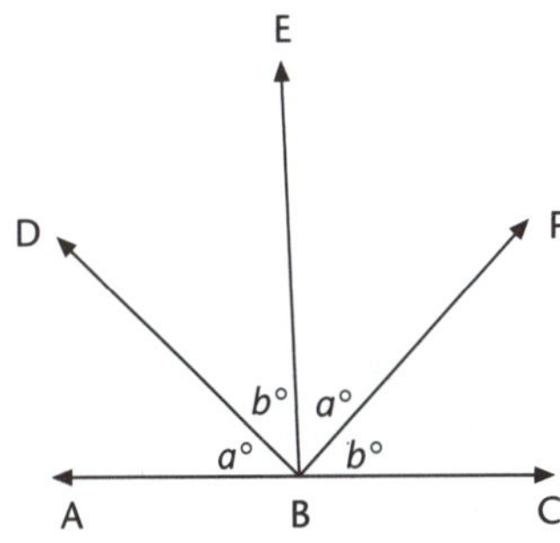


1 $a + b + a + b = 180$

$\therefore\ 2a + 2b = 180$

$2(a + b) = 180$

$\therefore\ a + b = 90$

But $\angle DBF = (a + b)°$

$\therefore\ \angle DBF = 90°$

## For Example

1 In the figure, AB || CD and AC || BD.

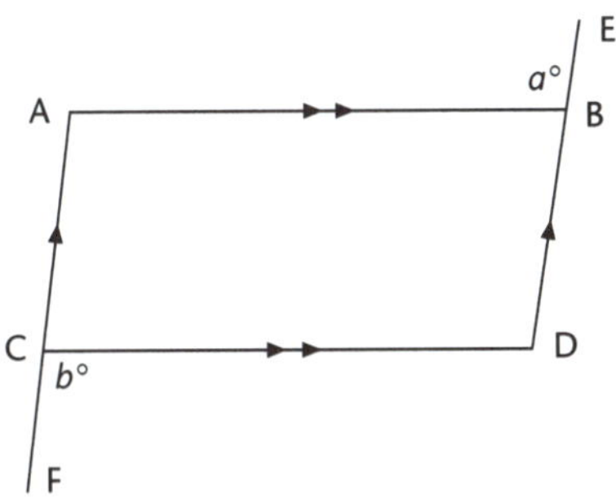


$\angle ABE = a°$ and $\angle DCF = b°$.
Prove that $a = b$.

1

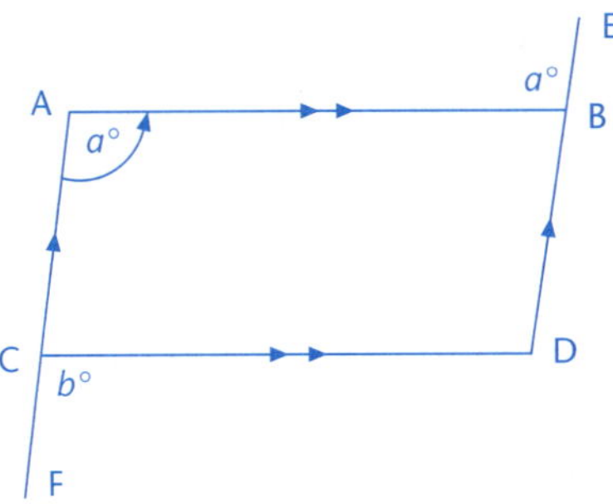


$\angle BAC = a°$ [Alternate to $\angle ABE$ and AC || BD.]

$\angle BAC = \angle DCF$

[Corresponding angles and AB || CD.]

But $\angle BAC = a°$ and $\angle DCF = b°$

$\therefore\ a = b$.

## For Example

1 In the diagram, BC || EF, $\angle ABC = a°$, $\angle BCA = b°$ and $\angle CAB = c°$.
Prove that $a + b + c = 180$.
(That is, prove that the angle sum of a triangle is 180°.)

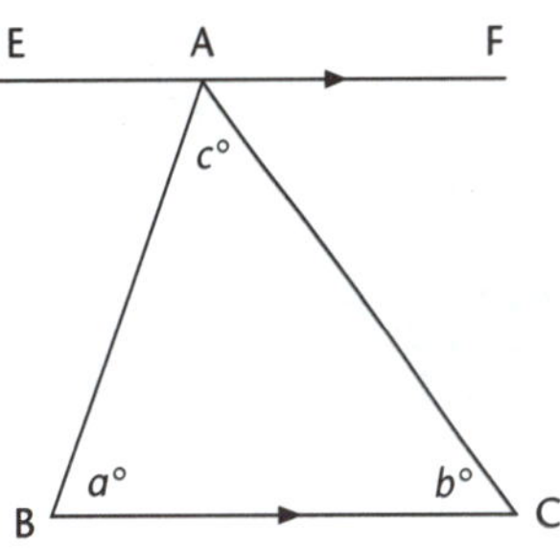


1 $\angle FAC = b°$ [Alternate to $\angle BCA$ and BC || EF.]

$\angle BAE = a°$ [Alternate to $\angle ABC$ and BC || EF.]

$\angle FAC + \angle CAB + \angle BAE = 180°$

[Angles in a straight line.]

$\therefore\ a + b + c = 180$.

# Congruence

Congruent figures are two or more figures that have the exact same size and shape: they are exact copies of each other.

These figures can be made to **superimpose** on one another if they were cut out.

The term 'superimpose' is used to describe the placement of one figure upon another in such a way that all the parts of one co-incide with the parts of the other.

Examples of pairs of congruent figures:

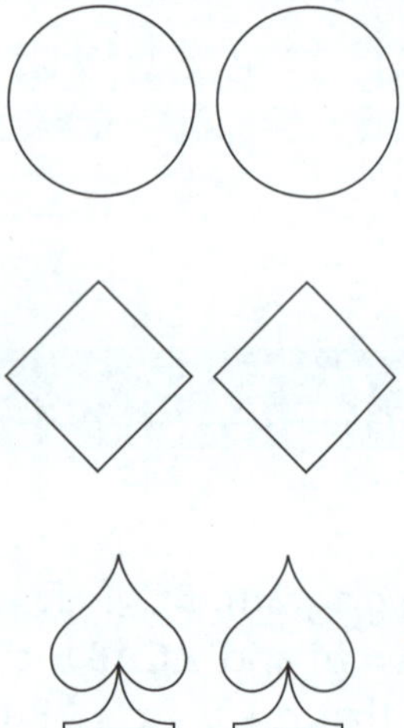

## Transformations and Congruence

A figure and its image under reflection, translation and rotation, are congruent.

**Figure 1**

Shape 1 Shape 2

In figure 1, shape 1 is reflected to shape 2. It can be seen that the parts of shape 1 coincide with the parts of shape 2 under the reflection.

**Figure 2**

Shape 1 Shape 2

In figure 2, shape 1 is translated to shape 2. It can be seen that the parts of shape 1 coincide with the parts of shape 2 under the translation.

**Figure 3**

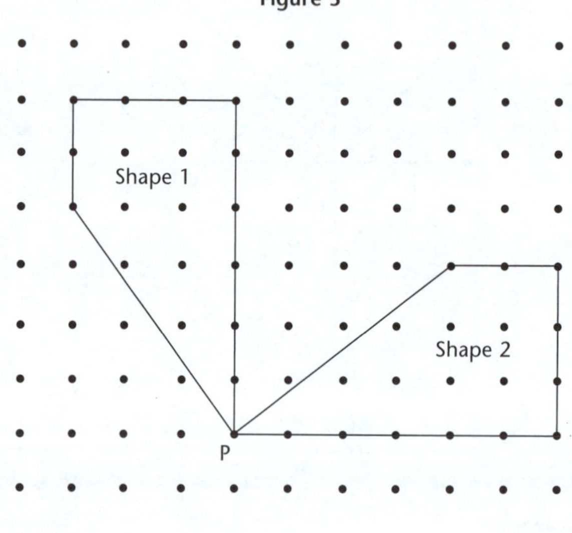


In figure 3, shape 1 is rotated to shape 2 about the point P. It can be seen that the parts of shape 1 coincide with the parts of shape 2 under the rotation.

## Corresponding Parts of Congruent Figures

We have learned that congruent figures have the same size and shape.

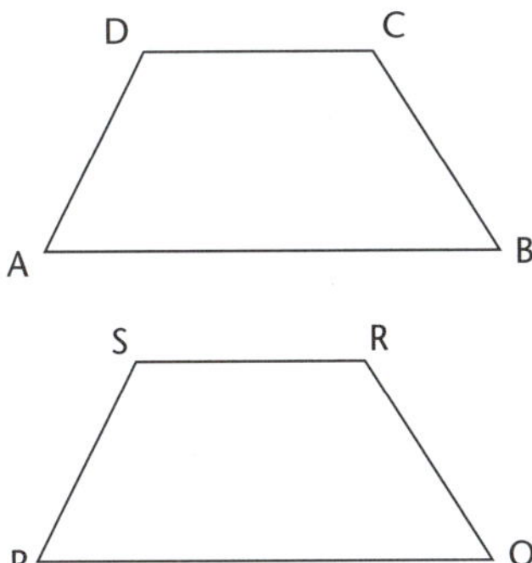


Figure ABCD and figure PQRS are congruent. If you cut figure ABCD you could fit it exactly over figure PQRS so that the vertices and sides coincide. In fact, we say that point A corresponds to point P, point B to point Q, point C corresponds to point R and point D corresponds to point S. In congruent figures we have equal corresponding sides and equal corresponding angles.

In the above congruent figures:

- Corresponding sides
  AB corresponds to PQ and AB = PQ
  BC corresponds to QR and BC = QR
  CD corresponds to RS and CD = RS
  DA corresponds to SP and DA = SP
- Corresponding angles
  $\angle$A corresponds to $\angle$P and $\angle A = \angle P$
  $\angle$B corresponds to $\angle$Q and $\angle B = \angle Q$
  $\angle$C corresponds to $\angle$R and $\angle C = \angle R$
  $\angle$D corresponds to $\angle$S and $\angle D = \angle S$

## Notation for Congruence

The symbol to denote 'is congruent to' is $\cong$.

To indicate that figure ABCD is congruent to figure PQRS you write ABCD $\cong$ PQRS.

## Naming Congruent Figures

When naming two congruent figures, you must list corresponding vertices in the same order.

### For Example

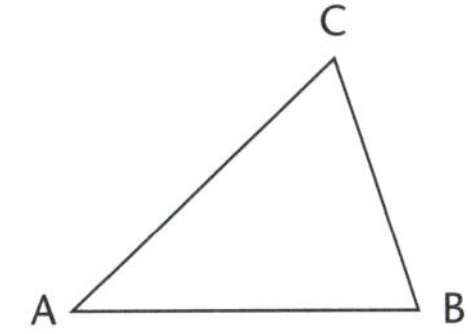


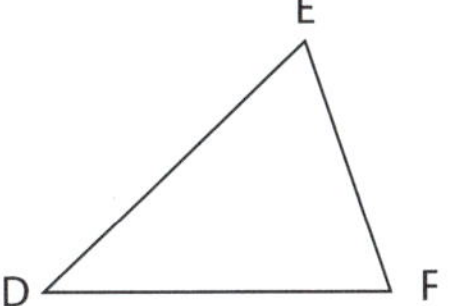


We write $\triangle ABC \cong \triangle DFE$ since A corresponds to D, B corresponds to F and C corresponds to E. Or, $\triangle CBA \cong \triangle EFD$ or $\triangle BCA \cong \triangle FED$, but not, say, $\triangle ABC \cong \triangle EFD$.

### For Example

$\triangle$ABC is congruent to $\triangle$FGH.

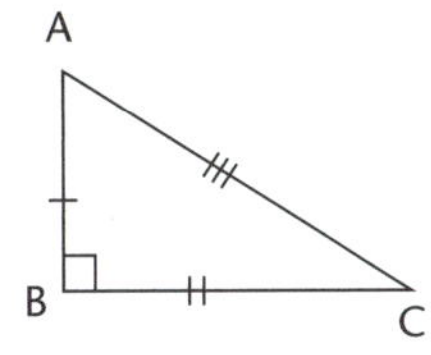


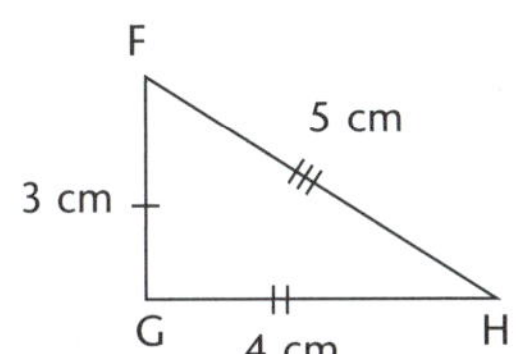


1 What side corresponds to side FH?

2 What is the measure of side AB?

3 What angle corresponds to $\angle$CAB?

4 What is the measure of $\angle$FGH?

1 Side AC corresponds to side FH.

2 Side AB = 3 cm because it corresponds to side FG = 3 cm.

3 $\angle$HFG corresponds to $\angle$CAB.

4 $\angle FGH = 90°$ because it corresponds to $\angle$ABC.

Note that in congruent figures:
- Corresponding sides are equal.
- Corresponding angles are equal.

For Example

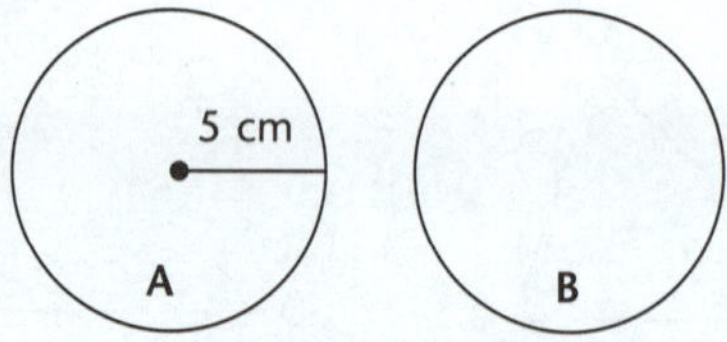


Circle A is congruent to circle B. The area of circle A is 79 $cm^2$, to the nearest whole.

1 What is the radius of circle B?

2 What is the area of circle B?

1 The radius of circle B is 5 cm since circle A is congruent to circle B.

2 The area of circle B is 79 $cm^2$ since circle A is congruent to circle B.

Note:
- Congruent circles have the same radius.
- Congruent figures have the same area.
- Congruent figures have the same perimeter.

## Congruent Triangles

Congruent triangles are equal in all respects.

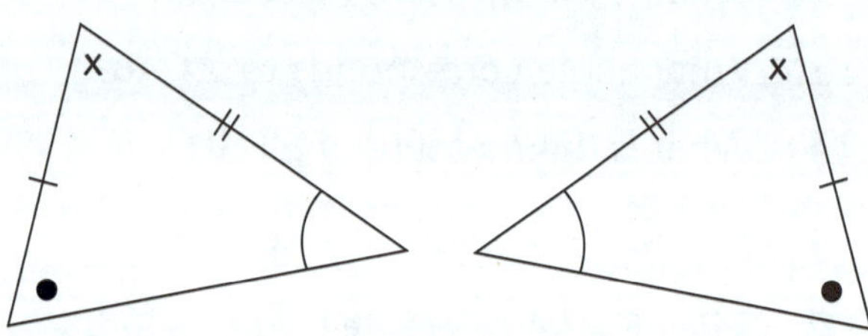


**Note:**

- Three sides of one equal to three sides of the other.
- Three angles of one equal to three angles of the other.
- Their areas are equal.
- Their perimeters are equal.

### Tests for Congruent Triangles

Two triangles are congruent if any one of the following four tests (conditions) is true.

**(SSS) Test**

Three sides of one triangle are respectively equal to the three sides of the other (SSS):

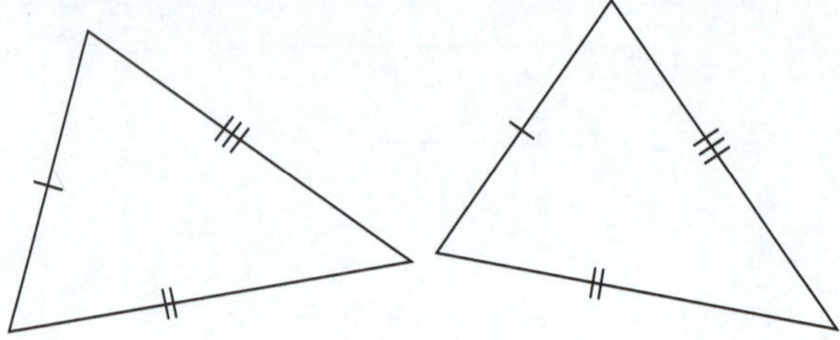

**(SAS) Test**

Two sides of one triangle and the included angle are respectively equal to two sides and the included angle of the other (SAS):

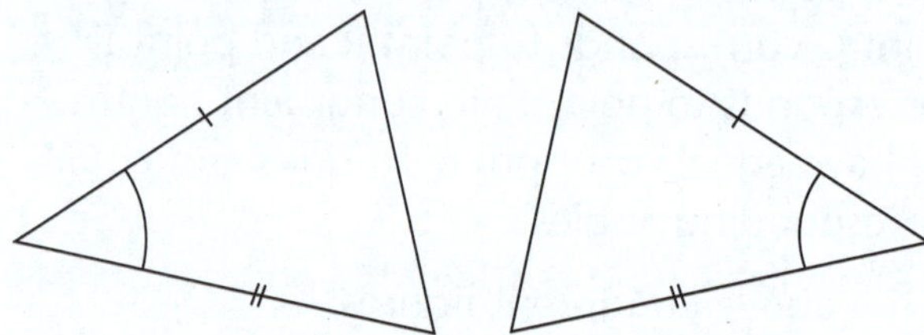

**(AAS) Test**

Two angles and one side of one triangle are respectively equal to two angles and the corresponding side of the other (AAS):

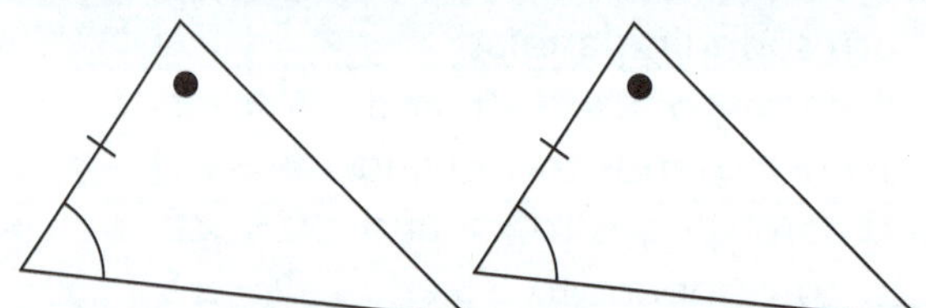

**(RHS) Test***

Two right-angled triangles are congruent if the hypotenuse and one side of one triangle are respectively equal to the hypotenuse and one side of the other (RHS):

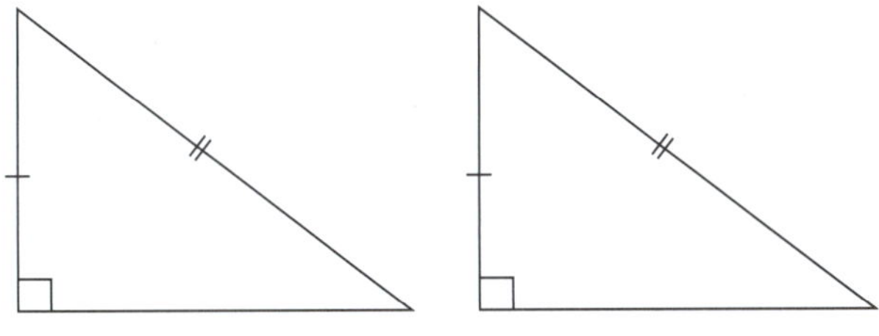
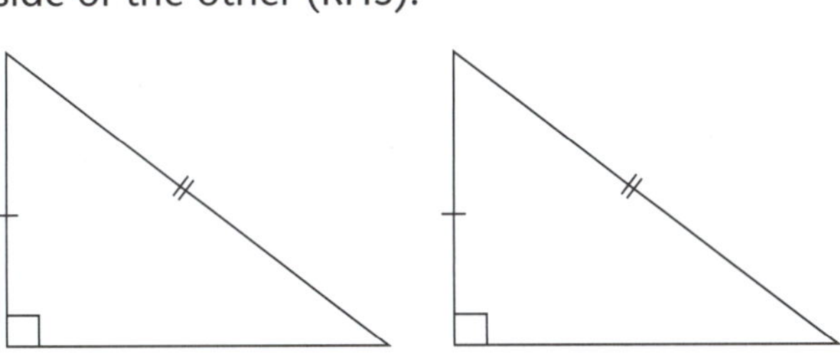

[*Note: only for right-angled triangles.]

For Example

1 In each case below, state which of the congruence tests would be used in proving each pair of triangles congruent:

**a**

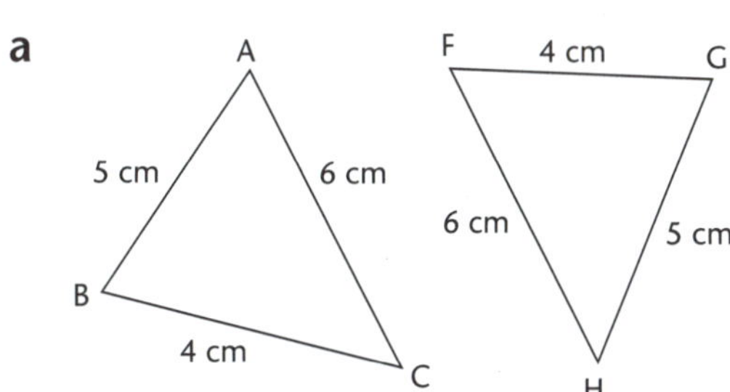


**b**

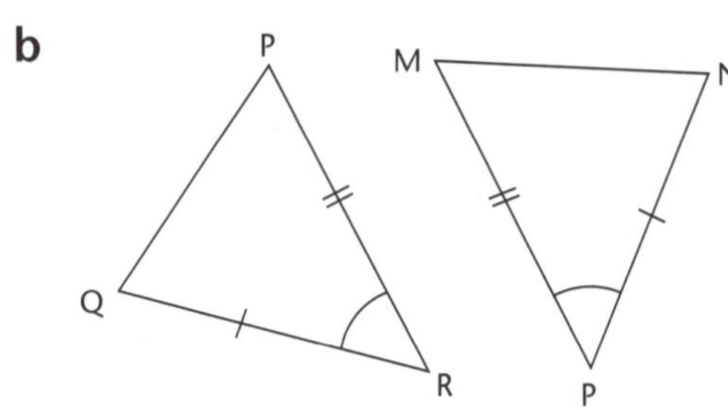


**c**

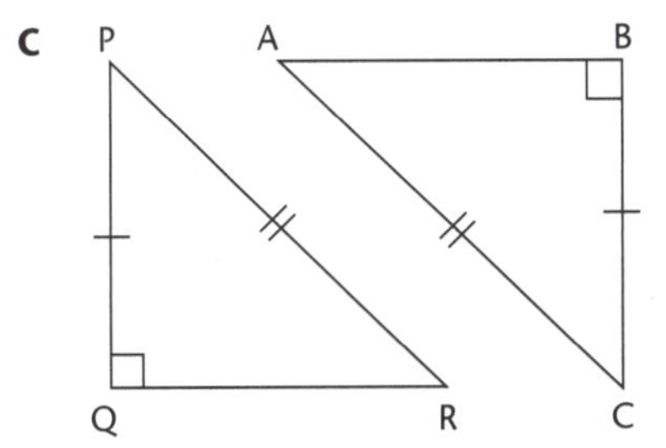


**d**

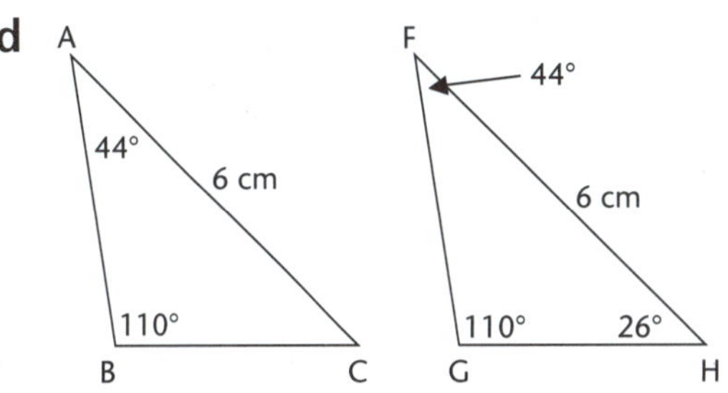


**1 a** △ABC ≡ △HGF (SSS)

[Note: ≡ means 'is congruent to'.]

**b** △QRP ≡ △NPM (SAS)

**c** △QRP ≡ △BAC (RHS)

**d** △ABC ≡ △FGH (AAS)

## Formal Proofs for Congruent Triangles

When proving congruent triangles:

- Mark all information given (data) on the diagram and then mark any other angles or sides that you can prove are equal.
- Remember that a congruence proof consists of three equality statements.
- Always give reasons.

For Example

1 Given that AC = EC and BC = CD, prove triangles ABC and EDC are congruent.

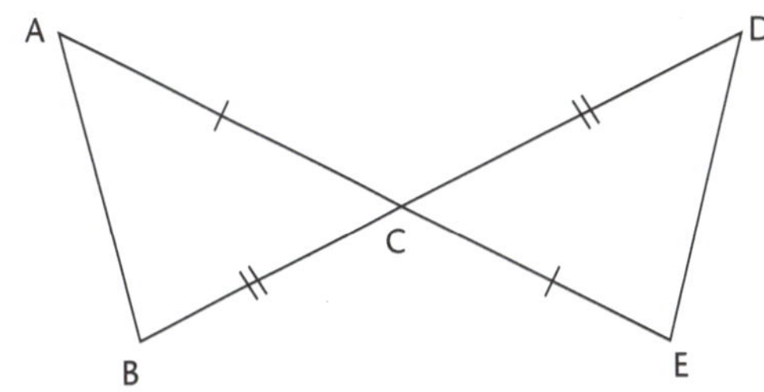

***Note:** the included angle is the one contained between the sides proved equal.

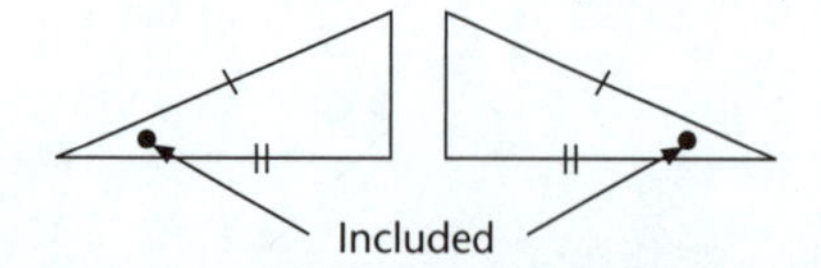


**2** Given that $\angle CDA = \angle ABC$, and also $\angle ACD$ and $\angle CAB$ are both right-angles, prove that $\triangle ACD \equiv \triangle CAB$.

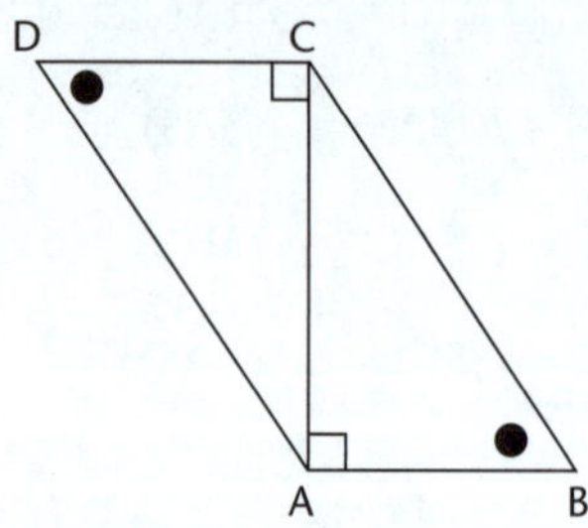


**3** Given that $AB \perp DC$ ($\angle CDB = 90°$) and $AC = BC$, prove that $\triangle ADC \equiv \triangle BDC$.

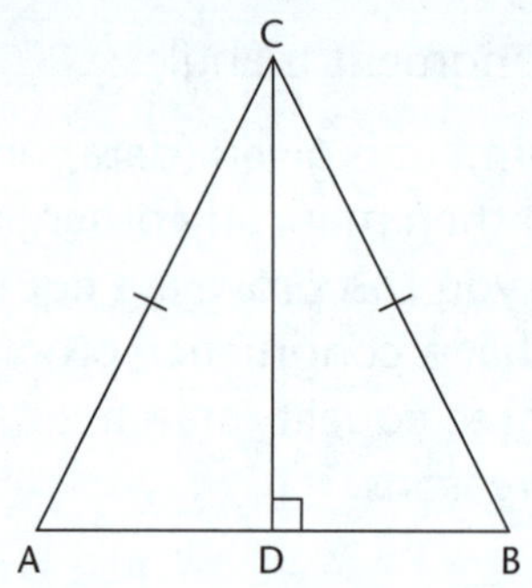


**1**

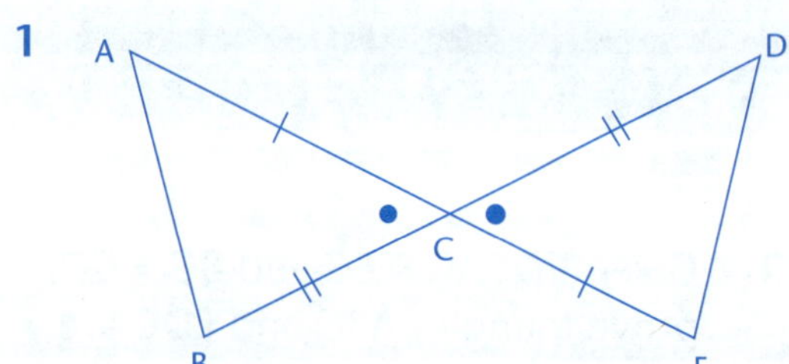


In the triangles ABC and EDC:

| | | |
|---|---|---|
| $AC = EC$ | (S) | (data) |
| $BC = DC$ | (S) | (data) |
| $\angle BCA = \angle DCE$ | (A) | |

[vertically opposite angles]

$\therefore \triangle ABC \equiv \triangle EDC$ (SAS)

**2**

This is how common sides are marked.

In $\triangle$s ACD and CAB:

| | | |
|---|---|---|
| $\angle CDA = \angle ABC$ | (A) | (data) |
| $\angle ACD = \angle CAB$ | (A) | (data, both 90°) |
| CA is common | (S) | |
| $\therefore \triangle ACD \equiv \triangle CAB$ | (AAS) | |

[Note: two right-angles are treated as equal angles in this example.]

**3**

In $\triangle$s ACD and BDC:

| | | |
|---|---|---|
| $\angle ADC = \angle BDC = 90°$ | (R) | ($AB \perp DC$) |
| $AC = BC$ | (H) | (data) |
| CD is common | (S) | |
| $\therefore \triangle ADC \equiv \triangle BDC$ | (RHS) | |

Go to p. 288 for quick answers, or to pp. 359–365 for worked solutions.

1 Complete the following statements: pp. 199–200

a Complementary angles add up to ______°.
b Supplementary angles add up to ______°.
c Angles on a straight line add up to ______°.
d Vertically opposite angles are always ______________.
e Angles at a point always add up to ______°.
f The angle sum of a triangle is ______°.
g The angle sum of a quadrilateral is ______°.
h The base angles of an isosceles triangle are ______________.
i Any angle of an equilateral triangle is equal to ______°.
j For a pair of parallel lines cut by a transversal:
  i alternate angles are ____________________.
  ii corresponding angles are ____________________.
  iii co-interior angles are ____________________.

2 In each of the following cases, evaluate the pronumeral, giving a brief reason: pp. 201–205

a
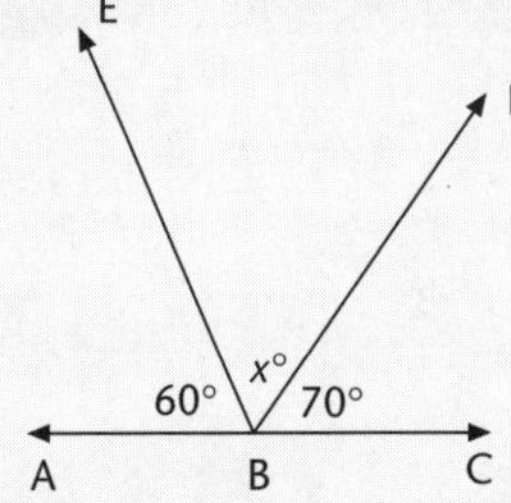


b
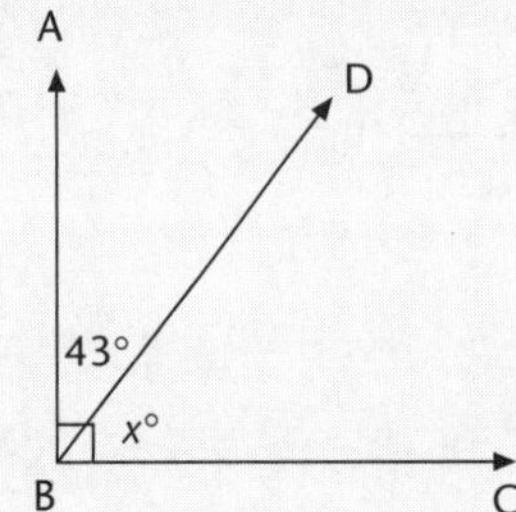


c
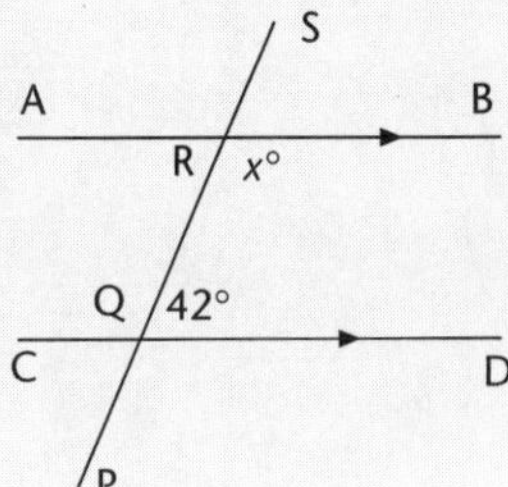


d
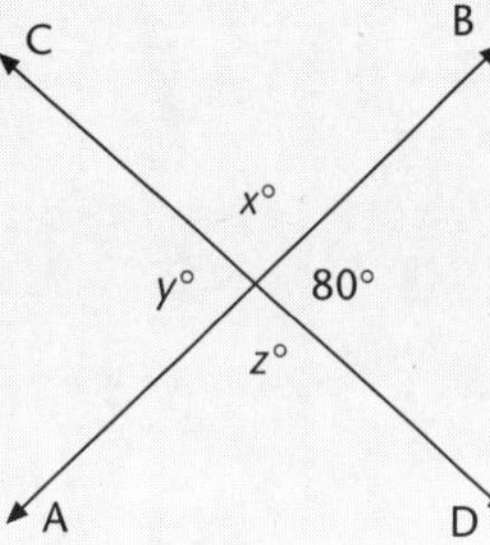


e
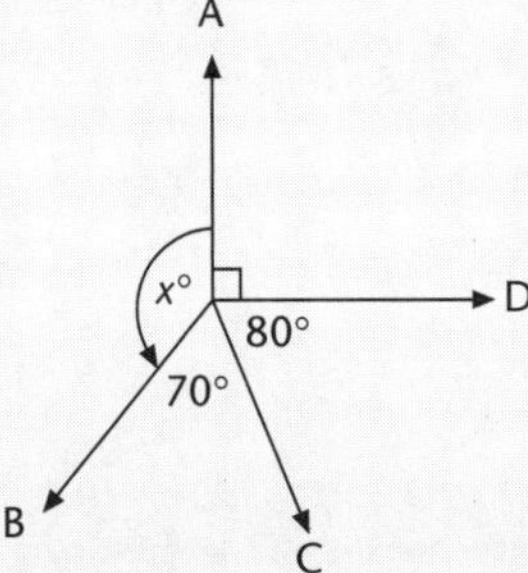


f
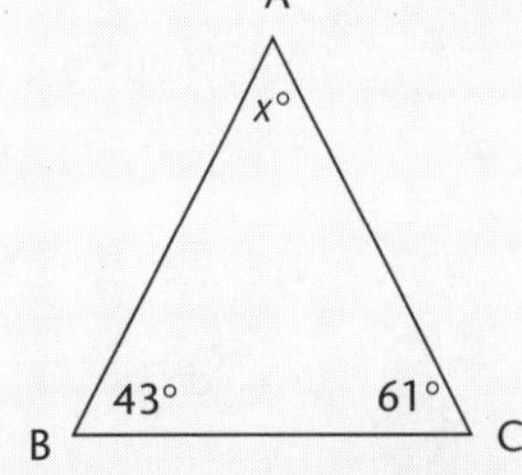

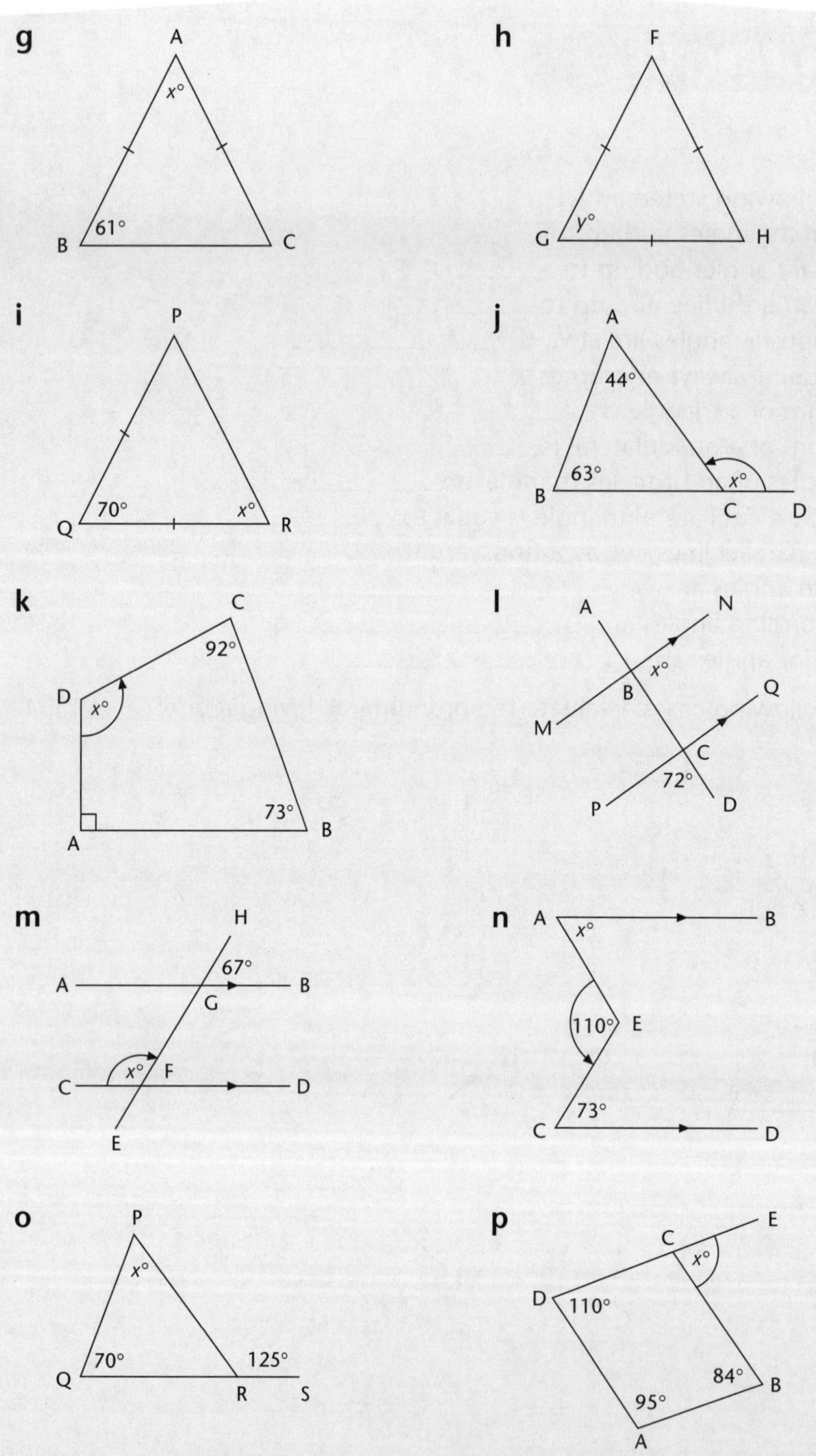
g
A
x°
61°
B
C
h
F
y°
G
H
i
P
70°
x°
Q
R
j
A
44°
63°
x°
B
C
D
k
C
92°
D
x°
73°
A
B
l
A
N
x°
B
Q
M
C
72°
P
D
m
H
67°
A
B
G
x°
F
C
D
E
n
A
x°
B
110°
E
73°
C
D
o
P
x°
70°
125°
Q
R
S
p
E
C
x°
D
110°
84°
B
95°
A

**3** **a** **i** Draw a trapezium and describe it in your own words.

pp. 200–201

**ii** Does it have any axis of symmetry?

**b** ABCD is a parallelogram.

**i** Using a ruler, find the lengths of AB, DC, AD and BC. In your own words, make a statement about the lengths of the sides of a parallelogram.

**ii** Using a protractor, find the size of ∠DAB, ∠BCD, ∠ABC and ∠CDA. In your own words, make a statement about the angles of a parallelogram.

**iii** Does a parallelogram have any axis of symmetry?

**c** XYZW is a rhombus.

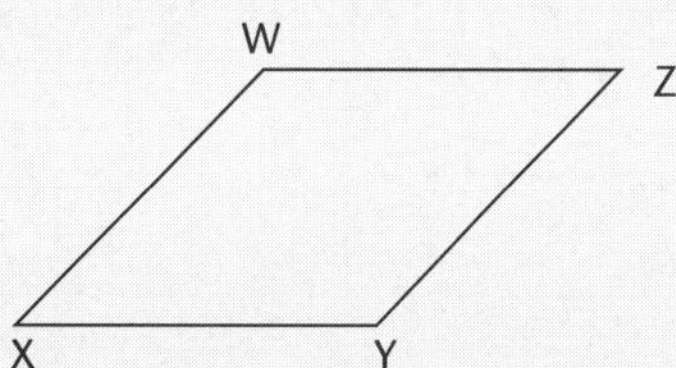


**i** Measure the length of the sides of rhombus XYZW.
In your own words make a statement about the lengths of sides of a rhombus.

**ii** What is the order of symmetry of a rhombus? (How many axes of symmetry does it have?)

**iii** Are the opposite angles of the rhombus equal? (Use a protractor.)

**iv** Is a rhombus a parallelogram?

**d** **i** Draw a square and label all the axes of symmetry. How many are there?

**ii** What do you know about the sides of a square?

**iii** What is the size of all the angles of a square?

**iv** Measure the diagonals of the square. What do you conclude?

**e** **i** Draw a rectangle and label the axes of symmetry. How many are there?

**ii** What do you know about the sides of a rectangle?

**iii** What is the size of all the angles of a rectangle?

**iv** Measure the diagonals of the rectangle. What do you conclude?

**f** **i** Draw a kite and label any axes of symmetry. How many are there?

**ii** Measure the lengths of all the sides of the kite. In your own words make a statement about the lengths of the sides of a kite.

**g** Which of the following quadrilaterals would be classed as trapeziums?

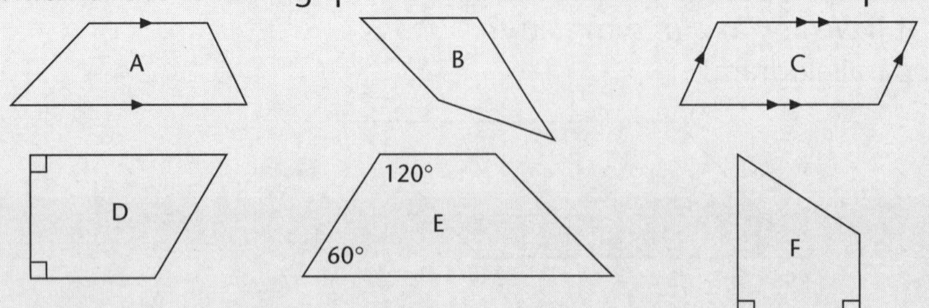


**Note:** a trapezium is a quadrilateral with one pair of opposite sides parallel.

**h** Which of the following quadrilaterals would be classed as parallelograms?
**Note:** a parallelogram is a quadrilateral with both pairs of opposite sides parallel.

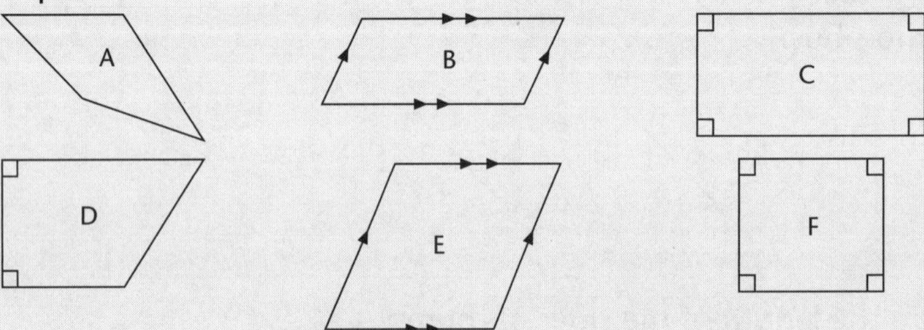


**i** Are the following statements true (T) or false (F)?

- **i** A trapezium is a parallelogram.
- **ii** A rhombus is a parallelogram.
- **iii** A square is a parallelogram.
- **iv** A rectangle is a parallelogram.
- **v** A parallelogram is a rhombus.
- **vi** All quadrilaterals have two diagonals.
- **vii** A square is a rhombus.
- **viii** A square is a rectangle.

**j**

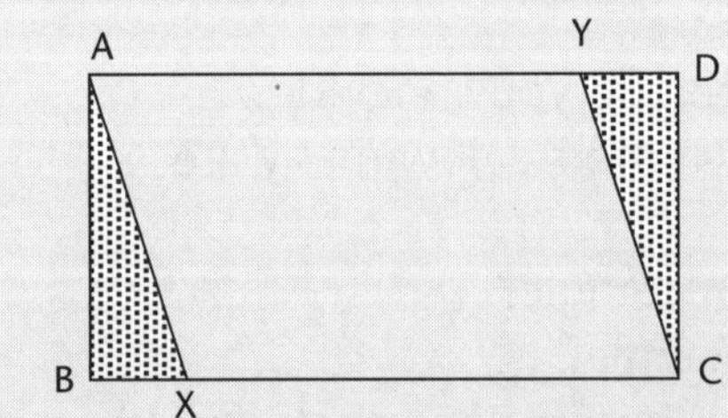


Take a rectangular piece of paper, ABCD. Label two points X and Y such that BX = DY as shown in the diagram. Cut along AX and CY to form the quadrilateral AXCY.

- **i** Name the quadrilateral AXCY.
- **ii** Show by folding that the quadrilateral AXCY has no axis of symmetry.

**4** **a** If AB || CD, find the value of $x$, giving reasons for your answer. p. 205

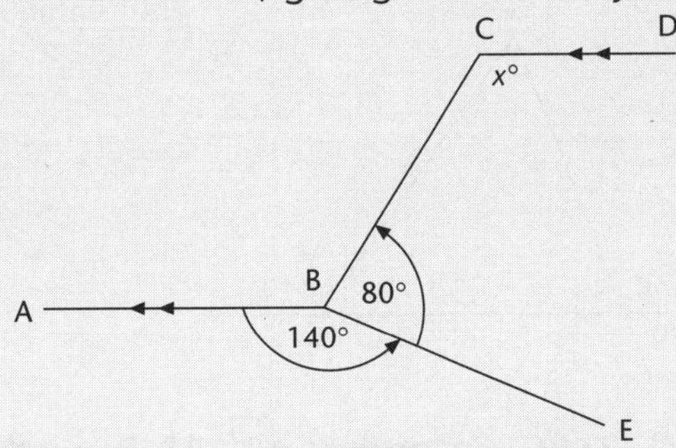


**b** In the figure AB || DC, DC = CE, ∠CBA = 70° and ∠BCE = 148°. Find the values of $x$ and $y$, giving reasons for your answer.

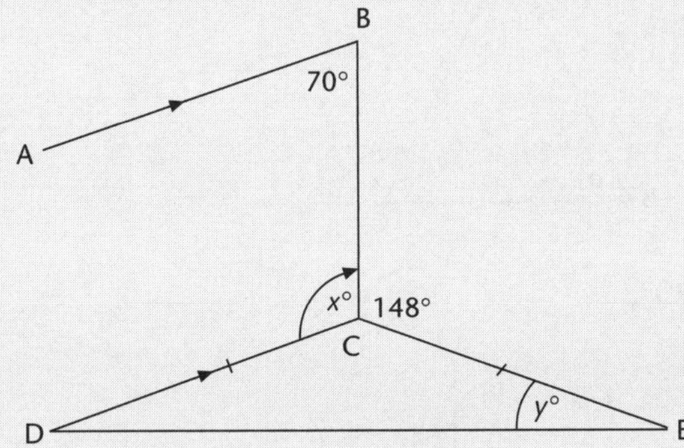


**c** In the figure, AB = AC and ∠ACD = 110°. Find the value of $x$, giving reasons for your answer.

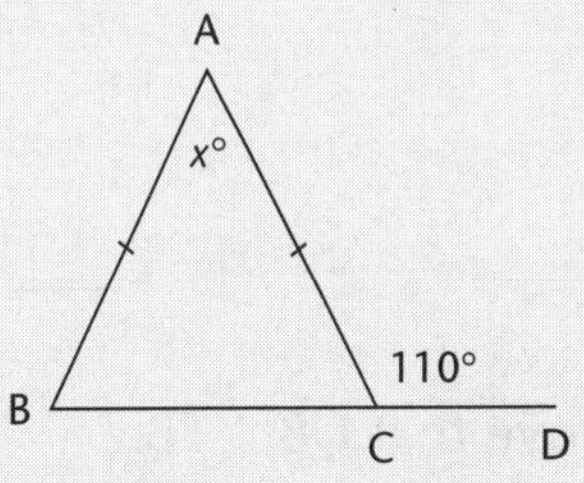


**d** In the figure, AB || PQ and ∠CRB = 48°. Find the size of ∠QSD.

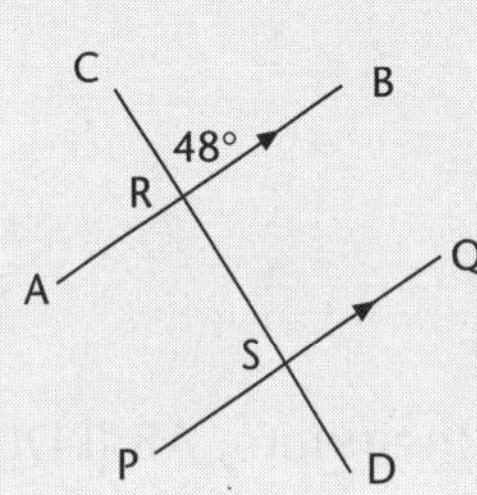


**5** **a** In the figure, AB || CD and ∠HGB = $a$°. Prove that ∠EFC = $a$°. p. 205

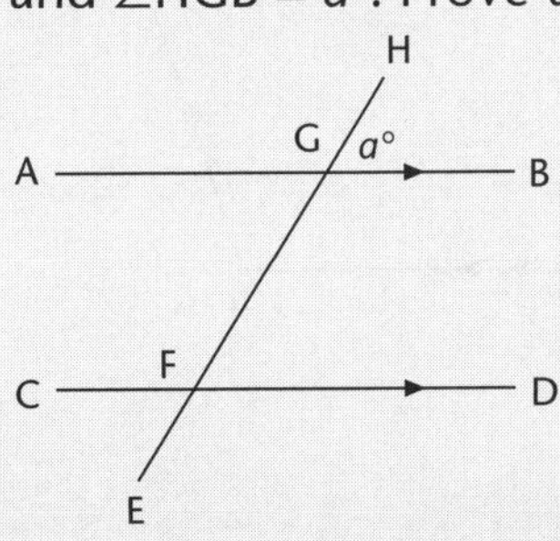

**b** Prove that AB = AC.

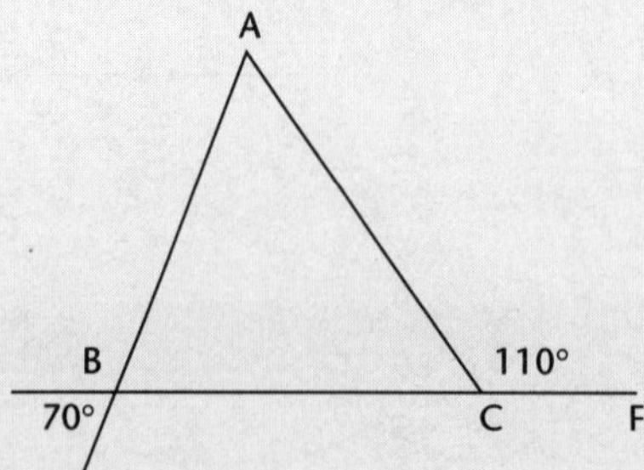


**c** In the diagram, EF || BG, ∠ABC = $a$° and ∠CAB = $b$°. Prove that $x = a + b$.

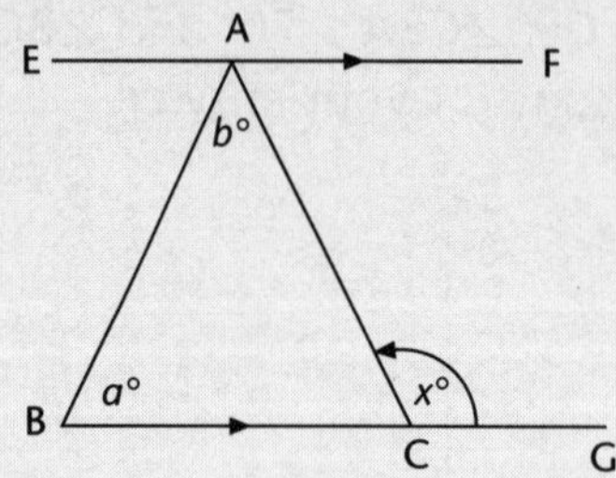


**d** Prove that ∠GBC = 90°.

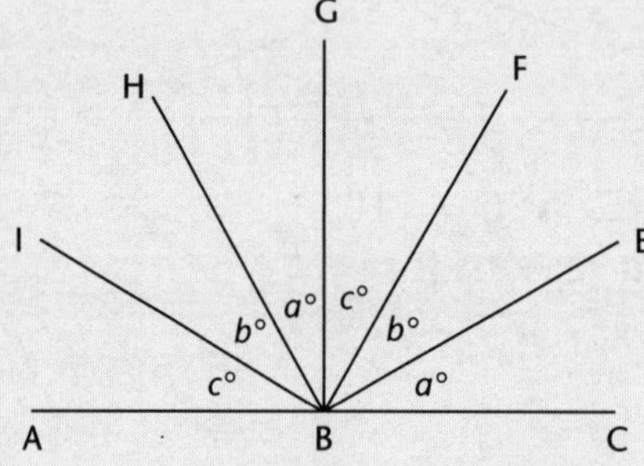


**e** Prove that DB ⊥ FB.

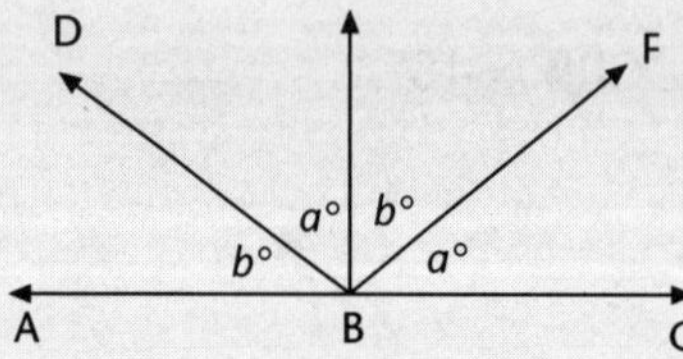


**f** In the figure, AB || DC, AD || BC and ∠DAB = $x$°. Prove that ∠BCD = $x$°.

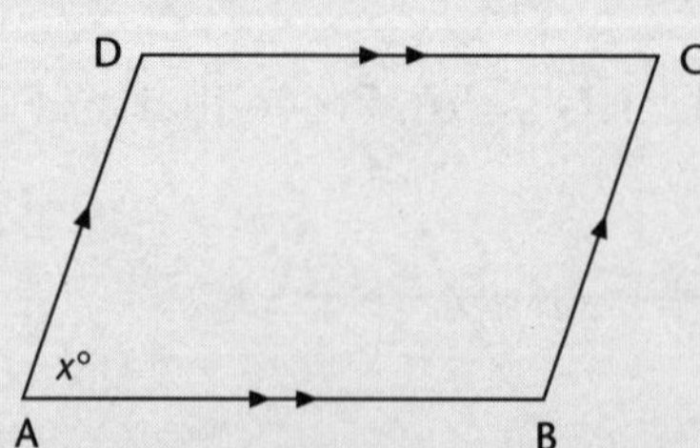

**6** p. 206

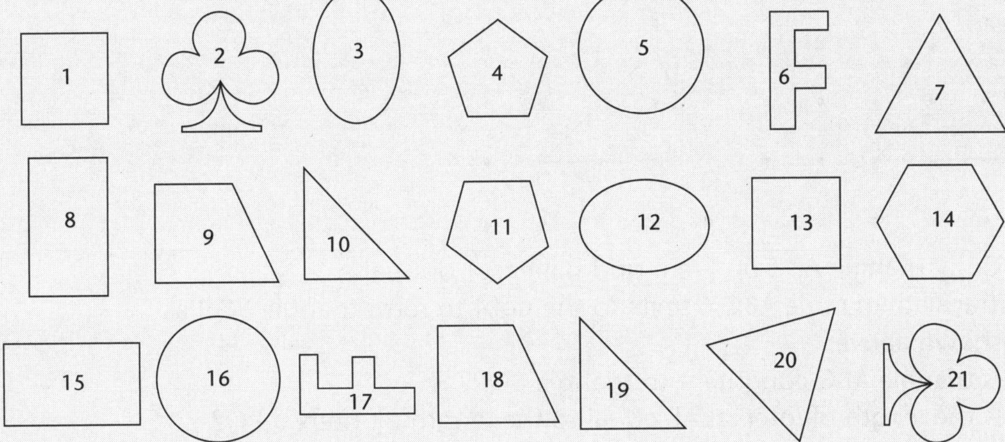


Complete the following table by listing the pairs of congruent figures from the above numbered shapes.

| Figure | | Figure |
|---|---|---|
| 1 | | 13 |
| 2 | | |
| 3 | | |
| 4 | | |
| 5 | is congruent to | |
| 6 | | |
| 7 | | |
| 8 | | |
| 9 | | |
| 10 | | |

**7** p. 206

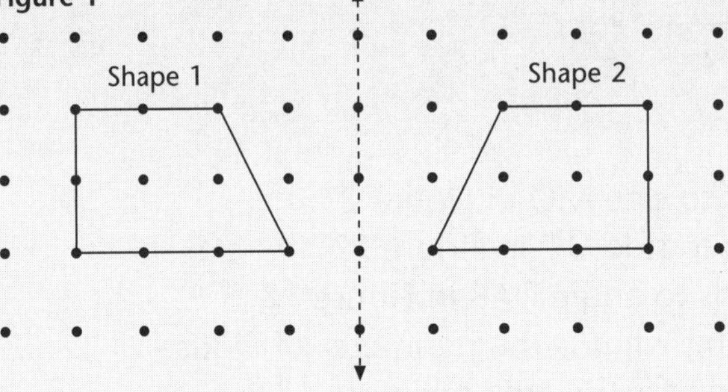


**a** Copy shape 1 using 1-centimetre grid paper (or dot paper).
**b** Reflect shape 1 to form shape 2 as shown in Figure 1.
**c** Cut out both shapes and place them on top of each other. Do they fit exactly?
**d** Is shape 1 congruent to shape 2?

**8** 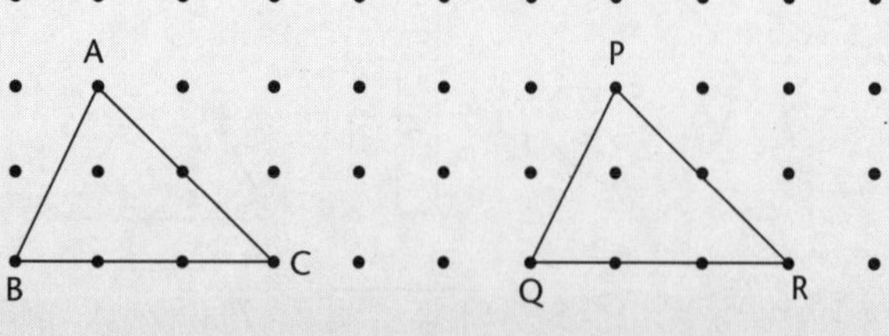


p. 206

**a** Copy triangle ABC on 1-cm grid paper (or dot paper).
**b** Translate triangle ABC 6 units to the right to form triangle PQR as shown above.
**c** Is triangle ABC congruent to triangle PQR?
**d** Is the length of interval AB equal to the length of interval PQ?
**e** Measure the lengths of intervals AC and PQ to the nearest millimetre. Are they equal?
**f** Does $\angle ABC = \angle PQR$? (Use a protractor.)

**9** $\triangle ABC$ below is translated to form $\triangle PQR$ and then $\triangle PQR$ is rotated about R to form $\triangle TSR$. p. 206

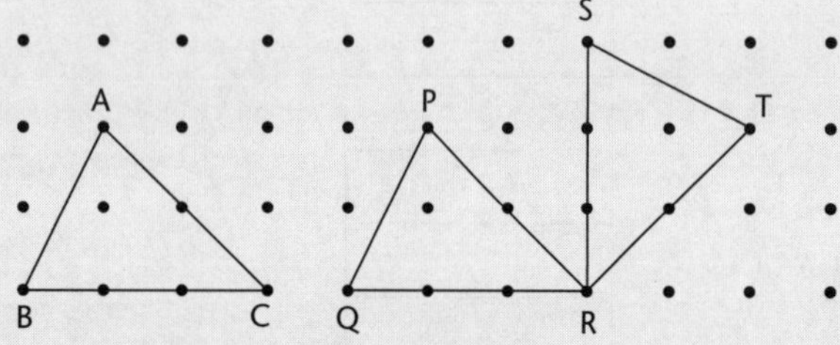


**a** Is $\triangle ABC$ congruent to $\triangle PQR$?
**b** Is $\triangle PQR$ congruent to $\triangle TSR$?
**c** Is $\triangle ABC$ congruent to $\triangle TSR$?

**10** 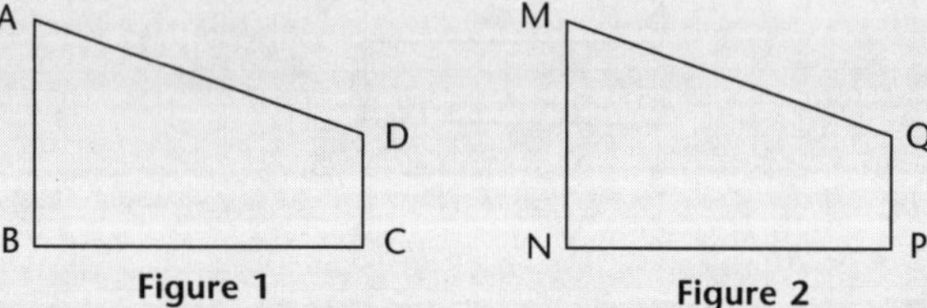


**Figure 1** **Figure 2**

p. 207

Figure 1 $\cong$ Figure 2.

**a** What side of Figure 1 corresponds to side MQ in Figure 2?
**b** What side of Figure 2 corresponds to side DC in Figure 1?
**c** What angle of Figure 2 corresponds to angle DAB in Figure 1?
**d** If the perimeter of Figure 1 is 25 cm, what is the perimeter of Figure 2?
**e** If the area of Figure 2 is 34 $cm^2$, what is the area of Figure 1?

**11** pp. 207–208

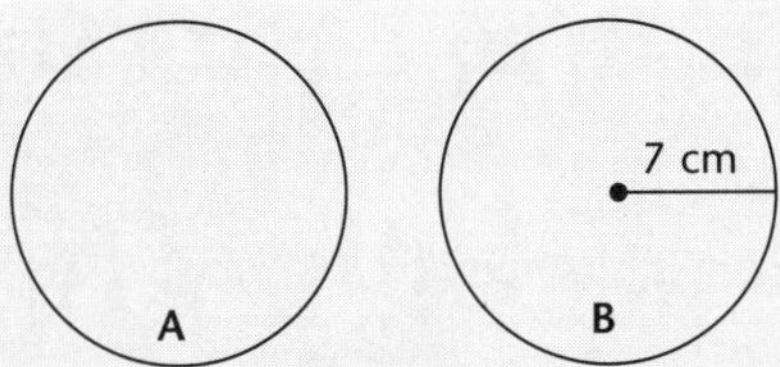


Circle A ≅ Circle B.

**a** What is the length of the radius of Circle A?
**b** If Circle A has an area of 154 $cm^2$, find the area of Circle B.

**12** pp. 207–208

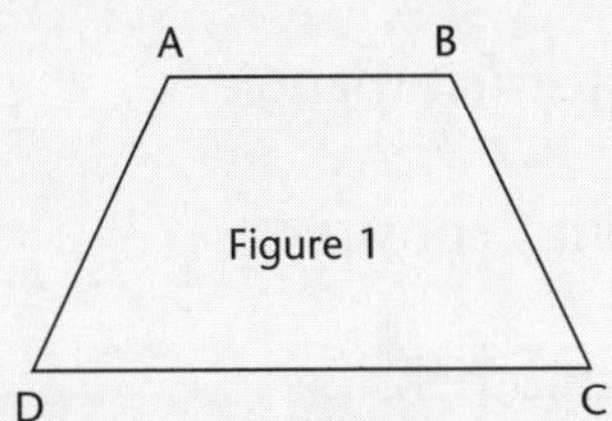


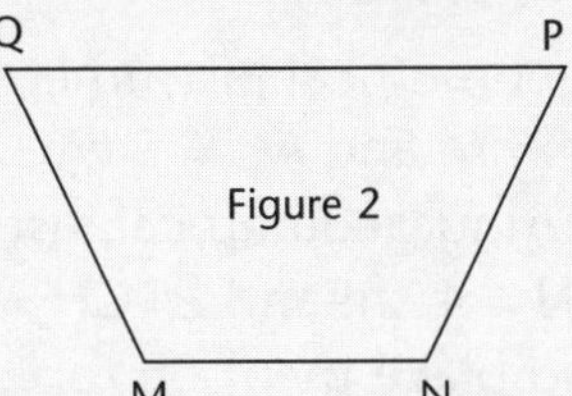


Figure 1 ≅ Figure 2.

**a** What side of Figure 2 corresponds to side AB of Figure 1?
**b** What side of Figure 1 corresponds to side MQ of Figure 2?
**c** What angle of Figure 2 corresponds to ∠ADC in Figure 1?
**d** What angle of Figure 1 corresponds to ∠MNP in Figure 2?

**13** pp. 207–208

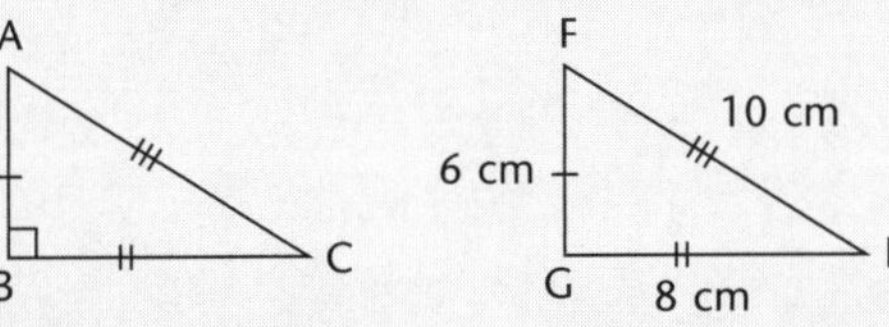


△ABC ≅ △FGH.

**a** What side corresponds to side GH?
**b** What is the measure of side AC?
**c** What angle corresponds to ∠BCA?
**d** What is the measure of ∠FGH?
**e** What is the perimeter of △ABC?
**f** What is the area of △ABC?

**14** pp. 207–208

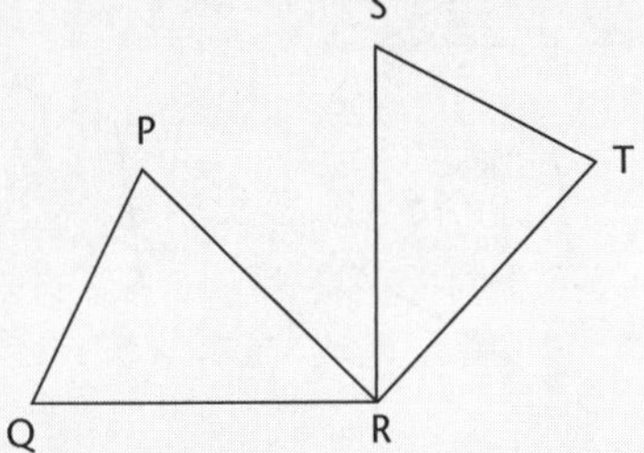


Triangle QPR is rotated about R to form triangle STR.

**a** What side of triangle QPR corresponds to side ST of triangle STR?
**b** What side of triangle STR corresponds to side QR of triangle QPR?

**15** pp. 207–208

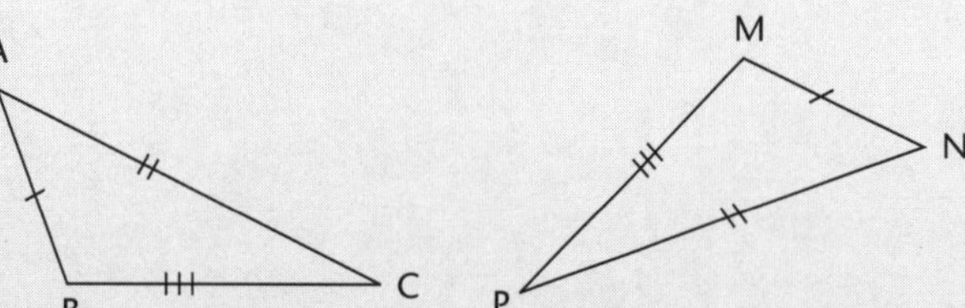


$\triangle ABC \cong \triangle NMP$.

**a** Name all pairs of corresponding sides.
**b** Name all pairs of corresponding angles.

**16** **a** On a piece of paper, using geometrical instruments, construct triangle ABC given AB = 5 cm, BC = 4 cm and AC = 3 cm. pp. 207–208
**b** On another piece of paper, using geometrical instruments, construct triangle given FG = 3 cm, GH = 4 cm and $\angle FGH = 90°$.
**c** Cut the two triangles out and place them on top of each other. You should find that they superimpose on each other.
**d** Is $\triangle ABC \cong \triangle FHG$?
**e** List all pairs of corresponding sides and corresponding angles of $\triangle ABC$ and $\triangle FHG$.

**17** Draw a square ABCD that has side 5 cm long. pp. 207–208
**a** Another square, FGHI, is congruent to square ABCD. What is the length of the side of square FGHI?
**b** What is the perimeter of square FGHI?
**c** What is the area of square FGHI?
**d** Are all squares congruent?

**18** Name the test you would use in proving each of the following pairs of triangles congruent: pp. 208–209

**a**

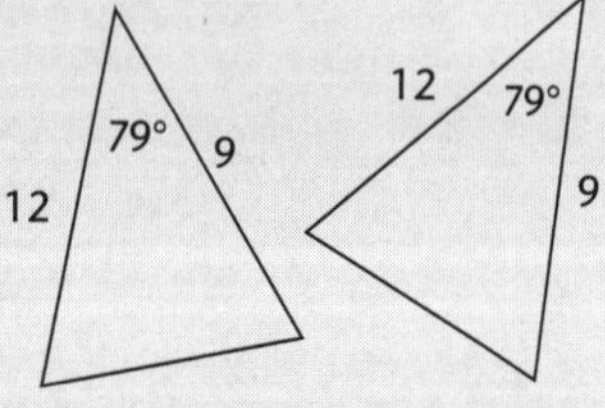


**b**

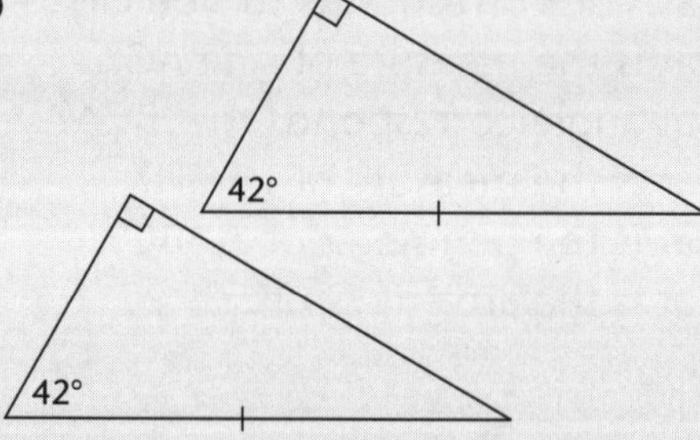


**c**

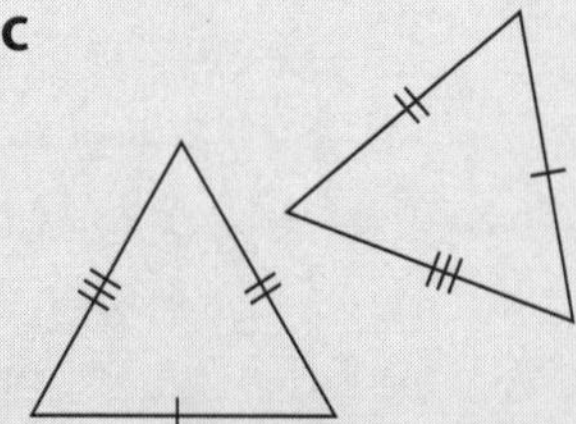

**d**

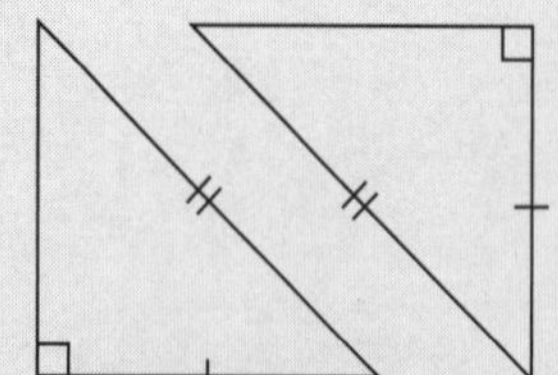

**e**

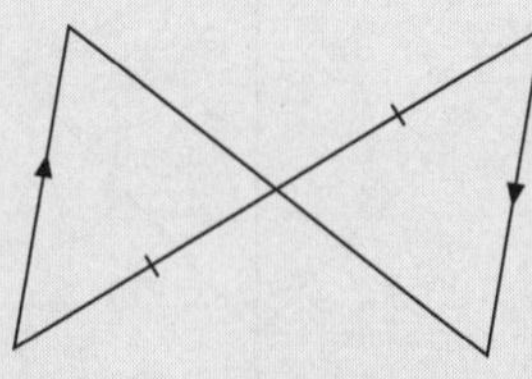

**19** In this question, prove these triangles are congruent, giving reasons: pp. 209–210

**a**

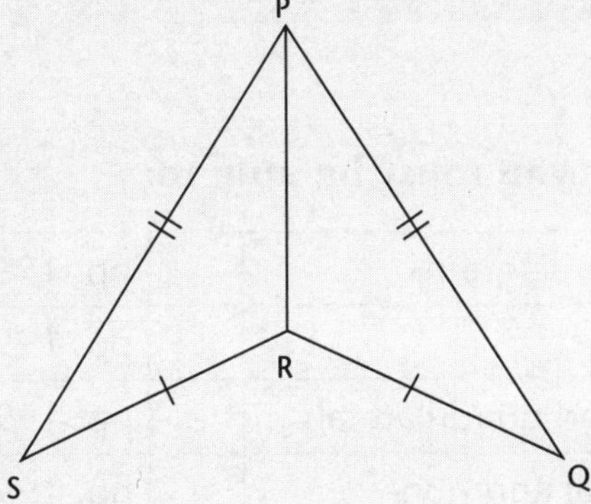


Given PS = PQ and SR = QR, prove that △PSR ≡ △PQR.

**b**

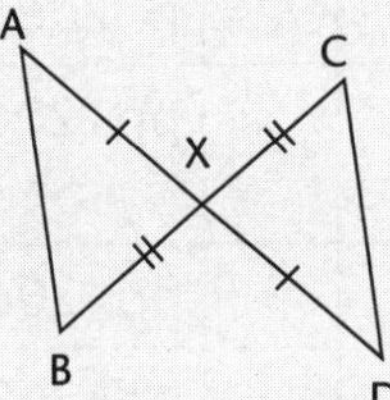


Given that AX = DX and BX = CX prove that △ABX ≡ △DCX.

**c**

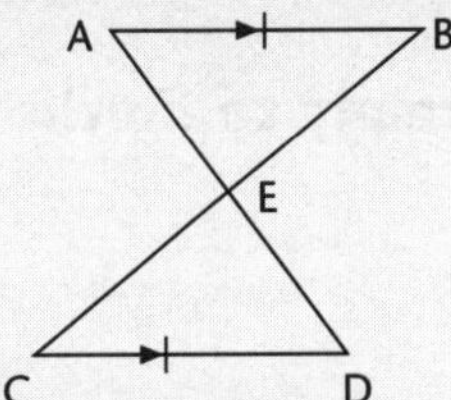


Given that AB || CD and AB = CD prove that △BAE ≡ △CDE.

**d**

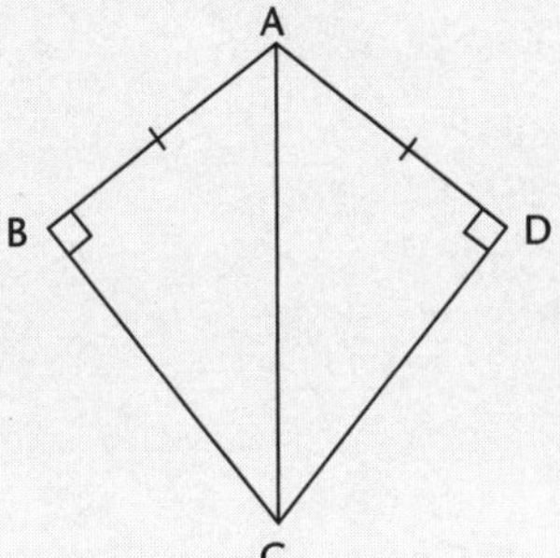


In the diagram AB = AD and ∠ABC = ∠ADC = 90°.
Prove that △ABC ≡ △ADC.

Go to p. 288 for quick answers, or to pp. 359–365 for worked solutions.

## YOUR CHECKLIST

**For a complete understanding of this topic you must be able to:**

| | | | |
|---|---|---|---|
| ✓ | Use angle properties to find angles in given diagrams | | pp. 199–204 |
| ✓ | Use angle properties to identify parallel lines | | pp. 199–204 |
| ✓ | Investigate properties of special triangles and quadrilaterals | | pp. 199–204 |
| ✓ | Use simple deductive reasoning in numerical and non-numerical problems in geometry | | pp. 199–205 |
| ✓ | Determine whether two figures are congruent | | p. 206 |
| ✓ | Identify congruent figures through a combination of transformations (reflection, translation and rotation) | | pp. 206–208 |
| ✓ | Match corresponding parts of congruent figures | | pp. 206–208 |
| ✓ | Use the four tests for congruent triangles | | pp. 208–209 |
| ✓ | Prove congruent triangles. | | pp. 209–210 |

**Now you are ready to do the tests!**

**(30 marks)**

1 Complete the following statements:

a Complementary angles add up to ______.

b Supplementary angles add up to ______.

c The angle sum of a triangle is ______.

d The angle sum of a quadrilateral is ______.

e Angles at a point add up to ______. (5 marks)

2 Name the following quadrilaterals:

a 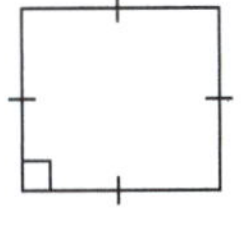

b 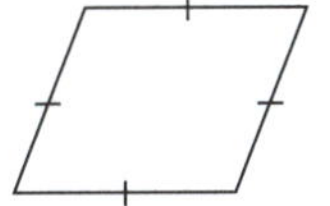

c 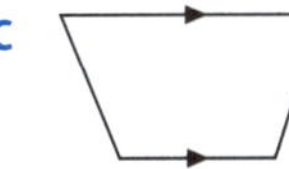

d 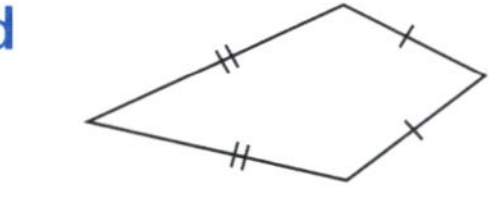

e 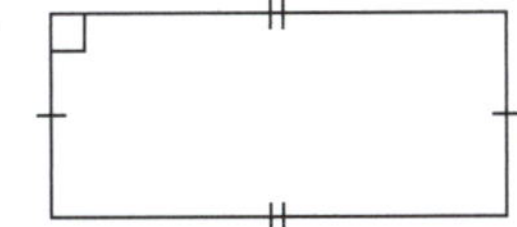

(5 marks)

3 Find the value of the pronumerals:

a 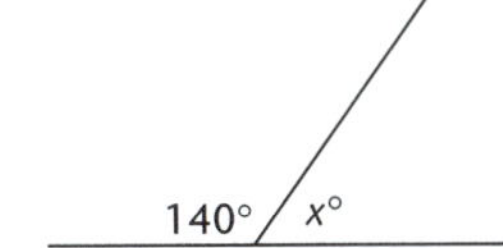


b 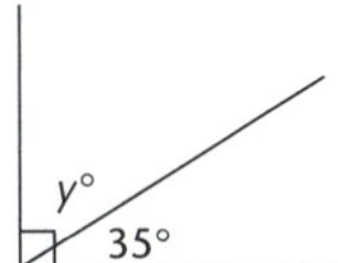


c 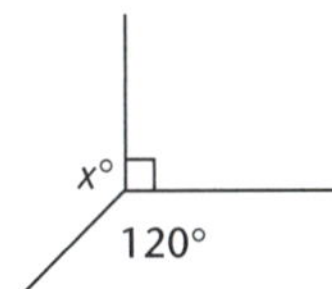


d 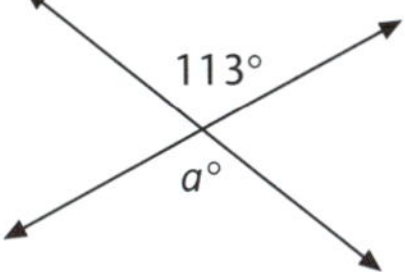


e 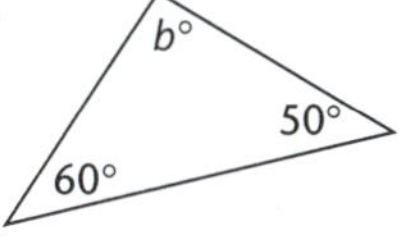


f 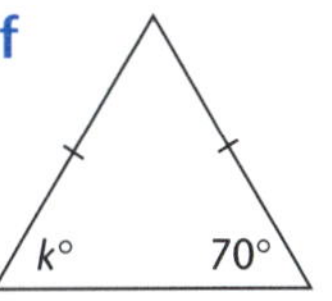


g 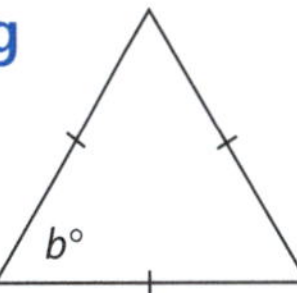


h 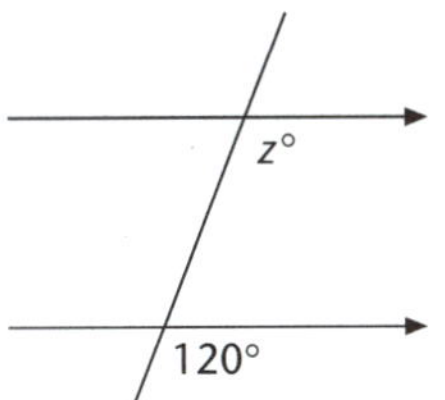


i 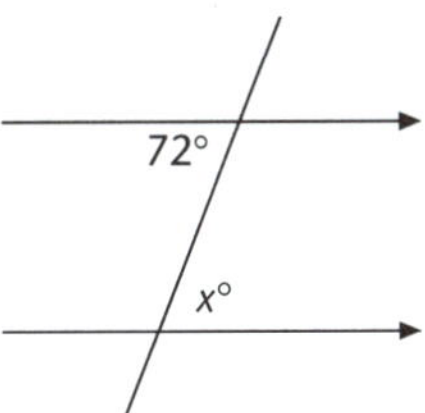


j 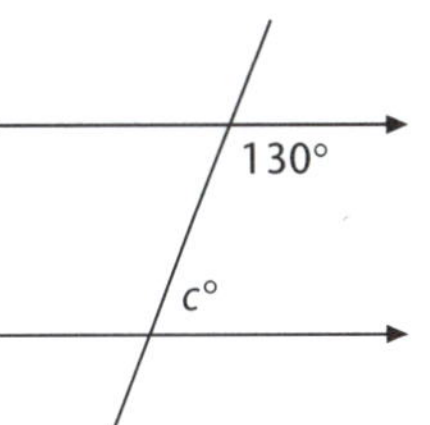


k 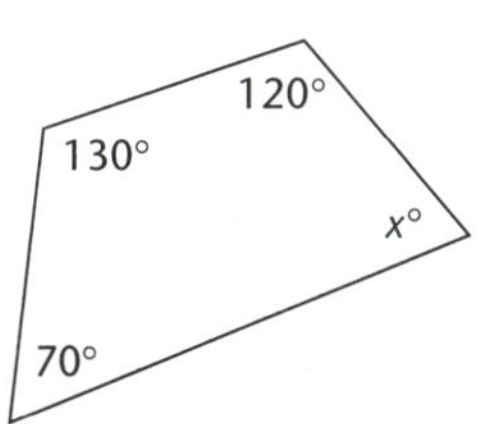


l 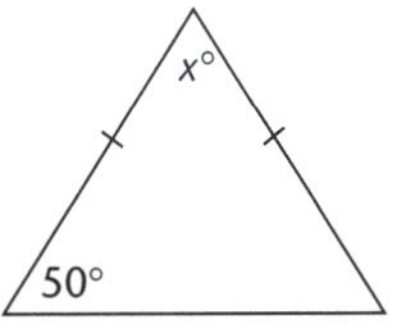


m 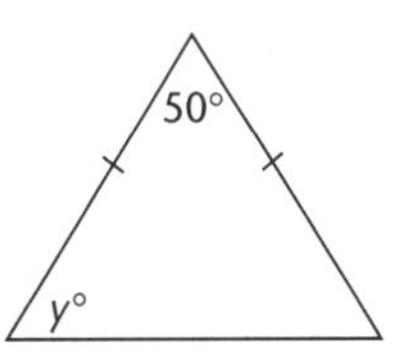


(13 marks)

*(cont.)*

**4** In this diagram triangle XYZ is congruent to triangle MNP.

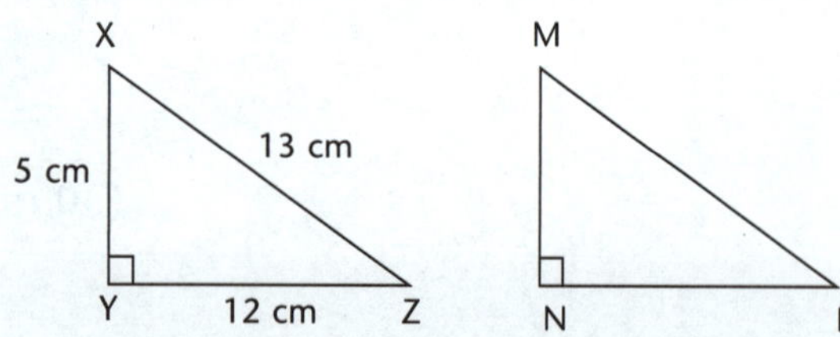


**a** What side corresponds to side XZ?

**b** How long is side MN?

**c** What angle corresponds to ∠ZXY? (3 marks)

**5** Figure ABCD is congruent to figure FGHI.

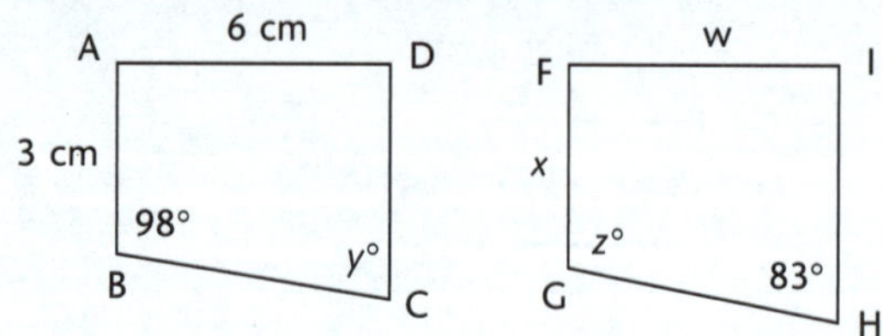


Find the value of $x$, $y$, $z$ and $w$. (4 marks)

**Your Feedback** $\frac{\square}{30} \times 100\% = \square\%$

☞ **Quick answers on page 293**
☞ **Worked solutions on page 394**

**(40 marks)**

1 Find the value of the pronumerals.

a 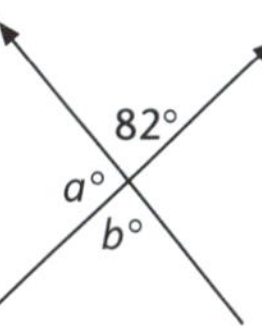


b 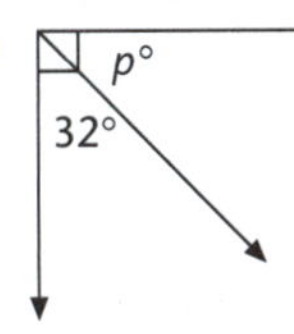


c 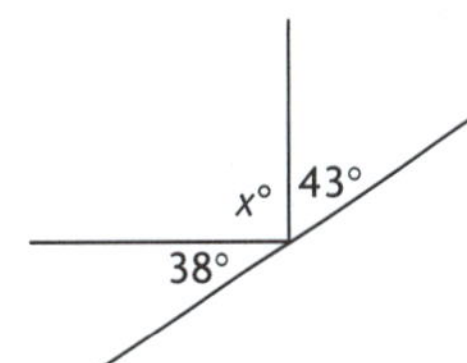


d 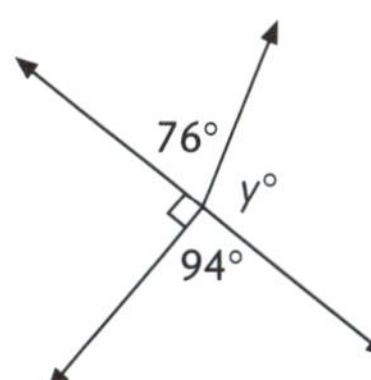


e 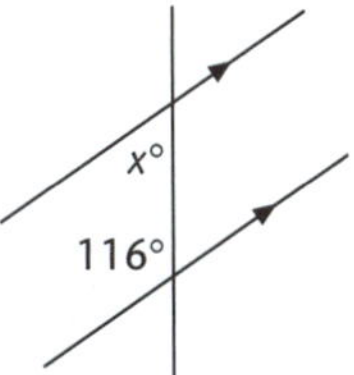


f 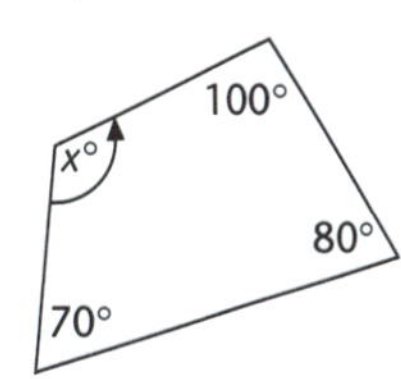


(7 marks)

2 Find the value of the pronumeral.

a 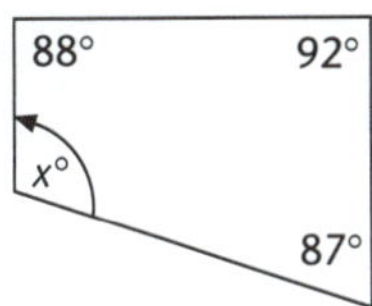


b 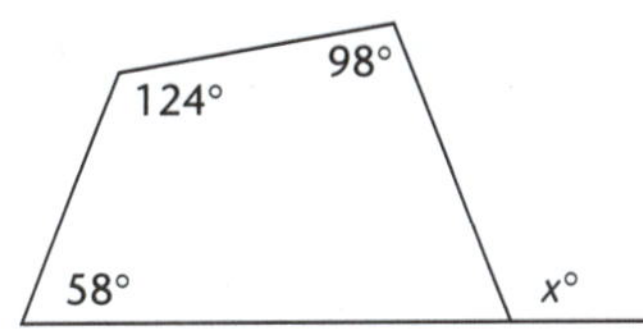


(3 marks)

3 Find the value of $x$.

a 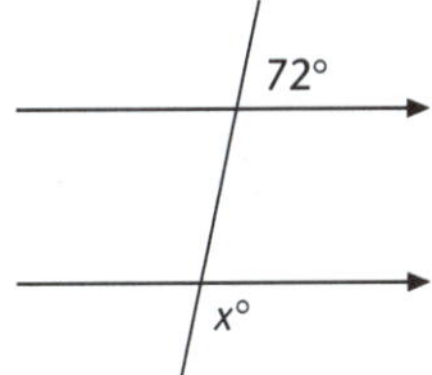


b 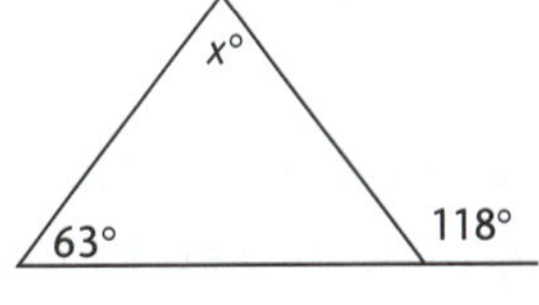


(3 marks)

4 Find the value of $x$.

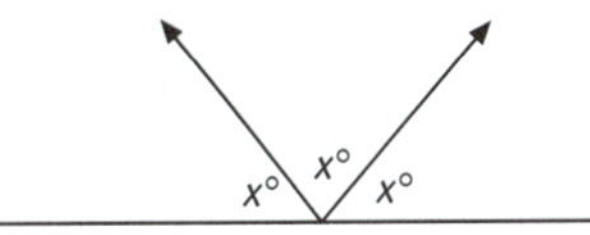


(1 mark)

5 Find the value of $x$.

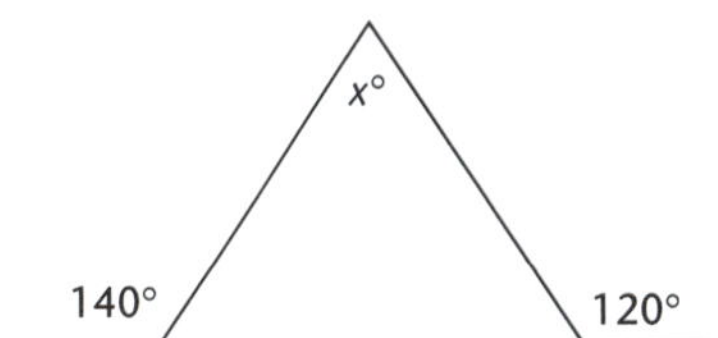


(2 marks)

*(cont.)*

**6** True or false? Line AB || line CD:

**a**

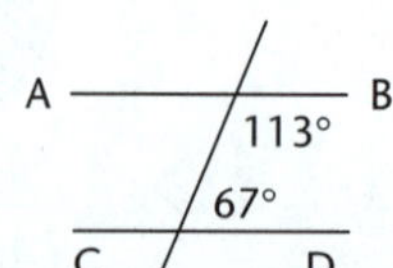


**b**

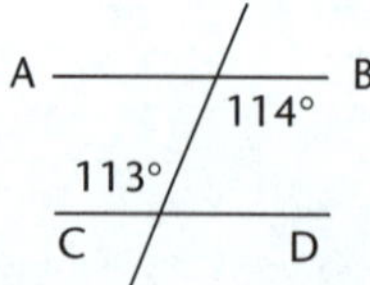


**c**

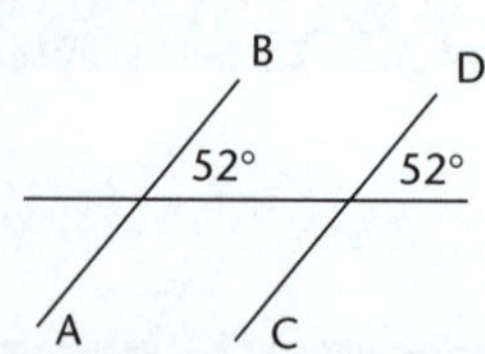


(3 marks)

**7** Find the value of $x$.

**a**

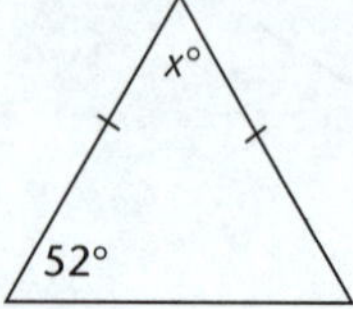


**b**

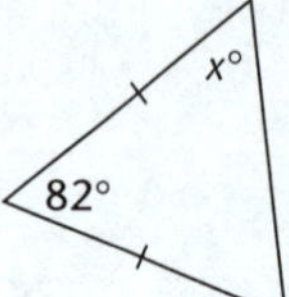


**c**

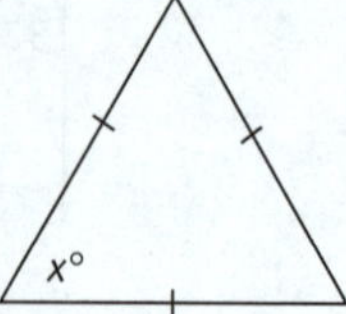


(3 marks)

**8** Find the value of $x$.

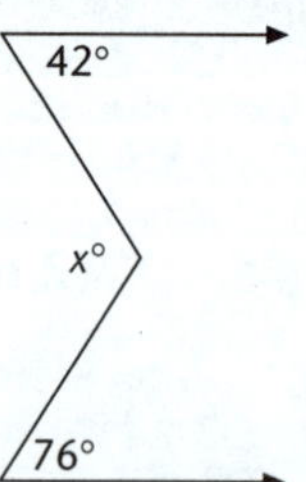


(2 marks)

**9**

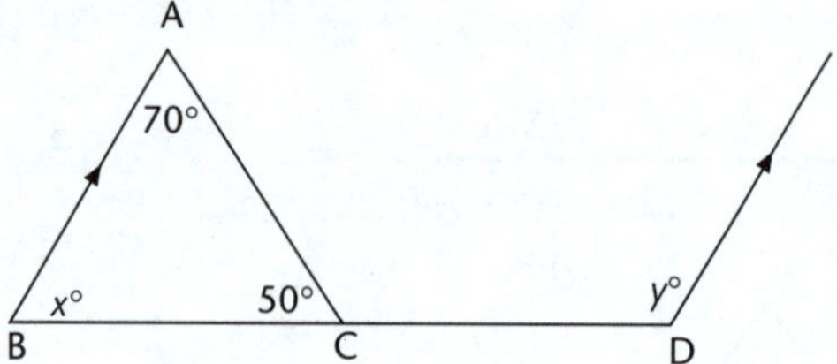


**a** $x =$ ______

Why? ____________________

**b** $y =$ ______

Why? ____________________ (4 marks)

**10** In the figure below, triangle A is reflected about line m, to form triangle B. Triangle B is translated, then rotated to form triangle C.

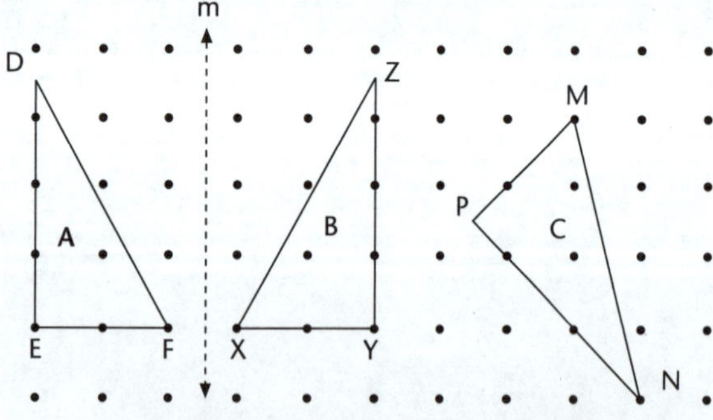


**a** Is triangle A congruent to triangle B?

**b** Is triangle B congruent to triangle C?

**c** Which side of triangle C corresponds to side DE of triangle A?

**d** Which angle of triangle C corresonds to angle DEF in triangle A?

**e** Which triangle has the largest area? (5 marks)

**11** Which of the following pairs of triangles are congruent?
If so, state the test used.

**a**

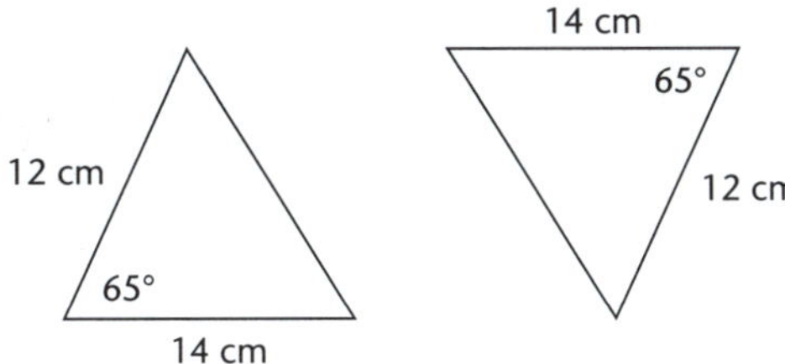


**b**

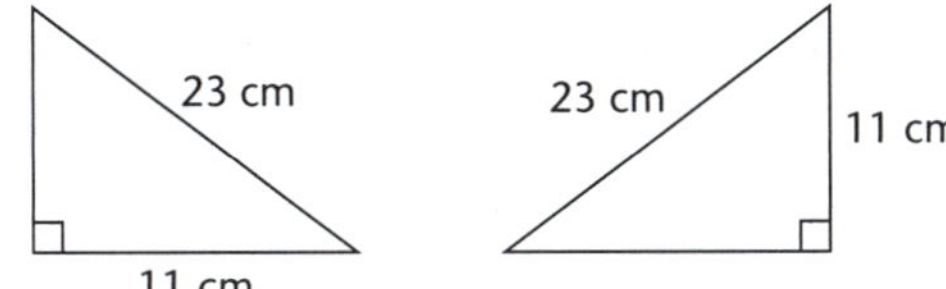


**c**

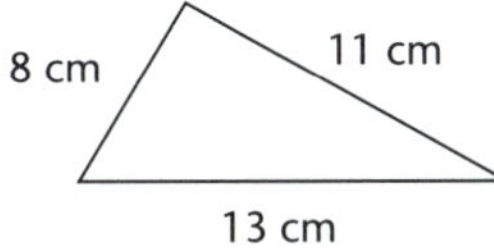


8 cm
13 cm
11 cm

**d**

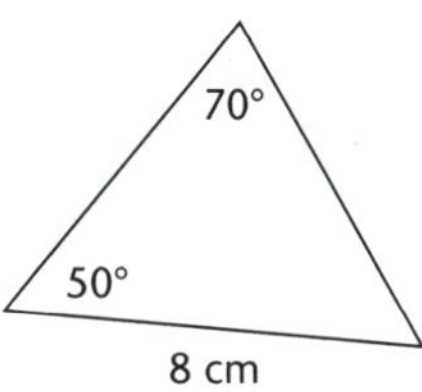


70°
60°
8 cm

(4 marks)

**12**

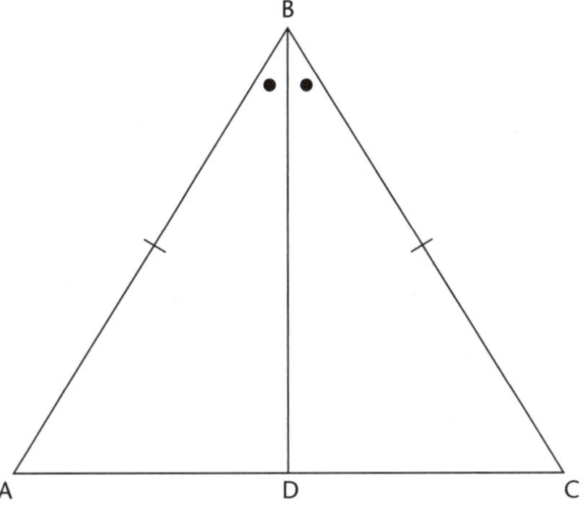


In $\triangle ABC$, $AB = CB$ and $\angle ABD = \angle CBD$.
Prove $\triangle ABD \cong \triangle CBD$. (3 marks)

☞ Quick answers on page 293
☞ Worked solutions on page 394

**Your Feedback** $\dfrac{\square}{40} \times 100\% = \square\%$

**(45 marks)**

1 True or false?

a All rectangles and squares are parallelograms.

b A trapezium is a parallelogram.

c A rhombus is a parallelogram. (3 marks)

2 Find the value of the pronumeral.

a 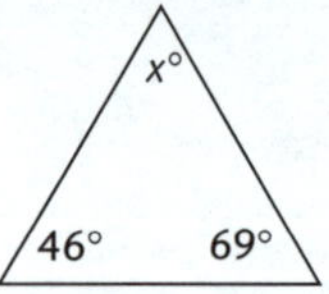


b 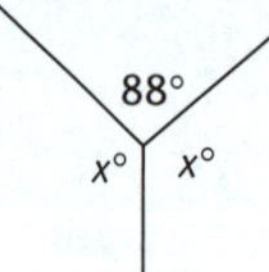


c 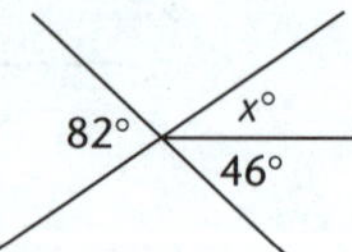


d 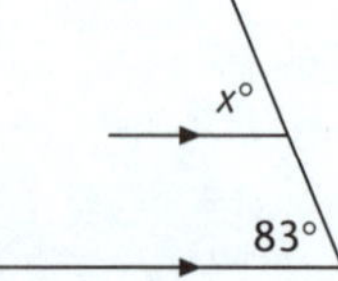


e 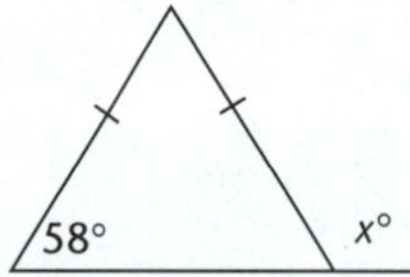


f 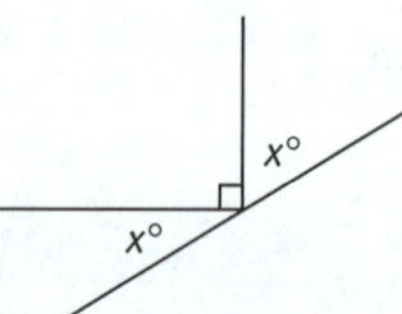


g 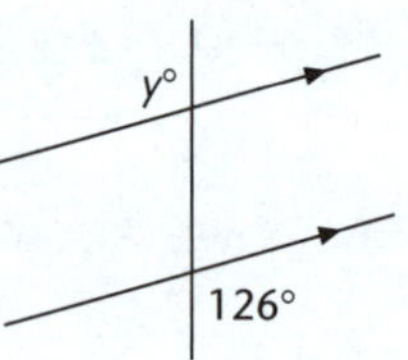


h 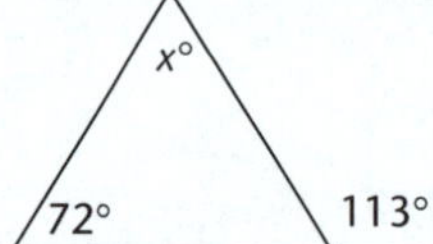


i 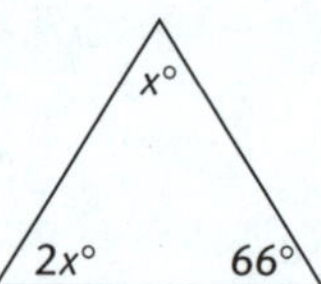


j 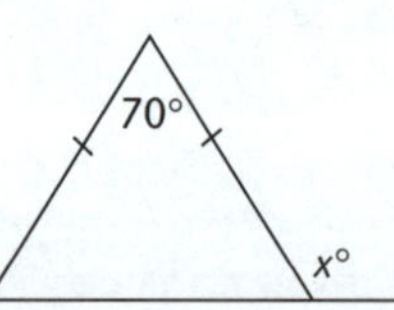


(10 marks)

3 Find the value of $x$.

a 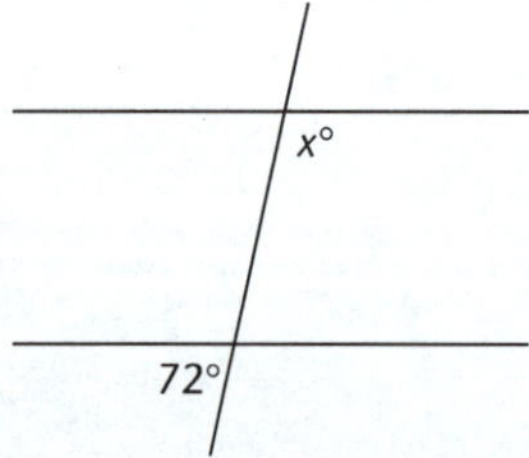


b 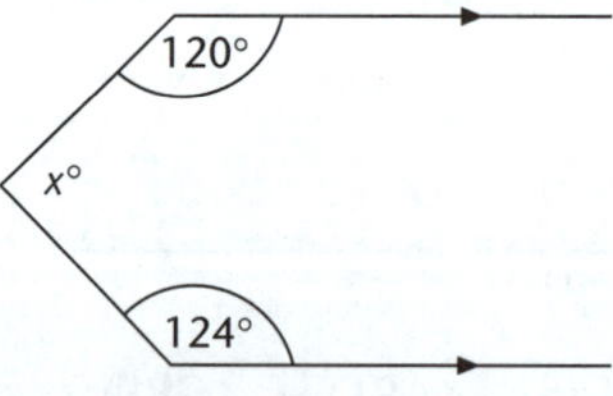


c 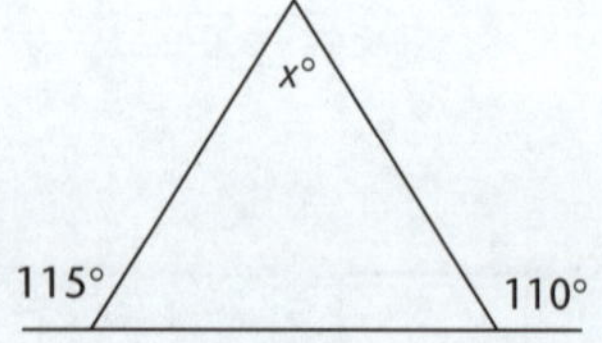


d 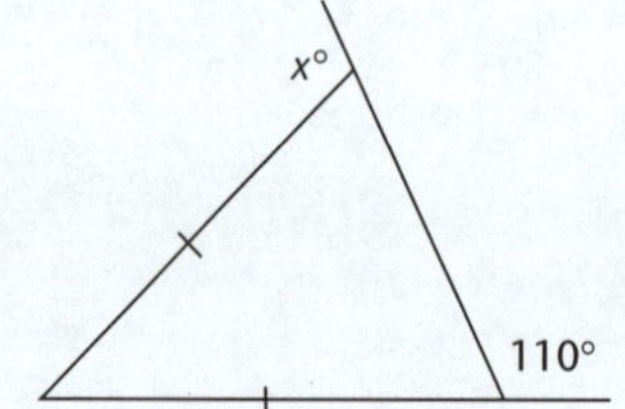


(6 marks)

4 Find the value of $a$.

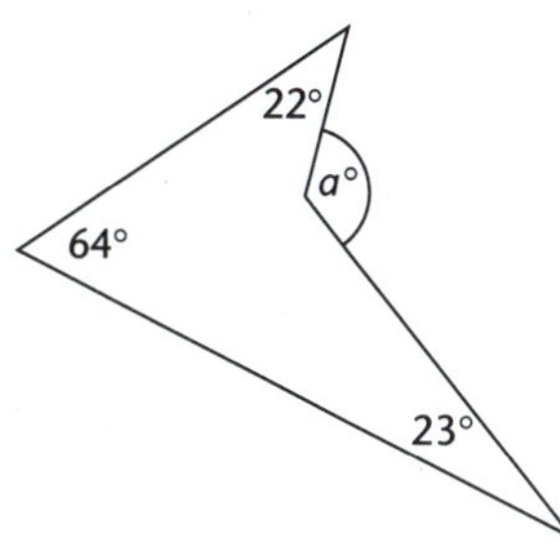


(2 marks)

5 Find the value of $x$, $y$ and $z$.

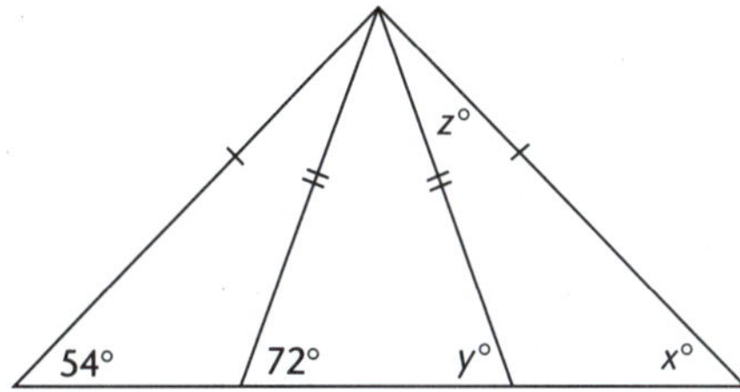


(3 marks)

6 Find the value of $a$ and $b$.

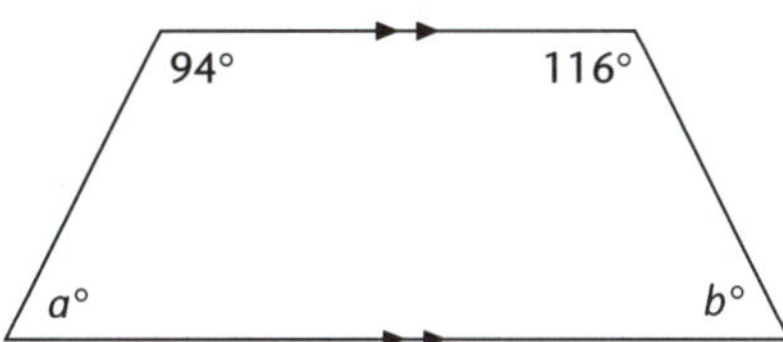


(2 marks)

7 In the diagram, XZ || PQ, ∠YPQ = 46° and ∠QYP = 81°.

Reasons:

a ∠PYX = ______

Why? ______________________________

b ∠ZYP = ______

Why? ______________________________

c ∠PQY = ______

Why? ______________________________

(6 marks)

8 AB || CD ∠BMI = 70° and ∠GQC = 70°.

a Giving reasons, find the size of ∠HND and ∠FQD.

b Is line FG parallel to line HI? Why?

(4 marks)

*(cont.)*

**9**

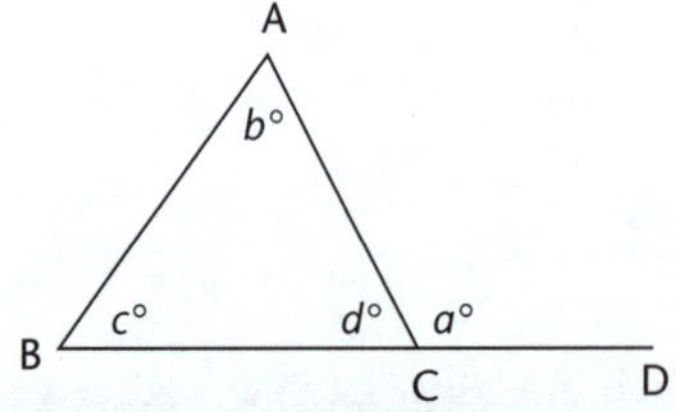


Prove that $a = b + c$. (3 marks)

**10** Determine whether the following pairs of triangles are congruent.
If so, state the test.

**a**

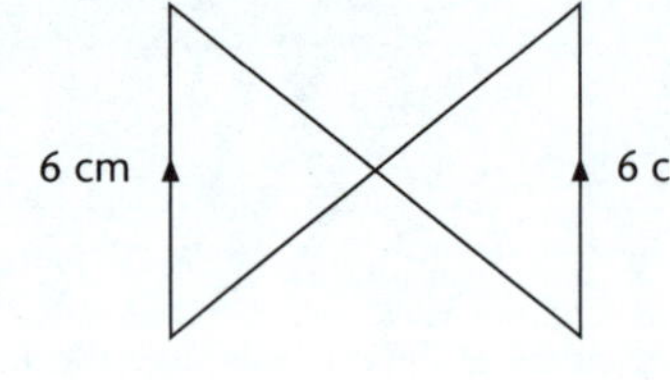


**b**

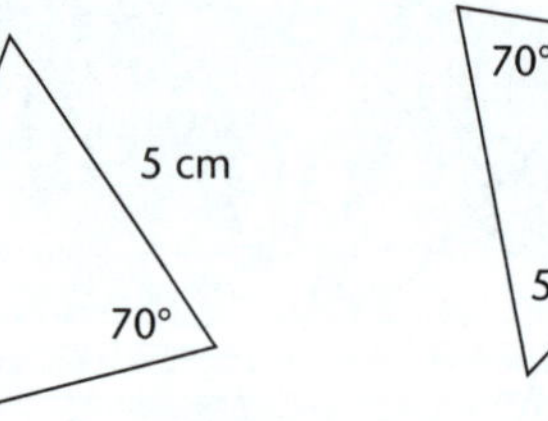


**c**

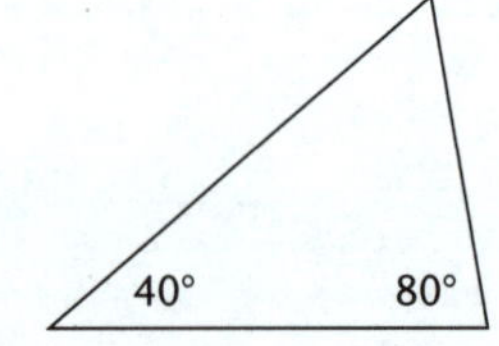


(3 marks)

**11**

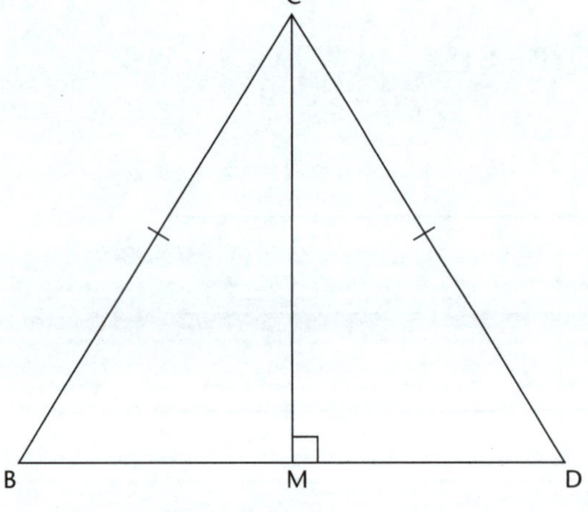


Prove $\triangle BCM \cong \triangle DCM$. (3 marks)

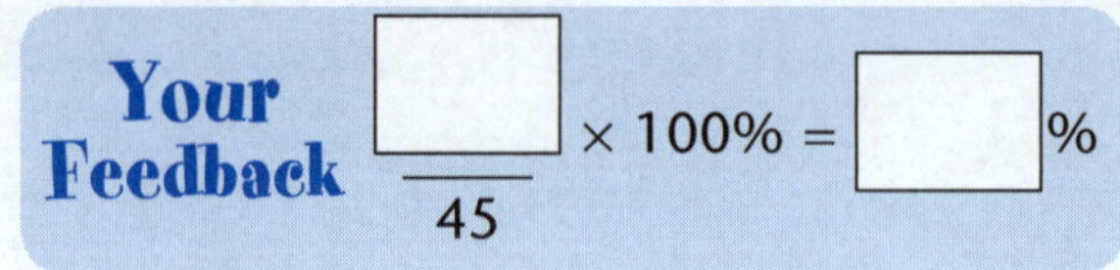


☞ Quick answers on page 294
☞ Worked solutions on page 394

# 11 STATISTICS

- Some Explanations
- Census versus Sample
- Biased and Random Samples
- Data Representation
- Data Analysis

## KEYWORDS

| | |
|---|---|
| **Analyse** | **Median** |
| **Bias** | **Mode** |
| **Categorical data** | **Organisation** |
| **Census** | **Outlier** |
| **Clustered** | **Polygon** |
| **Collection** | **Population sample** |
| **Complementary events** | **Probability event outcome** |
| **Continuous data** | **Quantitative data** |
| **Data** | **Random** |
| **Discrete data** | **Range** |
| **Dot plot** | **Scatter diagram** |
| **Frequency** | **Score** |
| **Grouped data** | **Spread** |
| **Histogram** | **Stem and leaf** |
| **Information** | **Tally** |
| **Location** | |
| **Mean** | |

## Some Explanations

Statistics involves the collection and organisation of information (**data**) so that:

**a** large amounts of information can be easily analysed, and

**b** predictions can be made, based on analysis of the data collected.

Tables and graphs allow information to be presented in a clear, concise form. The information can also be readily analysed from the table or graph.

## Census versus Sample

A **census** involves surveying every member of the target population, while a **sample** is a limited survey used to get a representative or typical view. The Australian Government conducts a regular census of all Australian households, asking questions about such things as number of residents, ages of residents, occupations, etc. The collected census information is analysed and used for future planning by government departments. Things such as new subdivisions, schools, infrastructure and hospitals can be planned in advance, based on information in the census.

## Biased and Random Samples

If a sample is **random**, then there is no **bias**. A biased sample is one that does not represent the whole population. For example, if a sample is conducted at a shopping centre on a Thursday morning to find the percentage of Australians who eat Weet Bix for breakfast, it would be a biased sample.

In a random sample participants are chosen 'at random'. For example, to randomly select from 1000 students a teacher might allocate a three-digit number to each student. She can then use her calculator to generate random numbers to find 20 students to interview.

## Data Representation

Data can be **quantitative** or **categorical**. Examples of quantitative data are shoe sizes (discrete) or height (continuous) while an example of categorical data is hair colour: brown, black, blonde, red. Collected data can be represented in tables or graphs.

## Frequency Distribution Table

A table is used to summarise listed data and allows easy analysis of the location and spread of the data.

### For Example

Consider the shoe sizes of 50 11-year-old boys:

| | | | | |
|---|---|---|---|---|
| 6 | $4\frac{1}{2}$ | 7 | 4 | 5 |
| $5\frac{1}{2}$ | 6 | 8 | 5 | 7 |
| 5 | $5\frac{1}{2}$ | 7 | $4\frac{1}{2}$ | 5 |
| 6 | $5\frac{1}{2}$ | 4 | 6 | 6 |
| 7 | $4\frac{1}{2}$ | 5 | 5 | 6 |
| 5 | 5 | $5\frac{1}{2}$ | $4\frac{1}{2}$ | 5 |
| $7\frac{1}{2}$ | $3\frac{1}{2}$ | 5 | $4\frac{1}{2}$ | $6\frac{1}{2}$ |
| $4\frac{1}{2}$ | $5\frac{1}{2}$ | 6 | $7\frac{1}{2}$ | 6 |
| 6 | 5 | 6 | 5 | 7 |
| 6 | 6 | 7 | $5\frac{1}{2}$ | 6 |

1. Draw up a frequency distribution table for this information.
2. How many boys wear a size $5\frac{1}{2}$ shoe?
3. Which size shoe is most commonly worn?
4. If this is a typical example of 11-year-old boys, what fraction of 11-year-old boys in Australia wear a size 6 shoe?
5. If the population of 11-year-olds in Hambelton is 200, how many would you predict wear a size 6 shoe?

**1**

| Score ($x$) | Tally | Frequency ($f$) | |
|---|---|---|---|
| $3\frac{1}{2}$ | \| | 1 | |
| 4 | \|\| | 2 | |
| $4\frac{1}{2}$ | ~~\|\|\|\|~~ \| | 6 | |
| 5 | ~~\|\|\|\|~~ ~~\|\|\|\|~~ \|\| | 12 | |
| $5\frac{1}{2}$ | ~~\|\|\|\|~~ \| | 6 | ← (2) |
| 6 | ~~\|\|\|\|~~ ~~\|\|\|\|~~ ~~\|\|\|\|~~ \|\|\| | 13 | ← (3) |
| $6\frac{1}{2}$ | \| | 1 | |
| 7 | ~~\|\|\|\|~~ \| | 6 | |
| $7\frac{1}{2}$ | \|\| | 2 | |
| 8 | \| | 1 | |
| | $\Sigma f$ | 50 | |

The completed table can then be used to answer the remaining questions.

**2** Six boys wear size $5\frac{1}{2}$ shoes.

**3** The most common size is 6 because 13 boys wear size 6.

**4** Fraction of 11-year-old boys with size 6 shoe:

$$= \frac{13}{50}\left(\frac{\text{Number wearing size 6}}{\text{Total number of boys in survey}}\right)$$

**5** Number wearing size 6 $\doteqdot \frac{13}{\cancel{50}_1} \times \frac{\cancel{200}^4}{1}$

$\doteqdot 52$

[$\doteqdot$ or $\approx$ mean 'approximately equal to']

You would expect about 52 11-year-olds in Hambelton to wear size 6 shoes.

## Frequency Distribution Table for Grouped Data

When data is continuous it is sensible to use class intervals.

The mass of 40 students was measured and the results listed below:

| | | | | |
|---|---|---|---|---|
| 53 | 51 | 62 | 64 | 63 |
| 71 | 75 | 79 | 56 | 60 |
| 53 | 48 | 61 | 64 | 63 |
| 67 | 59 | 63 | 68 | 44 |
| 53 | 55 | 59 | 52 | 64 |
| 67 | 68 | 72 | 75 | 79 |
| 61 | 65 | 48 | 41 | 47 |
| 48 | 55 | 57 | 56 | 61 |

1 Arrange the data in a frequency distribution table using class intervals of 41–45, 46–50, etc. Also, find the class centre for each interval.

2 What was the most common class interval of masses (the **modal** class)?

1 The completed table can then be used to answer the questions:

| Class interval | Class centre ($x$) | Tally | Frequency ($f$) |
|---|---|---|---|
| 41–45 | 43 | \|\| | 2 |
| 46–50 | 48 | \|\|\|\| | 4 |
| 51–55 | 53 | ~~\|\|\|\|~~ \|\| | 7 |
| 56–60 | 58 | ~~\|\|\|\|~~ \| | 6 |
| 61–65 | 63 | ~~\|\|\|\|~~ ~~\|\|\|\|~~ \| | 11 |
| 66–70 | 68 | \|\|\|\| | 4 |
| 71–75 | 73 | \|\|\|\| | 4 |
| 76–80 | 78 | \|\| | 2 |
| | | | 40 |

2 Most common class interval of masses was 61–65.

[Note: the scores are **clustered** in the low 60s.]

## Stem-and-Leaf Plot

The stem is the *first* digit or digits of a number, whereas the leaf is the *last* digit. The leaf is always a single digit.

1 The results of a mathematics test were recorded:

| | | | | | | | |
|---|---|---|---|---|---|---|---|
| 19 | 48 | 36 | 40 | 31 | 22 | 18 | 27 |
| 18 | 20 | 18 | 36 | 49 | 60 | 45 | 13 |
| 9 | 17 | 22 | 31 | 39 | 26 | 28 | 30 |
| 44 | 8 | 23 | 19 | 46 | 33 | | |

Draw a stem-and-leaf plot for this data.

1

| Stem | Leaf |
|---|---|
| 0 | 98 |
| 1 | 9888379 |
| 2 | 2702683 |
| 3 | 6161903 |
| 4 | 809546 |
| 5 | |
| 6 | 0 |

This is sometimes referred to as the initial plot. However, the plot can be refined by ordering each of the leaves:

| Stem | Leaf |
|---|---|
| 0 | 89 |
| 1 | 3788899 |
| 2 | 0223678 |
| 3 | 0113669 |
| 4 | 045689 |
| 5 | |
| 6 | 0 |

An ordered stem-and-leaf plot can be used to analyse the data. A back-to-back stem-and-leaf plot is used to compare two sets of data.

[Note: the score of 60 is called an **outlier**.]

## Frequency Histogram and Polygon

There are two basic types of frequency graphs, a **frequency histogram**, and a **frequency polygon**.

The table below has been used to graph a histogram and frequency polygon:

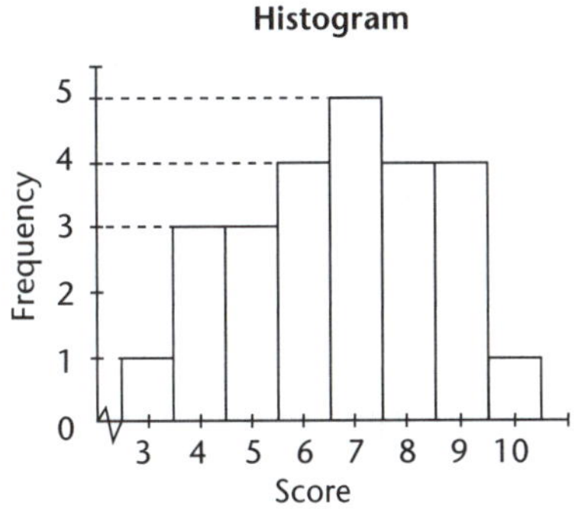


| $x$ | $f$ |
|---|---|
| 3 | 1 |
| 4 | 3 |
| 5 | 3 |
| 6 | 4 |
| 7 | 5 |
| 8 | 4 |
| 9 | 4 |
| 10 | 1 |

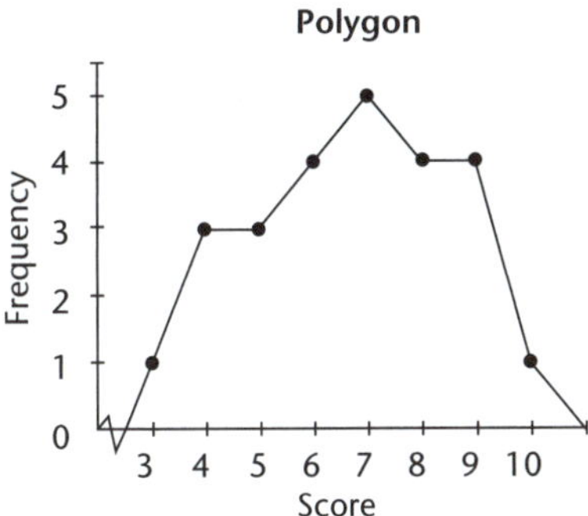


Note that the use of ⩗ to indicate that some numbers have been omitted (in this case 1 and 2). This cuts out unnecessary parts of the graph.

Columns on the frequency histogram are around the scores. Their height is their frequency.

Straight lines on the frequency polygon join dots at the height of the frequency value. The line meets the horizontal axis where the next score would be if there was one. The polygon can also be made by joining the mid-points of the tops of the columns in the histogram.

## Dot Plots

A dot plot is an alternative to drawing a histogram and can either be vertical or horizontal.

**1** The type of cars passing along Honus Avenue over a half-hour period was noted and the results recorded:

| Car manufacturer | Frequency |
|---|---|
| Ford | 8 |
| Holden | 11 |
| Toyota | 7 |
| Mitsubishi | 4 |
| Other | 6 |

Use a dot plot to represent this information.

**1**

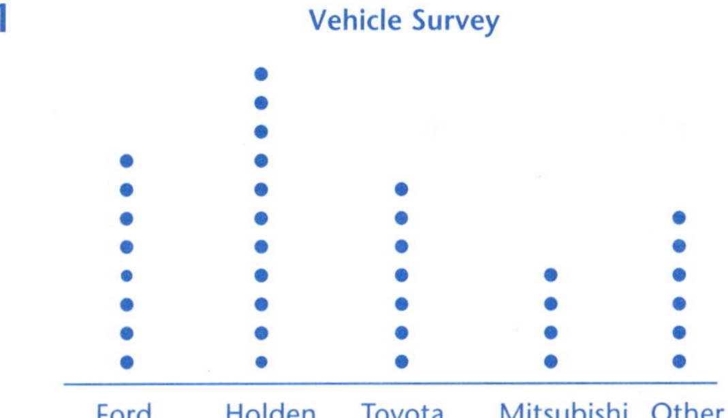


## Scatter Diagram

A **scatter diagram** is formed by using points to represent a pair of results.

The results of an English test and a Science test for ten students are detailed:

Student: A B C D E F G H I J

English: 13 17 14 11 18 17 15 10 20 16

Science: 15 16 13 15 17 18 16 11 14 19

**1** Draw a scatter diagram for the data.

**2** Which students scored the highest and lowest scores in each test.

1

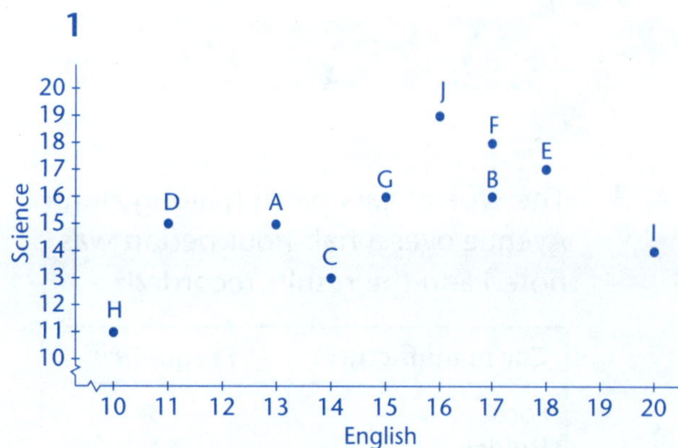


2 English highest: student I; lowest: student H.
Science highest: student J; lowest: student H.

## Data Analysis

### Measures of Spread

The **range** indicates the size of the distribution of scores, that is, the range = highest score – lowest score.

**For Example**

1 Find the range of the scores 4, 17, 6, –5, 4, 2, 12.

1 Range = 17 – (–5)
= 17 + 5 = 22

### Measures of Location

- **Mean** is the *average score*:

  i.e. Mean $= \dfrac{\text{Sum of scores}}{\text{Number of scores}}$

  [Notation: $\bar{x}$ = mean]

- **Mode** is most common score—the score with the highest frequency.
- **Median** is the middle score when the scores are arranged in order.

**For Example**

Consider these sets of scores for Alison's first season of cricket. Over the first seven innings she scored 17, 14, 10, 10, 9, 21 and 10. Over the next eight innings she scored 0, 19, 10, 26, 19, 35, 0 and 29.

1 For the first seven innings calculate the range, mode, mean and median.

2 For the next eight innings calculate the range, mode, mean and median.

3 Taking all scores together, find the season's range, mode, mean and median.

1 17, 14, 10, 10, (9), (21), 10

Range = 21 – 9
= 12

Mode = 10

Mean $= \frac{17+14+10+10+9+21+10}{7}$
$= \frac{91}{7}$
= 13

For median, rewrite the scores in ascending order:

9 10 10 10 14 17 21
↑
3 scores below ← **middle score** → 3 scores above

Median = 10

[A suggested method to locate the middle is to cross off the first and last numbers, then the remaining first and last numbers, etc., until only one or the two middle numbers are left.]

2 0, 19, 10, 26, 19, 35, 0, 29

Range = 35 – 0 = 35

Mode = 0 and 19

$$\text{Mean} = \frac{0 + 19 + 10 + 26 + 19 + 35 + 0 + 29}{8}$$

$$= \frac{138}{8}$$

$$= 17.25$$

For median: 0, 0, 10, 19, 19, 26, 29, 35

Median is middle of 4th and 5th scores, i.e. 19, 19

Median = 19

3 17, 14, 10, 10, 9, 21, 10, 0, 19, 10, 26, 19, 35, 0, 29

Range = 35 – 0 = 35

Mode = 10

$$\text{Mean} = \frac{17 + 14 + 10 + 10 + 9 + 21 + 10 + 0 + 19 + 10 + 26 + 19 + 35 + 0 + 29}{15}$$

$$= \frac{229}{15}$$

$$= 15.27 \text{ (to 2 decimal places)}$$

For median: 0, 0, 9, 10, 10, 10, 10, 14, 17, 19, 19, 21, 26, 29, 35

Median is the middle score.

Median = 14

## Using a Calculator to Find the Mean

A calculator can be placed in 'statistics' mode to find a number of statistical measures. Refer to your calculator manual for instructions.

## Measures of Location and Spread from a Table or Graph

Mean, mode, median and range can be found when data is arranged in a table or graph.

### For Example

| Score | Frequency |
|---|---|
| 12 | 5 |
| 13 | 7 |
| 14 | 8 |
| 15 | 7 |
| 16 | 3 |

1 Find the mean, mode, median and range.

1 To find the mean, another column is added to the table. In this column, we multiply each score by its frequency, i.e. the $x \times f$ or $fx$, column.

| Score ($x$) | Frequency ($f$) | $fx$ |
|---|---|---|
| 12 | 5 | 60 |
| 13 | 7 | 91 |
| 14 | 8 | 112 |
| 15 | 7 | 105 |
| 16 | 3 | 48 |
| **Total** | 30 | 416 |

- **Mean** $(\bar{x}) = \dfrac{\text{Sum of scores}}{\text{Number of scores}} = \dfrac{\text{Sum of } fx}{\text{Sum of } f} = \dfrac{\sum fx}{\sum f}$

  i.e. $\bar{x} = \dfrac{416}{30}$

  $= 13.87$ (to 2 decimal places)

- **Mode**: 14 (has highest frequency of 8)
- **Median**: As there are 30 scores, then the median is the average of the two middle scores:

  i.e. the average of 15th and 16th scores

  i.e. the average of 14 and 14

  ∴ Median = 14

- **Range**: 16 – 12 = 4

## For Example

1 The stem-and-leaf plot shows the results for the students in a maths test.

| Stem | Leaf |
|---|---|
| 0 | 9 |
| 1 | 35689 |
| 2 | 555899 |
| 3 | 0025 |

Find the mean, mode, median and range.

1 Mean: $(9 + 13 + 15 + \ldots) \div 16$

$= 23.625$

Mode: 25

Median:

| Stem | Leaf |
|---|---|
| 0 | 9 |
| 1 | 35689 |
| 2 | 555899 |
| 3 | 0025 |

Middle of 25 and 25 is 25.

Range: $35 - 9 = 26$

## For Example

1 A class is surveyed to find the number of mobile phones per household.

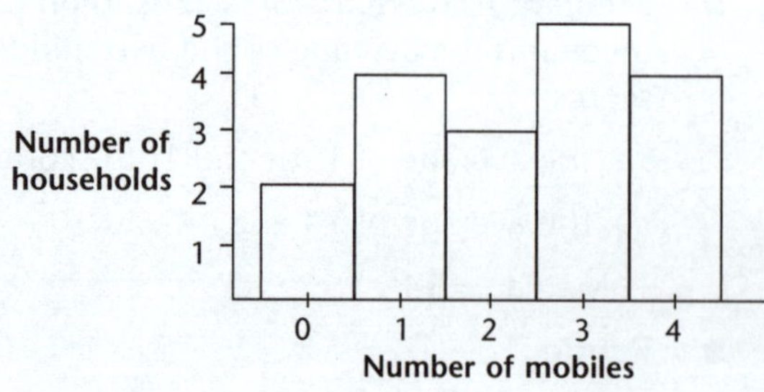


Find the mean, mode and median.

1 **Mean:** To help find the mean students can either complete a frequency table for the histogram or note the $fx$ data on the existing histogram.

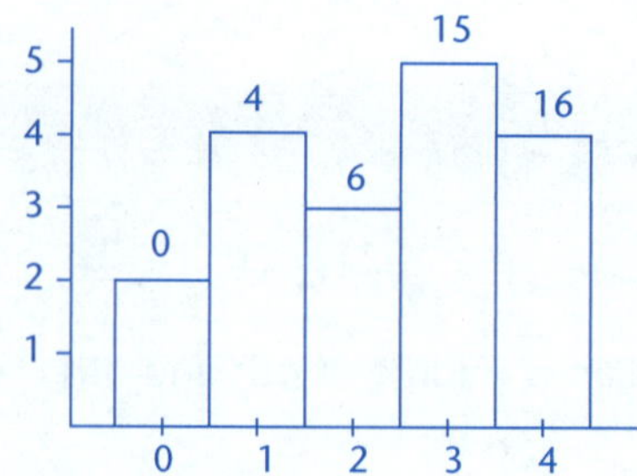


| Scores ($x$) | $f$ | $fx$ |
|---|---|---|
| 0 | 2 | 0 |
| 1 | 4 | 4 |
| 2 | 3 | 6 |
| 3 | 5 | 15 |
| 4 | 4 | 16 |
| | 18 | 41 |

$$\bar{x} = \frac{\sum fx}{\sum f}$$

$$= \frac{0 + 4 + 6 + 15 + 16}{2 + 4 + 3 + 5 + 4}$$

$$= \frac{41}{18}$$

$$= 2.2\dot{7}$$

**Mode:** 3 (highest frequency – tallest column).

**Median:** As there are 18 scores:

∴ median is middle of 9th and 10th score

i.e. 0, 0, 1, 1, 1, 1, 2, 2, (2, 3) . . .

∴ Median = 2.5

Go to p. 289 for quick answers, or to pp. 361–365 for worked solutions.

**1** Johann Bark surveyed the numbers of the letter t in each line of his book. pp. 233–235
His results were:

| 5 | 3 | 6 | 3 | 4 | 2 | 5 | 3 | 4 | 7 |
|---|---|---|---|---|---|---|---|---|---|
| 2 | 2 | 4 | 4 | 5 | 5 | 4 | 3 | 2 | 5 |

From this information, draw up a frequency distribution table and answer the following questions.

**a** How many lines contained:
  **i** 5 ts? **ii** 5 or fewer ts?
  **iii** more than 5 ts?

**b** What fraction of lines contained:
  **i** 5 ts? **ii** 5 or fewer ts?
  **iii** more than 5 ts?

**c** Find the range and mode.

**d** Draw a histogram to illustrate this information.

**2** The height of 30 students was measured and the results listed: pp. 233–237

| 142 | 148 | 156 | 158 | 167 | 169 |
|---|---|---|---|---|---|
| 171 | 163 | 158 | 149 | 148 | 155 |
| 158 | 150 | 153 | 163 | 164 | 169 |
| 148 | 149 | 157 | 158 | 164 | 165 |
| 173 | 161 | 165 | 149 | 148 | 155 |

Complete the frequency distribution table for the data:

| Class | Class centre | Tally | Frequency |
|---|---|---|---|
| 141–145 | | | |
| 146–150 | | | |
| … | | | |

**3** A survey of students' pulse rates are detailed: p. 235

| 59 | 63 | 82 | 76 | 88 | 83 | 76 | 72 | 65 |
|---|---|---|---|---|---|---|---|---|
| 58 | 71 | 69 | 74 | 79 | 81 | 69 | 67 | 68 |
| 73 | 75 | 81 | 80 | 71 | 69 | 59 | 63 | 71 |
| 78 | 81 | 76 | | | | | | |

Complete a stem-and-leaf plot for the data.

4 Goldilocks surveyed 25 bears in the forest to find out the number of lumps each had in their porridge that morning. This is the set of numbers that she wrote down: p. 235

7 4 0 2 3 6 4 5 3 2 1 3 7
6 6 4 5 2 4 5 5 4 6 7 6

Draw up a frequency distribution table for this information and answer the following questions.

a How many bears reported:
   i 5 lumps?
   ii 5 or fewer lumps?
   iii less than 5 lumps?
   iv more than 5 lumps?

b What fraction of the bears survey reported:
   i 5 lumps?
   ii 5 or fewer lumps?
   iii less than 5 lumps?
   iv more than 5 lumps?

c Draw a frequency polygon to illustrate this information.

5 The table shows the number of soccer goals scored by five friends during a season: p. 235

| Name | Mitchell | Brendan | Mathew | Jason | Eddie |
|---|---|---|---|---|---|
| Goals | 6 | 3 | 5 | 4 | 2 |

Draw a dot plot to represent the information.

6 Eight students completed a history and a geography quiz and the results are detailed below: pp. 235–236

| Student | A | B | C | D | E | F | G | H |
|---|---|---|---|---|---|---|---|---|
| History | 5 | 8 | 4 | 7 | 6 | 3 | 9 | 5 |
| Geography | 2 | 9 | 3 | 7 | 6 | 4 | 8 | 3 |

Draw a scatter diagram showing the information.

7 Find the range of the scores 8, 6, 4, 7, 6, 2, –1. p. 236

8 Calculate the mean, and find the mode, median and range of these sets of numbers. (Calculate the mean to 1 decimal place where necessary.) pp. 236–237

a 7, 4, 6, 6, 7, 6, 3
b 5, 10, 8, 7, 11, 6, 5, 9
c 7, 7, 4, 4, 4, 1, 1, 1, 7
d 0, 0, 2, 2, 1, 7, 3, 7, 7, 4, 9, 5
e 7, 11, 9, 4, 11, 7, 4, 11, 9, 3, 7, 4, 5, 7, 6

**9** George surveyed children buying joggers from his shoe store and discovered this information:

pp. 233–237

| Shoe size ($x$) | Number of children ($f$) | $x \times f$ |
|---|---|---|
| 3 | 8 | |
| $3\frac{1}{2}$ | 8 | |
| 4 | 6 | |
| $4\frac{1}{2}$ | 12 | |
| 5 | 10 | |
| $5\frac{1}{2}$ | 8 | |
| 6 | 6 | |
| $6\frac{1}{2}$ | 4 | |
| 7 | 2 | |
| Total: | | |

By completing this frequency table, answer the following questions:

**a** How many children were surveyed?

**b** How many children wore shoes of size:
  **i** 5 or less? **ii** greater than 5?

**c** What fraction (in simplest form) of the children wore shoes of size:
  **i** 5? **ii** 5 or less?
  **iii** greater than 5?

**d** Find the range and mode.

**e** Calculate the mean shoe size.

**f** Find the median.

**g** Draw a histogram to illustrate this information.

**h** George wishes to cater for the children of the region. Which size shoe would you advise George to be sure to keep in stock? Which measure (mean, median or mode) is relevant to this question?

**10** Professor Kowalski drew a frequency polygon to represent his students' grades out of 10.

pp. 233–237

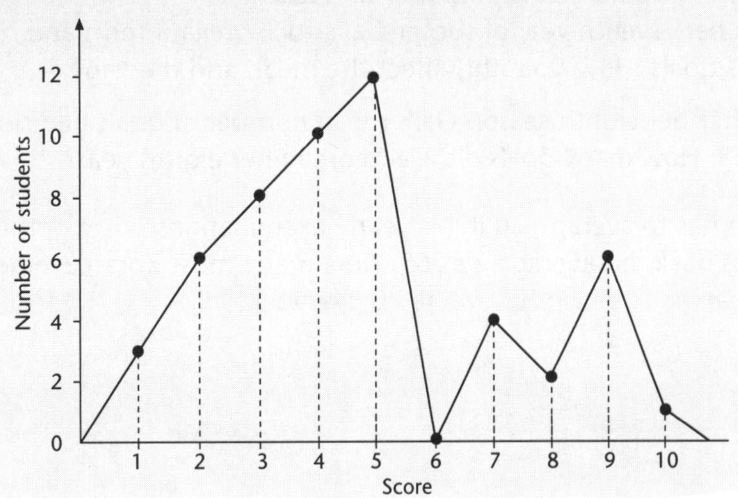

**a** Use the previous diagram to complete the frequency distribution table.

| Score ($x$) | Frequency ($f$) | $x \times f$ |
|---|---|---|
| 1 | | |
| 2 | | |
| 3 | | |
| 4 | | |
| 5 | 12 | 60 |
| 6 | | |
| 7 | | |
| 8 | | |
| 9 | | |
| 10 | | |
| | | |

**b** Write down the range and mode.

**c** How many students scored:

**i** 5 marks?

**ii** 5 or fewer marks?

**iii** fewer than 5 marks?

**iv** more than 5 marks?

**d** How many students were in Professor Kowalski's class?

**e** What fraction of the students scored:

**i** 5 marks?

**ii** 5 or fewer marks?

**iii** fewer than 5 marks?

**iv** more than 5 marks?

**f** Calculate the mean.

**g** Find the median score.

**11** Gai has been playing soccer for six years and has kept a record of her goal scoring. Over these years Gai scored 8, 6, 5, 9, 5 and 3 goals. pp. 236–237

**a** **i** Calculate Gai's mean number of goals.

**ii** Find the range and the mode.

**iii** Find the median number of goals.

**b** In her seventh year of soccer Gai struck brilliant form and scored 13 goals. How does this affect the mean and median?

**c** After her eighth season Gai's mean number of goals had slumped to 6.5. How many goals did Gai score in her eighth year?

**12** Zoran has to average 70 in his yearly examinations. After getting four results back, his average was 65. How many must Zoran average in his final two papers to attain his overall 70 average? pp. 236–237

**13** A survey of families in Nicholas Avenue was conducted to find the number of children in each family. The results were graphed as a histogram: pp. 233–237

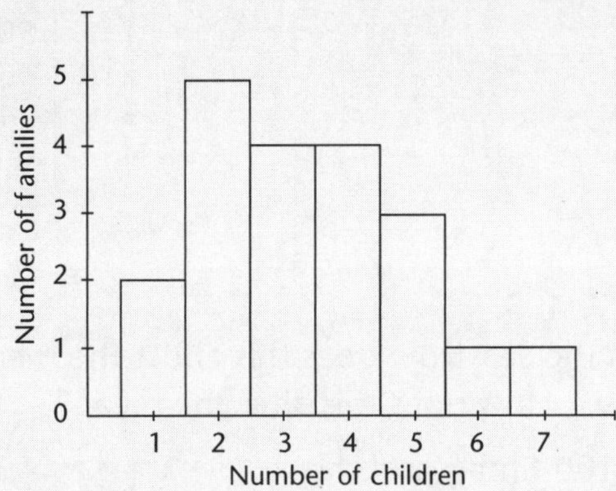


**a** **i** Complete the frequency distribution table using information from the histogram:

| Number of children (score) $x$ | Number of families (frequency) $f$ | $x \times f$ |
|---|---|---|
| 1 | | |
| 2 | | |
| 3 | | |
| 4 | | |
| 5 | | |
| 6 | | |
| 7 | | |
| | | |

**ii** How many families were surveyed?
**iii** How many children live in Nicholas Avenue?

**b** Find the range and the mode.

**c** How many families had:
**i** 4 children?
**ii** less than 4 children?
**iii** more than 4 children?

**d** What fraction of the families had:
**i** 4 children?
**ii** less than 4 children?
**iii** more than 4 children?

**e** Calculate the mean number of children.

**f** Find the median number of children.

**14** The stem-and-leaf plot shows the results of a survey of waiting times in a doctor's surgery. Find the:

| Stem | Leaf |
|---|---|
| 0 | 7 |
| 1 | 889 |
| 2 | 46789 |
| 3 | 44666889 |
| 4 | 0 |

pp. 237–238

**a** outlier
**b** range
**c** mode
**d** median
**e** mean.

**15** Take the set of numbers 8, 3, 7, 4, 3. pp. 236–237

**a** Add six to each number in the set. How does this affect the mean?
**b** Multiply each number by six. How does this alter the mean?

**16** A five-sector wheel was spun 100 times, and the number of times recorded that each sector stopped at the pointer. The results were then placed in a table. pp. 236–237

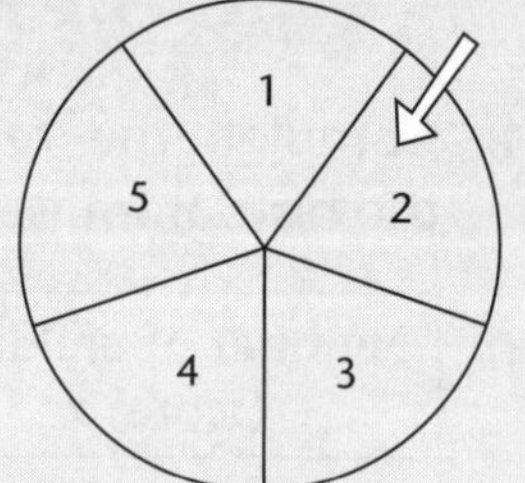


| Number ($x$) | Frequency ($f$) | $fx$ |
|---|---|---|
| 1 | 19 | |
| 2 | 21 | |
| 3 | 20 | |
| 4 | 19 | |
| 5 | 21 | |
| | 100 | |

**a** Complete the ($x \times f$) column in the table.
**b** Calculate the times each number occurs as a fraction and then as a decimal.

For example:

| Number | Fraction | Decimal |
|---|---|---|
| 1 | $\frac{19}{100}$ | 0.19 |

What do you notice about the decimals? Is there a reason for this?

**c** Calculate the mean for these scores.
**d** Find the mode and median.

**17** Greg's cricket average over his first five innings was 15.8, but over his next 10 innings he averaged 19.1 runs. Calculate Greg's total runs over the full 15 innings. Find Greg's average score over the 15 innings. If Greg scored 34 runs in his final innings, calculate his season average over 16 innings. pp. 236–237

Go to p. 289 for quick answers, or to pp. 361–365 for worked solutions.

## YOUR CHECKLIST

**For a complete understanding of this topic you must be able to:**

| | | | |
|---|---|---|---|
| ✓ | Create a frequency distribution table | | p. 233–234 |
| ✓ | Draw a stem-and-leaf plot | | p. 234 |
| ✓ | Draw a frequency histogram and polygon | | p. 235 |
| ✓ | Draw a dot plot | | p. 235 |
| ✓ | Draw a scatter diagram | | pp. 235–236 |
| ✓ | Find the range as a measure of spread | | p. 236 |
| ✓ | Find the mean, median, mode as measures of location. | | pp. 236–238 |

**Now you are ready to do the tests!**

(30 marks)

1

| | | | | | | | | | |
|---|---|---|---|---|---|---|---|---|---|
| 12 | 17 | 16 | 13 | 15 | 17 | 18 | 19 | 18 | 15 |
| 14 | 11 | 19 | 16 | 13 | 14 | 16 | 15 | 17 | 16 |
| 13 | 18 | 17 | 16 | 17 | 18 | 19 | 13 | 14 | 19 |
| 15 | 16 | 14 | 11 | 12 | 18 | 17 | 19 | 13 | 18 |

For the data above:

a Complete a frequency distribution table using column headings score ($x$), tally, frequency ($f$).

b Draw a frequency histogram and polygon. (5 marks)

2 a Complete the table. (2 marks)

b Find the mean. (2 marks)

c Find the mode, median and range. (3 marks)

d How many scores are less than 24? (1 mark)

e How many scores are even? (1 mark)

| Score ($x$) | Frequency ($f$) | $fx$ |
|---|---|---|
| 22 | 5 | |
| 23 | 7 | |
| 24 | 10 | |
| 25 | 8 | |
| 26 | 5 | |
| Total: | | |

3 For the scores 12, 18, 14, 3, 5, 19, 9, 11, 14, 20, find the:

a mean b median c mode d range. (6 marks)

4 Use the stem-and-leaf plot to find the:

a outlier b mean

c median d mode

e range.

| Stem | Leaf |
|---|---|
| 0 | 69 |
| 1 | 244789 |
| 2 | 3489 |
| 3 | |
| 4 | 5 |

(6 marks)

5 The diagram shows a dot plot.

Find the:

a outlier

b mode

c range

d median.

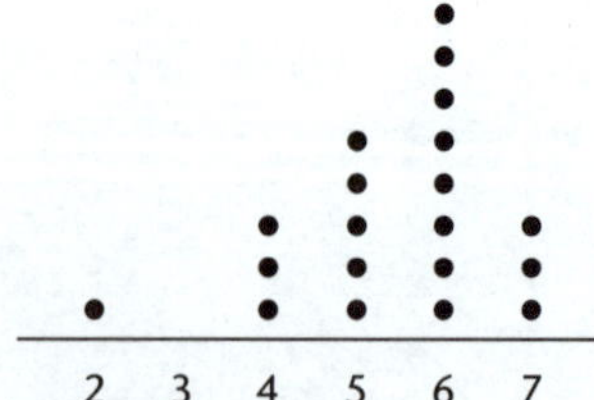


(4 marks)

☞ Quick answers on page 294
☞ Worked solutions on page 398

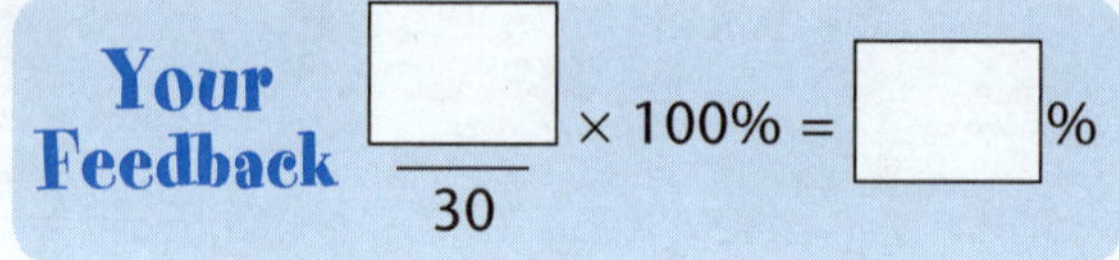

**(35 marks)**

1

| 68 | 73 | 76 | 61 | 62 | 58 | 54 | 75 | 71 | 68 |
|---|---|---|---|---|---|---|---|---|---|
| 63 | 61 | 72 | 71 | 68 | 59 | 54 | 73 | 70 | 54 |
| 69 | 64 | 65 | 73 | 69 | 62 | 65 | 75 | 60 | 55 |

**a** Complete the frequency distribution table for the data above:

| Class | Class centre ($x$) | Tally | Frequency ($f$) | $fx$ |
|---|---|---|---|---|
| 51–55 | | | | |
| 56–60 | | | | |
| 61–65 | | | | |
| 66–70 | | | | |
| 71–75 | | | | |
| 76–80 | | | | |
| | | **Sum** | | |

(3 marks)

**b** Find the mean. (1 mark)

**c** Find the modal class. (1 mark)

**d** Draw a frequency histogram and polygon. (3 marks)

2 For the scores 16, 3, 5, 4, 7, 8, 6, 5, 2, 11, 13, find:

**a** mean **b** median **c** mode **d** range. (4 marks)

3 The histogram shows the number of people living in each house in Donald Street.

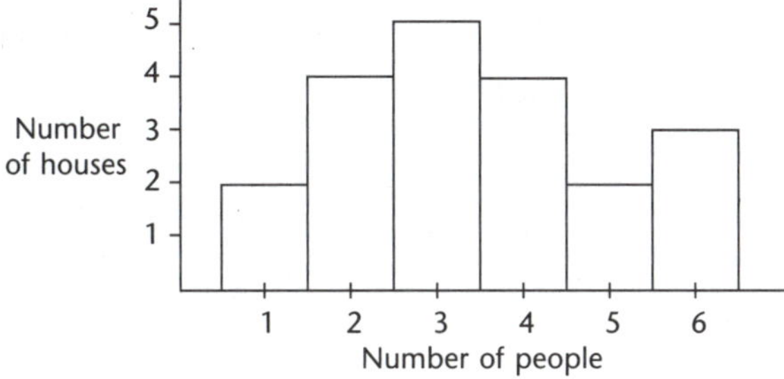


Use the data in the histogram to:

**a** complete a frequency table (2 marks)

**b** find the number of people living in the street (2 marks)

**c** find the number of houses in the street (1 mark)

**d** find the mean number of people living in each house. (1 mark)

*(cont.)*

4 The mean of five scores is 8.

a Find the sum of the five scores. (1 mark)

b When a sixth score is included the mean increases to 10.
What is the new score? (2 marks)

5

| Stem | Leaf |
|---|---|
| 2 | 2069 |
| 3 | 84257 |
| 4 | 3428681 |
| 5 | 19 |

a Draw an ordered stem-and-leaf plot using above the data. (2 marks)

b Find the:

i mean (2 marks)

ii mode iii median iv range. (3 marks)

6 Two student samples were taken from Class 8M to record the number of children living in students' homes.

| Sample A | |
|---|---|
| **Children** | **Number** |
| 1 | 2 |
| 2 | 6 |
| 3 | 1 |
| 4 | 1 |

| Sample B | |
|---|---|
| **Children** | **Number** |
| 1 | 1 |
| 2 | 3 |
| 3 | 5 |
| 4 | 1 |

a What was the total number of students in the samples? (1 mark)

b Compare the mode of both samples. (1 mark)

c What was the median of:

i Sample A? ii Sample B? (2 marks)

d Which sample had the greater mean, and by how much? (3 marks)

**Your Feedback** $\frac{\square}{35} \times 100\% = \square\%$

☞ **Quick answers on page 248**
☞ **Worked solutions on page 398**

**(35 marks)**

1 a Complete the frequency distribution table: (2 marks)

| Score ($x$) | Frequency ($f$) | $f \times x$ |
|---|---|---|
| 3 | 4 | |
| 5 | | 20 |
| 8 | | 48 |
| | 4 | 40 |
| Sum | 23 | 180 |

b Find the mean, mode, median and range. (4 marks)

c Draw a frequency polygon. (3 marks)

2

| Stem | Leaf |
|---|---|
| 3 | 24789 |
| 4 | 00246 |
| 5 | 258 |
| 6 | 36 |
| 7 | □ |

The stem-and-leaf plot shows the scores in a PDHPE test marked out of 80.
The range is the same as the median.

a Find the median. b Find the digit represented by □. (2 marks)

c Arrange the data in a grouped frequency table: (3 marks)

| Class | Class centre ($x$) | Frequency ($f$) | $fx$ |
|---|---|---|---|
| 31–35 | | | |
| 36–40 | | | |
| 41–45 | | | |
| 46–50 | | | |
| 51–55 | | | |
| 56–60 | | | |
| 61–65 | | | |
| 66–70 | | | |
| 71–75 | | | |

d Compare the mean calculated from the stem-and-leaf plot with the mean from the grouped frequency table. (2 marks)

3

| Score | Frequency |
|---|---|
| $a$ | 4 |
| $b$ | 6 |

In the frequency table, $a < b$. The mode is 4 and the range is 2. Find the mean of the distribution. (2 marks)

*(cont.)*

4 After five netball games the average number of goals scored by Laura's team is 22. After another game the average dropped by 3. How many goals were scored in the sixth game? (2 marks)

5 The stem-and-leaf plot records the heights of 10 students:

| Stem | Leaf |
|---|---|
| 14 | 3 |
| 15 | 68 |
| 16 | 02249 |
| 17 | 01 |

a Identify the outlier. (1 mark)

b If the outlier is ignored, comment on the change to the:

i mode  ii median  iii range

iv mean. (5 marks)

6 A survey was conducted at a supermarket checkout recording the number of items in baskets. The results were recorded in the histogram:

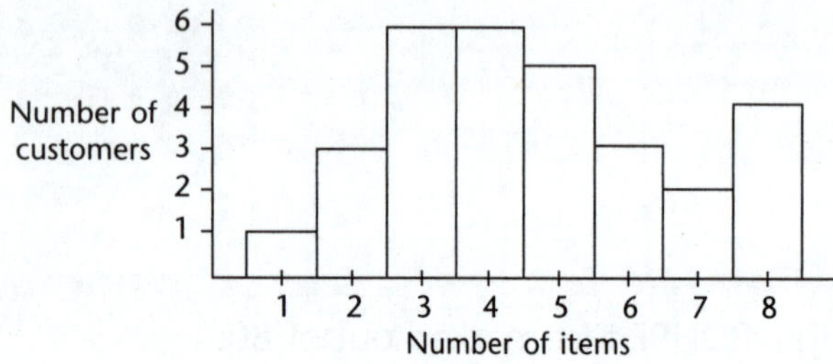


a Find the median, mode and range. (3 marks)

b Find the mean. (2 marks)

7 The results of a maths quiz were recorded as:

| | | | | |
|---|---|---|---|---|
| 16 | 12 | 15 | 19 | 18 |
| 17 | 20 | 16 | 17 | 15 |
| 13 | 18 | 17 | 17 | 15 |
| 16 | 17 | 20 | 19 | 19 |

a Record the results in a table.

b What percentage of the students scored the:

i mode?  ii median? (4 marks)

**Your Feedback** $\frac{\square}{35} \times 100\% = \square\,\%$

☞ **Quick answers on page 294**
☞ **Worked solutions on page 399**

# PROBABILITY 12

- Probability
- Relative Frequency
- Probability Using Two-way Tables
- Probability Using Venn Diagrams

## KEYWORDS

**Complementary** **Two-way**
**Outcome** **Venn Diagram**
**Probability**

# Probability

A coin is tossed once. There is one chance in two that it will land with a head facing up. We say that the probability of throwing a head is one out of two, or:

$P(\text{Head}) = \frac{1}{2}$.

Similarly, if a die (singular of dice) is thrown once, there is one chance in six of throwing a 2:

i.e. $P(2) = \frac{1}{6}$.

The probability of an event occurring is defined as:

$$\text{Probability(Event)} = \frac{\text{Number of favourable outcomes}}{\text{Number of possible outcomes}}$$

*or* $$P(E) = \frac{n(E)}{\text{Total number possible outocomes}}$$

Note: a certainty has a probability of one. An impossibility has a probability of zero. Therefore probability is expressed as a fraction from 0 and 1.

## For Example

A die is rolled once. Find the probability of rolling:

1 a 3

2 a 5

3 an odd number

4 an even number

5 a number larger than 4.

1 $P(3) = \frac{1}{6}$

[One 3 can occur out of six possible numbers.]

2 $P(5) = \frac{1}{6}$

3 $P(\text{Odd}) = \frac{3}{6} = \frac{1}{2}$ [Odd numbers = 3]

4 $P(\text{Even}) = \frac{3}{6} = \frac{1}{2}$

5 $P(>4) = \frac{2}{6} = \frac{1}{3}$ [Numbers greater than 4 are 5 and 6, i.e. two numbers out of six.]

### Total Probability

If there are only two possible things that can happen, then the sum of their probabilities is 1.

When a coin is tossed, for example, the two possible outcomes are head or tail:

$P(\text{Head}) = \frac{1}{2}$ $P(\text{Tail}) = \frac{1}{2}$

Now $P(\text{H}) + P(\text{T}) = \frac{1}{2} + \frac{1}{2} = 1$

This is called total probability. It is true for any number of possible outcomes. If you add up the probabilities of all possible outcomes, they will total one.

## For Example

1 When a die is thrown there are six possible outcomes: 1, 2, 3, 4, 5 or 6.

1 Now $P(1) = \frac{1}{6}$, $P(2) = \frac{1}{6}$, $P(3) = \frac{1}{6}$, $P(4) = \frac{1}{6}$, $P(5) = \frac{1}{6}$, $P(6) = \frac{1}{6}$

Total $P = \frac{1}{6} + \frac{1}{6} + \frac{1}{6} + \frac{1}{6} + \frac{1}{6} + \frac{1}{6} = 1$

### Complementary Events

Also, because the total of the probabilities of all possible outcomes is 1, then if there are two possible outcomes and we know the probability of one outcome, then we also know the probability of the other.

If the probability of rain on a Monday is $\frac{2}{5}$, then the probability of no rain on a Monday

$= 1 - \frac{2}{5}$

$= \frac{3}{5}$.

These are called **complementary events**.

1 If the probability of an archer hitting a tree is 0.8, calculate the probability of her not hitting the tree.

1 $P(\text{Not hitting}) = 1 - 0.8$

$= 0.2$

A die is thrown. Calculate the probability that the score on the uppermost face is:

1 not 6

2 greater than 2.

1 $P(6) = \frac{1}{6}$

$\therefore P(\text{Not } 6) = 1 - \frac{1}{6}$

$= \frac{5}{6}$

2 $P(1 \text{ or } 2) = \frac{2}{6}$

$= \frac{1}{3}$

$P(>2) = 1 - P(1 \text{ or } 2)$

$= 1 - \frac{1}{3}$

$= \frac{2}{3}$

## Relative Frequency

When data is collected it is often presented in a table that can be used to determine probabilities. An additional column titled 'Relative frequency' is used, and the sum of all relative frequencies is 1.

1 The number of muffins sold at a stall during the first hour of a school fete was recorded.

| Sales of muffins | | |
|---|---|---|
| **Type of muffin** | **Frequency** | **Relative frequency** |
| Blueberry | 6 | 0.3 |
| Lemon | 2 | |
| Poppy seed | | |
| Choc chip | 8 | |
| Total | 20 | |

a Complete the table.

b How many poppy seed muffins were sold?

c Estimate the probability that the next muffin sold will:

i be blueberry

ii not be choc chip.

1 a

| Sales of muffins | | |
|---|---|---|
| **Type of muffin** | **Frequency** | **Relative frequency** |
| Blueberry | 6 | 0.3 |
| Lemon | 2 | 0.1 |
| Poppy seed | 4 | 0.2 |
| Choc chip | 8 | 0.4 |
| Total | 20 | 1.0 |

b 4

c i P(Blueberry) = 0.3

ii P(Not choc chip)

= 1 − P(Choc chip)

= 1 − 0.4

= 0.6

## Probabilities Using Two-way Tables

When data is comprised of two variables a two-way table can be used.

### For Example

1 Students in Year 8 study two electives. A sample of 30 students is conducted and the numbers who study Japanese and Food Technology are recorded in the table:

| | | Japanese | | |
|---|---|---|---|---|
| | | **Yes** | **No** | **Total** |
| **Food Technology** | **Yes** | 8 | 12 | |
| | **No** | | 7 | |
| | **Total** | | | |

a Complete the table.

b Estimate the probability that a randomly selected student studies:

i Japanese and Food Technology

ii Food Technology but not Japanese

iii neither Japanese nor Food Technology.

2 A survey was conducted of Year 10 students who have a part-time job.

The results are recorded in the table.

**Students with Part-Time Jobs**

| | **Males** | **Females** | **Total** |
|---|---|---|---|
| **Have job** | 14 | 16 | |
| **Do not have job** | | 9 | |
| **Totals** | | | 50 |

a Complete the table.

b How many girls were surveyed?

c Find the number of students who have a part-time job.

d A student involved in the survey is chosen at random.

What is the probability the student is a:

i male?

ii male who does not have a part-time job?

1 a

| | | Japanese | | |
|---|---|---|---|---|
| | | **Yes** | **No** | **Total** |
| **Food Technology** | **Yes** | 8 | 12 | 20 |
| | **No** | 3 | 7 | 10 |
| | **Total** | 11 | 19 | 30 |

**b** **i** $P$(Japanese and Food Technology) $= \frac{8}{30} = \frac{4}{15}$

**ii** $P$(Food Technology but not Japanese) $= \frac{12}{30} = \frac{2}{5}$

**iii** $P$(Neither Japanese nor Food Technology) $= \frac{7}{30}$

**2** **a** **Students with Part-Time Job**

| | Males | Females | Total |
|---|---|---|---|
| **Have job** | 14 | 16 | 30 |
| **Do not have job** | 11 | 9 | 20 |
| **Totals** | 25 | 25 | 50 |

**b** 25

**c** 30

**d** **i** $\frac{25}{50} = \frac{1}{2}$ **ii** $\frac{11}{50}$

## Probability Using Venn Diagrams

Data can be represented visually using a Venn Diagram.

**For Example**

**1** In a class of 28 students, 21 own a mobile phone and 14 own an MP3 player. Three students own neither device.

**a** Show the information on a Venn Diagram.

**b** How many students own both devices?

**c** If a student is chosen at random, what is the probability that the student:

**i** owns a phone but not a MP3 player?

**ii** owns an MP3 player but not a phone?

**2** In one day a dentist saw 20 patients. All the patients needed either a filling, an X-ray or both. If 16 patients required a filling and 12 required an X-ray, draw a Venn Diagram. Use the diagram to find the probability that a patient chosen at random required a filling and an X-ray.

**3** Travis placed 10 numbers from 1 to 10 on the Venn Diagram shown below:

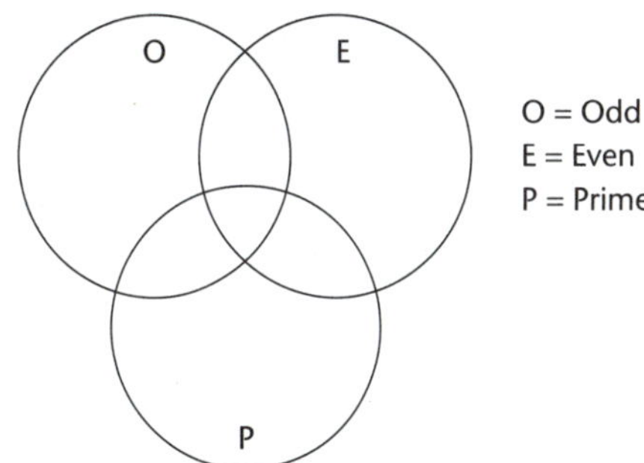


If a number is chosen at random ,what is the probability it is odd but not prime?

**1** **a** Firstly, consider that three students don't own devices:

$28 - 3 = 25$

$\therefore$ 25 students own at least 1 device.

As $21 + 14 - 25 = 10$, then 10 students own both devices.

As $21 - 10 = 11$, then 11 students own only a phone, and as $14 - 10 = 4$, then four students own only an MP3 player.

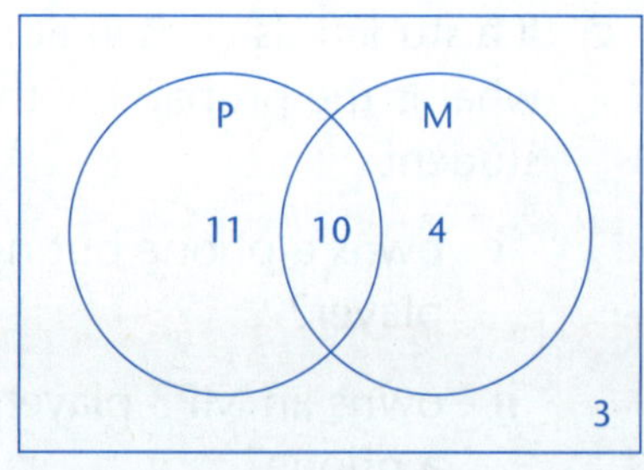


Let P = phone, M = MP3 player

Checking: 11 + 10 + 4 + 3 = 28

**b** 10 students

**c** **i** $P(\text{Phone, not MP3}) = \frac{11}{28}$

**ii** $P(\text{MP3, not phone}) = \frac{4}{28} = \frac{1}{7}$

**2**

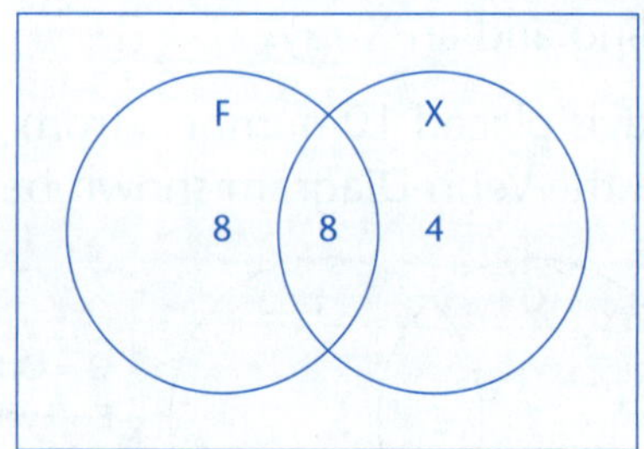


Let F = filling, X = X-ray

$P(\text{Filling and X-ray}) = \frac{8}{20}$

$= \frac{2}{5}$

**3**

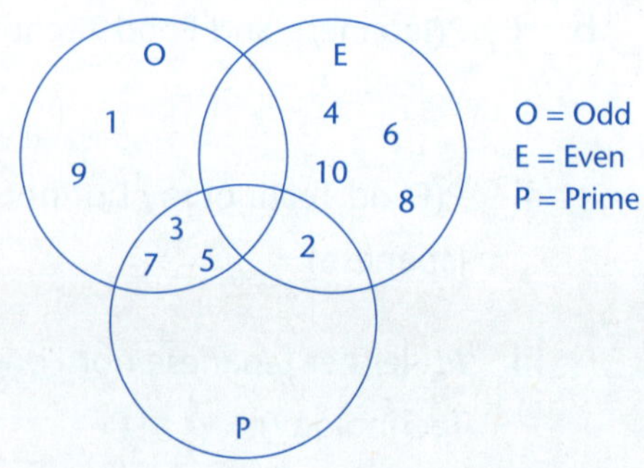


From the diagram, two numbers are odd and not primes: 1 and 9.

$\therefore P(\text{Odd and not prime}) = \frac{2}{10}$

$= \frac{1}{5}$

Go to p. 289 for quick answers, or to pp. 365–368 for worked solutions.

**1** A pack of 52 playing cards with four suits—diamonds and hearts (red), clubs and spades (black)—numbered from 2 to 10, and with King, Queen, Jack and Ace in each suit, is shuffled. One card is drawn at random from the pack. Find the probability that the card is: pp. 252–253

**a** an Ace  **b** a King
**c** a heart  **d** a spade
**e** a black card  **f** a picture card (Jack, Queen or King)
**g** not a picture card  **h** a number less than 7
**i** the 6 of hearts  **j** a red card or a 6
**k** a red 6.

**2** The probability of rain on the June long weekend is $\frac{7}{10}$. What is the probability that it will not rain on the June long weekend? p. 253

**3** A jar contains five red, six black and four white jelly babies. If one jelly baby is chosen at random from the jar, find the probability that it is: pp. 252–253

**a** red  **b** not red  **c** black
**d** red or white  **e** green  **f** white
**g** not white.

**4** It has been calculated that the probability of a male birth is 0.48. pp. 252–253

**a** Find the probability of a female birth.
**b** At a large Adelaide hospital, 1400 babies were born in 2001. How many male babies would you expect were born that year?

**5** Two dice are thrown (faces numbered 1 to 6). The sum of the two uppermost faces is noted. Find the probability that the sum is: pp. 252–253

**a** 8  **b** 5
**c** odd  **d** greater than 9
**e** 9 or greater  **f** not 9
**g** odd or less than 5  **h** odd and less than 5.

**6** A pair of four-sided dice (numbered 1–4) are thrown. The numbers face up are added. Find the probability that the sum is: pp. 252–253

**a** 5  **b** 7  **c** 1
**d** greater than 5  **e** at least 5.

**7** In a supermarket promotion 10 gold tokens numbered 1 to 10 are hidden. Rees found one token. Find the probability that Rees found the token with the number: pp. 252–253

**a** 9  **b** even  **c** less than 6  **d** 6 or more.

**8** The table shows the survey results of the destinations of domestic travellers at Adelaide Airport. pp. 253–254

| Destinations | | |
|---|---|---|
| **State/Territory** | **Frequency** | **Relative frequency** |
| SA | 18 | |
| Victoria | 20 | |
| NSW | 22 | |
| Qld | 18 | |
| Tasmania | 12 | |
| WA | 6 | |
| NT | 4 | |
| Totals | 100 | |

**a** Complete the relative frequency column.

**b** Estimate the probability that the next passenger interviewed will:

**i** be travelling to NSW

**ii** be travelling to WA

**iii** be travelling to either Victoria or Tasmania

**iv** not be travelling to Qld

**v** not be travelling to the NT or NSW.

**9** Complete the relative frequency table. pp. 253–254

| **Score** | **Frequency** | **Relative frequency** |
|---|---|---|
| 7 | 5 | 0.1 |
| 8 | | 0.3 |
| 9 | 20 | |
| 10 | | |
| Totals | | |

**10** A survey was used to find the number of 17 year olds who have a driver's licence. pp. 254–255

**Seventeen-Year-Old Drivers**

| | Licensed | Not licensed | Total |
|---|---|---|---|
| **Female** | 8 | 14 | |
| **Male** | 12 | 16 | |
| Totals | | | |

**a** Complete the table.

**b** How many people participated in the survey?

**c** If a 17 year old was chosen at random from those surveyed, what is the probability that they are:

- **i** a licensed female?
- **ii** an unlicensed male?

**d** If a 17-year-old female was chosen at random from those surveyed, what is the probability she is unlicensed?

**11** The students in 8P were asked whether or not they had completed their Geography and English assignments. The results are recorded below. pp. 254–255

**Completed Assignment**

| | English completed | English not completed | Total |
|---|---|---|---|
| **Geography completed** | 19 | | |
| **Geography not completed** | | 3 | |
| Totals | | 7 | 28 |

**a** Complete the table.

**b** If a student is chosen at random, what is the probability that they have:

- **i** completed Geography but not English?
- **ii** completed neither assignment?

**12** A Venn Diagram is drawn to show the number of people who like to watch Sport and Lifestyle television programs. pp. 255–256

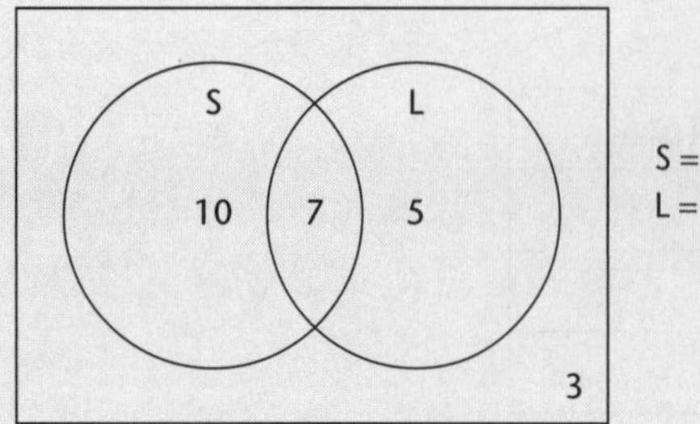


**a** How many people:

**i** like both Sport and Lifestyle programs?

**ii** like Sport programs, but not Lifestyle programs?

**iii** do not like Sport, nor Lifestyle programs?

**b** Find the number of people who participated in the survey.

**c** If a person is chosen at random, what is the probability that the person

**i** likes Lifestyles programs, but not Sport programs

**ii** likes both Sport and Lifestyle programs

**d** Use the Venn Diagram to complete the two-way table

| | **Likes Sport** | **Does not like Sport** | **Total** |
|---|---|---|---|
| **Likes Lifestyle** | | | |
| **Does not like Lifestyle** | | | |
| **Totals** | | | |

**13** Fifty students were surveyed to find whether they regularly drink coffee or tea. Twenty-five students said they regularly drink coffee, while 18 said they drink tea on a regular basis. Sixteen students said that they did not drink either beverage on a regular basis. pp. 255–256

**a** Draw a Venn Diagram to represent the information.

**b** A student is chosen at random. What is the probability that this student:

**i** drinks both coffee and tea regularly?

**ii** drinks tea but not coffee regularly?

**14** A group of young people was interviewed about their preference in take-away food. Twenty-two students enjoyed Indian while 34 enjoyed Thai. Seven people did not like Indian or Thai food. Sixteen people enjoyed both Indian and Thai food. pp. 255–256

**a** Represent the information on a Venn Diagram.

**b** How many people were interviewed?

**c** If a person who had been interviewed was chosen at random, what is the probability they enjoy Indian but not Thai food?

**15** A group of Year 12 students was surveyed to find the Science courses they study at a particular high school. Some of the results are recorded on the Venn Diagram. pp. 255–256

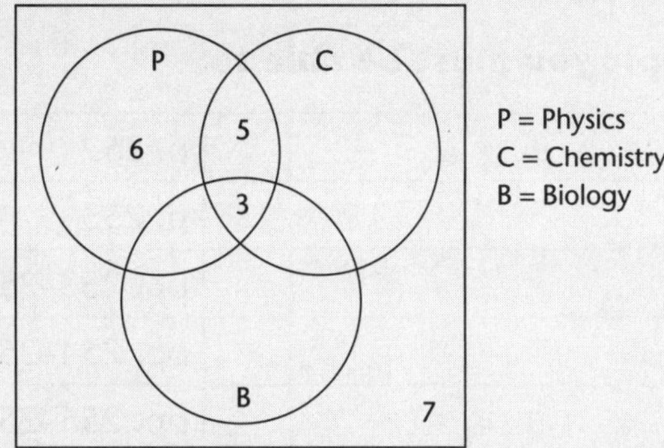


It is known that 50 students participated in the survey. Twenty-two students study Physics, 10 students study both Chemistry and Biology, and nine students study only Chemistry.

**a** Complete the Venn Diagram.

**b** If a student from the group B was chosen at random, what is the probability that they:

  **i** study Chemistry?

  **ii** study Biology and Physics?

  **iii** do not study Biology?

Go to p. 289 for quick answers, or to pp. 365–368 for worked solutions.

## YOUR CHECKLIST

**For a complete understanding of this topic you must be able to:**

| ✓ | Find the probability of simple events | | p. 252 |
|---|---|---|---|
| ✓ | Use complementary events | | p. 253 |
| ✓ | Use a table to find relative frequencies | | pp. 253–254 |
| ✓ | Find probabilities using two-way tables | | pp. 254–255 |
| ✓ | Find probabilities using a Venn Diagram. | | pp. 255–256 |

**Now you are ready to do the tests!**

**(25 marks)**

**1** A bag contains four red balls, three green balls and two blue balls. Find the probability that a randomly selected ball is:

**a** red
**b** green
**c** yellow
**d** red or green
**e** not green
**f** neither red nor green. (6 marks)

**2** From a deck of playing cards, one card is selected at random, find the probability that the card is:

**a** red
**b** not red
**c** green
**d** 9
**e** red 4
**f** 7 of clubs
**g** heart
**h** 7 or 8. (8 marks)

**3** Complete the relative frequency table.

| Score | Frequency | Relative frequency |
|---|---|---|
| 4 | 6 | |
| 5 | 7 | |
| 6 | 2 | |
| 7 | 5 | |

(3 marks)

**4** A survey was conducted of passengers waiting for a bus, and the results shown in the two-way table below.

| | Males | Females | Total |
|---|---|---|---|
| **Smokers** | 5 | 8 | |
| **Non-smokers** | 4 | 6 | |
| **Totals** | | | |

**a** Complete the table.
**b** How many males were surveyed?
**c** What fraction of the non-smokers were females?
**d** If a passenger was chosen at random what is the probability that they were a:
  **i** female smoker?
  **ii** male non-smoker? (5 marks)

*(cont.)*

5 The Venn Diagram shows the results of a questionnaire used to determine the number of students who had a dog or a cat as a family pet.

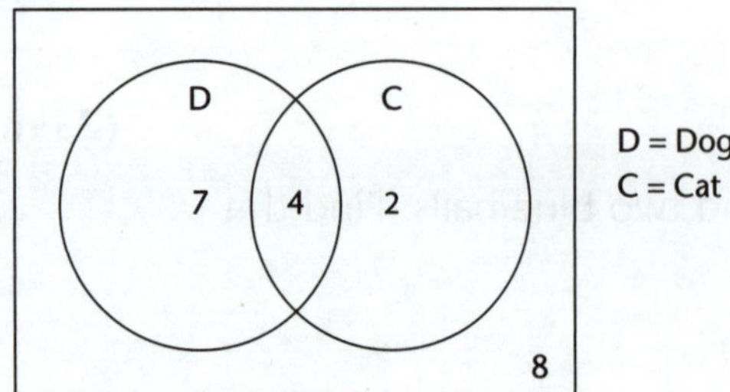


A student was chosen at random. What is the probability that they:

a have a dog and a cat?

b have a cat but not a dog?

c do not have a dog or a cat? (3 marks)

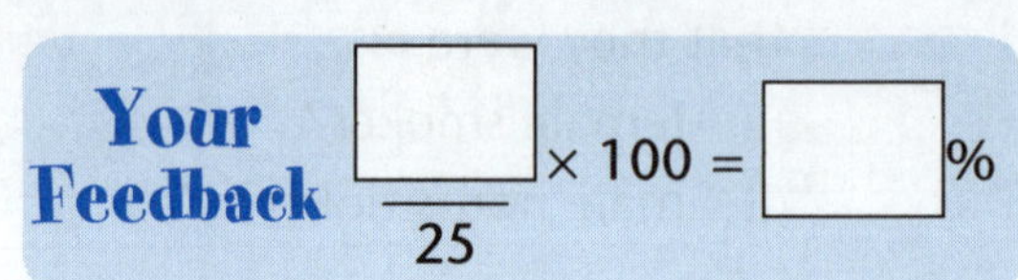


☞ Answers on page 294
☞ Worked solutions on page 400

**(25 marks)**

1 The numbers 1 to 10 are written on 10 identical balls and placed in a bag. If one ball is drawn at random from the bag, find the probability that the number is:

a a 6
b even
c less than 8
d composite
e a factor of 12
f a multiple of 3
g not a 7. (7 marks)

2 A die is rolled 60 times and the results recorded in the frequency table.

| Score | Frequency | Relative frequency |
|---|---|---|
| 1 | 9 | |
| 2 | 11 | |
| 3 | 7 | |
| 4 | | |
| 5 | 8 | |
| 6 | 13 | |
| Total: | | |

Complete the table, expressing the relative frequencies as fractions. (2 marks)

3 A box contains 15 identical balls. Six balls are blue, four balls are red and the remainder are green.

a A ball is chosen at random from the box.
Find the probability that the ball is:
  i green
  ii not blue
  iii red, blue or green.

b A ball is removed from the bag. The chance of selecting a blue ball is now twice the probability of selecting a red ball. If a ball is now selected at random from the box, find the probability that it is:
  i green
  ii blue or red. (5 marks)

*(cont.)*

4 In their Year 8 music class, students were asked whether they played the piano and guitar. The results were recorded on the table below.

| | Plays guitar | Does not play guitar | Total |
|---|---|---|---|
| **Plays piano** | | 5 | 11 |
| **Does not play piano** | | | 20 |
| | 17 | | |

a Complete the table.

b How many students play the piano?

c A student from the class is chosen at random. What is the probability that they play:

i guitar, but not piano?

ii both guitar and piano?

d A student who does not play guitar is chosen. What is the probability they do not play the piano, either? (5 marks)

5 A group of 20 girls play either hockey or netball or both. Twelve girls play hockey and 14 girls play netball.

a Draw a Venn Diagram to represent the information.

b If a girl is chosen at random, what is the probability that she plays:

i both sports?

ii netball but not hockey? (3 marks)

6 A group of people were surveyed to find the number who had travelled by train or ferry in the previous month and the results recorded in the Venn Diagram.

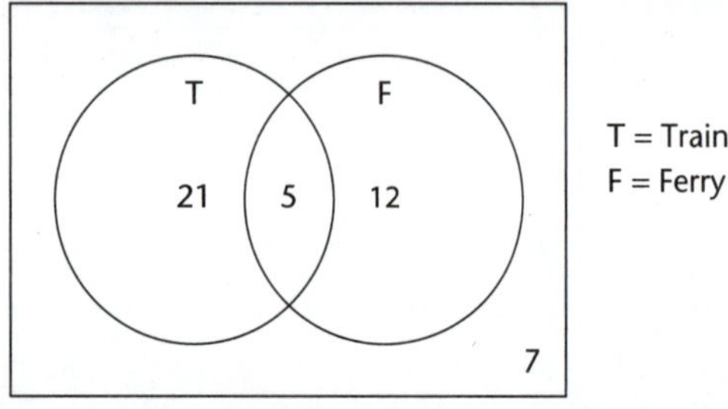


a How many people had travelled by ferry?

b A person was chosen at random. What is the probability that they had travelled by:

i both train and ferry?

ii train but not by ferry? (3 marks)

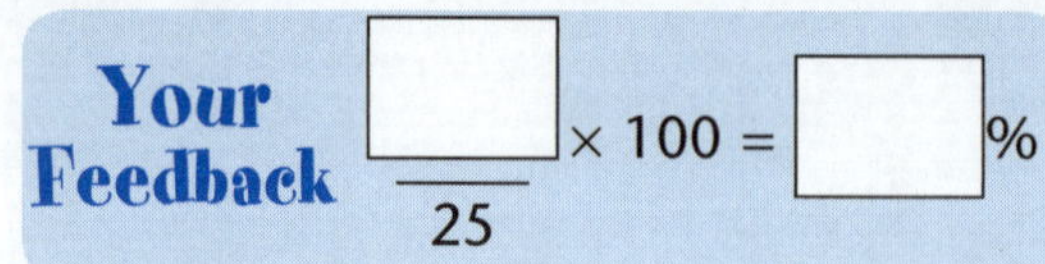


☞ Answers on page 294
☞ Worked solutions on page 401

**(30 marks)**

1 A normal deck of playing cards is shuffled and a card chosen at random. Find the probability that the card is:

a red
b not red
c not a 7
d a Jack or a Queen
e a black 8
f not a red Jack. (6 marks)

2 Identical cards are numbered 1 to 20. The 20 cards are turned face down on a table and a card selected at random. Find the probability the card is:

a less than 10
b greater than 16
c prime
d composite
e divisible by 3
f a multiple of 4
g a factor of 24
h even and a multiple of 3. (8 marks)

3 A survey of 200 university students found the number of people who regularly eat breakfast.

| | **Eat breakfast** | **Do not eat breakfast** | **Total** |
|---|---|---|---|
| **Male** | 42 | | 90 |
| **Female** | 46 | | |
| **Totals** | | | |

a Complete the table.
b If one of the university students was chosen at random, what is the probability that they are a:
  i male who does not eat breakfast?
  ii female who eats breakfast?
c If one of the university students was chosen at random, what is the probability that they:
  i are female?
  ii eat breakfast? (5 marks)

*(cont.)*

4 A newspaper conducted a survey regarding the increase in pension payments. The results are detailed in the table below.

**Support for Pension Increase**

| | | For | Against | No opinion | Total |
|---|---|---|---|---|---|
| Ages | **18–35** | 12 | | 5 | 25 |
| | **36–50** | | 9 | 3 | 25 |
| | **51–65** | | 6 | 0 | 25 |
| | **Over 65** | 23 | | 0 | 25 |
| | **Total** | | | | 100 |

a Complete the table.

b If a person who had participated in the survey was chosen at random, what is the probability that they:

i are aged 51–65?

ii were against the pension increase?

iii were aged 18–35 and were 'for' the pension increase? (4 marks)

5 Forty women at a netball gala were surveyed to find whether they had volunteered at the canteen this year and whether they had even played netball.

Three women had never volunteered or played. Sixteen had volunteered at the canteen while 32 had played netball.

a Record the information on a Venn Diagram.

b If a woman was chosen at random, what is the probability that she had volunteered at the canteen but never played?

c Record the information on a two-way table.

d If 120 women were present at the gala, estimate the number of women who had:

i never played netball

ii played netball but never volunteered. (7 marks)

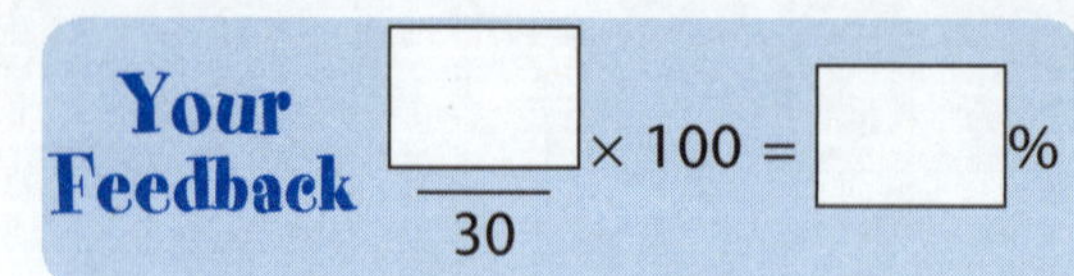


☞ Answers on page 294
☞ Worked solutions on page 402

# SAMPLE EXAMINATIONS 13

- Sample Examination 1
  - Part A
  - Part B
  - Part C
- Sample Examination 2
  - Part A
  - Part B
  - Part C

# SAMPLE EXAMINATION 1

**Time Allowed: 1 hour 15 minutes**

## Part A

**(30 marks)**

1 Simplify 28:18. (1 mark)

2 Find the area, correct to two decimal places, of a circle with radius 4.8 cm. (1 mark)

3 Does the line $y = 3x - 2$ pass through the point (2, 4)? (1 mark)

4 Simplify $4a - 2b - 3a + 4b$. (1 mark)

5 The maximum temperatures over a one-week period were recorded: 23°, 24°, 27°, 34°, 39°, 31°, 25°. Find the range. (1 mark)

6 Find the value of the pronumeral.

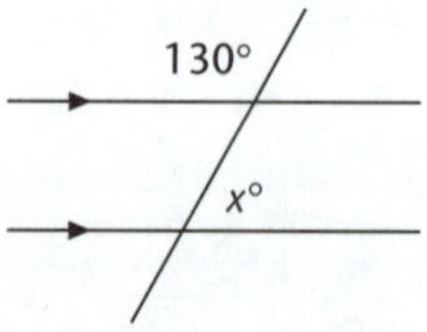

130°

x°

(1 mark)

7 What is 12% of $740? (1 mark)

8 Evaluate to two decimal places: $\dfrac{4.76}{3.91 - 1.007}$ (1 mark)

9 The temperature in Orange at 5 pm was 4°. By midnight it had fallen six degrees, and by 7 am had fallen another three degrees. What was the temperature at 7 am? (1 mark)

10 Simplify $1.4 - (0.2)^2$. (1 mark)

11 Increase $320 by 25%. (1 mark)

12 A car leaves Launceston at 2:30 pm and averages 80 km/h to arrive at its destination at 6 pm. How far has the car travelled? (1 mark)

13 Find the area of the quadrilateral.

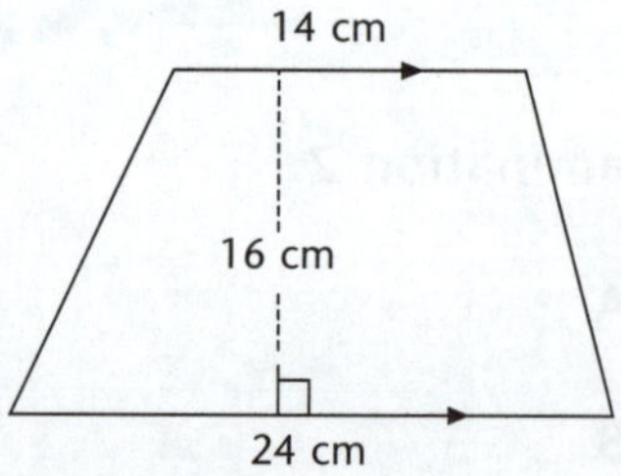

14 cm

16 cm

24 cm

(1 mark)

14 Expand $a(3 - 4a)$. (1 mark)

**15** Joanne is able to type 1680 words in 20 minutes. What is her typing rate per minute? (1 mark)

**16** A die is rolled. What is the probability of rolling a number that is a factor of 12? (1 mark)

**17** What test is used to prove these triangles are congruent?

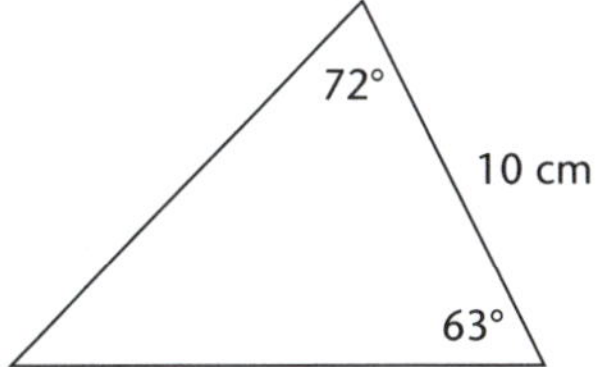


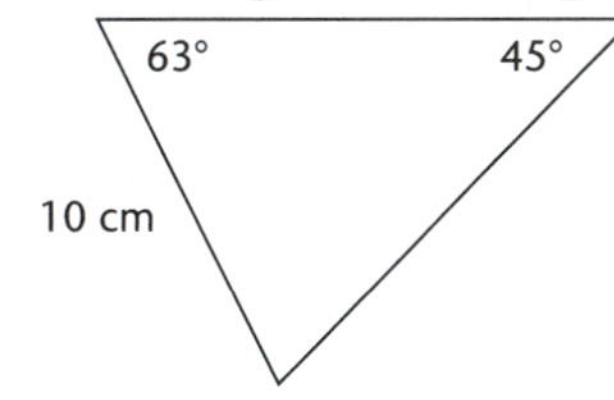


(1 mark)

**18** When the equations $y = -2$ and $x = 3$ are graphed on the same number plane they intersect. Find the coordinates of this point of intersection. (1 mark)

**19** Solve for $x$: $4x - 2 = 3x + 5$ (1 mark)

**20** Simplify: $4x^2 \times 3x^4$ (1 mark)

**21** Rewrite 24% as a simplified fraction. (1 mark)

**22** Find the pattern rule used:

| $x$ | $y$ |
|---|---|
| 1 | 5 |
| 2 | 8 |
| 3 | 11 |
| 4 | 14 |

(1 mark)

**23** Convert 43 670 m$^2$ to hectares. (1 mark)

**24** Raylene bought a watch 10 years ago for $200. She sold it this year for $240. Find her profit as a percentage of the cost price. (1 mark)

**25** Factorise $12x - 18$. (1 mark)

**26** A B C D E F G H I Find the ratio of AF:DI. (1 mark)

**27** A ballet teaches charges $440 for one term's lessons. How much GST is included in the cost? (1 mark)

**28** Find the value of $(2 \times 3)^0$. (1 mark)

**29** What is the outlier in the scores 38, 34, 28, 40, 3, 29, 45? (1 mark)

**30** Maria won $1200 in a lottery prize. She decided to put half of it in the bank and used two-thirds of the remainder to pay off her credit card debt. How much was her credit card debt? (1 mark)

## Part B

**(40 marks)**

**1**

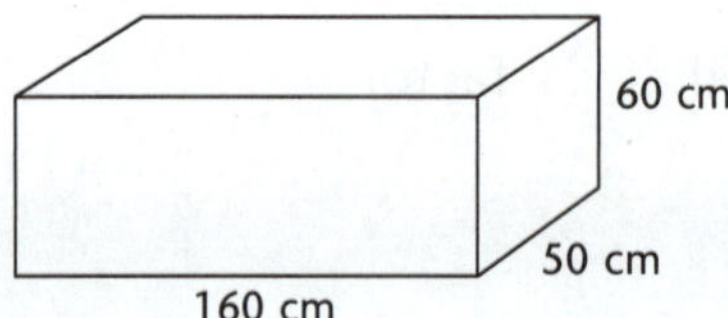


Find the:

**a** volume **b** capacity in L. (2 marks)

**2** Solve $3(2x - 1) = 4(x + 1)$. (2 marks)

**3** A plumber charges a call-out fee of $60 and then $70 per hour.
Using $c$ = cost and $n$ = number of hours, complete the table.

| $n$ | 0 | 1 | 2 | 3 |
|---|---|---|---|---|
| $c$ | | | | |

(2 marks)

**4** Find the area of the shaded region.

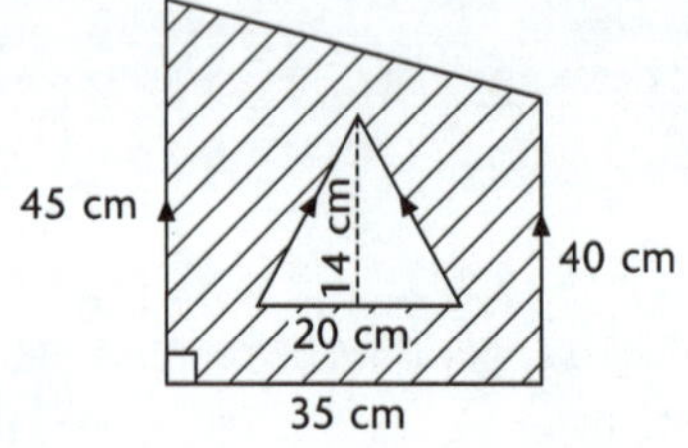


(2 marks)

**5** Expand and simplify $3(a - 4) - 2(2a - 5)$. (2 marks)

**6** By first completing the table, graph the line $y = 4 - 2x$ on a number plane.

| $x$ | 0 | 1 | 2 |
|---|---|---|---|
| $y$ | | | |

(2 marks)

**7** If $a = -3$, $b = -2$, $c = -4$, evaluate $a^2 - bc$. (2 marks)

**8** Find the volume of a cylinder with radius 5 cm and height 7 cm.
Write your answer correct to two decimal places. (2 marks)

**9** Find the value of the pronumeral. Write your answer as a surd.

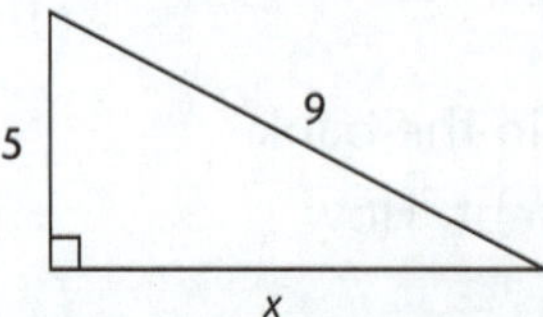


(2 marks)

**10** Ann and Trevor invest $40 000 and $50 000 respectively to start a computer store business. They plan to divide any profits in the ratio of their investments.

**a** Find the ratio of their investments.

**b** If in the first year they make $58 500, what is Trevor's share of the profit? (2 marks)

**11** **a** Complete the table of scores.

| Score ($x$) | Frequency ($f$) | $fx$ |
|---|---|---|
| 4 | 3 | |
| 5 | 2 | |
| 6 | 4 | |
| 7 | 3 | |
| Total | | |

**b** Find the mean of the scores. (2 marks)

**12** The scale on a map of NSW is recorded as 1:2 500 000.

**a** Rewrite this scale as 1 cm = ____________ km.

**b** If the distance from Nyngan to Sydney is 475 km, how far apart are they on the map? (2 marks)

**13** A car leaves Kununurra and travels to Halls Creek, a distance of 360 km. If the car averaged a speed of 80 km/h, find the:

**a** time taken for the journey in hours and minutes

**b** arrival time in Sydney, if it had left Moruya at 9:30 am. (2 marks)

**14** The sign below was displayed in a department store.

**15%**
**OFF**
**everything**
**in the store**

**a** What will be saved on a pair of jeans marked at $89?

**b** Find the price for a microwave, originally marked at $349. (2 marks)

**15** From her fortnightly pay, Fiona spent 36% on groceries, 8% on petrol, 14% on a dress and 18% on an electricity bill. If she had $336 remaining, what amount had she spent on groceries? (2 marks)

**16** A bushwalker leaves A and walks 4 km in a northerly direction then 6 km in an easterly direction. How far is he from A (to the nearest metre)? (2 marks)

**17** The figure shows a square surmounted by a semicircle. Find the perimeter of the figure, to the nearest centimetre.

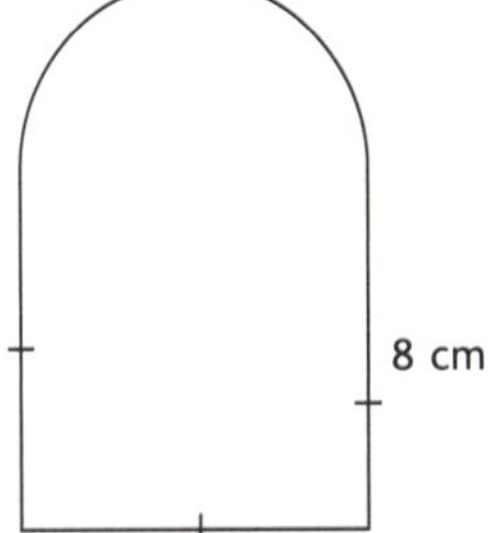


(2 marks)

**18** Find the values of the pronumerals.

**a**

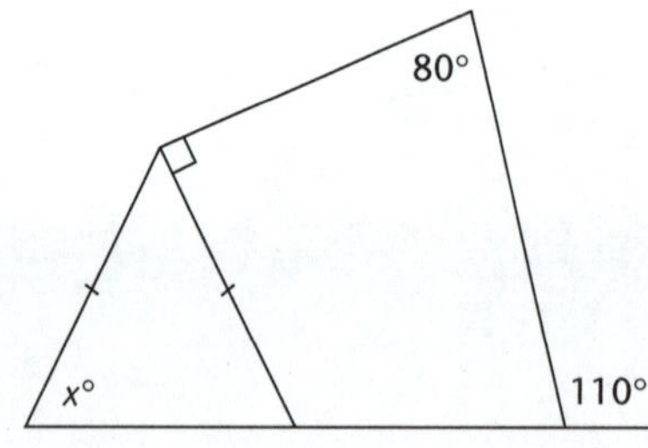


**b**

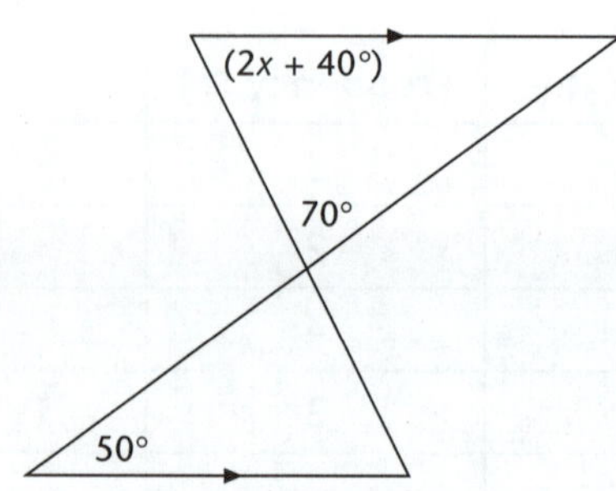


(2 marks)

**19** Handyman Ken bought a television for \$240 and spent \$80 repairing it. He then sold it for \$400.

**a** Find the ratio of his profit to cost price.

**b** Find his percentage profit. (2 marks)

**20** A survey was conducted to find the students who are paid an allowance by their parents.

**a** Complete the table.

| Student allowance (weekly) | | | | |
|---|---|---|---|---|
| | **None** | **\$0–\$10** | **More than \$10** | **Total** |
| Males | 3 | 7 | | 12 |
| Females | | | | |
| Total | 8 | 16 | | 30 |

**b** If one of the students is chosen at random, what is the probability that they are a female who is paid more than \$10 per week allowance? (2 marks)

## Part C

**(15 marks)**

**1** For the scores 4, 7, 2, 3, 8, 9, 2, 8, find the:

**a** mode **b** median **c** mean. (3 marks)

**2** The histogram shows the results of a survey of the number of children in the families of a group of Year 8 students.

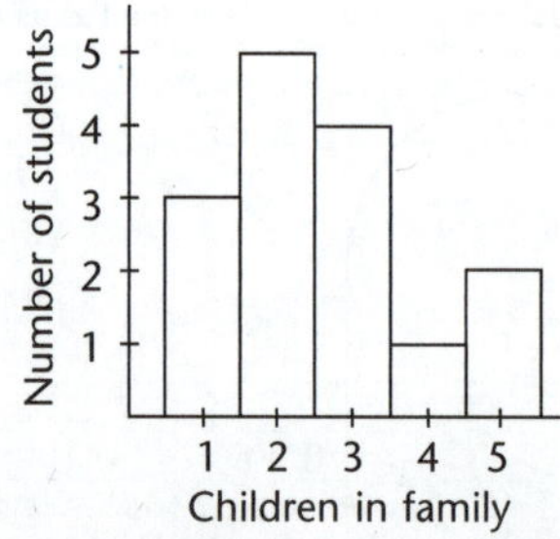


**a** How many students were surveyed?

**b** What was the most common-sized family?

**c** What was the average number of children in a family? (3 marks)

3 A class of students was used to sample teenagers' perceptions of their own weight, by asking: Do you consider yourself overweight, underweight or 'normal'?

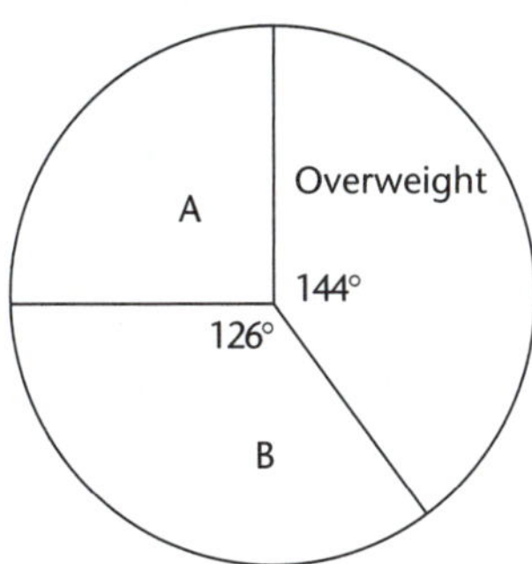


a What percentage of the class believed they were overweight?

b What sector represents the 35% who believed their weight was 'normal'?

c What was the ratio of 'overweight' students to 'underweight' students? (3 marks)

4 a Use Pythagoras' Theorem to find the length of AB on the triangular prism.

b If the volume of the triangular prism below is 672 $m^3$, find the length of CD.

c What would be the total area of the *five* faces of the triangular prism? (3 marks)

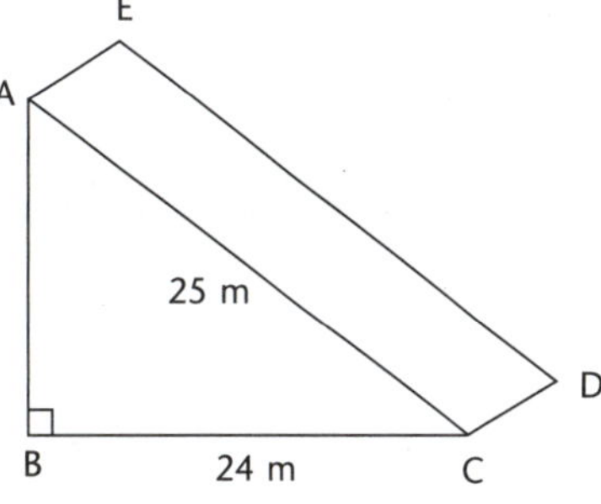


5 In the diagram, ∠ABC = 40°, AB || CD and AB = BC. Find the value of:

a ∠BCD

b ∠BAC

c ∠DCE. (3 marks)

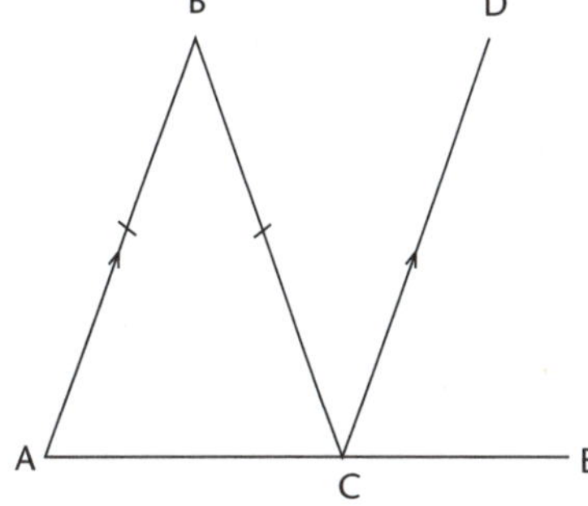

☞ Answers on page 295
☞ Worked solutions on page 403

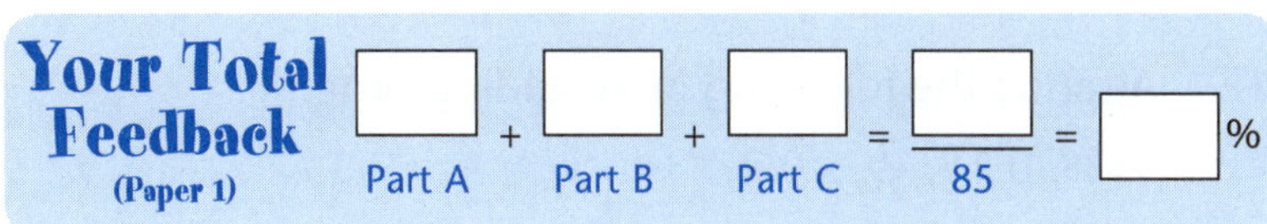

# SAMPLE EXAMINATION 2

1¼ hours

**Time Allowed: 1 hour 15 minutes**

## Part A

**(30 marks)**

1 Use the calculator to evaluate $\sqrt{26.89}$ correct to two decimal places. (1 mark)

2 Express 35% as a simple fraction. (1 mark)

3 Simplify $8a + 7 - 3a + 4$. (1 mark)

4 Simplify $-3 + 4 - 2$. (1 mark)

5 Expand $3(2x - 4)$. (1 mark)

6 How many centimetres in 2.4 metres? (1 mark)

7 Find the area of:

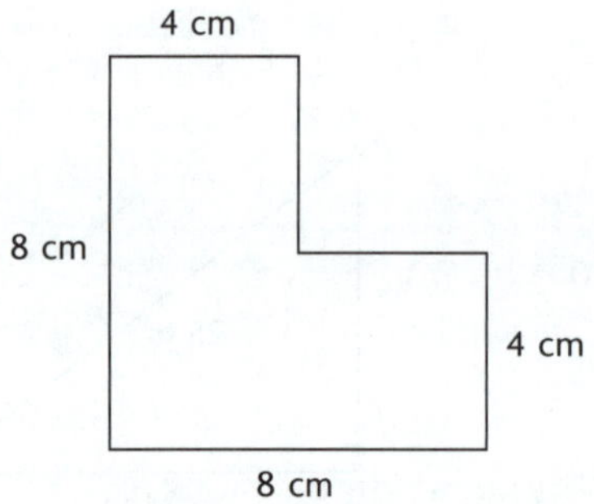

4 cm

8 cm

4 cm

8 cm

(1 mark)

8 Simplify 12:15. (1 mark)

9 Express $\frac{5}{8}$ as a decimal. (1 mark)

10 Simplify $(1\frac{1}{2})^2$. (1 mark)

11 What is the reciprocal of $2\frac{2}{3}$? (1 mark)

12 Evaluate $2a \times 3a$. (1 mark)

13 Evaluate $3b^2$ if $b = 2$. (1 mark)

14 Find 12% of $9.50. (1 mark)

15 Solve $x + 8 = 2$. (1 mark)

16

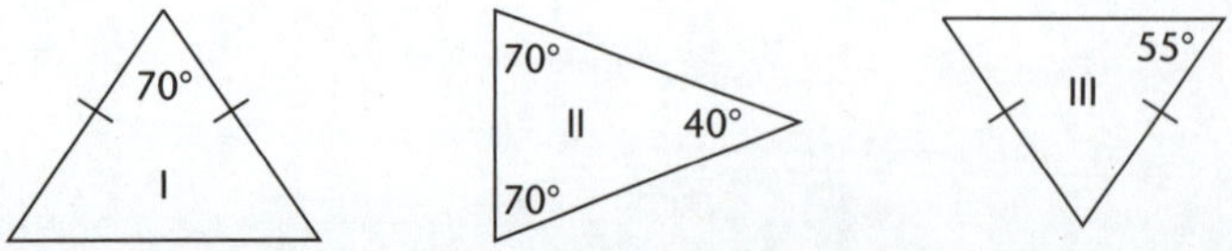

70° I

70° II 40° 70°

55° III

Which two triangles are congruent? (1 mark)

17 Arrange the following in ascending order:
13%, 0.15, $\frac{3}{25}$, $\sqrt{0.9}$. (1 mark)

18 Simplify $\frac{-5 + 9}{-2}$. (1 mark)

19 Find the average of 8, 7, 6, 5, 9, 7. (1 mark)

20 What percentage is 17 out of 20 marks? (1 mark)

21 A square has perimeter of 28 cm. Find its area. (1 mark)

22 A bag contains twice as many red balls as blue balls and half as many green balls as blue balls.

What is the probability that a red ball is randomly selected from the bag? (1 mark)

23 Simplify $15ab \div 5a$. (1 mark)

24 Increase $3a$ by 14. (1 mark)

25 How much time has elapsed between 11:45 am and 2:28 pm? (1 mark)

26 Simplify $\frac{5x + 15y}{5}$. (1 mark)

27 Find the value of $a$.

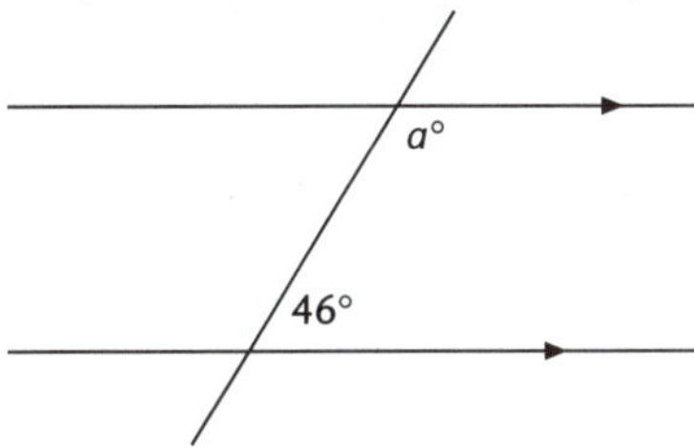


(1 mark)

28 How many 250 mL glasses of cordial could be poured from a 2 L bottle? (1 mark)

29 The graph shows the scores in a spelling test. How many students did the test?

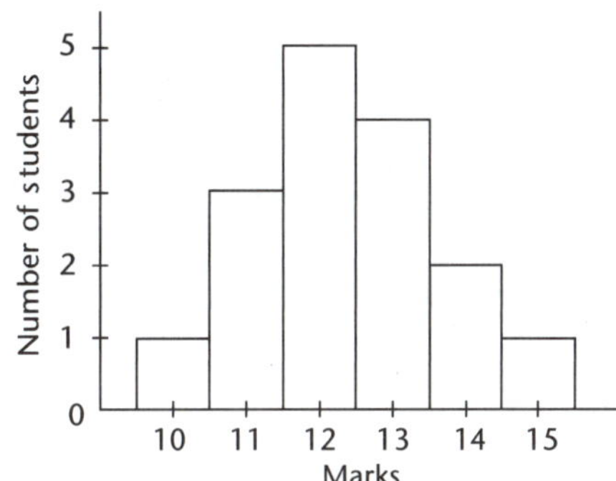


(1 mark)

30 Find the perimeter.

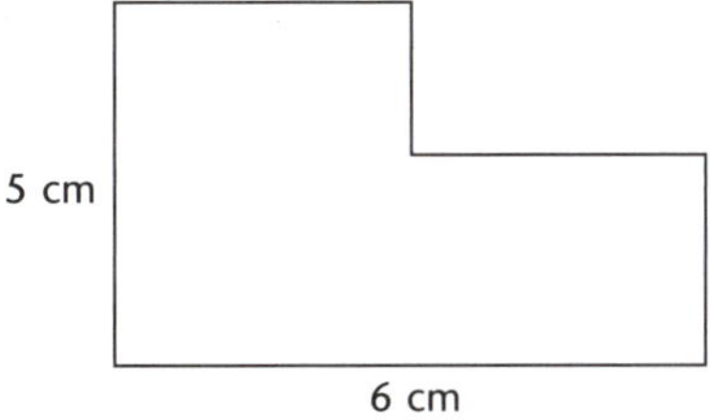


(1 mark)

## Part B

**(40 marks)**

Show all necessary working.

1 In a class the ratio of boys to girls is 4:3. If there are 12 girls, how many boys are there in the class? (2 marks)

2 Solve $\frac{2x - 1}{3} = 5$. (2 marks)

3 Expand and simplify $3(2x - 5) + 2(x + 3)$. (2 marks)

4 A shop offers '15% OFF EVERYTHING'. What would be paid for a washing machine originally marked at $830? (2 marks)

5 Find the area of a circle with a radius of 7 cm. Give your answer to the nearest square centimetre.

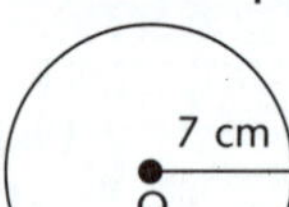


(2 marks)

6 On a number plane, graph the line $y = 2x - 1$. (2 marks)

7 Use Pythagoras' Theorem to find the value of $x$.

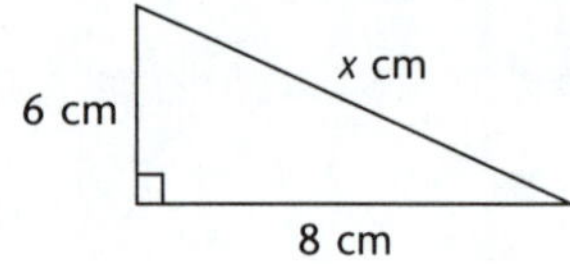


(2 marks)

8 Find the volume of:

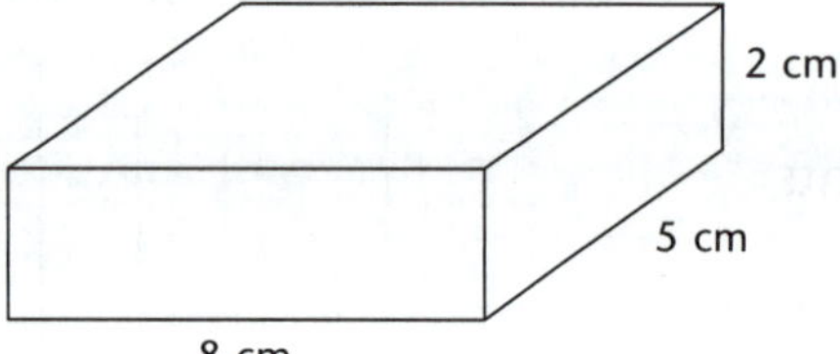


(2 marks)

9 The shape shown is a semicircle. Find its perimeter to the nearest centimetre.

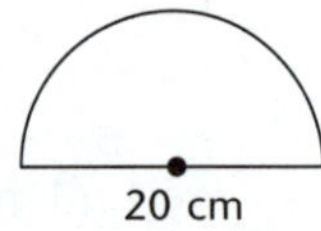


(2 marks)

10 Craig's flight from New York to Hawaii takes 10 hours 20 minutes and leaves New York at 7:00 am. If Hawaii is 6 hours behind New York, what time does he land in Hawaii (local time)? (2 marks)

11 A bag contains six white, four red and two black balls. If a ball is picked at random from the bag, what is the probability of selecting a:

**a** white ball? **b** pink ball? (2 marks)

**12** Given that $F = \frac{mv^2}{r}$, find the value of $F$ when $m = 12$, $v = 4$ and $r = 8$. (2 marks)

**13** The results of a mathematics test are recorded in the stem-and-leaf display.
Find the difference between the mode and the median.

| Stem | Leaf |
|---|---|
| 0 | 9 |
| 1 | 2578 |
| 2 | 33467 |
| 3 | 0 |

(2 marks)

**14**

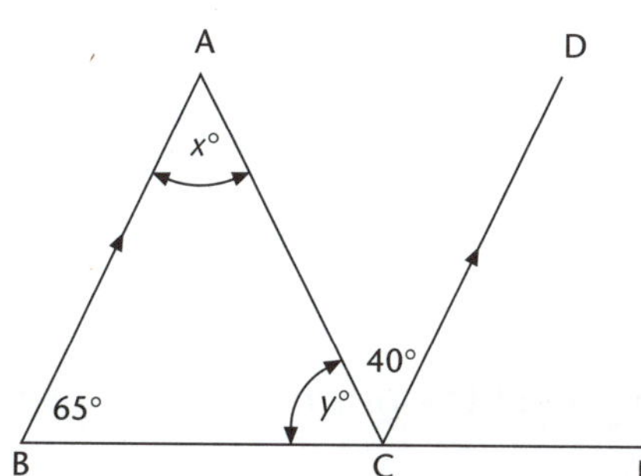


In the figure AB || DC.
Angle ABC = 65°
Angle ACD = 40°

**a** $x =$ __________. Why? ____________________

**b** $y =$ __________. Why? ____________________ (2 marks)

**15** Eleni earns $28 365 p.a. She spends $\frac{1}{3}$ on rent, $\frac{2}{5}$ on food and saves the rest.
How much money does Eleni save per year? (2 marks)

**16** A car travels 306 kilometres in $4\frac{1}{2}$ hours. Find the average speed of the car. (2 marks)

**17** Find the median of these scores:
1, 5, 1, 2, 3, 1, 4, 4, 3, 4 (2 marks)

**18** Solve the equation $3x + 5 = 8x - 35$. (2 marks)

**19** Find the volume of the cylinder. Give your answer to the nearest cubic centimetre. (Use the formula $V = \pi r^2 h$.)

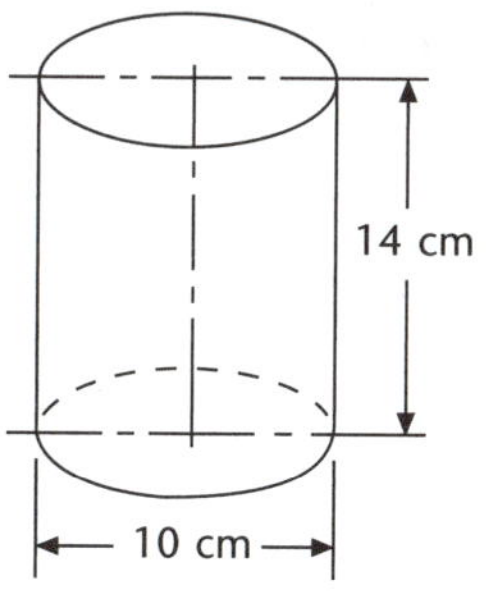


(2 marks)

20 In the figure, each brick has a length of 10 cm and a width of $m$ cm. Assume each higher brick rests half-way along the lower brick, i.e. 5 cm. Find the:

a perimeter of the figure

b area of the figure.

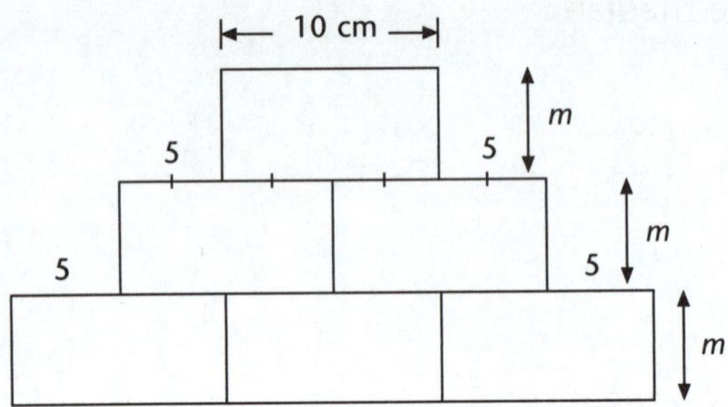


(2 marks)

## Part C

**(15 marks)**

Show all necessary working.

1

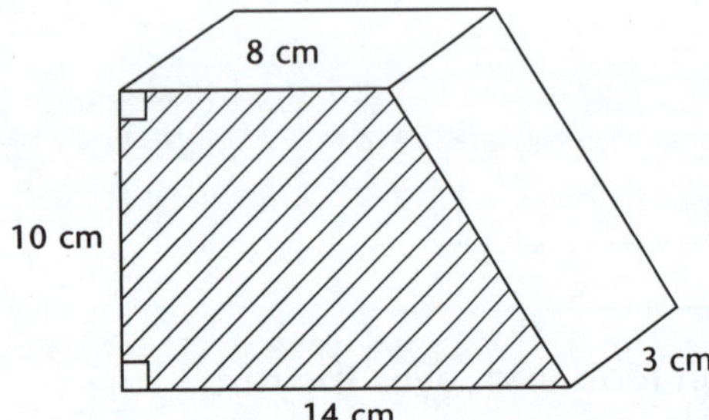


a Find the shaded area.

b Find the volume of the prism. (2 marks)

2 Thirty students were asked whether they had seen a movie ($M$) or gone to the beach ($B$) during the recent holidays. Twenty students had gone to the beach and 18 had seen a movie. Four students said they had not seen a movie, or gone to the beach.

a Record the information on a Venn Diagram. (1 mark)

b If a student who went to the beach was chosen at random, what is the probability they had also seen a movie? (1 mark)

3 The number of passengers in a taxi for each trip was recorded in one day, as shown:

| | | | | | | | | | |
|---|---|---|---|---|---|---|---|---|---|
| 3 | 0 | 6 | 2 | 0 | 2 | 5 | 2 | 6 | 1 |
| 5 | 1 | 0 | 4 | 5 | 1 | 2 | 6 | 4 | 5 |
| 3 | 5 | 3 | 2 | 3 | 0 | 5 | 5 | 3 | 1 |

a Complete the frequency distribution table.

| Number of passengers ($x$) | Tally | Frequency ($f$) | $f \times x$ |
|---|---|---|---|
| 0 | \|\|\|\| | | |
| 1 | \|\|\|\| | | |
| 2 | ~~\|\|\|\|~~ | | |
| 3 | | | |
| 4 | | | |
| 5 | | | |
| 6 | | | |
| | Totals | | |

(2 marks)

b Use this table to find the mean (average) of the number of passengers riding in this taxi. (1 mark)

c For the above distribution, find the:

i mode ii range. (2 marks)

4

a Write an expression for the perimeter of the rectangle. (1 mark)

b If the perimeter is 44 cm, find the value of $x$. (2 marks)

5 O is the centre of a circle.

AD and BC are diameters.

a Prove $\triangle ABO = \triangle DCO$.

b If $\angle ABO = 70°$, find $\angle COA$.

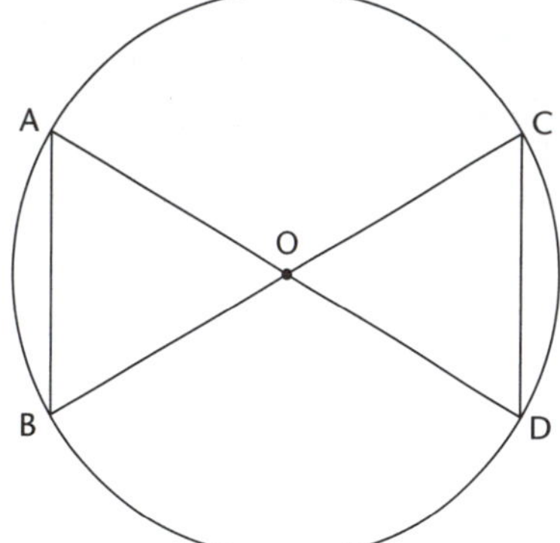


(3 marks)

☞ Answers on page 295
☞ Worked solutions on page 407

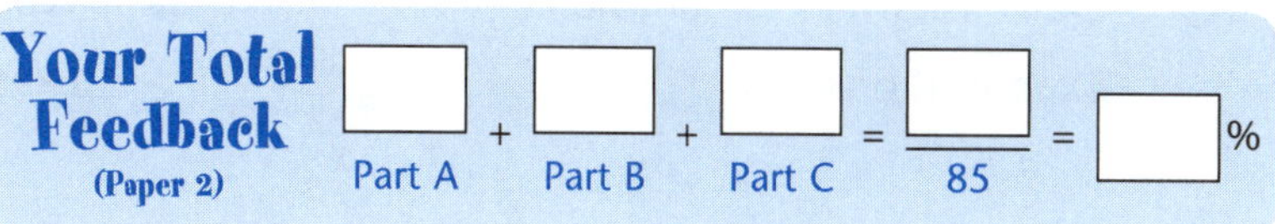

# 14 ANSWERS AND WORKED SOLUTIONS

- Quick Answers to Practise, Practise
- Quick Answers to Tests
- Quick Answers to Sample Examinations
- Worked Solutions to Practise, Practise
- Worked Solutions to Tests
- Worked Solutions to Sample Examinations

# QUICK ANSWERS to Chapter 1 **Practise, Practise**

## Chapter 1—Selected Revision of Year 7 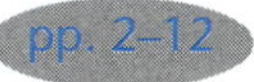



### Number

**1** a 28 b 3 c 12 d 58 e 84 f 32 g 12 h 63 i 108 j 14 k 20 l 14 **2** 48 **3** a 36 b 54 c 55 d 15 e 13 f 9 g 3 h 134 i 32 j 90 **4** \$30 **5** \$46.20 **6** 334 **7** 24 **8** 312 km **9** 13 trees

### Fractions

**1** a $\frac{1}{3}$ b $\frac{1}{2}$ c $\frac{2}{5}$ d $\frac{3}{4}$ e $\frac{3}{4}$ f $\frac{2}{3}$ g $\frac{2}{3}$ h 3 i $2\frac{1}{7}$ j $1\frac{1}{2}$ k $2\frac{1}{2}$ l $1\frac{1}{3}$ **2** a $\frac{10}{3}$ b $\frac{32}{5}$ c $\frac{19}{4}$ d $\frac{100}{11}$ e $\frac{25}{3}$ **3** a $\frac{1}{5}$ b $1\frac{1}{7}$ c $\frac{5}{12}$ d 2 e $\frac{3}{4}$ f $2\frac{3}{5}$ g $1\frac{29}{60}$ h $4\frac{5}{12}$ **4** a $\frac{12}{25}$ b $\frac{9}{32}$ c $\frac{7}{12}$ d $\frac{5}{6}$ e $1\frac{1}{8}$ f 2 g 48 h $\frac{5}{36}$ i $\frac{7}{9}$ j 32 k $3\frac{1}{3}$ l $1\frac{5}{7}$ m $1\frac{1}{3}$ n 4 o 40 p $2\frac{1}{5}$ q 12 **5** a $1\frac{1}{6}$ b $1\frac{17}{30}$ c $\frac{23}{30}$ d $\frac{1}{3}$ e $\frac{2}{15}$ f $\frac{4}{9}$ g $\frac{7}{36}$ h $\frac{7}{36}$ i $\frac{13}{15}$ j $\frac{9}{25}$ k $\frac{9}{25}$ l 7 **6** \$60 **7** \$40 **8** \$1600 **9** \$129.60 **10** a $\frac{4}{5}$ b $1\frac{3}{4}$ c $\frac{8}{15}$ d $1\frac{3}{40}$ e $2\frac{2}{3}$

### Decimals

**1** a $\frac{2}{5}$ b $\frac{11}{25}$ c $\frac{1}{25}$ d $1\frac{3}{5}$ e $2\frac{1}{2}$ f $3\frac{1}{20}$ g $2\frac{1}{8}$ h $1\frac{1}{200}$ **2** a 0.7 b 0.17 c 0.07 d 0.77 e 3.999 f 1.79 g 42.4 **3** a 23.64 b 2.0 or 2 c 2.6 d 1.155 e 470 f 73.4 g 5.7 h 1.17 i 0.8 j $32 \div 8 = 4$ k 60 l 600 m 7.6 n 6.67 **4** a 3.5 b 1.1 c 1.2 d 1 e $1.\dot{3}$ f 5 g 2.46 h 2.25 i 9 j 0.81 **5** 23.35 m **6** 33, 15 cm remaining **7** 36 kg **8** 6.72 m **9** \$149.10 **10** a 2.64 b 8.24 c 8 d 5.8 e 1.3 f 5.3

### Percentages

**1** a $\frac{17}{100}$ b $\frac{73}{100}$ c $\frac{7}{100}$ d $\frac{4}{5}$ e $\frac{3}{50}$ f $\frac{1}{4}$ g $\frac{3}{40}$ h $\frac{49}{400}$ **2** a 0.69 b 0.08 c 0.8 d 0.45 e 0.065 f 0.1725 g 1.4 h 3 **3** a 75% b 70% c 81% d 80% e 62% f 125% g 63% h 70% I 5% j 120% k 37.5% l 120.5% **4** a \$270 b \$1960 c 455 d 24 e 70 cm f 200 g **5** a 25% b 90% c 10% d 10% e 20%

### Basic Algebra

**1** a $8a$ b $10y$ c $7t$ d $5t$ e $4a$ f $7y$ g $1.8x$ h $\frac{7}{12}y$ i $-2y$ j $-3a$ k $-7m$ l $0.7n$ **2** a $4y$ b $14a$ c $ab$ d $15t$ e $35y$ f $28a^2$ g $54xy$ h $-18y$ i $28t^2$ j $3y$ k 3 l $3y$ m 7 n $-9q$ **3** a $10x + y$ b $16y + 6x$ c $37a + 10$ d $15t + 17x$ e $16y + 2x$ f 0 **4** a $x = 6$ b $a = 21$ c $y = 14$ d $a = 7$ e $t = 8$ f $a = 27$ g $y = 9$ h $m = 7$ i $x = 5$ j $x = 2$ **5** a $10x$ b $8y$ c $8a + 8b$ d $20a$ e $38a$

### Basic Geometry

**1** Rectangle **2** Parallelogram **3** Trapezium **4** Kite **5** Rhombus **6** Sphere **7** Cone **8** Cylinder **9** Quadrilateral **10** Scalene triangle **11** Isosceles triangle **12** Equilateral triangle **13** Right-angled triangle **14** Cube **15** Square prism **16** Triangular prism **17** Pentagon **18** Hexagon **19** Square **20** Octagon **21** Hemisphere **22** Concurrent lines **23** Collinear points **24** Parallel lines **25** Complementary angles **26** Vertically opposite angles **27** Supplementary angles **28** Exterior angle **29** Alternate angles **30** Co-interior angles **31** Corresponding angles **32** Revolution **33** Reflex angle **34** Obtuse angles **35** Acute angle **36** Straight angle **37** a 23° b 42° c $(90 - x)°$ **38** a 113° b 132° c $(180 - x)°$ d 32° **39** a Equal b 90° c 180° d Equal e 360° f Sum … interior g Transversal i Equal ii Corresponding iii Supplementary **40** 56 **41** 74 **42** 57 **43** 21 **44** 85 **45** 116, 64, 64 **46** 292

# QUICK ANSWERS to Chapters 1, 2, 3 and 4 **Practise, Practise**

**47** 149 **48** 150 **49** 30 **50** 36 **51** 73 **52** 30 **53** 103 **54** 118 **55** 65 **56** 83 **57** 68 **58** 111 **59** 65 **60** 114 **61** 94, 86, 94, 86 **62** 59, 59, 59 **63** 82, 98, 98, 98 **64** 140 **65** 60 **66** 52 **67** 69 **68** 52 **69** 64, 64, 68 **70** 75

## Statistics

**1** See solutions **2** See solutions **3** a 5 b 8 c 6 d $5\frac{6}{7}$ **4** a 12 b 8 c 9 d $9\frac{2}{3}$
**5** a See solutions b i 6 ii 4 iii 6.05 **6** Purple

## Probability

**1** a $\frac{1}{10}$ b $\frac{1}{2}$ c $\frac{1}{2}$ d $\frac{9}{10}$ e $\frac{2}{5}$ f $\frac{1}{2}$ g 0 h 0 i $\frac{1}{2}$ j $\frac{3}{10}$ k $\frac{3}{10}$ l $\frac{3}{10}$ m $\frac{7}{10}$ **2** a $\frac{1}{6}$ b $\frac{1}{2}$ c $\frac{1}{2}$ d $\frac{1}{3}$ e $\frac{1}{2}$ f $\frac{1}{3}$ g $\frac{1}{6}$

## Chapter 2—Number pp. 19–20

**1** a 8 b 15 c 12 **2** 12 **3** 4 **4** a 12 b 18 c 48 **5** 12 **6** 22 **7** a 2, 3, 5, 7, 11, 13, 17, 19, 23, 29 b 4, 6, 8, 9, 10, 12, 14, 15, 16, 18, 20, 21, 22, 24, 25, 26, 27, 28 **8** 2 **9** 8 **10** a $2\times2\times2\times3\times3$ b $2\times2\times2\times2\times2\times3$ c $5\times5\times5\times2$ **11** a $4^5$ b $3^2\times5^3$ c $10^5$ **12** $2^3\times3^3$ **13** a $2^6$ b $3^3$ c $6^7$ d 7 e $4^2$ f $2^{12}$ g $6^6$ **14** a 1 b 1 c 0 d 1 **15** a –5 b 8 c –6 d 11 e –12 f –2 **16** a –6 b 12 c –12 d –10 e –24 f 6 g 9 h 25 i 9 **17** a –2 b 4 c –5 d 9 e 3 f –2 g 12 h –8 i –3 **18** a –5 b –3 c 2 d –17 e –6 f –5 g –5 h 1 i 1 j 2 k 25 l 4 m –20 n 14 o 8 p –18 q 65 r 2 **19** a 0.6 b 0.55 c 0.17 d $0.\dot{6}$ e $0.\dot{3}\dot{6}$ f $0.\dot{8}5714\dot{2}$ **20** a $\frac{7}{10}$ b $\frac{7}{100}$ c $\frac{2}{25}$ d $\frac{31}{100}$ e $\frac{29}{200}$ f $\frac{9}{125}$

## Chapter 3—Percentages and Applications pp. 34–37

**1** a 40% b 45% c 52% d 17.5% e 41% f $33\frac{1}{3}$% g $83\frac{1}{3}$% h 325% 2 a 28.6% b 45.5% c 226.7% **3** a 30% b 3% c 0.3% d 19% e 57.6% f 112% g 470% h 7.5% **4** a $\frac{3}{5}$ b $\frac{3}{100}$ c $\frac{19}{20}$ d $1\frac{27}{100}$ e 3 f $\frac{11}{200}$ g $\frac{49}{400}$ h $\frac{19}{500}$ **5** a 0.16 b 0.07 c 0.6 d 0.076 e 0.125 f 1.76 g 1.046 h 0.0025 i 0.0775 j $0.\dot{3}$ **6** a 162 b \$22.40 c \$25.20 d 420 m e 360 mL f 30 min g 15 g h \$435 i \$35 200 j 635 **7** a \$22 b 826 c 58.5 L d 3240 g **8** a 208 b \$61.75 c 2625 d \$34.80 **9** \$91 **10** a 48% b 10% c 0.8% d 12.5% e $8\frac{1}{3}$% f 5% **11** a 10% b 0.175% c 200% d 6000% **12** a 3500 b 500 **13** 3500 g **14** a \$71.25 b \$28.50 **15** a \$2.75 b \$19.25 **16** 12.5% **17** \$54.60 **18** \$2549.85 **19** a 60c b 75% **20** a \$2400 b 70.6% **21** a \$0.96 b \$5.76 c 1.6c **22** 80% **23** \$384 000 **24** \$43 **25** \$6.50 **26** \$59.40 **27** \$32 **28** \$240 **29** \$863.64 **30** \$62.73 **31** 40% **32** a 36% b 8% **33** a 20% b 25% **34** \$626.08 **35** 80% **36** 900 **37** a \$170 b \$34 **38** 12.5% **39** a 68.89% b 1820

## Chapter 4—Algebra pp. 51–57

**1** a $2bc$ b $12a$ c $ab$ d $3b$ e $a^2$ f $10t$ g $3b^2$ h $6a^2$ i $30a^2$ j $3a^2$ k $21ab$ l $2a^3$ m $12xy^2$ n $12xy$ o $4x^2$ **2** a $a\times b$ b $2\times x\times x\times y$ c $3\times a\times a$ d $5\times a\times b\times c$ e $2\times a\times a\times b\times c\times c$ f $5\times5\times x\times x\times x$ **3** a 6 b 15 c 7 d 6 e 4 f 3 g 30 h 16 i 5 j 25 k 9 l 50 m 12 n 100 o 2 p 3 q 2 r 25 s 4 **4** a 1.32 b 1.2 c 0.432 d 3 e 2.28 f 0.5184 g 0.36 h 0.72 **5** a $1\frac{7}{15}$ b $1\frac{1}{5}$

# QUICK ANSWERS to Chapters 4 and 5 **Practise, Practise**

c $\frac{8}{15}$ d 2 e $1\frac{3}{5}$ f $\frac{16}{25}$ g $1\frac{1}{3}$ h $4\frac{7}{15}$ i $3\frac{3}{5}$ j 11 **6** a –12 b 16 c –10 d –6 e 7 f –5 g –2 h –20 i 144 j –1 k –7 **7** a 6 b 50 c –10 d 154 e 64 f $1\frac{5}{6}$ g $\frac{2}{3}$ h 5.04 i 3 j $3\frac{3}{7}$ **8** a $7a$ b $6a$ c $5a+2b$ d $3a$ e $10b-2$ f $x-4$ g $5a^2+2a$ h $8b$ i $3p-4$ j $2x+y$ k $3ab$ l $4m^2+5m$ m $3b+7a$ **9** a $5b$ b 5 c $3b$ d $\frac{5b}{c}$ e $2y$ f $\frac{3a}{2}$ g $\frac{2n}{3m}$ h $7x^6y$ i $\frac{3b}{7}$ j 2 **10** a $a^8$ b $a^8$ c $a^{10}$ d $a^{14}$ e $a^{17}$ f $a^{12}$ g $a^{10}$ h $a^9$ i $a^{12}$ j $a^{28}$ k $a^{10}$ l $a^3b^3$ m $9x^2$ n $16a^2b^2$ o $64a^3$ p $25a^8$ q $49x^2$ r $a^{12}b^6$ s $y^8$ t $a^8$ u $3a^3b$ v $3x$ w $2a^2$ x $8x$ y $3a^2b^4$ z $3a^4$ aa $2^{10}$ bb $5^5$ cc $10^7$ **11** a $5a+15$ b $4b-8$ c $3a-3b$ d $8x-12$ e $ab-2a$ f $-4x-8$ g $-3x+6$ h $8a-12b$ i $-4a+6b$ j $2a^2-ab$ k $-4a+12$ l $-a+2$ m $-2+3a$ n $6-12a$ o $-6-3a$ **12** a $-4a$ b $-6t$ c $-7a$ d $-4a$ e 0 f $3x-3y$ g $6x-4y$ h $a^2-4a$ i $4x-9$ j $2-x^2$ k $-2x-5$ l $11x-18$ m $11b-6$ n $2a-10b$ o $-4b$ p $2a-4b$ **13** a $8x+10$ b $4a+7b$ c $14+3a$ d $7a+17$ e $6x+10$ f $5x-8$ g $13-2x$ h $6-x$ i $-x-2$ j $13x-6$ k $8x-11$ l $-2-2x$ m $1+2a$ n $a^2+5a+3$ o $a^2-6a$ p $5b^2-4b$ q $10p^2-8p+6$ **14** a $-8a$ b $-6ab$ c $6ab$ d $4bc$ e $-2x^2$ f $-15b^2$ g $-4a$ h $3b$ i $10p^2$ j $-2x$ k 12 l –2 **15** a $9a-2$ b $-8t$ c $6a$ d 5 e $2p$ f $6a^2b$ g $3a$ h $a^3$ i $4a$ j $16x^2$ k $8a^{12}$ l $-7a$ m $-20a^2$ n $4a^2b^2$ o $6x^2-5x$ p $6x$ q $\frac{2x}{3y}$ **16** a $4(a+5)$ b $5(a+b)$ c $a(b-c)$ d $2(3a+2b)$ e $3(b^2-2)$ f $5x(x+5)$ g $2a(2a-1)$ h $5(x^2+2y+4)$ i $6(2-a)$ j $x(5x-6)$ k $7xy(5x-1)$ l $2b(2a-1)$ m $4a(2-a)$ n $4b(3-2b)$ **17** a $2a+c$ b $mn$ c $5x$ d $2b+a$ e $3x+8$ f $12y$ cents g $\$(x-6)$ h $a+1, a+2, a+3$ i $a+2, a+4$ j $9b^2$ cm$^2$ **18** a i $100b$ cm ii $60a$ sec iii $2000a$ mL iv $300a$ dollars v $2t$ min b i $(5x+3)$ cm ii $(4x+8)$ cm iii $36a$ cm c i $(4x-12)$ cm$^2$ ii $9x^2$ cm$^2$ iii $(6x-12)$ cm$^2$ **19** a $2a+3$ b $8a-12$ c $5b$ d $8c^2-2c$ e $6y^7$ f 16 g $6a^2$ h $7m-6n$ i 9 j –15 k $8x-9$ l –32 m –14 n $\frac{5}{6}$ o $6a^3b^2$ p $-ab+8a$ q 8 r $11-2a$ s $2(4b-a^2)$ t $2x^3y^3$ u $4x(x+4)$ v $5x^2-11x$ w $x+2, x+4$ x 0.32 y $P=(6x+14)$ cm, $A=(2x^2+14x)$ cm$^2$ **20** a 4, 5, 6, 7, 13, 103 b 1, 3, 5, 17, 77, 197 c –5, –2, 1, 4, 61, 151 d 11, 9, 7, 5, 3, 1 e 10, 5, 2, 1, 2, 5 f –3, 0, 3, 6, 24, 294 g $-1, -\frac{1}{2}, 0, 1, 1\frac{1}{2}, 2$ h –6, –5, –4, –3, –2, –1 **21** a $y=4x$ b $y=2x+1$ c $y=3x+1$ d $y=4-x$ e $y=2x+1$ f $y=x^2$ g $y=x^2+1$ h $y=3-2x$

## Chapter 5—Patterns and Linear Relationships pp. 76–79

**1** a (2, 3), b (–2, 1) c (2, –4) d (5, –2) e (0, 5) f (–6, 0) g (4, –6) h (–3, –2) i (–5, –5) j (1, 1) **2** See solution a Rectangle b 6, 5 c 22 units d 30 units$^2$ e 2nd **3** a K(3, 4), L(3, –3), M(–4, –3), N(–4, 4) b Square c 49 units$^2$ **4** See solution a Triangle b 7 c 4 d 14 units$^2$ **5** a i 5 ii 5 b Isosceles c 12.5 units$^2$ **6** See solution a (3, 0) or (0, 4) b 6 units$^2$ **7** See solution D(4, –4)

**8** a

| $y = 3x + 2$ | |
|---|---|
| $x$ | $y$ |
| 0 | 2 |
| 1 | 5 |
| 2 | 8 |
| 3 | 11 |

b

| $y = 4 - 2x$ | |
|---|---|
| $x$ | $y$ |
| 0 | 4 |
| 1 | 2 |
| 2 | 0 |
| 3 | –2 |

**9** a $y = 4x$ b $y = x - 2$ c $y = 3x - 1$ **10** a 0, 1, 2, 3 b 0, 2, 4, 6 c $y = 2x$ **11** See solution **12** Yes **13** No **14** D **15** C **16** a No b Yes c Yes d Yes B, C, D collinear **17** Yes, No, Yes, Yes, Concurrent lines are $y = 3x + 2$, $y = 6 - x$ and $y = 6x - 1$ **18** See solution **19** See solution a Rectangle b 12 units$^2$ **20** a (–1, 3) b (2, 3) c (2, 2) d (1, 2) **21** See solutions: a, d and f are parallel; b and e are parallel and perpendicular to a, d, and f; c is neither perpendicular nor parallel a 2 b $-\frac{1}{2}$ c $\frac{1}{2}$ d 2 e $-\frac{1}{2}$ f 2 **22** See solutions **23** See solution

## Chapter 6—Equations and Formulae pp. 99–104

**1** a Solution b Solution **2** a Solution b Solution c Not solution d Solution **3** a $y = 4$ b $y = -4$ c $y = 18$ d $y = 18$ e $y = -13$ f $t = -11$ g $m = 8$ h $t = 3$ i $p = 11$ j $m = 21$ k $n = -17$ l $t = 7$ m $y = -2$ n $a = 6$ o $t = 6$ p $x = 8$ q $a = -9$ r $t = -5$ s $y = -4$ t $t = 8$ u $t = 72$ v $z = 27$ w $a = -30$ x $n = -48$ y $m = 28$ z $x = 3$ **4** a $y = 2\frac{1}{4}$ b $a = 2\frac{1}{2}$ c $t = 2\frac{1}{3}$ d $a = 2\frac{4}{5}$ e $y = 1\frac{3}{5}$ f $m = 4\frac{2}{7}$ g $a = 2\frac{2}{3}$ h $a = 1\frac{3}{4}$ i $a = \frac{3}{4}$ j $t = \frac{2}{3}$ k $y = 2\frac{1}{2}$ l $a = -1\frac{1}{4}$ m $y = -3\frac{1}{3}$ n $t = 2\frac{1}{2}$ o $z = 1\frac{1}{4}$ p $m = 1\frac{2}{3}$ q $y = 0$ **5** a $a = 3$ b $x = 3$ c $y = 4$ d $t = 4$ e $t = 3$ f $y = 4$ g $t = 7$ h $y = 3$ i $n = 4$ j $a = 6$ k $t = 9$ l $y = -7$ m $m = -8$ n $y = 4$ o $a = 20$ p $a = -1$ q $t = -5$ r $n = -4$ s $t = -2$ t $n = 3$ **6** a $a = 1\frac{1}{5}$ b $y = 2\frac{1}{2}$ c $t = 1\frac{4}{7}$ d $a = 1\frac{1}{5}$ e $t = 3\frac{3}{4}$ f $t = 3\frac{1}{2}$ g $m = -1\frac{1}{2}$ h $a = 2\frac{1}{8}$ i $t = -3\frac{2}{3}$ j $y = 1\frac{5}{6}$ k $y = 2\frac{1}{2}$ l $t = 1\frac{1}{3}$ m $t = 1\frac{2}{5}$ n $m = -4\frac{1}{2}$ **7** a $a = -7$ b $y = 5$ c $y = 4\frac{1}{3}$ d $n = 5$ e $a = 5$ f $y = 4$ g $t = 3$ h $y = -5$ i $y = -1$ j $m = 6$ k $t = 3$ l $a = -4\frac{2}{3}$ m $y = 4\frac{1}{2}$ n $a = -\frac{2}{3}$ **8** a $x = 5$ b $a = 5$ c $t = 11$ d $a = 5$ e $y = 9$ f $m = 3$ g $t = 9$ h $a = -\frac{1}{3}$ i $y = 10\frac{1}{3}$ j $t = 2$ k $t = -5\frac{4}{5}$ l $y = 12$ m $a = -2\frac{2}{3}$ n $a = -2\frac{2}{3}$ o $a = 12$ p $a = 9\frac{4}{5}$ q $a = 3$ r $x = 13$ s $x = 2$ t $a = -6\frac{1}{4}$ **9** a $a = 17$ b $y = 10$ c $a = -8\frac{1}{2}$ d $a = 8$ e $a = 3\frac{1}{5}$ f $x = 4\frac{1}{2}$ **10** a $y = 30$ b $a = 90$ c $y = 20$ d $t = 4$ e $a = 12$ f $a = 8\frac{3}{4}$ g $m = -18$ h $t = 4\frac{1}{2}$ i $m = 5$ j $a = 11\frac{2}{7}$ k $y = 6\frac{1}{3}$ l $x = \frac{2}{7}$ **11** a $x = 12\frac{8}{11}$ b $x = 10$ c $x = 32\frac{10}{11}$ d $x = -\frac{120}{127}$ e $x = \frac{39}{44}$ **12** a $a + b$ b $ab$ c $y + 2$ d $y + 1$ e $2p + 1$ f $\dfrac{x}{1000}$ g $1000x$ h $2b - a$ **13** a 18 b 14 c 14 d 15 e 8 **14** a 3 b 13, 14, 15 c 31, 33 d 18 e 18, 54, 35 **15** a 10.5 b 6 c 9 d 4 e 5 f 8 **16** a 8 b 70.5 c 5 d 34 **17** 10 **18** 40 **19** 24 **20** 12 **21** 2 km **22** 38, 55, 68 **23** 54 km **24** a 118 b 11 c 16 d 4.5 e 157 f 176 g 135 h 806 i 18 j 560 k 10 l 2 m 24 n 50 o 401.92 **25** a $2\frac{1}{2}$ b $\frac{1}{2}$ c 4 d 6 e 2 f $-5\frac{1}{4}$ g 9 h 9

## Chapter 7—Pythagoras' Theorem pp. 119–128

**1** a $r^2 = p^2 + q^2$ b $n^2 = m^2 + p^2$ c $10^2 = a^2 + b^2$ d $PR^2 = PQ^2 + QR^2$ e $c^2 = 3^2 + 5^2$ f $d^2 = 4^2 + 7^2$ **2** a $d^2 = c^2 + e^2$ b $y^2 = x^2 + 5^2$ c $PQ^2 = PR^2 + RQ^2$ d $z^2 = 15^2 - 12^2$ e $w^2 + x^2 = y^2$ f $s^2 + 7^2 = t^2$ **3** a 49 b 121 c 289 d 841 e 961 f 12 g 30 h 16 i 25 j 130 **4** a $a = 10$ b $b = 2.5$ c $c = 17$ d $d = 15$ e $e = 25$ f $f = 25$ g $g = 41$ h $h = 26$ i $i = 61$ j $j = 30$ **5** a $x = 4$ b $y = 5$ c $t = 7$ d $n = 15$ e $m = 9$ f $z = 0.4$ g $p = 0.6$ h 30 cm **6** a $x = 20$ b $y = 24$ c $a = 8$ d $b = 35$ e $c = 0.9$ f $d = 1.2$ g $n = 32$ h $q = 48$ i BD is 2.5 cm j PQ is 40 cm **7** a △ is right-angled b △ is not right-angled c △ is not right-angled d △ is right-angled e △ is right-angled f △ is right-angled g △ is not right-angled h △ is right-angled **8** a Triad is {7, 24, 25} b Triad is {9, 12, 15} c Triad is {18, 24, 30} d Triad is {9, 40, 41} e Triad is {13, 84, 85} f Triad is {21, 72, 75} **9** a $n = 6.4$ b $w = 7.3$ c $x = 8.6$ d $m = 8.5$ e $p = 12.1$ f $q = 4.3$ g AC = 1.4 cm h QS = 5.8 cm **10** a $a = 5.7$ b $b = 4.1$ c $c = 2.6$ d $d = 8.5$ e $e = 7.9$ f $f = 11.2$ g $g = 14.4$ h $h = 13.4$ **11** a $p = 12.4$ b $q = 7.14$ c $p = 1.33$ d $q = 7.07$

# QUICK ANSWERS to Chapters 7 and 8 **Practise, Practise**

**12** a $n = \sqrt{50}$ b $m = \sqrt{301}$ c $r = \sqrt{132}$ d $s = \sqrt{297}$ **13** Length of diagonal is 20 cm **14** Diagonal is 4.2 cm **15** Brace reaches 28 cm up the wall **16** Foot of ladder is 2.8 m from the wall **17** Wires are attached 10.96 m from the top **18** Length of side is 12.7 cm **19** Width is 15 cm **20** Length of each side is 13 cm **21** $y = 11.2$ (to 3 significant figures) **22** Third side is 24 m **23** Perimeter is 36 m **24** Shortest distance AG is 25 cm **25** Shortest distance for Harry is 29 cm. Harry's extra distance = 29 − 25 = 4. Harry travels an extra 4 cm **26** Gradient of railway line is $\frac{11}{60}$ **27** Length of PR is 12 cm **28** Perimeter is 54 cm **29** Foot of ladder should be placed 4.8 m from the wall **30** Area of semicircle is 157 $cm^2$ **31** See solutions **32** Slant height of roof is 14.4 metres **33** Perimeter is 208 metres

## Chapter 8—Measurement pp. 155–167

**1** a 56.7 $mm^2$ b 324 000 $m^2$ c 76.5 $cm^2$ d 370 $cm^2$ e 250 $cm^2$ f 263 $cm^2$ **2** a 18 $cm^2$ b 102 $cm^2$ c 359 $cm^2$ **3** a 120 $cm^2$ b 405 $cm^2$ c 119 $cm^2$ d 448 $cm^2$ e 66 $cm^2$ **4** a 6.78 ha b 48 ha **5** a 3 ha b 47 200 $m^2$ c 6.54 $m^2$ d 115 700 $cm^2$ **6** a Centre b Radius c Sector d Segment e Diameter f Quadrant g Semicircle h Chord i Circumference j Arc **7** a 9 cm b 15 m c 7.5 mm d 2.1 m e 0.05 cm **8** a 14 cm b 30 cm c 7 m d 9 mm e 4.1 m **9** 52.5 m **10** a $\frac{1}{4}$ b $\frac{3}{4}$ c $\frac{1}{12}$ d $\frac{2}{3}$ e $\frac{5}{6}$ **11** a 31.4 cm b 176 cm c 31.42 cm d 110 m e 12.568 m f $12\pi$ cm g $16\pi$ cm **12** a 88 cm b 440 mm c 264 cm d 132 cm e 11 m f 11 m **13** a 628 mm b 12.56 cm c 125.6 cm d 28.26 m e 94.2 m f 94.2 cm g 9.42 m **14** a 6.28 m b 3.14 cm c 125.66 cm d 314.16 m e 942.48 m f 628.32 mm g 15.71 m h 282.74 m **15** 75.4 cm **16** $140\pi$ m **17** a 41.1 cm b 35.7 cm c 17.9 cm d 36.6 cm e 30.5 cm f 33.6 cm **18** a 12.7 cm b 79.6 cm c 1.4 cm d 318.3 m e 2037.2 km **19** a 8.0 cm b 0.4 m c 0.1 cm d 15.9 m e 50.0 m **20** 32 m **21** $3749 **22** a 616 $m^2$ b 1386 $cm^2$ c $38\frac{1}{2}$ $cm^2$ d 154 $m^2$ e $38\frac{1}{2}$ $m^2$ f $3\frac{1}{7}$ $cm^2$ **23** a 314.2 $m^2$ b 78.5 $cm^2$ c 31 415.9 $cm^2$ d 314.2 $cm^2$ e 50.3 $cm^2$ f 113.1 $m^2$ g 1963.5 $cm^2$ h 19.6 $cm^2$ i 1017.9 $cm^2$ j 1590.4 $m^2$ **24** a 3.34 cm b 5.64 cm c 7.04 mm d 4.35 cm **25** a 11.28 mm b 7.14 cm **26** a 13.1 $cm^2$ b 235.6 $cm^2$ c 39.3 $cm^2$ d 12.6 $m^2$ e 94.2 $cm^2$ f 457.1 $cm^2$ g 78.5 $cm^2$ h 50.3 $cm^2$ i 23.1 $cm^2$ j 125.7 $m^2$ k 87.5 $cm^2$ l 370.2 $cm^2$ m 34.6 $cm^2$ n 102.5 $cm^2$ **27** a 54.9 $cm^2$ b 201.1 $cm^2$ c 27.5 $cm^2$ d 28.5 $cm^2$ **28** 1514 $m^2$ **29** 963 $cm^2$ **30** a 5.365 045 9 $cm^2$ b 20 c 8000 d 42 920 $cm^2$ **31** a 282 $m^3$ b 3744 $cm^3$ c 3375 $cm^3$ d 180 000 $cm^3$ e 1600 $cm^3$ f 1232 $cm^3$ g 920 $cm^3$ h 1680 $cm^3$ i 9000 $cm^3$ j 840 $cm^3$ **32** 396 $m^3$ 10 mm 40 $mm^2$ **33** 5 cm **34** 8 cm **35** a 100.5 $m^3$ b 502.7 $cm^3$ c 14 726.2 $cm^3$ d 54.3 $cm^3$ e 0.6 $m^3$ **36** a 3080 $cm^3$ b 1100 $m^3$ c 9504 $cm^3$ d 770 $cm^3$ e 61 600 $mm^3$ f 577.5 $m^3$ **37** 1.54 kL **38** 678.6 $m^3$ **39** 17 996.8 $cm^3$ **40** 62 168 $cm^3$ **41** 1.42 L **42** a 3.5 L b 28 mL c 760 $cm^3$ d 7.85 L e 58 000 $cm^3$ f 1 000 000 $cm^3$ **43** 1 $m^3$ = 1 kL **44** a 13.5 L b 150.48 L c 8000 L **45** a 5 h b 96 h c 390 sec d 45 min e 12 sec f $2\frac{1}{2}$ days g 20 centuries h 6 decades **46** a 35 minutes b 3 h 55 min c 12 h 35 min **47** a 142 days b 114 days

**48**

| | |
|---|---|
| 4:17 am | **0417** |
| **11:42** am | 1142 |
| 3:47 pm | **1547** |
| **9:48** pm | 2148 |

# QUICK ANSWERS to Chapters 8, 9 and 10 **Practise, Practise**

**49** 3:56 **50** a 9:00 pm b Wakes at 6:20 am **51** a $8\frac{1}{2}$ hours b $102 **52** a 3:10 pm in Adelaide b 6:52 pm in Sydney c 1:13 am in Adelaide **53** a 6:15 b 9:58 **54** a i 20 min ii 48 min b Arrive at 5:41 c Arrive at 8:16 **55** a 2 hours ahead b 7 hours behind c 8 am d Finishes 11:30 am Monday morning

## Chapter 9—Ratio and Rates pp. 188–193

**1** a 5:3 b 3:7 c 1:2:3 d 2:7 e 1:1000 f 3:5 **2** a $x=3$ b $x=3$ c $x=2$ d $x=49$ e $x=39$ **3** a 7:3 b 10:7 c 2:1 d 13:17 e 3:2 f 4:5 g 2:5 h 9:10 i 4:11 **4** a 15:4 b 3:2 c 1:6 d 1:10:1000 e 7:2 f 2:1 g 1:1 000 000 h 3:20 **5** a $2b$:1 b 1:5 c 1:2 d 1:$p^2$ **6** a $200 b i 7:9 ii 2:7 **7** a 25 b i 31:25 ii 56:25 **8** 5:12 **9** a 3:5 b 1:1 **10** a i 3 cm ii 8 cm b i 3:8 ii 2:3 iii 8:7 **11** 1:2 **12** a 1:3 b 1:9 c 1:27 **13** 5:1 **14** a 60° b i 3:5 ii 1:1 iii 1:4 **15** 16 **16** a 18 cm b 60 cm **17** $840 **18** 10 cm **19** a 80:4:1 b 960 students, 12 assistants **20** $120 000 **21** $55, $20 **22** $120 **23** a F b B c G d J **24** a $549 000 b $594 000 c 61:66 **25** 40 minutes **26** a 3:7 b 12 000 **27** $12.45/h **28** a $29.70 b 1.32 kL/day c 99c/day **29** a $411.51 b 1914 km c 8 km/L d 21.5c/km **30** $1762.25 **31** 800 **32** a 12 h b 90 min **33** 375 g **34** 1 kg **35** a 72 km/h b 0.384 kL/day c 342.72 m/week **36** a $23\frac{1}{3}$ m/sec b 1 L/min **37** 1 min 12 s **38** $136.88 **39** 5h, 156 km, 95 km/h, $2\frac{1}{2}$ h, 20 km, 240 km/h **40** 6 pm **41** 75 km/h **42** 8:15 pm **43** A arrives 10 min after B **44** a 1:25 b 1:300 000 c 1:200 000 d 3:200 **45** a 4.2 m b 3.1 m c 6.2 m **46** a i 75 km ii 240 km b i 3 cm ii 5 cm **47** a 1:700 b 24.5 m c 15.4 m × 11.2 m **48** a 1 cm = 5 m b 17.5 m c No room

## Chapter 10—Geometry and Congruence pp. 211–221

**1** a 90° b 180° c 180° d Equal e 360° f 180° g 360° h Equal i 60° j i Equal ii Equal iii Supplementary **2** a $x=50$ b $x=47$ c $x=138$ d $y=80, x=100, z=100$ e $x=120$ f $x=76$ g $x=58$ h $y=60$ i $x=55$ j $x=107$ k $x=105$ l $x=108$ m $x=113$ n $x=37$ o $x=55$ p $x=109$ **3** a i A trapezium is a quadrilateral with one pair of opposite sides parallel ii Not necessarily b i Opposite sides are equal ii Opposite angles are equal iii Not unless it's a rhombus c i All sides are equal ii 2 iii Yes iv Yes, a special type of parallelogram d i 4 ii Equal iii 90° iv They are equal e i 2 ii Opposite sides are equal iii 90° iv They are equal f i 1 ii 2 pairs of adjacent sides equal g A, C, D, E, F h B, C, E, F i i F ii T iii T iv T v F vi T vii T viii T j i Parallelogram **4** a $x=140$ b $x=110, y=39$ c $x=40$ d 132° **5** See solutions **6** 1 – 13, 2 – 21, 3 – 12, 4 – 11, 5 – 16, 6 – 17, 7 – 20, 8 – 15, 9 – 18, 10 – 19 **7** c Yes d Yes **8** c Yes d Yes e No f Yes **9** a Yes b Yes c Yes **10** a AD b QP c ∠QMN d 25 cm e 34 cm$^2$ **11** a 7 cm b 154 cm$^2$ **12** a MN or NM b AD or BC depending on if you took Figure 2 as a reflection or rotation of Figure 1 c ∠MQP or ∠NPQ d ∠ABC or ∠BAD **13** a BC b 10 cm c ∠GHF d 90° e 24 cm f 24 cm$^2$ **14** a QP b SR **15** a AB = NM, AC = NP, BC = MP b ∠A = ∠N, ∠B = ∠M, ∠C = ∠P **16** d Yes e AB = FH, AC = FG, BC = HG, ∠A = ∠F, ∠B = ∠H, ∠C = ∠G **17** a 5 cm b 20 cm c 25 cm$^2$ d No **18** a SAS b AAS c SSS d RHS e AAS **19** a △PSR ≡ △PQR (SSS) b △ABX ≡ △DCX (SAS) c △BAE ≡ △CDE (AAS) d △ABC ≡ △ADC (RHS)

# QUICK ANSWERS to Chapters 11 and 12 **Practise, Practise**

## Chapter 11—Statistics pp. 239–244

**1** a i 5 ii 18 iii 2 b i $\frac{1}{4}$ ii $\frac{9}{10}$ iii $\frac{1}{10}$ c 5 4 and 5 d See solutions **2** See solutions **3** See solutions **4** a i 4 ii 17 iii 13 iv 8 b i $\frac{4}{25}$ ii $\frac{17}{25}$ iii $\frac{13}{25}$ iv $\frac{8}{25}$ c See solutions **5** See solutions **6** See solutions **7** 9 **8** a 5.6, 6, 6, 4 b 7.625, 5, 7.5, 6 c 4, 1 and 4 and 7, 4, 6 d 3.9, 7, 3.5, 9 e 7, 7, 7, 8 **9** a 64 b i 44 ii 20 c i $\frac{5}{32}$ ii $\frac{11}{16}$ iii $\frac{5}{16}$ d 4, $4\frac{1}{2}$ e 4.6875 f $4\frac{1}{2}$ g See solutions h $4\frac{1}{2}$ **10** a See solutions b 9, 5 c i 12 ii 39 iii 27 iv 13 d 52 e i $\frac{3}{13}$ ii $\frac{3}{4}$ iii $\frac{27}{52}$ iv $\frac{1}{4}$ f 4.75 g 4 **11** a i 6 ii 6, 5 iii 5.5 b Mean increased by 1, median increased by $\frac{1}{2}$ c 3 goals **12** 80 **13** a i See solutions ii 20 iii 68 b 6, 2 c i 4 ii 11 iii 5 d i $\frac{1}{5}$ ii $\frac{11}{20}$ iii $\frac{1}{4}$ e 3.4 f 3 **14** a 7 b 33 c 36 d 31.5 e 29.28 (2 dec. pl.) **15** a Increase by 6 b Multiplied by 6 **16** a, b See solutions c 3.02 d 2 and 5, 3 **17** 18, 19

## Chapter 12—Probability pp. 257–261

**1** a $\frac{1}{13}$ b $\frac{1}{13}$ c $\frac{1}{4}$ d $\frac{1}{4}$ e $\frac{1}{2}$ f $\frac{3}{13}$ g $\frac{10}{13}$ h $\frac{5}{13}$ i $\frac{1}{52}$ j $\frac{7}{13}$ k $\frac{1}{26}$ **2** $\frac{3}{10}$ **3** a $\frac{1}{3}$ b $\frac{2}{3}$ c $\frac{2}{5}$ d $\frac{3}{5}$ e 0 f $\frac{4}{15}$ g $\frac{11}{15}$ **4** a 0.52 b 672 **5** a $\frac{5}{36}$ b $\frac{1}{9}$ c $\frac{1}{2}$ d $\frac{1}{6}$ e $\frac{5}{18}$ f $\frac{8}{9}$ g $\frac{11}{18}$ h $\frac{1}{18}$ **6** a $\frac{1}{4}$ b $\frac{1}{8}$ c 0 d $\frac{3}{8}$ e $\frac{5}{8}$ **7** a $\frac{1}{10}$ b $\frac{1}{2}$ c $\frac{1}{2}$ d $\frac{1}{2}$ **8** a See solutions b i 0.22 ii 0.06 iii 0.32 iv 0.82 v 0.74 **9** See solutions **10** a See solutions b 50 i $\frac{4}{25}$ ii $\frac{8}{25}$ c $\frac{7}{11}$ **11** a See solutions b i $\frac{1}{7}$ ii $\frac{3}{28}$ **12** a i 7 ii 10 iii 3 b 25 c i $\frac{1}{5}$ ii $\frac{7}{25}$ d See solutions **13** a See solutions b i $\frac{9}{50}$ ii $\frac{9}{50}$ **14** a See solutions b 47 c $\frac{6}{47}$ **15** a See solutions b i $\frac{12}{25}$ ii $\frac{11}{50}$ iii $\frac{27}{50}$

# QUICK ANSWERS to Chapters 2, 3 and 4 **Tests**

## Chapter 2—Number

### Level 1 Test p. 22

**1** a 1, 2, 4, 8, 16 b 1, 5, 7, 35 c 1, 2, 5, 10, 25, 50 **2** a 2 b 4 c 6 **3** a 3, 6, 9, 12, 15 b 5, 10, 15, 20, 25 c 7, 14, 21, 28, 35 **4** a 12 b 15 c 24 **5** a 4, 6, 8, 9, 10 b 2, 3, 5, 7 **6** a $2^4$ b $6^2 \times 7^3$ **7** a 6 b −8 c −4 **8** a −18 b −8 c 15 **9** a −4 b −3 c 5 **10** a 0.8 b $0.\dot{3}$ **11** a $\frac{3}{100}$ b $\frac{4}{5}$

### Level 2 Test p. 23

**1** a 4 b 5 c 20 **2** a 20 b 75 c 36 **3** a $2 \times 2 \times 2 \times 2 \times 5$ b $2 \times 2 \times 2 \times 3 \times 3$ c $2 \times 2 \times 2 \times 5 \times 5$ **4** a $3^6$ b $2^2$ c $5^6$ **5** a 1 b 1 c 1 **6** a −8 b −2 c 6 **7** a −16 b 11 c 2 d 12 **8** a $0.\dot{1}4285\dot{7}$ b $0.\dot{2}\dot{7}$ c 0.15 **9** a $\frac{3}{100}$ b $\frac{33}{1000}$ c $\frac{3003}{10000}$

### Level 3 Test p. 24

**1** a 20 b 5 **2** 20 **3** a 120 b 48 **4** 60 **5** a 40% b 50% **6** a $4^4$ b 1 **7** a 0 b 1 c 0 **8** a −21 b −2 c 3 d 43 **9** a −1 b 1 **10** a $2^6 \times 5$ b $2^8$ c $2^2 \times 19^2$ **11** a −14 b −7 **12** $0.\dot{0}\dot{1}$

## Chapter 3—Percentages and Applications

### Level 1 Test p. 39

**1** a 60% b 150% c 45% d 80% e 6% f 35.5% **2** a $\frac{27}{100}$ b $\frac{2}{25}$ c $\frac{2}{5}$ **3** a 0.04 b 0.075 c 2.15 **4** a 25% b 80% **5** a \$3 b \$48 **6** a \$8.80 b \$96.80 **7** a \$600 b 25% **8** 6000 L **9** \$80.75 **10** \$7560

### Level 2 Test p. 40

**1** a $\frac{2}{5}$, 41%, 0.42 b 37%, $\frac{3}{8}$, 0.4 **2** a \$11.60 b \$57.60 c \$595 d 210 mm **3** a 10% b 0.25% c 5% d 1.75% **4** a \$27.50 b \$702 **5** a \$598 b \$3620 **6** a \$6.75 b \$496 **7** a \$68 b \$15.30 **8** 25% **9** 38.46% (to 2 decimal places)

### Level 3 Test p. 41

**1** a $66.\dot{6}\%$ b 87.5% **2** a 0.085 b 0.034 c 0.1275 **3** a \$8.75 b \$0.40 c 345 **4** 118.8 **5** \$109 **6** 6000 L **7** \$8000 **8** \$226.10 **9** $72.\dot{2}\%$ **10** \$78.41

## Chapter 4—Algebra

### Level 1 Test pp. 59–61

**1** a $4x$ b $2n$ c $12y$ d $6a$ e $ab$ f $a^2$ g $8mn$ h $6bc$ i $2a^2$ j $10a^2$ **2** a $4 \times a$ b $2 \times a \times b$ c $3 \times a \times a$ **3** a $3a$ b $2t$ c $2t$ d 5 **4** a 6 b 2 c 8 d 12 e 4 **5** a $8x$ b $5x$ c $5y$ **6** a $2a + 6$ b $5x - 10$ c $6a - 12$ **7** a $3b$ b $a + b$ c $x - 2$

**8**

| x | 0 | 1 | 2 | 3 | 4 | 5 |
|---|---|---|---|---|---|---|
| y | 2 | 3 | 4 | 5 | 6 | 7 |

**9** a $2(b + 3)$ b $3(x + 2y + 4)$ c $4(3x + 2)$ **10** a $6a + 3b$ b $6x + 5y$ c $4x + 4y$ **11** a 14 b 8 **12** a $6x + 6$ b $5x + 9$

# QUICK ANSWERS to Chapters 4 and 5 **Tests**

## Level 2 Test (pp. 61–62)

### Part A

**1** $7x - 3$ **2** $3 \times a \times b \times b$ **3** $2mn$ **4** 18 **5** 32 **6** $x = 17$ **7** $-6ab$ **8** 1 **9** 1 **10** $3(a - 4b)$
**11** $-3x - 6$ **12** $-4x$ **13** $F = 30$ **14** 100 **15** $15a^2$

### Part B

**1** a $6ab$ b $12b^2$ c $-6y$ **2** a $8x - 2$ b $9a + 3b$ c $8x + 6y$ **3** a $2b$ b $3a$ c $\frac{4x}{3y}$ **4** a $3x + 2$
b $6x + 8$ c $x + 4$ **5** a 7 b 9 c 4 **6** a $\$(a - b)$ b $2x + 16$ **7** a $2(3a - 2)$ b $a(a - 4)$ c $4x(3y - 1)$
**8** a $x^2 + 7x$ b $12a - 4a^2$ c $-2y + 7$ **9** a $5x + 7$ b $7x + 3$ c $5x + 15$

**10**

| x | −1 | 0 | 1 | 2 |
|---|---|---|---|---|
| y | −5 | −2 | 1 | 4 |

**11** $y = x + 2$

## Level 3 Test (pp. 63–64)

### Part A

**1** $8a^2$ **2** $3a + 5$ **3** 47 **4** $4x - 5y$ **5** 9 **6** $x$ **7** $4a$ **8** $6x - 3x^2$ **9** $5t(t - 6)$ **10** 16
**11** $x + 1, x + 3$ **12** $25x^2$ **13** $\frac{4x}{3y^2}$ **14** $\frac{4}{5}$ **15** $2y^2$

### Part B

**1** a $4a^2$ b $3 - a$ c $5y$ **2** 14 **3** a $-5(x + 5)$ b $3a(y^2 - 2a)$ c $t^2(t - 2)$ **4** a $x^3y^5$ b $x^2y^3$ c $y^3$
**5** a $7x - 2$ b $5x + 2$

**6**

| −1 | 0 | 1 | 2 |
|---|---|---|---|
| −1 | −2 | −1 | 2 |

**7** 40 **8** a $3x + 7$ b $x - 2$ **9** a $36y^2$ b $9y$ **10** a $2a^2 - 5a$ b $7xy - 4x$
**11** a $y = 3x + 1$ b $y = 3 - 2x$

## Chapter 5—Patterns and Linear Relationships

### Level 1 Test (pp. 81–82)

**1** a A(2, 2) B(−2, 1) C(−1, −3) D(3, −1) E(1, 0) F(0, 3) b See solutions

**2**

| x | y |
|---|---|
| 0 | 11 |
| 2 | 17 |
| 4 | 23 |

**3** $y = 3x$ **4** See solutions **5** a Yes b Yes c No d Yes e No **6** See solutions (−2, 5)
**7** See solutions **8** a See solutions b See solutions c \$15 d 7 kg

### Level 2 Test (pp. 83–84)

**1** a A(−3, −2) B(3, −2) C(0, 4) b Isosceles triangle c 18 units$^2$

**2**

| x | y |
|---|---|
| −1 | 14 |
| 1 | 10 |
| 3 | 6 |
| 5 | 2 |

**3** a

| x | −3 | −2 | −1 | 0 | 1 | 2 |
|---|---|---|---|---|---|---|
| y | −1 | 0 | 1 | 2 | 3 | 4 |

b $y = x + 2$

**4** a See solutions b (2, 1) **5** A, B, C collinear **6** a 3 b $-\frac{1}{3}$, perpendicular
**7** a See solutions b See solutions c i \$970 ii 7 h

# QUICK ANSWERS to Chapters 5, 6 and 7 Tests

## Level 3 Test (pp. 85–86)

**1** a, b See solution, D(3, –4) c E($-1\frac{1}{2}, -\frac{1}{2}$) F($3\frac{1}{2}, -1\frac{1}{2}$) d Yes, parallel; see solutions

**2** a

| $x$ | $y$ |
|---|---|
| –1 | 9 |
| 0 | 7 |
| 1 | 5 |

b

| $x$ | $y$ |
|---|---|
| 0 | 2 |
| 2 | 1 |
| 4 | $\frac{2}{3}$ |

**3** $y = x^2 + 3$ **4** a See solutions b (3, –1) c Concurrent **5** All points lie on line **6** a See solutions b See solutions c 8 days time d 15 days time

## Chapter 6—Equations and Formulae

### Level 1 Test (p. 106)

**1** Solution **2** a $x = 5$ b $y = -3$ c $t = 8$ d $m = 20$ **3** a $x = 7$ b $y = 2$ c $m = 5$ d $a = 6$ e $x = 23$ **4** a $y = 12$ b $t = 4$ **5** a $m + n$ b $100y$ c $a + 1$ **6** 16 **7** 120 **8** a $y = -2$ b $y = 5$ c $x = 6$ **9** $x = -3\frac{2}{3}$ **10** $x = 4$; see solutions

### Level 2 Test (p. 107)

**1** a Not solution b Solution **2** a $x = 1\frac{1}{2}$ b $y = -2$ c $m = 4$ d $t = 32$ **3** a $y = 3\frac{1}{2}$ b $m = 6$ c $m = -1\frac{1}{6}$ **4** a $y = 9$ b $m = 4$ c $m = 30$ d $m = 4\frac{4}{5}$ **5** Width = 3 m, length = 8 m **6** $v = 3$ **7** a $b = 8$ b $p = 64$ c $p = 4$ **8** Marcus 17, Rebecca 11 **9** 21 and 29

### Level 3 Test (p. 108)

**1** a Solution b Not solution c Solution **2** a $x = -\frac{4}{5}$ b $y = -4$ c $t = -15$ **3** a $m = 5\frac{1}{3}$ b $y = -1\frac{1}{5}$ c $a = 4\frac{1}{2}$ d $m = -1\frac{2}{5}$ e $t = 1\frac{4}{5}$ **4** a $y = 1\frac{9}{10}$ b $x = 20\frac{1}{3}$ **5** 23 years **6** $a = 8$ **7** 70°, 50°, 60° **8** 10 cm, 16 cm

## Chapter 7—Pythagoras' Theorem

### Level 1 Test (pp. 130–131)

**1** $l$ **2** $l^2 = m^2 + n^2$ **3** a 361 b 441 c 17 **4** 7.07 **5** a 1.7 b 34 **6** a 5.8 b $\sqrt{58}$ c 8.77 **7** 27 cm **8** a Yes b No **9** Yes **10** 1.6 m

### Level 2 Test (pp. 132–133)

**1** $PR^2 = PQ^2 + QR^2$ **2** a 1369 b 41 **3** 9.7 **4** a 20 b 2.9 **5** a 8.06 b 9.43 c 10.6 **6** a $\sqrt{104}$ b $\sqrt{55}$ **7** a No b Yes **8** 35 **9** 8.5 cm

### Level 3 Test (pp. 134–135)

**1** $BD^2 = BC^2 + DC^2$ **2** a 808 b 45 **3** 6.08 **4** a 30 b 10 c 3.5 d 30 **5** a 9.4 b 5.4 c $y = 2$, $x = 5.4$ **6** 30 cm **7** 30.4 cm **8** 120.1 metres

# QUICK ANSWERS to Chapters 8, 9 and 10 **Tests**

## Chapter 8—Measurement

### Level 1 Test (pp. 169–170)

**1** a 210 cm$^2$ b 144 cm$^2$ **2** a 420 cm$^3$ b 1344 cm$^3$ **3** 0.96 L **4** a 75.40 cm b 50.27 cm **5** a 201.062 cm$^2$ b 78.540 cm$^2$ **6** $4\pi$ cm$^2$, $4\pi$ cm **7** 9047.79 cm$^3$ **8** a London ahead 8 h b 11 am c 7 am d 6 am Wednesday **9** a 0225 b 0.37 m c 13 h 7 min d 0829

### Level 2 Test (pp. 171–172)

**1** a $(9x^2 + 15x)$ cm$^2$ b $(9x + 20)$ cm$^2$ **2** a 638.65 cm$^3$ b 180.25 cm$^3$ **3** a 1.944 L b 416 kL **4** a 58.850 cm b 107.398 cm c 5.142 m d 59.982 cm **5** 38.2 cm **6** a 75.40 cm$^2$ b 449.10 cm$^2$ **7** 25 447 cm$^3$ **8** 109 L **9** a 43 min b 9 min c 21 min d 1305

### Level 3 Test (pp. 173–174)

**1** a 100 cm$^2$ b 192 cm$^2$ **2** a 138 cm$^2$ b 2608 cm$^2$ **3** \$72 **4** 42 cm$^2$ **5** a 6 cm b 37.70 cm$^2$ c 12.57 cm **6** 885 **7** a 74.8407 cm b 62.8319 cm **8** a 117.81 cm$^2$ b 100.53 cm$^2$ c 37.70 cm$^2$ d 265.46 cm$^2$ e 164.53 cm$^2$ f 9.13 cm$^2$ **9** 904.8 L **10** 5:54 pm

## Chapter 9—Ratio and Rates

### Level 1 Test  (p. 195)

**1** a 1:4 b 9:7 c 10:1 **2** a 1:1 b 1:4 c 1:1 **3** a 8:7 b 7:15 **4** a 2:1 b 3:2 **5** 90 mL **6** 290 **7** 18 m **8** 240 km **9** 70 km/h **10** a 1:40 000 b 5 cm c 2.56 km

### Level 2 Test  (p. 196)

**1** a 4:5 b 5:4:2 c 1:10 000 **2** a 4 b 10 **3** 3:2 **4** 8 **5** 420 g, 120 g **6** 0.6 m **7** 200 ha **8** 34 min 10 s **9** 11 h 40 min **10** 80 km/h **11** 2 h 30 min **12** 4.5 km

### Level 3 Test  (p. 197)

**1** a 3:4 b 1:4 c 8:9 **2** a $x = 3$ b $x = 14$ **3** 2:1 **4** 3:20 **5** \$72, \$108, \$180 **6** 16 cm, 24 cm **7** 4.2 kg **8** 4 h 4 min 10 s **9** 2167 min **10** 8 h 20 min **11** 9 cm **12** a 15:10:4 b 180

## Chapter 10—Geometry and Congruence

### Level 1 Test (pp. 223–224)

**1** a 90° b 180° c 180° d 360° e 360° **2** a Square b Rhombus c Trapezium d Kite e Rectangle **3** a $x = 40$ b $y = 55$ c $x = 150$ d $a = 113$ e $b = 70$ f $k = 70$ g $b = 60$ h $z = 120$ i $x = 72$ j $c = 50$ k $x = 40$ l $x = 80$ m $y = 65$ **4** a MP b 5 cm c ∠PMN **5** $x = 3$, $y = 83$, $z = 98$, $w = 6$

### Level 2 Test (pp. 225–227)

**1** a $a = 98$, $b = 82$ b $p = 58$ c $x = 99$ d $y = 100$ e $x = 64$ f $x = 110$ **2** a $x = 93$ b $x = 100$ **3** a $x = 108$ b $x = 55$ **4** $x = 60$ **5** $x = 80$ **6** a True b False c True **7** a $x = 76$ b $x = 49$

# QUICK ANSWERS to Chapters 10, 11 and 12 **Tests**

c $x = 60$ **8** $x = 118$ **9** a $x = 60$ (angle sum of $\triangle ABC$) b $y = 120$ (co-interior to $\angle ABD$ and AB || ED)
**10** a Congruent b T c NP d $\angle NPM$ e All the same areas **11** a SAS b RHS c SSS d AAS
**12** See solutions

### Level 3 Test pp. 228–230

**1** a T b F c T **2** a $x = 65$ b $x = 136$ c $x = 36$ d $x = 83$ e $x = 122$ f $x = 45$ g $y = 126$ h $x = 41$
i $x = 38$ j $x = 125$ **3** a $x = 108$ b $x = 116$ c $x = 45$ d $x = 110$ **4** $a = 109$ **5** $x = 54, y = 72, z = 18$
**6** $a = 86, b = 64$ **7** a 46°, alternate to $\angle YPQ$ and XZ || PQ b 134°, supplementary to $\angle PXY$ c 53°, angle sum of $\triangle PQY$ **8** a $\angle HND = 110°$ (co-interior to $\angle BMI$ and AB || CD); $\angle FQD = 70°$ (vertically opposite to $\angle GQC$)
b No, because a pair of co-interior angles HND and FQD are not supplementary **9** $b + c + d = 180$ (angle sum of $\triangle ABC$); $a + d = 180$ (supplementary angles); $b + c + d = a + d \therefore a = b + c$ **10** a Yes, ASA b No c No
**11** See solutions

## Chapter 11—Statistics

### Level 1 Test p. 246

**1** See solutions **2** a See solutions b 24.03 c 24, 24, 4 d 12 e 20 **3** a 12.5 b 13 c 14 d 17
**4** a 45 b 19.85 c 18 d 14 e 39 **5** a 2 b 6 c 5 d 6

### Level 2 Test pp. 247–248

**1** a See solutions b $65.\dot{3}$ c 61–65 and 71–75 d See solutions **2** a $7.\dot{2}\dot{7}$ b 6 c 5 d 14
**3** a See solutions b 69 c 20 d 3.45 **4** a 40 b 20 **5** a See solutions b i $38.6\dot{1}$ ii 48 iii 39.5
iv 39 **6** a 20 b B is higher than A c i 2 ii 3 d B is higher than A

### Level 3 Test pp. 249–250

**1** a See solutions b 7.83, 8, 8, 9 c See solutions **2** a 43 b 5 c See solutions d See solutions
**3** 3.2 **4** 4 goals **5** a 143 b i Unchanged ii Unchanged iii Drops by 13 iv Mean increases by 2.1
**6** a 4, 3 and 4, 7 b 4.6 **7** a See solutions b i 25% ii 25%

## Chapter 12—Probability

### Level 1 Test pp. 263–264

**1** a $\frac{4}{9}$ b $\frac{1}{3}$ c 0 d $\frac{7}{9}$ e $\frac{2}{3}$ f $\frac{2}{9}$ **2** a $\frac{1}{2}$ b $\frac{1}{2}$ c 0 d $\frac{1}{13}$ e $\frac{1}{26}$ f $\frac{1}{52}$ g $\frac{1}{4}$ h $\frac{2}{13}$
**3** See solutions **4** a See solutions b 9 c $\frac{3}{5}$ d i $\frac{8}{23}$ ii $\frac{4}{23}$ **5** a i $\frac{4}{21}$ ii $\frac{2}{21}$ iii $\frac{8}{21}$

### Level 2 Test pp. 265–266

**1** a $\frac{1}{10}$ b $\frac{1}{2}$ c $\frac{7}{10}$ d $\frac{1}{2}$ e $\frac{1}{2}$ f $\frac{3}{10}$ g $\frac{9}{10}$ **2** See solutions **3** a i $\frac{1}{3}$ ii $\frac{3}{5}$ iii 1 b i $\frac{5}{14}$ ii $\frac{9}{14}$
**4** a See solutions b 11 c i $\frac{11}{31}$ ii $\frac{6}{31}$ d $\frac{9}{14}$ **5** a See solutions b i $\frac{3}{10}$ ii $\frac{2}{5}$ **6** a 17 b i $\frac{1}{9}$ ii $\frac{7}{15}$

### Level 3 Test pp. 267–268

**1** a $\frac{1}{2}$ b $\frac{1}{2}$ c $\frac{12}{13}$ d $\frac{2}{13}$ e $\frac{1}{26}$ f $\frac{25}{26}$ **2** a $\frac{9}{20}$ b $\frac{1}{5}$ c $\frac{2}{5}$ d $\frac{11}{20}$ e $\frac{3}{10}$ f $\frac{1}{4}$ g $\frac{7}{20}$ h $\frac{3}{20}$
**3** a See solutions b i $\frac{6}{25}$ ii $\frac{23}{100}$ c i $\frac{11}{20}$ ii $\frac{11}{25}$ **4** a See solutions b i $\frac{1}{4}$ ii $\frac{1}{4}$ iii $\frac{12}{25}$
**5** a See solutions b $\frac{1}{8}$ c See solutions d i 24 ii 63

# QUICK ANSWERS to Sample Examinations

## Sample Examination 1

### Part A pp. 270–272

**1** 14:9 **2** 72.38 cm$^2$ **3** Yes **4** $a + 2b$ **5** 16° **6** $x = 50$ **7** $88.80 **8** 1.64 **9** –5° **10** 1.36
**11** $400 **12** 280 km **13** 304 cm$^2$ **14** $3a - 4a^2$ **15** 84 words/min **16** $\frac{5}{6}$ **17** ASA **18** (3, – 2)
**19** $x = 7$ **20** $12x^6$ **21** $\frac{6}{25}$ **22** $y = 3x + 2$ **23** 4.367 ha **24** 20% **25** $6(2x - 3)$ **26** 1:1 **27** $40
**28** 1 **29** 3 **30** $400

### Part B pp. 272–274

**1** a 480 000 cm$^3$ b 480 L **2** $x = 3\frac{1}{2}$ **3** See solutions **4** 1347.5 cm$^2$ **5** $-a - 2$ **6** See solutions
**7** 1 **8** 549.78 cm$^3$ **9** $\sqrt{56}$ **10** a 4:5 b $32 500 **11** a See solutions b $5\frac{7}{12}$ **12** a 1 cm:25 km
b 19 cm **13** a 4 h 30 min b 2:00 pm **14** a $13.35 b $296.65 **15** $504 **16** 7 km 211 m
**17** 37 cm **18** a $x = 60$ b $x = 10$ **19** a 1:4 b 25% **20** a See solutions b $\frac{2}{15}$

### Part C pp. 274–275

**1** a 2, 8 b $5\frac{1}{2}$ c $5\frac{3}{8}$ **2** a 15 b 2 c $2\frac{3}{5}$ **3** a 40% b 35% c 8:5 **4** a 7 cm b 8 cm
c 616 m$^2$ **5** a 40° b 70° c 70°

## Sample Examination 2

### Part A pp. 276–277

**1** 5.19 **2** $\frac{7}{20}$ **3** $5a + 11$ **4** –1 **5** $6x - 12$ **6** 240 cm **7** 48 cm$^2$ **8** 4:5 **9** 0.625 **10** $2\frac{1}{4}$
**11** $\frac{3}{8}$ **12** $6a^2$ **13** 12 **14** $1.14 **15** $x = -6$ **16** I and III **17** $\frac{3}{25}$, 13%, 0.15, $\sqrt{0.9}$ **18** –2 **19** 7
**20** 85% **21** 49 cm$^2$ **22** $\frac{4}{7}$ **23** $3b$ **24** $3a + 14$ **25** 2 h 43 min **26** $x + 3y$ **27** $a = 134$ **28** 8
**29** 16 **30** 22 cm

### Part B pp. 278–280

**1** 16 **2** $x = 8$ **3** $8x - 9$ **4** $705.50 **5** 154 cm$^2$ **6** See solutions **7** $x = 10$ **8** 80 cm$^3$ **9** 51 cm
**10** 11:20 am **11** a $\frac{1}{2}$ b 0 **12** 24 **13** 0 **14** a $x = 40$ b $y = 75$ **15** $7564 **16** 68 km/h **17** 3
**18** $x = 8$ **19** 1100 cm$^3$ **20** a $(60 + 6m)$ cm b $60m$ cm$^2$

### Part C pp. 280–281

**1** a 110 cm$^2$ b 330 cm$^3$ **2** a See solutions b $\frac{3}{5}$ **3** a See solutions b 3 c i 5 ii 6 **4** a $6x + 14$
b $x = 5$ **5** a See solutions b 140°

# WORKED SOLUTIONS to Chapter 1 **Practise, Practise**

## Chapter 1—Selected Revision of Year 7

pp. 2–12

### Number

**1** **a** 28 **b** $19 - 16 = 3$
**c** 12 **d** $4 + 54 = 58$
**e** $35 + 49 = 84$ **f** $5 + 18 + 9 = 32$
**g** $15 - 3 = 12$ **h** $9 \times 7 = 63$
**i** $9 \times 12 = 108$ **j** $7 \times 2 = 14$
**k** $15 + [17 - 12]$
$= 15 + 5$
$= 20$
**l** $\{[2 + 5] \times 8\} \div 4$
$= \{7 \times 8\} \div 4$
$= 56 \div 4$
$= 14$

**2** Number of balls $= 8 \times 6$
$= 48$

**3** **a** Sum $= 19 + 17$
$= 36$

**b** Product $= 9 \times 6$
$= 54$

**c** Amount $= 84 - 29$
$= 55$

**d** Average $= \frac{19+11+17+13+15}{5}$
$= \frac{75}{5}$
$= 15$

**e** Difference $= 31 - 18 = 13$

**f** Quotient $= 63 \div 7$
$= 9$

**g** $75 \div 8 = 9$ remainder 3
Remainder $= 3$

**h** Total $= 16 + 19 + 41 + 34 + 24$
$= 134$

**i** Product $= (9 + 7) \times (9 - 7)$
$= 16 \times 2$
$= 32$

**j** Sum $= (6 \times 9) + (4 \times 9)$
$= 90$

**4** 8 cost \$48
$\therefore$ 1 costs \$48 ÷ 8 = \$6
$\therefore$ 5 cost \$6 × 5 = \$30

**5** Hourly rate = \$61.60 ÷ 8
= \$7.70
Wage (6 hrs) = \$7.70 × 6
= \$46.20

**6** Number of cards $= 144 + 132 + 24 + 19 + 15$
$= 334$

**7** Average $= \frac{19 + 26 + 42 + 17 + 35 + 5}{6}$
$= \frac{144}{6}$
$= 24$
The average score is 24.

**8** Distance $= 26 \times 12$
$= 312$ km [Note return trips]

**9** Remainder when $247 \div 18$ needed.

$$\begin{array}{r} 13 \\ 18\overline{)247} \\ \underline{18} \\ 67 \\ \underline{54} \\ 13 \end{array}$$

Remainder $= 13$
Thirteen trees will be left over.

### Fractions

**1** **a** $\frac{1}{3}$ **b** $\frac{1}{2}$ **c** $\frac{2}{5}$
**d** $\frac{3}{4}$ **e** $\frac{3}{4}$ **f** $\frac{2}{3}$
**g** $\frac{\frac{1}{2}}{\frac{3}{4}} = \frac{\frac{1}{2} \times 4}{\frac{3}{4} \times 4} = \frac{2}{3}$ **h** 3
**i** $2\frac{1}{7}$ **j** $1\frac{1}{2}$ **k** $2\frac{1}{2}$
**l** $1\frac{1}{3}$

**2** **a** $\frac{10}{3}$ **b** $\frac{32}{5}$ **c** $\frac{19}{4}$
**d** $\frac{100}{11}$ **e** $\frac{25}{3}$

**3** **a** $\frac{7}{10} - \frac{5}{10}$
$= \frac{2}{10}$
$= \frac{1}{5}$
**b** $\frac{8}{7} = 1\frac{1}{7}$
**c** $\frac{9}{12} - \frac{4}{12} = \frac{5}{12}$ **d** 2
**e** $2\frac{1}{4} - 2 + \frac{1}{2}$
$= \frac{1}{4} + \frac{1}{2}$
$= \frac{3}{4}$
**f** $6 - 4 + \frac{3}{5} = 2\frac{3}{5}$

**g** $\frac{24}{60} + \frac{20}{60} + \frac{45}{60}$
$= \frac{89}{60}$
$= 1\frac{29}{60}$

**h** $(6 + 3 - 5) + \frac{1}{4} + \frac{2}{3} - \frac{1}{2}$
$= 4 + \frac{3}{12} + \frac{8}{12} - \frac{6}{12}$
$= 4\frac{5}{12}$

**4** **a** $\frac{12}{25}$

**b** $\frac{9}{32}$

**c** $\frac{7}{12}$

**d** $\frac{5}{\cancel{8}_2} \times \frac{\cancel{4}^1}{3} = \frac{5}{6}$

**e** $\frac{3}{2} \times \frac{3}{4} = \frac{9}{8} = 1\frac{1}{8}$

**f** $\frac{7}{\cancel{3}_1} \times \frac{\cancel{6}^2}{7} = 2$

**g** $\frac{8}{1} \times \frac{\cancel{8}^2}{\cancel{3}_1} \times \frac{\cancel{9}^3}{\cancel{4}_1} = 48$

**h** $\frac{5}{6} \times \frac{1}{6} = \frac{5}{36}$

**i** $\frac{7}{\cancel{12}_3} \times \frac{\cancel{4}^1}{3} = \frac{7}{9}$

**j** $\frac{\cancel{20}^4}{1} \times \frac{8}{\cancel{5}_1} = 32$

**k** $\frac{\cancel{25}^5}{\cancel{4}_1} \times \frac{\cancel{8}^2}{\cancel{15}_3}$
$= \frac{10}{3}$
$= 3\frac{1}{3}$

**l** $\frac{3}{\cancel{4}_1} \times \frac{\cancel{8}^2}{\cancel{5}_1} \times \frac{\cancel{10}^2}{7} = \frac{12}{7}$
$= 1\frac{5}{7}$

**m** $\frac{2}{3} + \left(\frac{\cancel{3}^1}{\cancel{4}_1} \times \frac{\cancel{8}^2}{\cancel{9}_3}\right) = \frac{2}{3} + \frac{2}{3}$
$= \frac{4}{3}$
$= 1\frac{1}{3}$

**n** $10 - \left(\frac{\cancel{10}^2}{\cancel{3}_1} \times \frac{\cancel{9}^3}{\cancel{5}_1}\right) = 10 - 6$
$= 4$

**o** $10 \div \left(\frac{3}{4} - \frac{1}{2}\right) = 10 \div \frac{1}{4}$
$= 10 \times 4$
$= 40$

**p** $\left(\frac{2}{3} + \frac{1}{4}\right) \div \left(\frac{2}{3} - \frac{1}{4}\right)$
$= \left(\frac{11}{12}\right) \div \left(\frac{5}{12}\right)$
$= \frac{11}{\cancel{12}_1} \times \frac{\cancel{12}^1}{5}$
$= \frac{11}{5}$
$= 2\frac{1}{5}$

**q** $\left(\frac{\cancel{25}^5}{\cancel{4}_2} \times \frac{\cancel{6}^3}{\cancel{5}_1}\right) + \left(\frac{\cancel{15}^3}{\cancel{4}_2} \times \frac{\cancel{6}^3}{\cancel{5}_1}\right)$
$= \frac{15}{2} + \frac{9}{2}$
$= \frac{24}{2}$
$= 12$

**5** **a** $\frac{2}{3} + \frac{1}{2} = \frac{7}{6}$
$= 1\frac{1}{6}$

**b** $\frac{2}{3} + \frac{1}{2} + \frac{2}{5} = 1\frac{1}{6} + \frac{2}{5}$
$= 1 + \left(\frac{5}{30} + \frac{12}{30}\right)$
$= 1\frac{17}{30}$

**c** $\frac{2}{3} + \frac{1}{2} - \frac{2}{5} = \frac{20}{30} + \frac{15}{30} - \frac{12}{30}$
$= \frac{23}{30}$

**d** $\frac{\cancel{2}^1}{3} \times \frac{1}{\cancel{2}_1} = \frac{1}{3}$

**e** $\frac{2}{3} \times \frac{1}{2} \times \frac{2}{5} = \frac{1}{3} \times \frac{2}{5}$
$= \frac{2}{15}$

**f** $\frac{2}{3} \times \frac{2}{3} = \frac{4}{9}$

**g** $\frac{4}{9} - \left(\frac{1}{2} \times \frac{1}{2}\right) = \frac{4}{9} - \frac{1}{4}$
$= \frac{16}{36} - \frac{9}{36}$
$= \frac{7}{36}$

**h** $\left(\frac{2}{3} + \frac{1}{2}\right) \times \left(\frac{2}{3} - \frac{1}{2}\right) = \frac{7}{6} \times \frac{1}{6}$
$= \frac{7}{36}$

**i** $\frac{2}{3} + \left(\frac{1}{\cancel{2}_1} \times \frac{\cancel{2}^1}{5}\right) = \frac{2}{3} + \frac{1}{5}$
$= \frac{10}{15} + \frac{3}{15}$
$= \frac{13}{15}$

**j** $\frac{2}{5} \times \frac{2}{5} + \frac{1}{\cancel{2}} \times \frac{\cancel{2}}{5} = \frac{4}{25} + \frac{1}{5}$
$= \frac{4}{25} + \frac{5}{25}$
$= \frac{9}{25}$

**k** $\frac{2}{5} \times \left(\frac{1}{2} + \frac{2}{5}\right) = \frac{2}{5} \times \left(\frac{5}{10} + \frac{4}{10}\right)$
$= \frac{\cancel{2}^1}{5} \times \frac{9}{\cancel{10}_5}$
$= \frac{9}{25}$

**l** $\left(\frac{2}{3} + \frac{1}{2}\right) \div \left(\frac{2}{3} - \frac{1}{2}\right)$
$= \left(\frac{4}{6} + \frac{3}{6}\right) \div \left(\frac{4}{6} - \frac{3}{6}\right)$
$= \frac{7}{6} \div \frac{1}{6}$
$= \frac{7}{\cancel{6}_1} \times \frac{\cancel{6}^1}{1}$
$= 7$

## WORKED SOLUTIONS to Chapter 1 **Practise, Practise**

**6** $\frac{3}{4}$ costs \$45

$\therefore \frac{1}{4}$ costs \$45 ÷ 3 = \$15

$\therefore \frac{4}{4}$ costs \$15 × 4 = \$60

1 tonne costs \$60.

**7** New salary $1\frac{1}{5}$ times old salary

i.e. $\frac{6}{5}$ = \$240

$\therefore \frac{1}{5}$ = \$240 ÷ 6 = \$40

$\therefore \frac{5}{5}$ = \$40 × 5 = \$200

Old salary was \$200.
The increase was \$40.

**8** Discount = $\frac{1}{3}$ of \$2400
= \$800
New price = \$2400 – \$800
= \$1600

**9** Payment = Hourly rate × hours × $2\frac{1}{2}$
$= \$8.64 \times \cancel{6}^{3} \times \frac{5}{\cancel{2}_{1}}$
= \$129.60

**10** **a** $\frac{7}{10} + \triangle = \frac{15}{10}$

$\therefore \triangle = \frac{8}{10} = \frac{4}{5}$

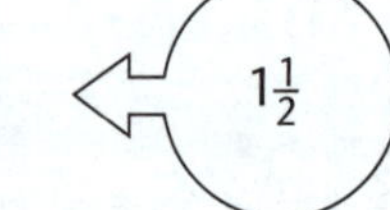


**b** $4 - * = 2\frac{1}{4}$

$\frac{16}{4} - * = \frac{9}{4}$

$\therefore * = \frac{7}{4}$

$= 1\frac{3}{4}$

**c** $\frac{4}{5} + * = 1\frac{1}{3}$

$\therefore \frac{12}{15} + * = \frac{20}{15}$

$\therefore * = \frac{8}{15}$

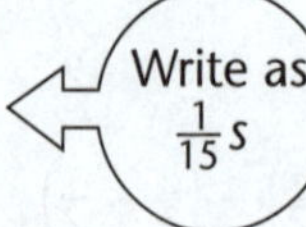


**d** $\frac{7}{10} + \left(\frac{3}{10} \times \frac{5}{4}\right) = \square$

$\therefore \frac{7}{10} + \frac{3}{8} = \square$

$\therefore \frac{28}{40} + \frac{15}{40} = \square$

$\therefore \square = \frac{43}{40}$

$= 1\frac{3}{40}$

**e** $\frac{3}{4} \times \square = 2$

i.e. $\frac{3}{4} \times \square = \frac{2}{1}$

By trial and error we need an 8 on top and a 3 on the bottom

i.e. $\square = \frac{8}{3} = 2\frac{2}{3}$

Because $\frac{\cancel{3}^{1}}{\cancel{4}_{1}} \times \frac{\cancel{8}^{2}}{\cancel{3}_{1}} = 2\frac{2}{3}$

### Decimals

**1** **a** $\frac{2}{5}$ **b** $\frac{11}{25}$ **c** $\frac{1}{25}$
**d** $1\frac{3}{5}$ **e** $2\frac{1}{2}$ **f** $3\frac{1}{20}$
**g** $2\frac{1}{8}$ **h** $1\frac{1}{200}$

**2** **a** 0.7 **b** 0.17 **c** 0.07
**d** 0.77 **e** 3.999 **f** 1.79
**g** 42.4

**3** **a** 23.64 **b** 2.0 or 2
**c** 2.6 **d** 1.155
**e** 470 **f** 73.4
**g** 5.7 **h** 1.17
**i** 0.8 **j** 32 ÷ 8 = 4
**k** 420 ÷ 7 = 60 **l** 5400 ÷ 9 = 600
**m** 7.6
**n** 6.666666…
= 6.67 (to 2 decimal places)

**4** **a** $x + y + z = 1.5 + 0.8 + 1.2$
$= 3.5$

**b** $x + y - z = 1.5 + 0.8 - 1.2$
$= 1.1$

**c** $xy = 1.5 \times 0.8$
$= 1.2$

**d** $xy \div z = 1.2 \div 1.2$
$= 1$

**e** $(y + z) \div x = (0.8 + 1.2) \div 1.5$
$= 2 \div 1.5$
$= 4 \div 3$
$= 1.\dot{3}$

**f** $(z + y) \div (z - y) = (1.2 + 0.8) \div (1.2 - 0.8)$
$= 2 \div 0.4$
$= 5$

**g** $x + yz = 1.5 + (0.8 \times 1.2)$
$= 1.5 + 0.96$
$= 2.46$

**h** $x^2 = (1.5)^2 = 2.25$

**i** $4x^2 = 4 \times (1.5)^2$
$= 9$

**j** $x^2 - z^2 = (1.5)^2 - (1.2)^2$
$= 2.25 - 1.44$
$= 0.81$

**5** Length = 19.45 + 3.9
= 23.35
The length is 23.35 metres.

**6** Pieces = 15 m ÷ 45 cm
= 1500 ÷ 45
= 33 remainder 15
Number of pieces cut is 33 with 15 cm remaining.

$$\begin{array}{r} 33 \\ 45\overline{)1500} \\ \underline{135}^{\times} \\ 150 \\ \underline{135} \\ 15 \end{array}$$

**7** Total weight = $8 \times 4.5$
= 36 kg

**8** Remainder = 10 − 3.28
= 6.72
The length of cloth remaining is 6.72 m.

**9** Payment = $6 \times \$14.20 + 3 \times \$14.20 \times 1.5$
= $149.10

**10** **a** $* = 9.34 - 6.7$
$= 2.64$

**b** $\triangle = 12 - 3.76$
$= 8.24$

**c** $\square = 75.2 \div 9.4$
$= 8$

**d** $\square = 92.8 \div 16$
$= 5.8$

**e** $* = 0.95 + 0.35$
$= 1.3$

**f** $(4.7 + \triangle) \times 6.7 = 67$
Now $10 \times 6.7 = 67$
$\therefore \quad 4.7 + \triangle = 10$
$\therefore \quad \triangle = 10 - 4.7$
$= 5.3$

## Percentages

**1** **a** $17\% = \frac{17}{100}$
**b** $73\% = \frac{73}{100}$
**c** $7\% = \frac{7}{100}$
**d** $80\% = \frac{80}{100} = \frac{4}{5}$
**e** $6\% = \frac{6}{100} = \frac{3}{50}$
**f** $25\% = \frac{25}{100} = \frac{1}{4}$
**g** $7\frac{1}{2}\% = \frac{7\frac{1}{2}}{100} = \frac{15}{200} = \frac{3}{40}$
**h** $12\frac{1}{4}\% = \frac{12\frac{1}{4}}{100} = \frac{49}{400}$

**2** **a** $69\% = 0.69$
**b** $8\% = 0.08$
**c** $80\% = 0.8$
**d** $45\% = 0.45$
**e** $6\frac{1}{2}\% = 0.065$
**f** $17\frac{1}{4}\% = 0.1725$
**g** $140\% = 1.4$
**h** $300\% = 3$

**3** **a** $\frac{3}{4} \times 100\% = 75\%$
**b** $\frac{7}{10} \times 100\% = 70\%$
**c** $\frac{81}{100} \times 100\% = 81\%$
**d** $\frac{4}{5} \times 100\% = 80\%$
**e** $\frac{31}{50} \times 100\% = 62\%$
**f** $1\frac{1}{4} \times 100\% = 125\%$
**g** $0.63 \times 100\% = 63\%$
**h** $0.7 \times 100\% = 70\%$
**i** $0.05 \times 100\% = 5\%$
**j** $1.2 \times 100\% = 120\%$
**k** $0.375 \times 100\% = 37.5\%$
**l** $1.205 \times 100\% = 120.5\%$

**4** **a** $0.45 \times 600 = 270$
$\therefore$ \$270
**b** $0.28 \times 7000 = 1960$
$\therefore$ \$1960
**c** $0.07 \times 6500 = 455$
$\therefore$ 455
**d** $0.04 \times 600 = 24$
$\therefore$ 24
**e** $0.35 \times 200 = 70$
$\therefore$ 70 cm
**f** $0.05 \times 4000 = 200$
$\therefore$ 200 g

**5** **a** $\frac{4}{16} \times 100\% = 25\%$
**b** $\frac{27}{30} \times 100\% = 90\%$
**c** $\frac{12}{120} \times 100\% = 10\%$
**d** $\frac{10}{100} \times 100\% = 10\%$
**e** $\frac{12}{60} \times 100\% = 20\%$

## Basic Algebra

**1** **a** $8a$ **b** $10y$ **c** $7t$
**d** $5t$ **e** $4a$ **f** $7y$
**g** $1.8x$ **h** $\frac{7}{12}y$ **i** $-2y$
**j** $-3a$ **k** $-7m$ **l** $0.7n$

**2** **a** $4y$ **b** $14a$ **c** $ab$
**d** $15t$ **e** $35y$ **f** $28a^2$

# WORKED SOLUTIONS to Chapter 1 **Practise, Practise**

g $54xy$ h $-18y$ i $28t^2$
j $3y$ k 3 l $3y$
m 7 n $-9q$

3 a $10x + y$ b $16y + 6x$
c $37a + 10$ d $15t + 17x$
e $16y + 2x$ f 0

4 a $x = 6$ b $a = 21$ c $y = 14$
d $a = 7$ e $t = 8$ f $a = 27$
g $y = 9$ h $m = 7$ i $x = 5$
j $x = 2$

5 a $P = 4x + x + 4x + x$
$= 10x$
b $P = 2y + 3y + 3y$
$= 8y$
c $P = 4a + 4b + 4a + 4b$
$= 8a + 8b$
d $P = 4 \times 5a$
$= 20a$
e $P = 12a + 7a + 5a + a + 7a + 6a$
$= 38a$

## Basic Geometry

1 Rectangle
2 Parallelogram
3 Trapezium
4 Kite
5 Rhombus
6 Sphere
7 Cone
8 Cylinder
9 Quadrilateral
10 Scalene triangle
11 Isosceles triangle
12 Equilateral triangle
13 Right-angled triangle
14 Cube
15 Square prism
16 Triangular prism
17 Pentagon
18 Hexagon
19 Square
20 Octagon
21 Hemisphere
22 Concurrent lines
23 Collinear points
24 Parallel lines
25 Complementary angles
26 Vertically opposite angles
27 Supplementary angles
28 Exterior angle
29 Alternate angles
30 Co-interior angles
31 Corresponding angles
32 Revolution
33 Reflex angle
34 Obtuse angle
35 Acute angle
36 Straight angle

37 a 23° b 42° c $(90 - x)°$

38 a 113° b 132°
c $(180 - x)°$ d 32°

39 a Equal b 90° c 180°
d Equal e 360°
f Sum … interior
g Transversal
i Equal ii Corresponding
iii Supplementary

**40** $a = 180 - 124$
$= 56$

**41** $b = 90 - 16$
$= 74$

**42** $c = 57$

**43** $d = 180 - (90 + 69)$
$= 21$

**44** $e = 180 - (46 + 49)$
$= 85$

**45** $f = 116$
$g = 180 - 116$
$= 64$
$h = 64$

**46** $i = 360 - 68$
$= 292$

**47** $k = 360 - (119 + 92)$
$= 149$

**48** $m = 360 - (88 + 16 + 106)$
$= 150$

**49** $n + n + 120 = 180$
$\therefore \quad 2n = 60$
$n = 30$

**50** $5p = 180$
$p = 36$

**51** $q = 180 - (39 + 68)$
$= 73$

**52** $6x = 180$
$x = 30$

**53** $r = 74 + 29$
$= 103$

**54** $s = 180 - 62$
$= 118$

**55** $t + 59 = 124$
$\therefore \quad t = 124 - 59$
$= 65$

**56** $u = 360 - (94 + 98 + 85)$
$= 83$ (360° in quadrilateral)

**57** $v = 68$

**58** $w = 180 - 69$
$= 111$

**59** $x = 65$

**60** $y = 114$

**61** $a = 94$ $\quad c = 94$ $\quad d = 86$
$b = 180 - 94$
$= 86$

**62** $e = 59$ $\quad g = 59$ $\quad f = 59$

**63** $h = 82$ $\quad m = 98$ $\quad n = 98$
$k = 180 - 82$
$= 98$

**64**

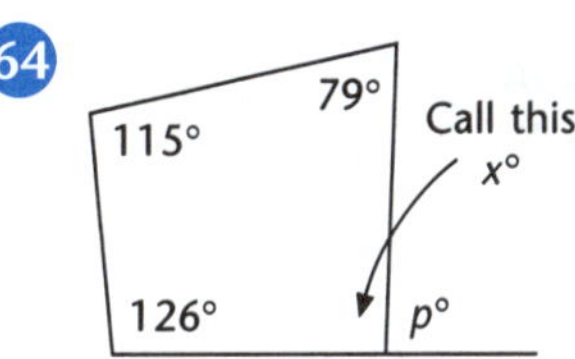


$x = 360 - (126 + 115 + 79)$
$= 40$

Then $p + x = 180$
$\therefore \quad p = 180 - 40$
$= 140$

**65** $q = 60$

**66** $t + 64 + 64 = 180$
$\therefore \quad t = 180 - 128$
$= 52$

**67** $u + u + 42 = 180$
$2u = 180 - 42$
$2u = 138$
$u = 69$

**68**

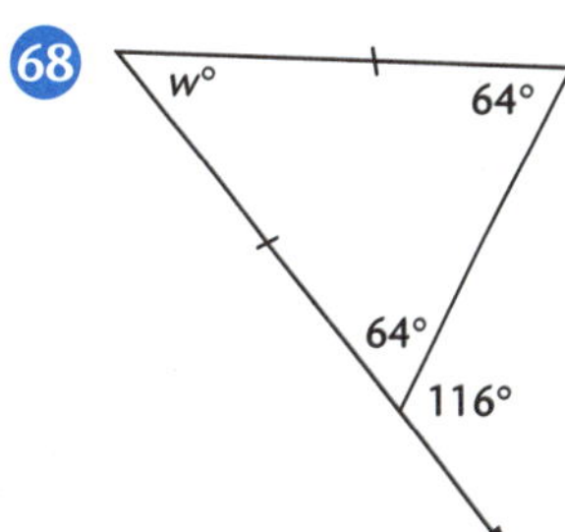

# WORKED SOLUTIONS to Chapter 1 **Practise, Practise**

Work out all of the angles.

$w + 64 + 64 = 180$
$w = 180 - 128$
$= 52$

**69** $z = 68$

$y + 48 + 68 = 180$
$\therefore y = 180 - 116$
$= 64$

$x = y = 64$

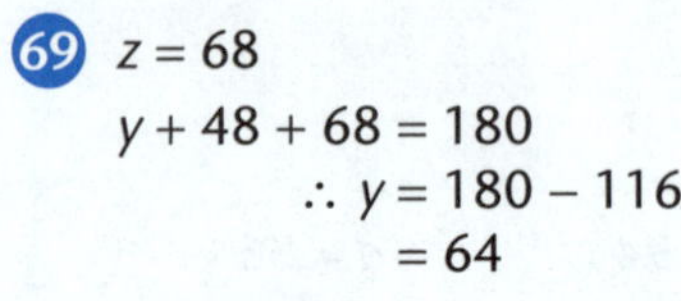

**70**

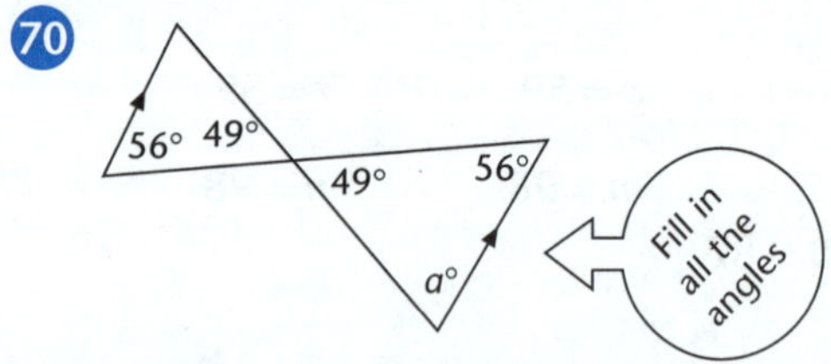


$a + 49 + 56 = 180$ ($\angle$ sum of $\triangle$)
$a = 180 - 105$
$a = 75$

## Statistics

**1** **a**

| Score | Tally | Frequency |
|---|---|---|
| 4 | \| | 1 |
| 5 | \| | 1 |
| 6 | \|\| | 2 |
| 7 | ~~\|\|\|\|~~ | 5 |
| 8 | \|\|\|\| | 4 |
| 9 | \|\|\| | 3 |
| 10 | \|\| | 2 |
| 11 | \|\| | 2 |
| | Total | 20 |

**b**

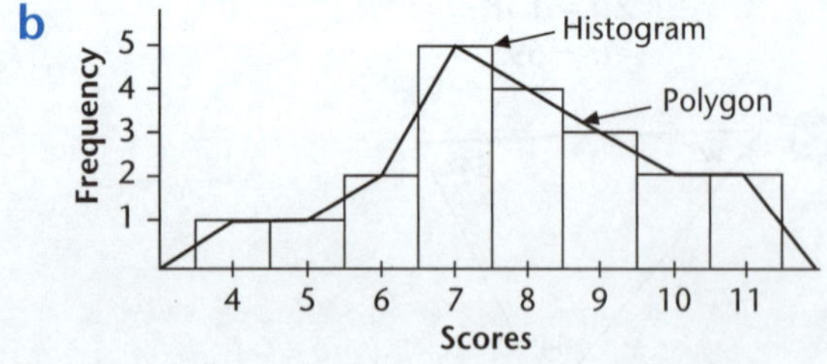


**2**

| Stem | Leaf |
|---|---|
| 4 | 0017789 |
| 5 | 23789 |
| 6 | 1338 |
| 7 | 03 |

**3** **a** range $= 8 - 3$
$= 5$

**b** mode = 8

**c** median = ~~3~~, ~~4~~, ~~5~~, 6, ~~7~~, ~~8~~, ~~8~~
$\therefore$ median is 6

**d** mean $= \frac{4+7+8+3+5+8+6}{7}$
$= \frac{41}{7}$
$= 5\frac{6}{7}$

**4** **a** range $= 16 - 4$
$= 12$

**b** mode = 8

**c** median: ~~4~~, ~~8~~, 8, 10, ~~12~~, ~~16~~
middle of 8 and 10 is 9
$\therefore$ median is 9

**d** mean $= \frac{16+4+8+8+10+12}{6}$
$= \frac{58}{6}$
$= 9\frac{2}{3}$

**5**

| Score ($x$) | Frequency ($f$) | $f \times x$ |
|---|---|---|
| 4 | 2 | 8 |
| 5 | 4 | 20 |
| 6 | 7 | 42 |
| 7 | 5 | 35 |
| 8 | 2 | 16 |
| Total | 20 | 121 |

**i** mode = 6

**ii** range $= 8 - 4 = 4$

**iii** mean $= \frac{\text{sum of } f \times x}{\text{sum of } f}$
$= \frac{121}{20}$
$= 6.05$

$\therefore$ mean is 6.05

**6** purple

# WORKED SOLUTIONS to Chapters 1 and 2 **Practise, Practise**

## Probability

**1** a $P(7) = \frac{1}{10}$

b $P(\text{Odd}) = \frac{5}{10} = \frac{1}{2}$

c $P(\text{Even}) = \frac{5}{10} = \frac{1}{2}$

d $P(\text{Not } 3) = \frac{9}{10}$

e $P(2, 3, 5, 7) = \frac{4}{10} = \frac{2}{5}$

f $P(4, 6, 8, 9, 10) = \frac{5}{10} = \frac{1}{2}$

g $P(13) = 0$

h $P(0) = 0$

i $P(1, 2, 3, 4, 6) = \frac{5}{10} = \frac{1}{2}$

j $P(3, 6, 9) = \frac{3}{10}$

k $P(1, 2, 3) = \frac{3}{10}$

l $P(8, 9, 10) = \frac{3}{10}$

m $P(4, 5, 6, 7, 8, 9, 10) = \frac{7}{10}$

**2** a $P(4) = \frac{1}{6}$

b $P(\text{Even}) = \frac{3}{6} = \frac{1}{2}$

c $P(\text{Not even}) = \frac{1}{2}$

d $P(2 \text{ or } 4) = \frac{2}{6} = \frac{1}{3}$

e $P(2, 4, 6) = \frac{3}{6} = \frac{1}{2}$

f $P(1 \text{ or } 4) = \frac{2}{6} = \frac{1}{3}$

g $P(1) = \frac{1}{6}$

## Chapter 2—Number

**1** a 16: 1, 2, 4, (8), 16
24: 1, 2, 3, 4, 6, (8), 12, 24
∴ Highest Common Factor is 8

b 45: 1, 3, 5, 9, (15), 45
75: 1, 3, 5, (15), 25, 75
∴ Highest Common Factor is 15

c 24: 1, 2, 3, 4, 6, 8, (12), 24
36: 1, 2, 3, 4, 6, 9, (12), 18, 36
∴ Highest Common Factor is 12

**2** 48: 1, 2, 3, 4, 6, 8, (12), 16, 24, 48
36: 1, 2, 3, 4, 6, 9, (12), 18, 36
∴ Highest Common Factor is 12
∴ 12 groups can be formed
[Each group will have 4 girls and 3 boys.]

**3** 20: 1, 2, (4), 5, 10, 20
16: 1, 2, (4), 8, 16
∴ Highest Common Factor is 4
∴ 4 practise groups can be formed
[Each group will have 5 altos and 4 sopranos.]

**4** a 4: 4, 8, (12), 16, …
6: 6, (12), 18, …
∴ Lowest Common Multiple is 12

b 6: 6, 12, (18), 24, …
9: 9, (18), 27, …
∴ Lowest Common Multiple is 18

c 12: 12, 24, 36, (48), 60, …
16: 16, 32, (48), 64, …
∴ Lowest Common Multiple is 48

**5** 6: 6, (12), 18, …
4: 4, 8, (12), 16, …
∴ Lowest Common Multiple is 12
∴ It will take 12 seconds until they flash again together.

**6** 3: 3, 6, 9, 12, 15, 18, (21), 24, …
7: 7, 14, (21), 28, …
∴ Lowest Common Multiple is 21
∴ It will take another 21 days until they ride together.
∴ Ride on the 22nd of June.

**7** a Primes: 2, 3, 5, 7, 11, 13, 17, 19, 23, 29

b Composites: 4, 6, 8, 9, 10, 12, 14, 15, 16, 18, 20, 21, 22, 24, 25, 26, 27, 28

**8** Primes: 53, 59
∴ 2 primes

**9** Composites: 62, 63, 64, 65, 66, 68, 69, 70
∴ 8 composites

# WORKED SOLUTIONS to Chapter 2 **Practise, Practise**

**10** **a**

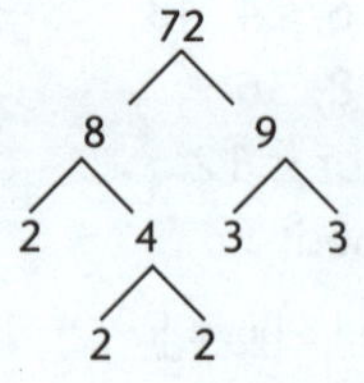


$\therefore 72 = 2 \times 2 \times 2 \times 3 \times 3$

**b**

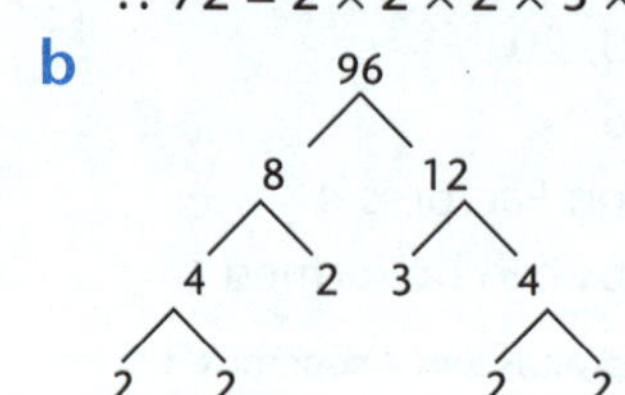


$\therefore 96 = 2 \times 2 \times 2 \times 2 \times 2 \times 3$

**c**

250
25 10
5 5 5 2

$\therefore 250 = 5 \times 5 \times 5 \times 2$

**11** **a** $4 \times 4 \times 4 \times 4 \times 4 = 4^5$
**b** $3 \times 3 \times 5 \times 5 \times 5 = 3^2 \times 5^3$
**c** $10 \times 10 \times 10 \times 10 \times 10 = 10^5$

**12**

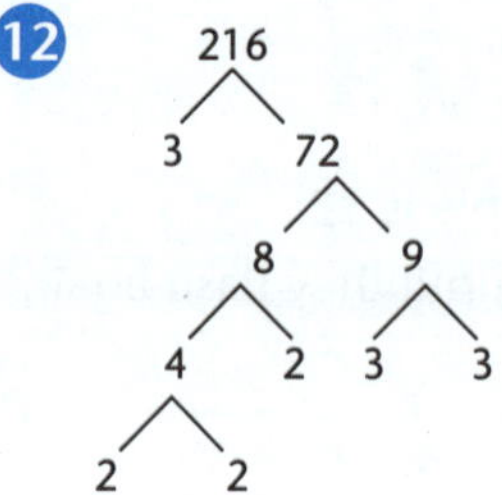


$\therefore 216 = 2 \times 2 \times 2 \times 3 \times 3 \times 3 = 2^3 \times 3^3$

**13** **a** $2^4 \times 2^2 = 2^6$ **b** $3^5 \div 3^2 = 3^3$
**c** $6^4 \times 6^1 \times 6^2 = 6^7$ **d** $7^4 \div 7^3 = 7^1 = 7$
**e** $4^3 \div 4^1 = 4^2$ **f** $(2^4)^3 = 2^{12}$
**g** $(6^3)^2 = 6^6$

**14** **a** $3^0 = 1$
**b** $(4 \times 2)^0 = 8^0 = 1$
**c** $5^0 - 2^0 = 1 - 1 = 0$
**d** $(5 - 2)^0 = 3^0 = 1$

**15** **a** $-2 - 3 = -5$
**b** $6 - (-2) = 6 + 2 = 8$
**c** $-5 + (-1) = -5 - 1 = -6$
**d** $10 - (-1) = 10 + 1 = 11$
**e** $-5 - 7 = -12$
**f** $-6 + 4 = -2$

**16** **a** $-2 \times 3 = -6$
**b** $-6 \times -2 = 12$
**c** $3 \times -4 = -12$
**d** $-5 \times -2 \times -1 = -10$
**e** $-4 \times -3 \times -2 = -24$
**f** $6 \times -1 \times -1 = 6$
**g** $(-3)^2 = -3 \times -3 = 9$
**h** $(-5)^2 = -5 \times -5 = 25$
**i** $(-2 - 1)^2 = (-3)^2 = 9$

**17** **a** $-12 \div 6 = -2$ **b** $-16 \div -4 = 4$
**c** $15 \div -3 = -5$ **d** $-18 \div -2 = 9$
**e** $\frac{-12}{-4} = 3$ **f** $\frac{-16}{8} = -2$
**g** $\frac{-120}{-10} = 12$ **h** $\frac{16}{-2} = -8$
**i** $\frac{15}{-5} = -3$

**18** **a** $3 - 4 \times 2 = 3 - 8 = -5$
**b** $12 - 5 \times 3 = 12 - 15 = -3$
**c** $-6 + 4 \times 2 = -6 + 8 = 2$
**d** $-15 - 4 \div 2 = -15 - 2 = -17$
**e** $16 \div 4 - 10 \div 1 = 4 - 10 = -6$
**f** $\frac{30}{6} - \frac{20}{2} = 5 - 10 = -5$
**g** $\frac{6+4}{-2} = \frac{10}{-2}$ $= -5$ **h** $\frac{10-14}{-4} = \frac{-4}{-4}$ $= 1$
**i** $\frac{12-15}{6-9} = \frac{-3}{-3}$ $= 1$ **j** $\frac{10-14}{6-8} = \frac{-4}{-2}$ $= 2$
**k** $(-3 - 2)^2 = (-5)^2 = 25$
**l** $(10 - 3 \times 4)^2 = (10 - 12)^2 = (-2)^2 = 4$
**m** $4(2 - 7) = 4 \times -5 = -20$
**n** $-2(-3 - 4) = -2 \times -7 = 14$
**o** $(4 - 6)(10 - 14) = -2 \times -4 = 8$
**p** $3(2 - 4 \times 2) = 3(2 - 8) = 3(-6) = -18$
**q** $-5(3 - 8 \times 2) = -5(3 - 16) = -5(-13) = 65$
**r** $\frac{12 \div 6 - 8 \times 2}{3 - 5 \times 2} = \frac{2-16}{3-10}$
$= \frac{-14}{-7}$
$= 2$

**19** a $\frac{3}{5}$ $\quad \begin{array}{r} 0.6 \\ 5\overline{)3.0} \end{array}$

$\therefore 0.6$

b $\frac{11}{20}$ $\quad \begin{array}{r} 0.55 \\ 20\overline{)11.00} \end{array}$

$\therefore 0.55$

c $\frac{17}{100} = 0.17$

d $\frac{2}{3}$ $\quad \begin{array}{r} 0.66 \\ 3\overline{)2.00\ldots} \end{array}$

$\therefore 0.\dot{6}$

e $\frac{4}{11}$ $\quad \begin{array}{r} 0.3636 \\ 11\overline{)4.000\ldots} \end{array}$

$\therefore 0.\dot{3}\dot{6}$

f $\frac{6}{7}$ $\quad \begin{array}{r} 0.85714285 \\ 7\overline{)6.000000\ldots} \end{array}$

$\therefore 0.\dot{8}5714\dot{2}$

**20** a $0.7 = \frac{7}{10}$

b $0.07 = \frac{7}{100}$

c $0.08 = \frac{8}{100} = \frac{2}{25}$

d $0.31 = \frac{31}{100}$

e $0.145 = \frac{145}{1000} = \frac{29}{200}$

f $0.072 = \frac{72}{1000} = \frac{9}{125}$

## Chapter 3—Percentages and Applications

pp. 34–37

**1** a $\frac{2}{5} \times 100\% = 40\%$

b $\frac{9}{20} \times 100\% = 45\%$

c $\frac{13}{25} \times 100\% = 52\%$

d $\frac{7}{40} \times 100\% = 17.5\%$

e $\frac{41}{100} \times 100\% = 41\%$

f $\frac{1}{3} \times 100\% = 33\frac{1}{3}\%$

g $\frac{5}{6} \times 100\% = 83\frac{1}{3}\%$

h $3\frac{1}{4} \times 100\% = 325\%$

**2** a $\frac{2}{7} \times 100\% = 28.\dot{5}7142\dot{8}$
$= 28.6\%$ (to 1 decimal place)

b $\frac{5}{11} \times 100\% = 45.\dot{4}\dot{5}$
$= 45.5\%$ (to 1 decimal place)

c $2\frac{4}{15} \times 100\% = 226.\dot{6}$
$= 226.7\%$ (to 1 decimal place)

**3** a $0.3 \times 100\% = 30\%$

b $0.03 \times 100\% = 3\%$

c $0.003 \times 100\% = 0.3\%$

d $0.19 \times 100\% = 19\%$

e $0.576 \times 100\% = 57.6\%$

f $1.12 \times 100\% = 112\%$

g $4.7 \times 100\% = 470\%$

h $0.075 \times 100\% = 7.5\%$

**4** a $\frac{60}{100} = \frac{3}{5}$

b $\frac{3}{100} = \frac{3}{100}$

c $\frac{95}{100} = \frac{19}{20}$

d $\frac{127}{100} = 1\frac{27}{100}$

e $\frac{300}{100} = 3$

f $\frac{5\frac{1}{2}}{100} = \frac{\frac{11}{2}}{100}$ [Numerator as improper fraction.]
$= \frac{11}{200}$ [Multiply numerator and denominator by 2.]

g $\frac{12\frac{1}{4}}{100} = \frac{\frac{49}{4}}{100}$ [Numerator as improper fraction.]
$= \frac{49}{400}$ [Multiply numerator and denominator by 4.]

h $\frac{3\frac{4}{5}}{100} = \frac{\frac{19}{5}}{100}$ [Numerator as improper fraction.]
$= \frac{19}{500}$ [Multiply numerator and denominator by 5.]

**5** a $16 \div 100 = 0.16$

b $7 \div 100 = 0.07$

c $60 \div 100 = 0.6$

d $7.6 \div 100 = 0.076$

e $12.5 \div 100 = 0.125$

f $176 \div 100 = 1.76$

g $104.6 \div 100 = 1.046$

h $\frac{1}{4} \div 100 = 0.0025$

i $7\frac{3}{4} \div 100 = 0.0775$

j $33\frac{1}{3} \div 100 = 0.\dot{3}$

## WORKED SOLUTIONS to Chapter 3 **Practise, Practise**

**6** a $0.27 \times 600 = 162$

b $0.16 \times 140 = 22.4$

$\therefore$ \$22.40

c $0.6 \times 42 = 25.2$

$\therefore$ \$25.20

d $0.07 \times 6000 = 420$ [Change to metres]

$\therefore$ 420 metres

e $0.045 \times 8000 = 360$ [Change to mL]

$\therefore$ 360 mL

f $0.125 \times 240 = 30$

$\therefore$ 30 minutes

g $0.0025 \times 6000 = 15$ [Change to grams]

$\therefore$ 15 g

h $1.45 \times 300 = 435$

$\therefore$ \$435

i $0.352 \times 100\,000 = 35\,200$

$\therefore$ \$35 200

j $0.158\,75 \times 4000 = 635$

**7** a Increase = 10% of \$20

= $0.1 \times 20$

= 2

$\therefore$ increase of \$2

$\therefore$ new amount = \$20 + \$2

$\therefore$ The increase was \$22.

Alternative method used for rest of question:

b 118% of 700 = $1.18 \times 700$ [100 + 18]

= 826

$\therefore$ The new amount is 826.

c 130% of 45 = $1.3 \times 45$ [100 + 30]

= 58.5

$\therefore$ The new amount is 58.5 L.

d 108% of 3000 = $1.08 \times 3000$

= 3240 [in grams]

$\therefore$ The new amount is 3240 g.

**8** a Decrease = 20% of 260

= 52

$\therefore$ decrease of 52

$\therefore$ new amount = 260 − 52 = 208

$\therefore$ The new amount is 208.

Alternative method used for rest of question:

b 95% of 65 = $0.95 \times 65$

= 61.75 [100 − 5]

$\therefore$ The new amount is \$61.75.

c $87\frac{1}{2}$% of 3000 = $0.875 \times 300$ [$100 - 12\frac{1}{2}$]

= 2625

$\therefore$ The new amount is 2625

d 58% of \$60 = $0.58 \times 60$

= 34.8 [100 − 42]

$\therefore$ The new amount is \$34.80.

**9** ↑ ed by 30% = 130% [100 + 30]

↓ ed by 30% = 70% [100 − 30]

130% of 100 = $1.3 \times 100$

= 130

Now, 70% of 130 = $0.7 \times 130$

= 91

$\therefore$ The result is \$91.

[Note that the two percentages do not cancel each other out.]

**10** a $\frac{12}{25} \times 100\% = 48\%$

b 20c, \$2 = 200c

$\therefore \frac{20}{200} \times 100\% = 10\%$

c 4 cm, 500 cm

$\therefore \frac{4}{500} \times 100\% = 0.8\%$

d 250 mL, 2000 mL

$\therefore \frac{250}{2000} \times 100\% = 12.5\%$

e 20 seconds, 240 seconds

$\therefore \frac{20}{240} \times 100\% = 8\frac{1}{3}\%$

f 45c, 900c

$\therefore \frac{45}{900} \times 100\% = 5\%$

**11** a $\frac{4}{40} \times 100\% = 10\%$

b 3.5 g, 2000 g

$\frac{3.5}{2000} \times 100\% = 0.175\%$

c 400 cm, 200 cm

$\frac{400}{200} \times 100\% = 200\%$

d 180 seconds, 3 seconds

$\frac{180}{3} \times 100\% = 6000\%$

## WORKED SOLUTIONS to Chapter 3 **Practise, Practise**

**12** **a** 20% of amount = 700
100% of amount = 700 × 5
= 3500 [100% = 5 × 20%]
∴ The amount is 3500.
[This was quicker than going down to 1% and then back up to 100%.]

**b** 9% of amount = 45
1% of amount = 45 ÷ 9
= 5
100% of amount = 5 × 100
= 500
∴ The amount is 500.

**13** 15% of object = 525
5% of object = 525 ÷ 3
= 175
100% of object = 175 × 20
= 3500
∴ The whole object weighs 3500 g.

**14** 25% off
∴ discounted price = 75% of original price
**a** 75% of 95 = 0.75 × 95
= 71.25
∴ The discounted price is $71.25.
**b** 75% of 38 = 0.75 × 38
= 28.5
∴ The discounted price is $28.50.

**15** **a** Discount = $12\frac{1}{2}$% of 22
= 0.125 × 22
= 2.75
∴ It gives a saving of $2.75.
**b** Cost of shovel = $22 – $2.75
= $19.25
∴ The shovel will cost $19.25.

**16** Discount = $800 – $700 = $100
∴ percentage discount $= \frac{\text{discount}}{\text{original Price}} \times 100\%$
$= \frac{100}{800} \times 100\%$
= 12.5%
∴ The percentage discount is 12.5%.

**17** Discount = 35%
∴ new price is 65% of old price
∴ new price = 65% of 84
= 54.6
∴ The new price is $54.60.

**18** Savings = 15% of 16 999
= 0.15 × 16 999
= 2549.85
∴ The savings are $2549.85.

**19** **a** Profit = $1.40 – $0.80
= $0.60
∴ The profit of 60 cents.
**b** $\%\ \text{profit} = \frac{\text{profit}}{\text{cost price}} \times 100\%$
$= \frac{60}{80} \times 100\%$
= 75%
∴ The profit as percentage cost is 75%.

**20** **a** Loss = $3400 – $1000
= $2400
∴ His loss is $2400.
**b** $\%\ \text{loss} = \frac{\text{loss}}{\text{cost price}} \times 100\%$
$= \frac{2400}{3400} \times 100\%$
= 70.588 235%
= 70.6% (to one decimal place)
∴ The percentage loss is 70.6%.

**21** **a** Profit = 20% of 4.80
= 0.2 × 4.8
= 0.96
∴ Profit is $0.96
**b** Selling price = Cost price + profit
= $4.80 + $0.96
= $5.76
∴ The oranges sold for $5.76 per box.
**c** $\frac{\text{Profit}}{\text{No. of oranges}} = \frac{96}{60}$
= 1.6
∴ 1.6 cents profit on each orange

## WORKED SOLUTIONS to Chapter 3 **Practise, Practise**

**22** Loss = \$1200 – \$240
= \$960

Percentage loss = $\frac{\text{loss}}{\text{cost price}} \times 100\%$
$= \frac{960}{1200} \times 100\%$
= 80%

∴ There was an 80% loss on lounge suite.

**23** If cost price = 100% and profit = 25%
∴ selling price is 125% (of cost price)
∴ 125% of cost price = 480 000
1% of cost price = 480 000 ÷ 125
= 3840
∴ 100% of cost price = 3840 × 100
= 384 000
∴ Maria had bought the house for \$384 000.

**24** GST = 473 ÷ 11 = 43
∴ GST of \$43.

**25** GST = 71.5 ÷ 11 = 6.5
∴ GST of \$6.50.

**26** GST = 0.1 × 54 = 5.4
∴ New price = 54 + 5.4 = 59.4
∴ New price of \$59.40.

**27** GST = 35.2 ÷ 11 = 3.2
∴ Old price = 35.2 – 3.2 = 32
∴ Pre-GST price of \$32.

**28** GST = 264 ÷ 11 = 24
∴ Old price = 264 – 24 = 240
∴ Pre-GST price of \$240.

**29** GST = 9500 ÷ 11 = 863.64 (2 dec. places)
∴ GST of \$863.64.

**30** GST = 69 ÷ 11 = 6.27 (2 dec. places)
∴ Old price = 69 – 6.27 = 62.73
∴ Pre-GST price of \$62.73.

**31** Lose = 20 – 12
= 8
∴ % Lose = $\frac{8}{20} \times 100\%$
= 40%
∴ Team lost 40% of games.

**32** Total students = 50
**a** $\frac{18}{50} \times 100\% = 36\%$
**b** $\frac{4}{50} \times 100\% = 8\%$

**33** **a** Yellow = 40 – (8 + 7 + 7 + 4 + 6)
= 40 – 32
= 8
% yellow = $\frac{8}{40} \times 100\%$
= 20%

**b** 40 – 8 = 32
% yellow = $\frac{8}{32} \times 100\%$
= 25%

**34** New pay = 104% of 602
= 1.04 × 602
= 626.08
∴ \$626.08

**35** 80 – 16 = 64
% failed = $\frac{64}{80} \times 100\%$
= 80%
∴ 80% failed to finish.

**36** 22% of total = 198
100% of total = 198 ÷ 22 × 100
= 900
∴ Enrolment of 900.

**37** **a** 40% of cost = 68
100% of cost = 68 ÷ 40 × 100
= 170
∴ Total cost is \$170.

**b** If 40% = \$68
then 20% = \$34
∴ Ken contributes \$34.

**38** Increase = 270 – 240
= 30
% increase = $\frac{30}{240} \times 100\%$
= 12.5%
∴ Increased by 12.5%.

**39** **a** Decrease = 4500 – 1400
= 3100
% decrese = $\frac{3100}{4500} \times 100\%$
= 68.89%

**b** New total = 130% of 1400
$= 1.3 \times 1400$
$= 1820$
$\therefore$ Trevor had 1820 sheep.

## Chapter 4—Algebra

pp. 51–57

**1** **a** $2 \times b \times c = 2bc$ **b** $3a \times 4 = 12a$
**c** $a \times b = ab$ **d** $b \times 3 = 3b$
**e** $a \times a = a^2$ **f** $5 \times t \times 2 = 10t$
**g** $3 \times b \times b = 3b^2$
**h** $2a \times 3a = 6a^2$
**i** $5 \times a \times 2a \times 3 = 30a^2$
**j** $3a \times a = 3a^2$
**k** $7a \times 3b = 21ab$
**l** $2a \times a \times a = 2a^3$
**m** $3y \times 4xy = 12xy^2$
**n** $4x \times 3 \times y = 12xy$
**o** $(2x)^2 = 2x \times 2x = 4x^2$

**2** **a** $ab = a \times b$
**b** $2xy = 2 \times x \times y$
**c** $3a^2 = 3 \times a \times a$
**d** $5abc = 5 \times a \times b \times c$
**e** $2a^2bc^2 = 2 \times a \times a \times b \times c \times c$
**f** $(5x)^2 = 5x \times 5x$
$= 5 \times x \times 5 \times x$

**3** **a** $bc = 2 \times 3$
$= 6$
**b** $3a = 3 \times 5$
$= 15$
**c** $a + b = 5 + 2$
$= 7$
**d** $c - b + a = 3 - 2 + 5$
$= 6$
**e** $b^2 = 2^2$
$= 2 \times 2$
$= 4$
**f** $3b - c = 3 \times 2 - 3$
$= 6 - 3$
$= 3$
**g** $abc = 5 \times 2 \times 3$
$= 30$
**h** $a^2 - c^2 = 5^2 - 3^2$
$= 25 - 9$
$= 16$
**i** $\sqrt{5a} = \sqrt{5 \times 5}$
$= \sqrt{25}$
$= 5$
**j** $a(b + c) = 5(2 + 3)$
$= 5 \times 5$
$= 25$
**k** $4c - a + b = 4 \times 3 - 5 + 2$
$= 12 - 5 + 2$
$= 9$
**l** $2a^2 = 2 \times 5^2$
$= 2 \times 25$
$= 50$
**m** $3b^2 = 3 \times 2^2$
$= 3 \times 4$
$= 12$
**n** $(2a)^2 = (2 \times 5)^2$
$= 10^2$
$= 100$
**o** $\frac{4c}{6} = \frac{4 \times 3}{6}$
$= \frac{12}{6}$ $\left[\frac{12}{6} = 12 \div 6\right]$
$= 2$
**p** $\frac{abc}{a + b + c} = \frac{5 \times 2 \times 3}{5 + 2 + 3}$
$= \frac{30}{10}$
$= 3$
**q** $bc - b^2 = 2 \times 3 - 2^2$
$= 6 - 4$
$= 2$
**r** $\frac{3a^2}{c} = \frac{3 \times 5^2}{3}$
$= \frac{75}{3}$
$= 25$
**s** $\sqrt{a^2 - c^2} = \sqrt{5^2 - 3^2}$
$= \sqrt{25 - 9}$
$= \sqrt{16}$
$= 4$

## WORKED SOLUTIONS to Chapter 4 **Practise, Practise**

**4** **a** $a + b = 0.72 + 0.6$
$= 1.32$

**b** $\frac{a}{b} = \frac{0.72}{0.6}$
$= 1.2$

**c** $ab = 0.72 \times 0.6$
$= 0.432$

**d** $5b = 5 \times 0.6$
$= 3$

**e** $4a - b = 4 \times 0.72 - 0.6$
$= 2.28$

**f** $a^2 = (0.72)^2$
$= 0.72 \times 0.72$
$= 0.5184$

**g** $3b - 2a = 3 \times 0.6 - 2 \times 0.72$
$= 0.36$

**h** $2b^2 = 2 \times (0.6)^2$
$= 2 \times 0.6 \times 0.6$
$= 0.72$

**5** [Use calculator $\boxed{a\,b/c}$]

**a** $m + n = \frac{4}{5} + \frac{2}{3}$
$= 1\frac{7}{15}$

**b** $\frac{m}{n} = \frac{\frac{4}{5}}{\frac{2}{3}} = \frac{4}{5} \div \frac{2}{3}$
$= 1\frac{1}{5}$

**c** $mn = \frac{4}{5} \times \frac{2}{3}$
$= \frac{8}{15}$

**d** $3n = 3 \times \frac{2}{3}$
$= 2$

**e** $2m = 2 \times \frac{4}{5}$
$= 1\frac{3}{5}$

**f** $m^2 = \left(\frac{4}{5}\right)^2$
$= \frac{4}{5} \times \frac{4}{5}$
$= \frac{16}{25}$

**g** $3n^2 = 3\left(\frac{2}{3}\right)^2$
$= 3 \times \frac{2}{3} \times \frac{2}{3}$
$= 1\frac{1}{3}$

**h** $5 - mn = 5 - \frac{4}{5} \times \frac{2}{3}$
$= 5 - \frac{8}{15}$
$= 4\frac{7}{15}$

**i** $2m + 3n = 2 \times \frac{4}{5} + 3 \times \frac{2}{3}$
$= \frac{8}{5} + 2$
$= 3\frac{3}{5}$

**j** $\frac{m+n}{m-n} = \frac{\frac{4}{5}+\frac{2}{3}}{\frac{4}{5}-\frac{2}{3}}$
$= \frac{1\frac{7}{15}}{\frac{2}{15}}$
$= \frac{22}{15} \div \frac{2}{15}$
$= 11$

**6** **a** $xy = -4 \times 3$ $[- \times + = -]$
$= -12$

**b** $x^2 = (-4)^2$
$= -4 \times -4$ $[- \times - = +]$
$= 16$

**c** $3x + z = 3 \times -4 + 2$
$= -12 + 2$
$= -10$

**d** $x - z = -4 - 2$
$= -6$

**e** $y - x = 3 - -4$ $[3 - -4 = 3 + 4]$
$= 7$

**f** $y + 2x = 3 + 2 \times -4$
$= 3 + -8$
$= -5$

**g** $\frac{x}{z} = \frac{-4}{2}$ $[- \div + = -]$
$= -4 \div 2$
$= -2$

**h** $x(y + z) = -4(3 + 2)$
$= -4 \times 5$
$= -20$

**i** $(3x)^2 = (3 \times -4)^2$
$= (-12)^2$
$= -12 \times -12$ $[- \times - = +]$
$= 144$

**j** $-y - z - x = -3 - 2 - (-4)$
$= -5 + 4$
$= -1$

**k** $y^2 - x^2 = 3^2 - (-4)^2$
$= 9 - 16$
$= -7$

**7** **a** $V = lbh$
$= 1 \times b \times h$
$= 4 \times 3 \times \frac{1}{2}$
$= 6$

**b** $T = 2R^2$
$= 2 \times (-5)^2$
$= 2 \times 25$
$= 50$

**c** $V = u + at$
$= -4 + -2 \times 3$
$= -4 + -6$
$= -10$

**d** $A = \frac{22}{7}R^2$
$= \frac{22}{\not{7}_1} \times 7 \times \not{7}^1$
$= 154$

# WORKED SOLUTIONS to Chapter 4 **Practise, Practise**

**e** $A = 4xy + x^2$
$= 4 \times 4 \times 3 + 4^2$
$= 48 + 16$
$= 64$

**f** $D = \frac{M}{V}$
$= \frac{2\frac{3}{4}}{1\frac{1}{2}}$
$= 2\frac{3}{4} \div 1\frac{1}{2}$
$= 1\frac{5}{6}$

**g** $T = \frac{10}{A + B}$
$= \frac{10}{4+11}$
$= \frac{10}{15}$
$= \frac{2}{3}$

**h** $P = 2(l + b)$
$= 2(1.72 + 0.8)$
$= 2 \times 2.52$
$= 5.04$

**i** $d = \sqrt{a^2 - b^2}$
$= \sqrt{(-5)^2 - 4^2}$
$= \sqrt{25 - 16}$
$= \sqrt{9}$
$= 3$

**j** $N = \frac{4B}{A + C}$
$= \frac{4 \times \frac{1}{2}}{\frac{1}{3} + \frac{1}{2}}$
$= \frac{2}{\frac{7}{12}} = 2 \div \frac{7}{12}$
$= \frac{24}{7}$
$= 3\frac{3}{7}$

**8** **a** $6a + 4a - 3a = 7a$
**b** $7a - a = 6a$ [$a = 1a$]
**c** $3a + 4b + 2a - 2b$
$= (3a + 2a) + (4b - 2b)$ [Collect like terms.]
$= 5a + 2b$ [Simplify.]
**d** $2a + a = 3a$ [$a = 1a$]
**e** $7b - 2 + 3b = 10b - 2$
**f** $8x - 4 - 7x = (8x - 7x) - 4$
$= x - 4$
**g** $3a^2 + 4a + 2a^2 - 2a = 5a^2 + 2a$
[$a^2$ and $a$ are different terms.]
**h** $a + 6b - a + 2b = 8b$
**i** $7p + 3 - 4p - 7 = (7p - 4p) + (3 - 7)$
$= 3p + (-4)$
$= 3p - 4$
[Adding –ve means subtracting.]
**j** $7x - 3x - 2x + y = 2x + y$
**k** $5ab - 2ba = 3ab$
**l** $m^2 + m + 3m^2 + 4m = 4m^2 + 5m$
**m** $5b + 6a - 2b + a = 3b + 7a$

**9** **a** $15b \div 3 = 5b$
**b** $15a \div 3a = 5$
**c** $12ab \div 4a = 3b$
**d** $20ab \div 4ac = \frac{\cancel{20}^5 \cancel{a}^1 b}{\cancel{4}_1 \cancel{a}_1 c}$
$= \frac{5b}{c}$
**e** $14xy^2t \div 7xyt = \frac{14xy^2t}{7xyt}$
$= \frac{\cancel{14}^2 \cancel{x}^1 y \cancel{y}^1 \cancel{t}^1}{\cancel{7}_1 \cancel{x}_1 \cancel{y}_1 \cancel{t}_1}$
$= 2y$
**f** $12a^2b \div 8ab = \frac{\cancel{12}^3 \cancel{a}^1 a \cancel{b}^1}{\cancel{8}_2 \cancel{a}_1 \cancel{b}_1}$
$= \frac{3a}{2}$
**g** $10mn^2 \div 15m^2n = \frac{10mn^2}{15m^2n}$
$= \frac{\cancel{10}^2 \cancel{m}^1 n \cancel{n}^1}{\cancel{15}_3 m \cancel{m}_1 \cancel{n}_1}$
$= \frac{2 \times n}{3 \times m}$
$= \frac{2n}{3m}$
**h** $35x^8y^4 \div 5x^2y^3 = 7x^6y$
**i** $3ab \div 7a = \frac{3\cancel{a}b}{7\cancel{a}}$
$= \frac{3b}{7}$
**j** $6a \div 3a = \frac{\cancel{6}^2 \cancel{a}}{\cancel{3}_1 \cancel{a}}$
$= \frac{2}{1}$ [$\frac{2}{1} = 2$]
$= 2$

**10** **a** $a^5 \times a^3 = a^8$ [To multiply: *add* indices.]
**b** $a^2 \times a^6 = a^8$
**c** $a \times a^9 = a^{10}$
**d** $a^{12} \times a^2 = a^{14}$
**e** $a^{15} \times a^2 = a^{17}$

**f** $a^{16} \div a^4 = a^{12}$ [To divide: *subtract* indices.]

**g** $a^{15} \div a^5 = a^{10}$

**h** $a^{12} \div a^3 = a^9$

**i** $(a^4)^3 = a^{12}$

**j** $(a^7)^4 = a^{28}$

**k** $(a^5)^2 = a^{10}$

**l** $(ab)^3 = a^3b^3$

**m** $(3x)^2 = 3x \times 3x$
$= 9x^2$

**n** $(4ab)^2 = 4^2a^2b^2 = 16a^2b^2$ Or $4ab \times 4ab = 16a^2b^2$

**o** $(4a)^3 = 4^3a^3$
$= 64a^3$

**p** $(5a^4)^2 = 5a^4 \times 5a^4$
$= 25a^8$

**q** $(7x)^2 = 7x \times 7x$
$= 49x^2$

**r** $(a^4b^2)^3 = (a^4)^3\,(b^2)^3$
$= a^{12}b^6$

**s** $y \times y^5 \times y^2 = y^8$

**t** $\dfrac{a^5 \times a^6}{a^3} = \dfrac{a^{11}}{a^3}$ $\left[\dfrac{a^{11}}{a^3} = a^{11} \div a^3\right]$
$= a^8$

**u** $15a^6b^3 \div 5a^3b^2 = 3a^3b$

**v** $12x^2y \div 4xy = 3x$

**w** $\dfrac{4 \times a^2 \times a^4}{a \times 2 \times a^3} = \dfrac{4a^6}{2a^4}$
$= 2a^2$

**x** $(2x)^3 \div x^2 = 2^3x^3 \div x^2$
$= 8x^3 \div x^2$
$= 8x$

**y** $\dfrac{15a^4b^6}{5a^2b^2} = 3a^2b^4$

**z** $\dfrac{(3a^4)^2}{3a^4} = \dfrac{9a^8}{3a^4}$
$= 3a^4$

**aa** $2^7 \times 2^3 = 2^{10}$

**bb** $5 \times 5^4 = 5^1 \times 5^4$
$= 5^5$

**cc** $10^{14} \div 10^7 = 10^7$

**11** **a** $5(a + 3) = 5a + 15$

**b** $4(b - 2) = 4b - 8$

**c** $3(a - b) = 3a - 3b$

**d** $4(2x - 3) = 8x - 12$

**e** $a(b - 2) = ab - 2a$

**f** $-4(x + 2) = -4x - 8$

**g** $-3(x - 2) = -3x + 6$

**h** $4(2a - 3b) = 8a - 12b$

**i** $-2(2a - 3b) = -4a + 6b$

**j** $a(2a - b) = 2a^2 - ab$

**k** $-4(a - 3) = -4a + 12$

**l** $-(a - 2) = -a + 2$

**m** $-(2 - 3a) = -2 + 3a$

**n** $3(2 - 4a) = 6 - 12a$

**o** $-3(2 + a) = -6 - 3a$

**12** **a** $7a - 2a - 3a - 6a = -4a$

**b** $-8t + 2t = -6t$

**c** $-3a - 4a = -7a$

**d** $-2a - 2a = -4a$

**e** $-5a + 5a = 0$

**f** $2x - 7y + x + 4y = 3x - 3y$

**g** $7x - 3y - x - y = 6x - 4y$

**h** $4a^2 + 6a - 3a^2 - 10a = a^2 - 4a$

**i** $6x - 5 - 2x - 4 = 4x - 9$

**j** $5 - x^2 - 3 = 2 - x^2$

**k** $2x - 8 - 4x + 3 = -2x - 5$

**l** $8x - 12 - 6 + 3x = 11x - 18$

**m** $8b - 12 + 3b + 6 = 11b - 6$

**n** $a - 7b + a - 3b = 2a - 10b$

**o** $a^2 - 6b - a^2 + 2b = -4b$

**p** $a - 3b + a - b = 2a - 4b$

**13** **a** $5(x + 2) + 3x$
$= 5x + 10 + 3x$ [Expand.]
$= 8x + 10$ [Collect like terms.]

**b** $4(a + b) + 3b = 4a + 4b + 3b$
$= 4a + 7b$

**c** $2 + 3(a + 4) = 2 + 3a + 12$
$= 14 + 3a$

**d** $5(a + 3) + 2(a + 1) = 5a + 15 + 2a + 2$
$= 7a + 17$

**e** $4(x + 3) + 2(x - 1) = 4x + 12 + 2x - 2$
$= 6x + 10$

**f** $3(x - 2) + 2(x - 1) = 3x - 6 + 2x - 2$
$= 5x - 8$

**g** $7 - 2(x - 3) = 7 - 2x + 6$
$= 13 - 2x$

**h** $4 - 1(x - 2) = 4 - x + 2$
$= 6 - x$

**i** $3(x - 2) - 4(x - 1) = 3x - 6 - 4x + 4$
$= -x - 2$

**j** $4x - 3(2 - 3x) = 4x - 6 + 9x$
$= 13x - 6$

**k** $5(2x - 3) - 2(x - 2) = 10x - 15 - 2x + 4$
$= 8x - 11$

**l** $4(1 - x) - 2(3 - x) = 4 - 4x - 6 + 2x$
$= -2 - 2x$

**m** $3 - 2(1 - a) = 3 - 2 + 2a$
$= 1 + 2a$

**n** $a(a + 2) + 3(a + 1) = a^2 + 2a + 3a + 3$
$= a^2 + 5a + 3$

**o** $2a(a - 4) - a(a - 2) = 2a^2 - 8a - a^2 + 2a$
$= a^2 - 6a$

**p** $2b(3b - 1) - b(b + 2) = 6b^2 - 2b - b^2 - 2b$
$= 5b^2 - 4b$

**q** $5p(2p - 1) - 3(p - 2)$
$= 10p^2 - 5p - 3p + 6$
$= 10p^2 - 8p + 6$

**14**

**a** $-4 \times 2a = -8a$ $[- \times + = -]$

**b** $-3a \times 2b = -6ab$

**c** $-2a \times -3b = 6ab$ $[- \times - = +]$

**d** $4 \times -b \times -c = 4bc$

**e** $-2x \times x = -2x^2$

**f** $5b \times -3b = -15b^2$

**g** $-12a \div 3 = -4a$ $[- \div + = -]$

**h** $-12b \div -4 = 3b$ $[- \div - = +]$

**i** $5 \times -2p \times -p = 10p^2$

**j** $-8x^2y \div 4xy = -2x$

**k** $-12b \div -b = 12$

**l** $4t \div -2t = -2$

**15**

**a** $6a - 4 + 3a + 2 = 9a - 2$

**b** $2t \times -4 = -8t$

**c** $18a \div 3 = 6a$

**d** $25a \div 5a = 5$

**e** $12p^2 \div 6p = 2p$

**f** $-2a \times -3a \times b = 6a^2b$

**g** $a + a + a = 3a$

**h** $a \times a \times a = a^3$

**i** $5a - a = 4a$

**j** $(4x)^2 = 4x \times 4x$
$= 16x^2$

**k** $(2a^4)^3 = 2^3(a^4)^3$ or $2a^4 \times 2a^4 \times 2a^4$
$= 8a^{12}$

**l** $5a - 12a = -7a$

**m** $5a \times -4a = -20a^2$

**n** $(-2ab)^2 = (-2)^2a^2b^2$
$= 4a^2b^2$

**o** $5x^2 - 6x + x^2 + x = 6x^2 - 5x$

**p** $18x^2y \div 3xy = 6x$

**q** $2x^4y \div 3x^3y^2 = \dfrac{2x^4y}{3x^3y^2}$
$= \dfrac{2x}{3y}$

**16**

**a** $4a + 20 = 4(a + 5)$

**b** $5a + 5b = 5(a + b)$

**c** $ab - ac = a(b - c)$

**d** $6a + 4b = 2(3a + 2b)$

**e** $3b^2 - 6 = 3(b^2 - 2)$

**f** $5x^2 + 25x = 5x(x + 5)$

**g** $4a^2 - 2a = 2a(2a - 1)$

**h** $5x^2 + 10y + 20 = 5(x^2 + 2y + 4)$

**i** $12 - 6a = 6(2 - a)$

**j** $5x^2 - 6x = x(5x - 6)$

**k** $35x^2y - 7xy = 7xy(5x - 1)$

**l** $4ab - 2b = 2b(2a - 1)$

**m** $8a - 4a^2 = 4a(2 - a)$

**n** $12b - 8b^2 = 4b(3 - 2b)$

**17**

**a** $2a + c$

**b** $mn$

**c** Average $= \dfrac{5x + 7x + 3x}{3}$
$= \dfrac{15x}{3}$
$= 5x$

**d** $2b + a$

**e** $(5x + 8) - 2x = 3x + 8$

**f** $y \times 12$ cents $= 12y$ cents

**g** Change $= \$(x - 6)$

**h** $a + 1, a + 2, a + 3$

**i** $a + 2, a + 4$

## WORKED SOLUTIONS to Chapter 4 **Practise, Practise**

**j**

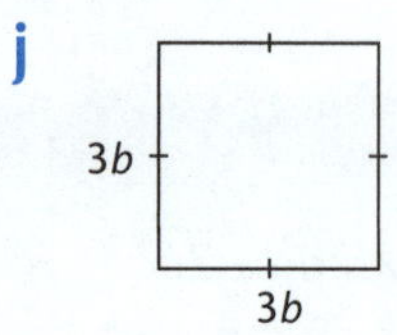


$A = 3b \times 3b$
$= 9b^2$

$\therefore$ The area is $9b^2$ cm$^2$.

**18** **a** **i** $b \times 100 = 100b$ cm

**ii** $a \times 60 = 60a$ seconds

**iii** $2a \times 1000 = 2000a$ mL

**iv** $3a \times 100 = 300a$ dollars

**v** $120t \div 60 = 2t$ minutes

**b** **i** $P = 2x + 2x + x + 3$
$= 5x + 3$

$\therefore$ The perimeter is $(5x + 3)$ cm.

**ii** $P = 2x - 1 + 5 + 2x - 1 + 5$
$= 4x + 8$

$\therefore$ The perimeter is $(4x + 8)$ cm.

**iii**

5a
3a
7a
6a
3a
12a

$P = 12a + 3a + 7a + 3a + 5a + 6a$
$= 36a$

$\therefore$ The perimeter is $36a$ cm.

**c** **i** $A = 4 \times (x - 3)$
$= 4(x - 3)$
$= 4x - 12$

$\therefore$ The area is $(4x - 12)$ cm$^2$.

**ii** $A = 3x \times 3x$
$= 9x^2$

$\therefore$ The area is $9x^2$ cm$^2$.

**iii** $A = \frac{1}{2}bh$
$= \frac{1}{2} \times 6 \times (2x - 4)$
$= 3 \times (2x - 4)$
$= 3(2x - 4)$
$= 6x - 12$

$\therefore$ The area is $(6x - 12)$ cm$^2$.

**19** **a** $2a + 3$

**b** $4(2a - 3) = 8a - 12$

**c** $15ab \times 3a = \frac{\cancel{15}^{5}\,\cancel{a}^{1}b}{\cancel{3}_{1}\,\cancel{a}_{1}}$
$= \frac{5b}{1}$
$= 5b$

**d** $3c^2 - 2c + 5c^2 = 8c^2 - 2c$

**e** $3y^4 \times 2y^3 = 6y^7$

**f** $d = -4,\ d^2 = (-4)^2$
$= -4 \times -4$
$= 16$ $\quad [- \times - = +]$

**g** $-3a \times -2a = 6a^2$ $\quad [a \times a = a^2]$

**h** $4m - 8n + 3m + 2n = 7m - 6n$

**i** $5 - a = 5 - (-4)$
$= 5 + 4$
$= 9$ $\quad$ [Note: $- - = +$]

**j** $4a - b = (4 \times -2) - 7$
$= -8 - 7$
$= -15$

**k** $3(2x - 5) + 2(x + 3) = 6x - 15 + 2x + 6$
$= 8x - 9$

**l** $2y^2 - (2y)^2 = 2(-4)^2 - (2x - 4)^2$
$= 32 - 64$
$= -32$

**m** $D = \frac{M}{V}$
$= -\frac{42}{3}$
$= -14$

**n** $M = \frac{x}{y}$
$= \frac{3}{\cancel{4}_{1}} \times \frac{\cancel{8}^{2}}{5}$
$= \frac{6}{5}$

$\therefore \frac{1}{M} = \frac{5}{6}$

**o** $2ab^2 \times 3a^2 = 6a^3b^2$

**p** $3ab + 2a - 4ab + 6a = -ab + 8a$

**q** $a^2 - b = 2^2 - (-4)$
$= 4 + 4$
$= 8$

**r** $3 - 2(a - 4) = 3 - 2a + 8$
$= 11 - 2a$

**s** $8b - 2a^2 = 2(4b - a^2)$

**t** $12x^4y^6 \div 6y^3x = \frac{12x^4y^6}{6y^3x}$

$= \frac{\cancel{12}^2 xxx\cancel{x}^1 yyy\cancel{y}^1\cancel{y}^1\cancel{y}^1}{\cancel{6}_1\cancel{y}_1\cancel{y}_1\cancel{y}_1\cancel{x}_1}$

$= 2xxxyyy$

$= 2x^3y^3$

**u** $4x^2 + 16x = 4x(x + 4)$

**v** $4x(x - 2) - x(3 - x)$
$= 4x^2 - 8x - 3x + x^2$
$= 5x^2 - 11x$

**w** $x + 2, x + 4$

**x** $2a^2 = 2 \times (0.4)^2$
$= 2 \times 0.16$
$= 0.32$

**y**

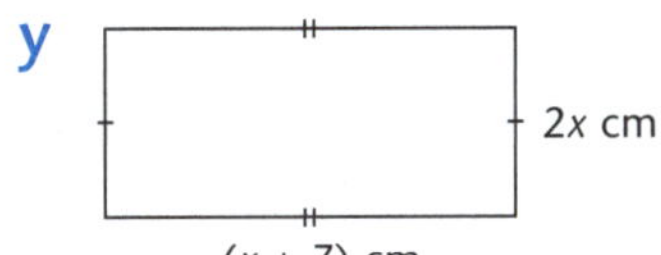


$P = x + 7 + 2x + x + 7 + 2x$
$= 6x + 14$

∴ The perimeter is $(6x + 14)$ cm.

$A = 2x \times (x + 7)$
$= 2x(x + 7)$
$= 2x^2 + 14x$

∴ The area is $(2x^2 + 14x)$ cm$^2$.

**20** **a**

| $x$ | 1 | 2 | 3 | 4 | 10 | 100 |
|---|---|---|---|---|---|---|
| $y$ | 4 | 5 | 6 | 7 | 13 | 103 |

**b**

| $x$ | 2 | 3 | 4 | 10 | 40 | 100 |
|---|---|---|---|---|---|---|
| $y$ | 1 | 3 | 5 | 17 | 77 | 197 |

**c**

| $x$ | −2 | −1 | 0 | 1 | 20 | 50 |
|---|---|---|---|---|---|---|
| $y$ | −5 | −2 | 1 | 4 | 61 | 151 |

**d**

| $x$ | −3 | −2 | −1 | 0 | 1 | 2 |
|---|---|---|---|---|---|---|
| $y$ | 11 | 9 | 7 | 5 | 3 | 1 |

**e**

| $x$ | −3 | −2 | −1 | 0 | 1 | 2 |
|---|---|---|---|---|---|---|
| $y$ | 10 | 5 | 2 | 1 | 2 | 5 |

**f**

| $d$ | 1 | 2 | 3 | 4 | 10 | 100 |
|---|---|---|---|---|---|---|
| $T$ | −3 | 0 | 3 | 6 | 24 | 294 |

**g**

| $q$ | 0 | $\frac{1}{4}$ | $\frac{1}{2}$ | 1 | $1\frac{1}{4}$ | $1\frac{1}{2}$ |
|---|---|---|---|---|---|---|
| $p$ | −1 | $-\frac{1}{2}$ | 0 | 1 | $1\frac{1}{2}$ | 2 |

**h**

| $t$ | 0 | 0.2 | 0.4 | 0.6 | 0.8 | 1 |
|---|---|---|---|---|---|---|
| $s$ | −6 | −5 | −4 | −3 | −2 | −1 |

**21** **a** $y = 4x$ **b** $y = 2x + 1$
**c** $y = 3x + 1$ **d** $y = 4 - x$
**e** $y = 2x + 1$ **f** $y = x^2$
**g** $y = x^2 + 1$ **h** $y = 3 - 2x$

## Chapter 5—Patterns and Linear Relationships

**1** **a** A(2, 3) **b** B(−2, 1)
**c** C(2, −4) **d** D(5, −2)
**e** E(0, 5) **f** F(−6, 0)
**g** G(4, −6) **h** H(−3, −2)
**i** I(−5, −5) **j** J(1, 1)

**2**

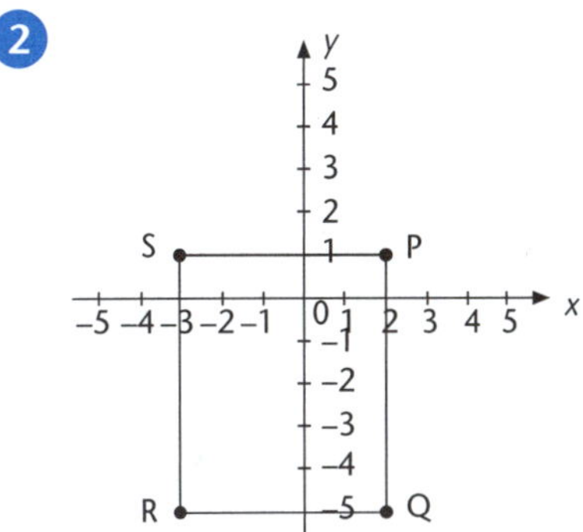


**a** Rectangle

**b** PQ = 6 units
PS = 5 units

**c** Perimeter = 2(6 + 5)
= 2(11)
= 22

∴ The perimeter is 22 units.

**d** Area = 6 × 5
= 30

∴ The area is 30 units$^2$.

**e** 2nd quadrant

## WORKED SOLUTIONS to Chapter 5 **Practise, Practise**

**3** a K(3, 4), L(3, –3), M(–4, –3), N(–4, 4)

b Square (all sides are 7 units)

c Area $= 7 \times 7$

$= 49$

$\therefore$ The area = 49 units$^2$.

**4**

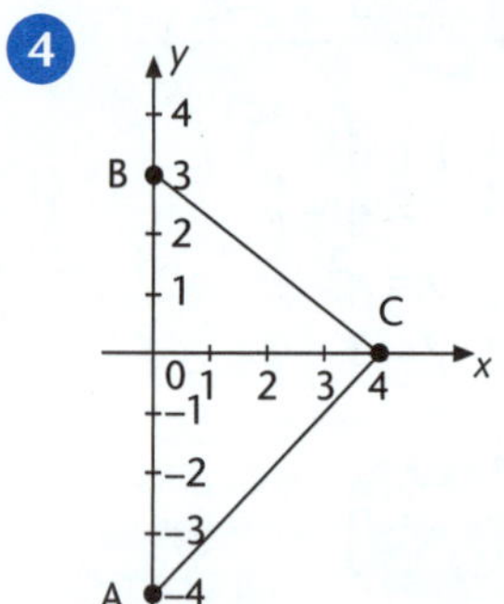

a ABC is a triangle.

b AB = 7 units

c Distance is 4 units

d Area $= \frac{1}{2} \times \text{base} \times \text{height}$

$= \frac{1}{2} \times 7 \times 4 = 14$

$\therefore$ the area is 14 units$^2$.

**5** a i $XY = 5$ units

ii $YZ = 5$ units

b isosceles (XY = YZ)

c Area $= \frac{1}{2} \times 5 \times 5 = 12.5$

$\therefore$ the area is 12.5 units$^2$

**6**

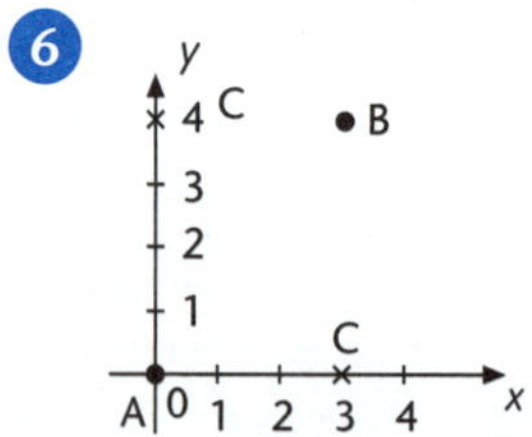

a C could be either (3, 0) or (0, 4)

b If C(3, 0) $\therefore$ Area $= \frac{1}{2} \times 3 \times 4$

$= 6$

If C(0, 4) $\therefore$ Area $= \frac{1}{2} \times 4 \times 3$ $\therefore$

$= 6$

$\therefore$ the area will always be 6 units$^2$.

**7**

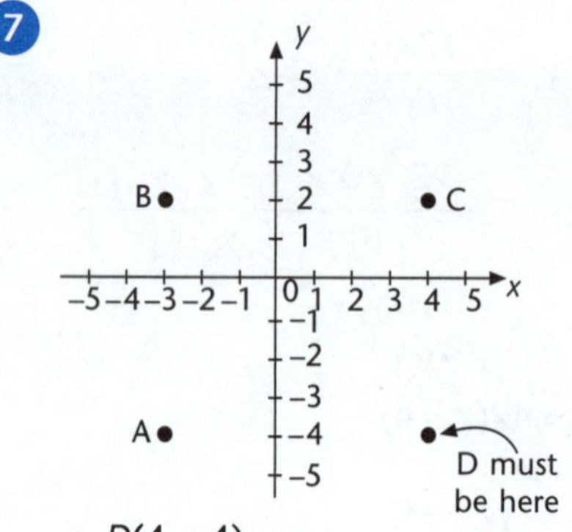


$\therefore D(4, -4)$

**8** a

| $y = 3x + 2$ | |
|---|---|
| $x$ | $y$ |
| 0 | 2 |
| 1 | 5 |
| 2 | 8 |
| 3 | 11 |

b

| $y = 4 - 2x$ | |
|---|---|
| $x$ | $y$ |
| 0 | 4 |
| 1 | 2 |
| 2 | 0 |
| 3 | –2 |

**9** a $y$ is 4 times $x$

$\therefore y = 4x$

b $y$ is 2 less than $x$

$\therefore y = x - 2$

c $y$ is 3 times $x$ minus 1

$\therefore y = 3x - 1$

**10** a Domain = 0, 1, 2, 3

b Range = 0, 2, 4, 6

c $y$ is 2 times $x$

$\therefore y = 2x$

# WORKED SOLUTIONS to Chapter 5 **Practise, Practise**

**11** a $y = 3x$

| $x$ | 0 | 1 | 2 |
|---|---|---|---|
| $y$ | 0 | 3 | 6 |

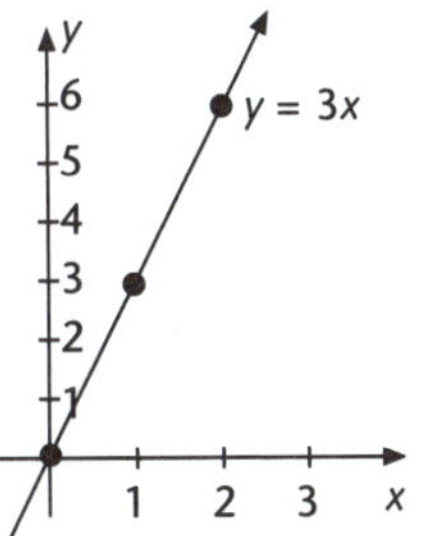


b $y = 2 - x$

| $x$ | 0 | 1 | 2 |
|---|---|---|---|
| $y$ | 2 | 1 | 0 |

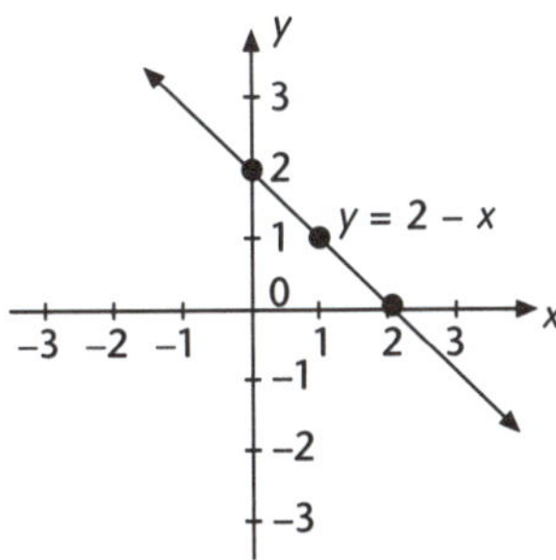


c $y = x + 3$

| $x$ | 0 | 1 | 2 |
|---|---|---|---|
| $y$ | 3 | 4 | 5 |

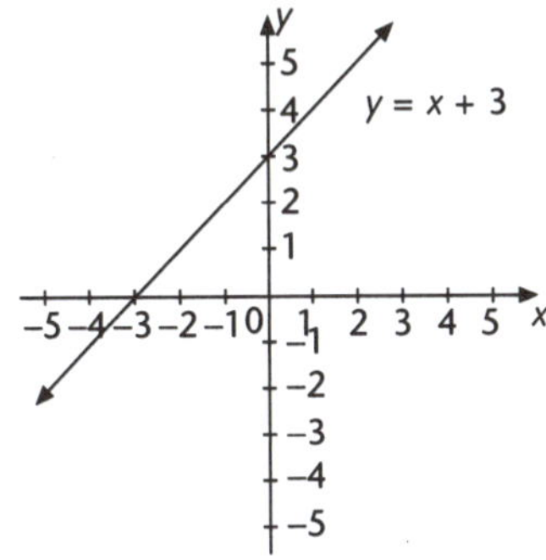


d $y = 4 - 2x$

| $x$ | 0 | 1 | 2 |
|---|---|---|---|
| $y$ | 4 | 2 | 0 |

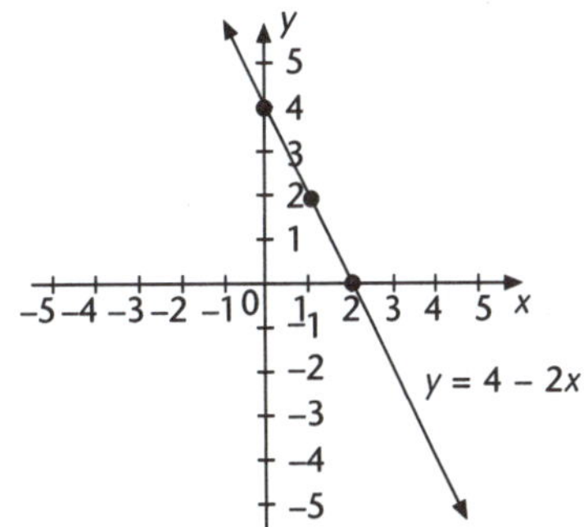


e $x + y = 3$

| $x$ | 0 | 1 | 2 |
|---|---|---|---|
| $y$ | 3 | 2 | 1 |

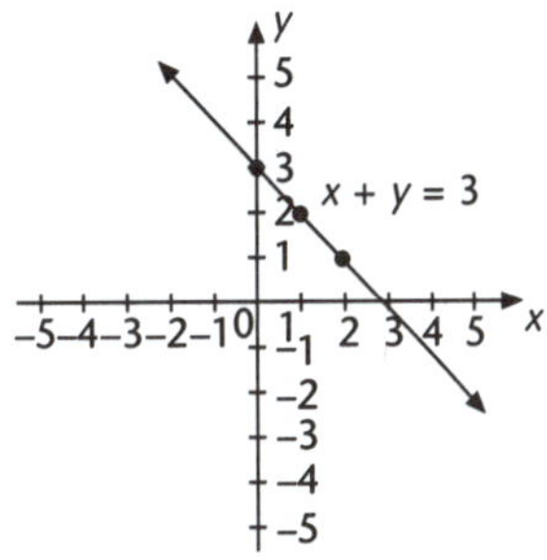


f $x - y = 0$

| $x$ | 0 | 1 | 2 |
|---|---|---|---|
| $y$ | 0 | 1 | 2 |

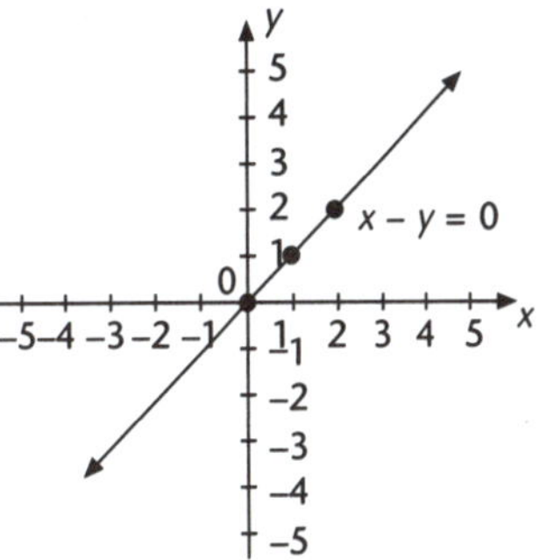


**12** Substitute $x = 3$, $y = 1$ in $y = x - 2$

$\therefore y = x - 2$

$1 = 3 - 2$

$1 = 1$ Yes!

$\therefore (3, 1)$ lies on $y = x - 2$

# WORKED SOLUTIONS to Chapter 5 **Practise, Practise**

**13** Substitute $x = 2$, $y = 4$ in $y = 3x + 1$

$\therefore y = 3x + 1$
$4 = 3(2) + 1$
$4 = 6 + 1$
$4 = 7$ No!

$\therefore$ (2, 4) does not lie on $y = 3x + 1$

**14** Substitute each point into $y = 3x - 2$:

**A** For (4, 1)
$\therefore 1 = 3(4) - 2$
$1 = 12 - 2$
$1 = 10$ No!

**B** For (2, 0):
$\therefore 0 = 3(2) - 2$
$0 = 6 - 2$
$0 = 4$ No!

**C** For (–1, 1):
$\therefore 1 = 3(-1) - 2$
$1 = -3 - 2$
$1 = -5$ No!

**D** For (3, 7):
$\therefore 7 = 3(3) - 2$
$7 = 9 - 2$
$7 = 7$ Yes!

$\therefore$ The answer is *D*.

Note: much of this could be done in your head.

**15** Substitute (1, 5) into each equation:

**A** $y = x + 2$
$5 = 1 + 2$ No!

**B** $y = 3 - x$
$5 = 3 - 1$ No!

**C** $y = 2x + 3$
$5 = 2(1) + 3$
$5 = 2 + 3$
$5 = 5$ Yes!

**D** We'll just check!
$y = 3x - 1$
$5 = 3(1) - 1$
$5 = 3 - 1$ No!

$\therefore$ The answer is *C*.

**16** For *A*(1, 2):
$y = 2x + 1$
$2 = 2(1) + 1$
$2 = 2 + 1$ No!

For *B*(2, 5):
$y = 2x + 1$
$5 = 2(2) + 1$
$5 = 4 + 1$
$5 = 5$ Yes !

For –(0, 1):
$y = 2x + 1$
$1 = 2(0) + 1$
$1 = 0 + 1$
$1 = 1$ Yes!

For *D*(1, 3):
$y = 2x + 1$
$3 = 2(1) + 1$
$3 = 2 + 1$
$3 = 3$ Yes!

$\therefore$ The collinear points are *B*, *C*, *D*.

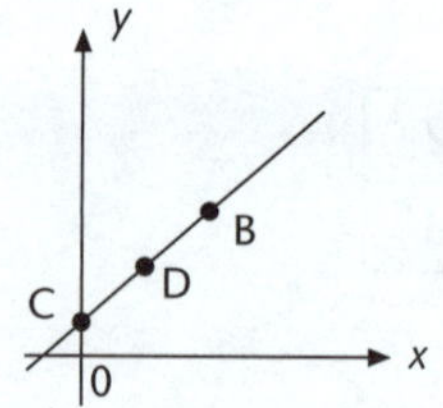


**17** Substitute (1, 5) in each line:

$y = 3x + 2$
$5 = 3(1) + 2$
$5 = 3 + 2$
$5 = 5$ Yes!

$y = 4x$
$5 = 4(1)$ No!

$y = 6 - x$
$5 = 6 - 1$
$5 = 5$ Yes!

$y = 6x - 1$
$5 = 6 - 1$
$5 = 5$ Yes!

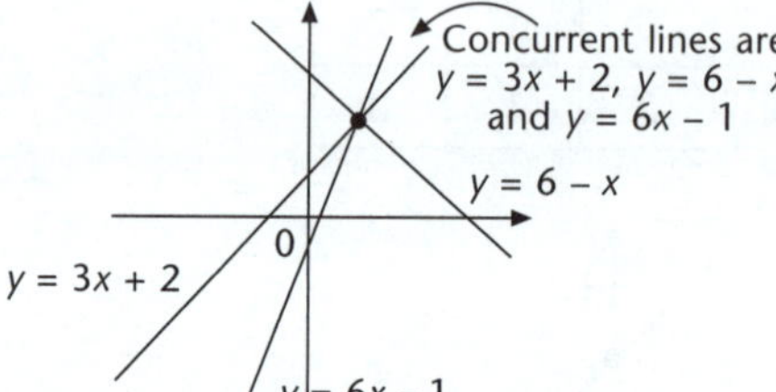


**18**

For $x = -5$: (–5, 1), (–5, 3), … etc.
For $y = 3$: (2, 3), (0, 3), … etc.

**19**

**a** Rectangle

**b** Area $= 3 \times 4$
$= 12$

$\therefore$ The area is 12 units$^2$.

**20** **a**

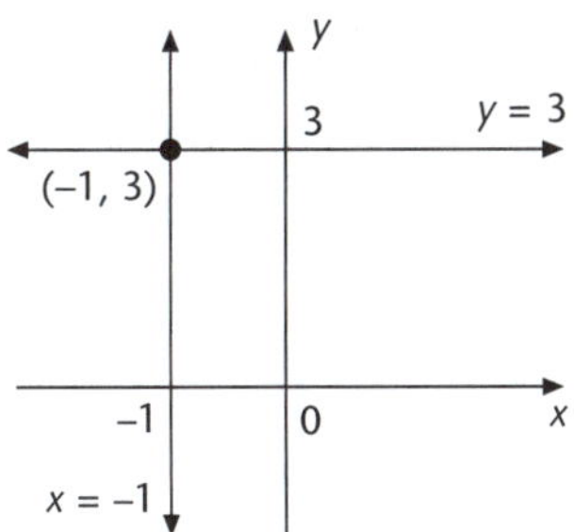


The point of intersection is (−1, 3).

**b** $y = x + 1$

| $x$ | 0 | 1 | 2 |
|---|---|---|---|
| $y$ | 1 | 2 | 3 |

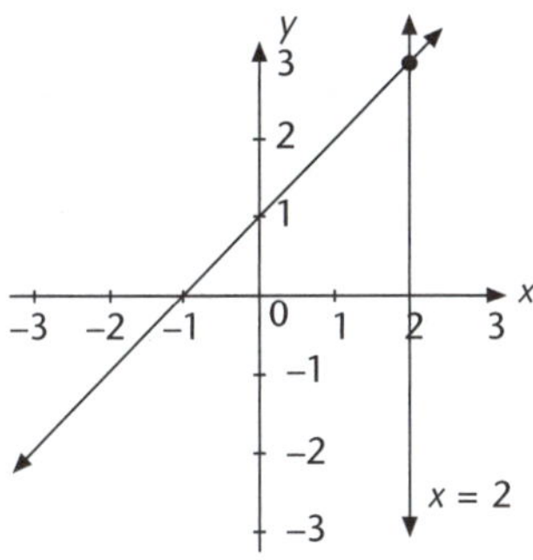


The point of intersection is (2, 3).

**c** $y = x$, $y = 2x - 2$

$y = x$

| $x$ | 0 | 1 | 2 |
|---|---|---|---|
| $y$ | 0 | 1 | 2 |

$y = 2x - 2$

| $x$ | 0 | 1 | 2 |
|---|---|---|---|
| $y$ | −2 | 0 | 2 |

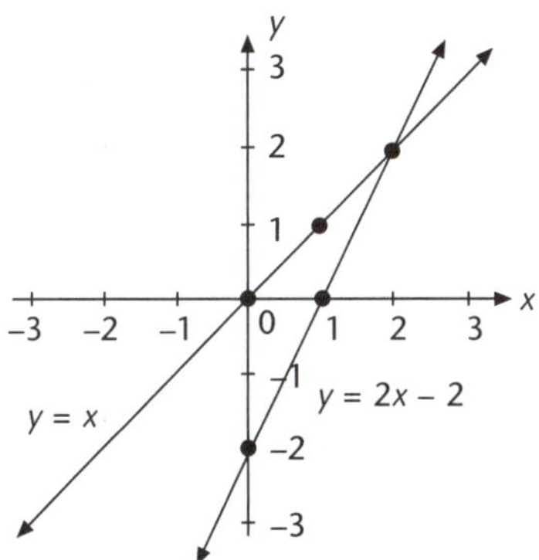


∴ The point of intersection is (2, 2).

**d** $y = 3x - 1$, $y = 3 - x$

$y = 3x - 1$

| $x$ | 0 | 1 | 2 |
|---|---|---|---|
| $y$ | −1 | 2 | 5 |

$y = 3 - x$

| $x$ | 0 | 1 | 2 |
|---|---|---|---|
| $y$ | 3 | 2 | 1 |

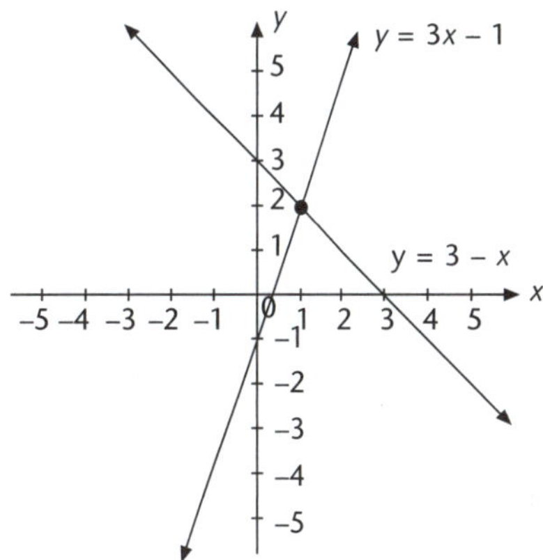


∴ The point of intersection is (1, 2).

**21** **a** $y = 2x$

| $x$ | 0 | 1 | 2 |
|---|---|---|---|
| $y$ | 0 | 1 | 2 |

**b** $y = -\frac{1}{2}x$

| $x$ | 0 | 1 | 2 |
|---|---|---|---|
| $y$ | 0 | $-\frac{1}{2}$ | 2 |

**c** $y = \frac{1}{2}x$

| $x$ | 0 | 1 | 2 |
|---|---|---|---|
| $y$ | 0 | $\frac{1}{2}$ | 1 |

**d** $y = 2x - 1$

| $x$ | 0 | 1 | 2 |
|---|---|---|---|
| $y$ | −1 | 1 | 3 |

**e** $y = -\frac{1}{2}x + 2$

| $x$ | 0 | 1 | 2 |
|---|---|---|---|
| $y$ | 0 | $1\frac{1}{2}$ | 2 |

**f** $2x - y = 2$

| $x$ | 0 | 1 | 2 |
|---|---|---|---|
| $y$ | −2 | 0 | 2 |

- Lines **a**, **d** and **f** are parallel to each other.
- Lines **b**, **e** are parallel to each other.
- Also lines **b** and **e** are perpendicular to **a**, **d** and **f**, while **c** is neither parallel nor perpendicular.

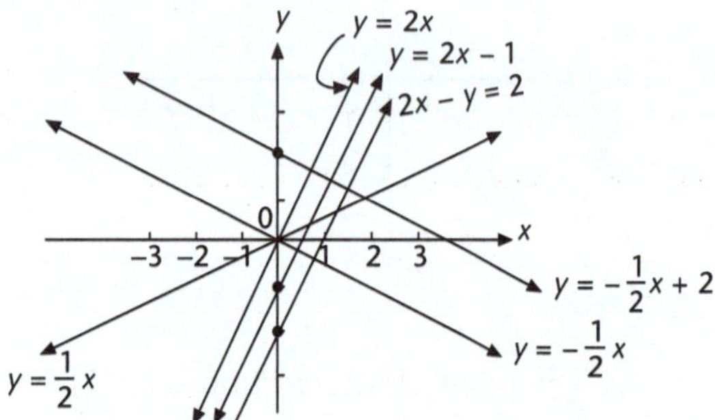


**a** Gradient = 2

**b** Gradient = $-\frac{1}{2}$

**c** Gradient = $\frac{1}{2}$

**d** Gradient = 2

**e** Gradient = $-\frac{1}{2}$

**f** $2x - y = 2$
$\therefore y = 2x - 2$
Gradient = 2

**22** **a**

| $n$ | 0 | 4 | 8 | 12 | 16 | 20 |
|---|---|---|---|---|---|---|
| $s$ | 0 | 320 | 640 | 960 | 1280 | 1600 |

**b**

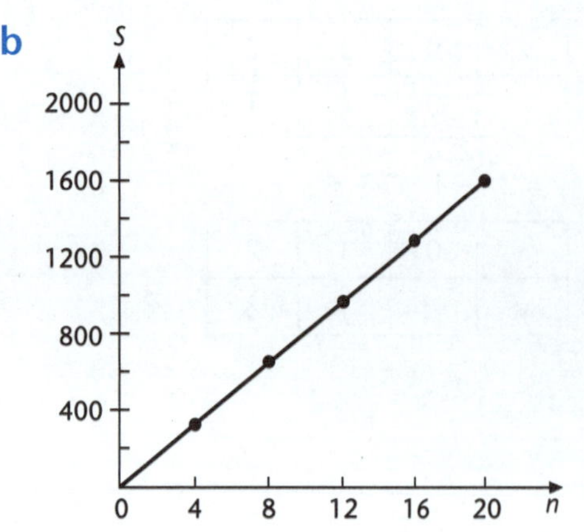


**23** **a**

| $n$ | 0 | 10 | 20 | 30 | 40 |
|---|---|---|---|---|---|
| $c$ | 200 | 300 | 400 | 500 | 600 |

**b**

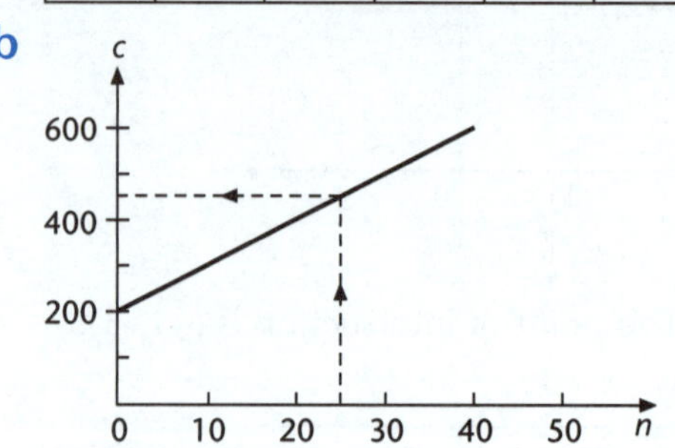


**c** Cost is \$450

## Chapter 6—Equations and Formulae

pp. 99–104

**1** **a** $7 \times (3) - 15 = 21 - 15$
$= 6$
$\therefore x = 3$ is a solution.

**b** LHS $= 11 - 2 \times 3$
$= 5$
RHS $= 7 \times 3 - 16$
$= 5$
LHS = RHS
$\therefore x = 3$ is a solution.

**2** **a** $8y - 5 = 8 \times 2 - 5$
$= 16 - 5$
$= 11$
$y = 2$ is a solution.

**b** $3(y + 2) = 3 \times (7 + 2)$
$= 27$
$\therefore y = 7$ is a solution.

**c** $30 - 4x = 30 - 4 \times (6)$
$= 30 - 24$
$= 6 \neq 8$
$\therefore x = 6$ is not a solution.

**d** $\frac{x}{5} - 11 = \frac{10}{5} - 11$
$= 2 - 11$
$= -9$
$\therefore x = 10$ is a solution.

**3** **a** $y + 7 = 11$
$y = 11 - 7$
$y = 4$

**b** $y + 11 = 7$
$y = 7 - 11$
$y = -4$

**c** $y - 11 = 7$
$y = 7 + 11$
$y = 18$

**d** $y - 7 = 11$
$y = 11 + 7$
$y = 18$

**e** $y + 17 = 4$
$y = 4 - 17$
$y = -13$

**f** $15 + t = 4$
$t = 4 - 15$
$t = -11$

**g** $19 = 11 + m$
$19 - 11 = m$
$8 = m$
$m = 8$

**h** $t - 5 = -2$
$t = -2 + 5$
$t = 3$

**i** $p + 13 = 24$
$p = 24 - 13$
$p = 11$

**j** $m - 17 = 4$
$m = 4 + 17$
$m = 21$

**k** $n + 31 = 14$
$n = 14 - 31$
$n = -17$

**l** $11 - t = 4$
$-t = 4 - 11$
$-t = -7$ [× both sides by –1]
$t = 7$

**m** $15 - y = 17$
$-y = 17 - 15$
$-y = 2$ [× both sides by –1]
$\therefore\ y = -2$

**n** $4a = 24$
$a = \frac{24}{4}$
$= 6$

**o** $12t = 72$
$t = \frac{72}{12}$
$= 6$

**p** $6x = 48$
$x = \frac{48}{6}$
$= 8$

**q** $3a = -27$
$a = \frac{-27}{3}$
$= -9$

**r** $7t = -35$
$t = \frac{-35}{7}$
$t = -5$

**s** $-4y = 16$
$y = \frac{16}{-4}$
$y = -4$

**t** $-8t = -64$
$t = \frac{-64}{-8}$
$t = 8$

**u** $\frac{t}{6} = 12$
$t = 12 \times 6$
$= 72$

**v** $\frac{z}{3} = 9$
$z = 9 \times 3$
$z = 27$

**w** $\frac{a}{5} = -6$
$a = -6 \times 5$
$a = -30$

**x** $\frac{n}{-8} = 6$
$n = 6 \times -8$
$n = -48$

**y** $\frac{m}{-7} = -4$
$m = -4 \times -7$
$m = 28$

**z** $27 = 9x$
$9x = 27$
$x = \frac{27}{9}$
$x = 3$

**4** **a** $4y = 9$
$y = \frac{9}{4}$
$= 2\frac{1}{4}$

**b** $6a = 15$
$a = \frac{15}{6}$
$= 2\frac{1}{2}$

**c** $9t = 21$
$t = \frac{21}{9}$
$= 2\frac{1}{3}$

**d** $5a = 14$
$a = \frac{14}{5}$
$= 2\frac{4}{5}$

**e** $5y = 8$
$y = \frac{8}{5}$
$= 1\frac{3}{5}$

**f** $7m = 30$
$m = \frac{30}{7}$
$= 4\frac{2}{7}$

**g** $9a = 24$
$a = \frac{24}{9}$
$= 2\frac{2}{3}$

**h** $8a = 14$
$a = \frac{14}{8}$
$= 1\frac{3}{4}$

**i** $4a = 3$
$a = \frac{3}{4}$

**j** $15t = 10$
$t = \frac{10}{15}$
$= \frac{2}{3}$

**k** $12y = 30$
$y = \frac{30}{12}$
$= 2\frac{1}{2}$

**l** $-8a = 10$
$a = \frac{10}{-8}$
$= -1\frac{1}{4}$

**m** $6y = -20$
$y = \frac{-20}{6}$
$= -3\frac{1}{3}$

**n** $-8t = -20$
$t = \frac{-20}{-8}$
$= 2\frac{1}{2}$

**o** $24z = 30$
$z = \frac{30}{24}$
$= 1\frac{1}{4}$

**p** $15m = 25$
$m = \frac{25}{15}$
$= 1\frac{2}{3}$

**q** $3y = 0$
$y = \frac{0}{3}$
$= 0$

**5** **a** $8a + 3 = 27$
$8a = 27 - 3$
$8a = 24$
$a = \frac{24}{8}$
$a = 3$

**b** $6x - 3 = 15$
$6x = 15 + 3$
$6x = 18$
$x = \frac{18}{6}$
$x = 3$

## WORKED SOLUTIONS to Chapter 6 **Practise, Practise**

**c** $4y + 13 = 29$
$4y = 29 - 13$
$4y = 16$
$y = \frac{16}{4}$
$y = 4$

**d** $7t - 15 = 13$
$7t = 13 + 15$
$7t = 28$
$t = \frac{28}{7}$
$t = 4$

**e** $11t - 3 = 30$
$11t = 30 + 3$
$11t = 33$
$t = \frac{33}{11}$
$t = 3$

**f** $9y + 17 = 53$
$9y = 53 - 17$
$9y = 36$
$y = \frac{36}{9}$
$y = 4$

**g** $2t + 7 = 21$
$2t = 21 - 7$
$2t = 14$
$t = \frac{14}{2}$
$t = 7$

**h** $13 + 9y = 40$
$9y = 40 - 13$
$9y = 27$
$y = \frac{27}{9}$
$y = 3$

**i** $11n - 31 = 13$
$11n = 13 + 31$
$11n = 44$
$n = \frac{44}{11}$
$n = 4$

**j** $7a = 3a + 24$
$7a - 3a = 24$
$4a = 24$
$a = \frac{24}{4}$
$a = 6$

**k** $5t = t + 36$
$5t - t = 36$
$4t = 36$
$t = \frac{36}{4}$
$t = 9$

**l** $13y = 8y - 35$
$13y - 8y = -35$
$5y = -35$
$y = \frac{-35}{5}$
$y = -7$

**m** $3m = 7m + 32$
$3m - 7m = 32$
$-4m = 32$
$m = \frac{32}{-4}$
$m = -8$

**n** $5y = 9y - 16$
$5y - 9y = -16$
$-4y = -16$
$y = \frac{-16}{-4}$
$y = 4$

**o** $\frac{a}{2} - 3 = 7$
$\frac{a}{2} = 7 + 3$
$\frac{a}{2} = 10$
$a = 10 \times 2$
$= 20$

**p** $7a + 21 = 14$
$7a = 14 - 21$
$7a = -7$
$a = \frac{-7}{7}$
$a = -1$

**q** $3t + 17 = 2$
$3t = 2 - 17$
$3t = -15$
$t = \frac{-15}{3}$
$t = -5$

**r** $11n + 47 = 3$
$11n = 3 - 47$
$11n = -44$
$n = \frac{-44}{11}$
$n = -4$

**s** $15 - 7t = 29$
$-7t = 29 - 15$
$-7t = 14$
$t = \frac{14}{-7}$
$t = -2$

**t** $11 - 3n = 2$
$-3n = 2 - 11$
$-3n = -9$
$n = \frac{-9}{-3}$
$n = 3$

**6** **a** $5a - 2 = 4$
$5a = 4 + 2$
$5a = 6$
$a = \frac{6}{5}$
$a = 1\frac{1}{5}$

**b** $2y + 7 = 12$
$2y = 12 - 7$
$2y = 5$
$y = \frac{5}{2}$
$= 2\frac{1}{2}$

**c** $7t + 13 = 24$
$7t = 24 - 13$
$7t = 11$
$t = \frac{11}{7}$
$= 1\frac{4}{7}$

**d** $5a + 11 = 17$
$5a = 17 - 11$
$5a = 6$
$a = \frac{6}{5}$
$= 1\frac{1}{5}$

**e** $4t - 9 = 6$
$4t = 6 + 9$
$4t = 15$
$t = \frac{15}{4}$
$= 3\frac{3}{4}$

**f** $11 + 4t = 25$
$4t = 25 - 11$
$4t = 14$
$t = \frac{14}{4}$
$= 3\frac{1}{2}$

**g** $27 + 2m = 24$
$2m = 24 - 27$
$2m = -3$
$m = \frac{-3}{2}$
$= -1\frac{1}{2}$

**h** $8a - 13 = 4$
$8a = 4 + 13$
$8a = 17$
$a = \frac{17}{8}$
$a = 2\frac{1}{8}$

**i** $6 - 3t = 17$
$-3t = 17 - 6$
$-3t = 11$
$t = \frac{11}{-3}$
$= -3\frac{2}{3}$

**j** $12y - 15 = 7$
$12y = 7 + 15$
$12y = 22$
$y = \frac{22}{12}$
$= 1\frac{5}{6}$

**k** $30 - 2y = 25$
$-2y = 25 - 30$
$-2y = -5$
$y = \frac{-5}{-2}$
$= 2\frac{1}{2}$

**l** $24t + 15 = 47$
$24t = 47 - 15$
$24t = 32$
$t = \frac{32}{24}$
$= 1\frac{1}{3}$

**m**
$$\begin{aligned} 13 &= 8t + 7 \\ 13t - 8t &= 7 \\ 5t &= 7 \\ t &= \tfrac{7}{5} \\ &= 1\tfrac{2}{5} \end{aligned}$$

**n**
$$\begin{aligned} 5m &= 7m + 9 \\ 5m - 7m &= 9 \\ -2m &= 9 \\ m &= \tfrac{9}{-2} \\ &= -4\tfrac{1}{2} \end{aligned}$$

**7** **a**
$$\begin{aligned} 5a + 14 &= 3a \\ 5a &= 3a - 14 \\ 5a - 3a &= -14 \\ 2a &= -14 \\ a &= \tfrac{-14}{2} \\ &= -7 \end{aligned}$$

**b**
$$\begin{aligned} 3y - 2 &= y + 8 \\ 3y &= y + 8 + 2 \\ 3y &= y + 10 \\ 3y - y &= 10 \\ 2y &= 10 \\ y &= \tfrac{10}{2} \\ &= 5 \end{aligned}$$

**c**
$$\begin{aligned} 8y + 13 &= 5y + 26 \\ 8y &= 5y + 26 - 13 \\ &= 5y + 13 \\ 8y - 5y &= 13 \\ 3y &= 13 \\ y &= \tfrac{13}{3} \\ &= 4\tfrac{1}{3} \end{aligned}$$

**d**
$$\begin{aligned} 9n - 4 &= 5n + 16 \\ 9n &= 5n + 16 + 4 \\ &= 5n + 20 \\ 9n - 5n &= 20 \\ 4n &= 20 \\ n &= \tfrac{20}{4} \\ &= 5 \end{aligned}$$

**e**
$$\begin{aligned} 11a - 17 &= 6a + 8 \\ 11a &= 6a + 8 + 17 \\ &= 6a + 25 \\ 11a - 6a &= 25 \\ 5a &= 25 \\ a &= \tfrac{25}{5} \\ &= 5 \end{aligned}$$

**f**
$$\begin{aligned} 8y - 19 &= 5y - 7 \\ 8y &= 5y - 7 + 19 \\ &= 5y + 12 \\ 8y - 5y &= 12 \\ 3y &= 12 \\ y &= \tfrac{12}{3} \\ &= 4 \end{aligned}$$

**g**
$$\begin{aligned} 13t - 19 &= 7t - 1 \\ 13t &= 7t - 1 + 19 \\ &= 7t + 18 \\ 13t - 7t &= 18 \\ 6t &= 18 \\ t &= \tfrac{18}{6} \\ &= 3 \end{aligned}$$

**h**
$$\begin{aligned} 10y + 21 &= 6y + 1 \\ 10y &= 6y + 1 - 21 \\ &= 6y - 20 \\ 10y - 6y &= -20 \\ 4y &= -20 \\ y &= \tfrac{-20}{4} \\ &= -5 \end{aligned}$$

**i**
$$\begin{aligned} 30y + 27 &= 15y + 12 \\ 30y &= 15y + 12 - 27 \\ &= 15y - 15 \\ 30y - 15y &= -15 \\ 15y &= -15 \\ y &= \tfrac{-15}{15} \\ &= -1 \end{aligned}$$

**j**
$$\begin{aligned} 13m - 17 &= 6m + 25 \\ 13m &= 6m + 25 + 17 \\ &= 6m + 42 \\ 13m - 6m &= 42 \\ 7m &= 42 \\ m &= \tfrac{42}{7} \\ &= 6 \end{aligned}$$

**k**
$$\begin{aligned} 5 + 3t &= 17 - t \\ 3t &= 17 - 5 - t \\ &= 12 - t \\ 3t + t &= 12 \\ 4t &= 12 \\ t &= \tfrac{12}{4} \\ &= 3 \end{aligned}$$

## WORKED SOLUTIONS to Chapter 6 **Practise, Practise**

**l** $3 - 5a = 17 - 2a$
$-5a = 17 - 3 - 2a$
$= 14 - 2a$
$-5a + 2a = 14$
$-3a = 14$
$a = \frac{14}{-3}$
$= -4\frac{2}{3}$

**m** $5y - 32 = 4 - 3y$
$5y = 4 + 32 - 3y$
$= 36 - 3y$
$5y + 3y = 36$
$8y = 36$
$y = \frac{36}{8}$
$= 4\frac{1}{2}$

**n** $25 - 7a = 27 - 4a$
$-7a = 27 - 25 - 4a$
$= 2 - 4a$
$-7a + 4a = 2$
$-3a = 2$
$a = -\frac{2}{3}$

**8** **a** $6x - 18 = 12$
$6x = 12 + 18$
$6x = 30$
$x = \frac{30}{6}$
$= 5$

**b** $5a + 10 = 35$
$5a = 35 - 10$
$5a = 25$
$a = \frac{25}{5}$
$= 5$

**c** $4t - 20 = 24$
$4t = 24 + 20$
$4t = 44$
$\therefore \quad t = \frac{44}{4}$
$t = 11$

**d** $4a + 12 = 32$
$4a = 32 - 12$
$4a = 20$
$a = \frac{20}{4}$
$= 5$

**e** $6y - 42 = 12$
$6y = 12 + 42$
$6y = 54$
$y = \frac{54}{6}$
$= 9$

**f** $6m + 15 = 33$
$6m = 33 - 15$
$6m = 18$
$m = \frac{18}{6}$
$= 3$

**g** $5t - 30 = 15$
$5t = 15 + 30$
$5t = 45$
$t = \frac{45}{5}$
$= 9$

**h** $6a + 18 - 11 = 5$
$6a + 7 = 5$
$6a = 5 - 7$
$6a = -2$
$a = \frac{-2}{6}$
$= -\frac{1}{3}$

**i** $8y - 16 = 5y + 15$
$8y = 5y + 15 + 16$
$= 5y + 31$
$8y - 5y = 31$
$3y = 31$
$y = \frac{31}{3}$
$= 10\frac{1}{3}$

**j** $6t + 18 + 2t - 8 = 26$
$8t + 10 = 26$
$8t = 26 - 10$
$= 16$
$t = \frac{16}{8}$
$= 2$

**k** $9(t + 3) - 4(t - 2) = 6$
$9t + 27 - 4t + 8 = 6$
$5t + 35 = 6$
$5t = 6 - 35$
$= -29$
$t = \frac{-29}{5}$
$= -5\frac{4}{5}$

**l** $8(y-6)-6(y-8)=24$
$8y-48-6y+48=24$
$2y=24$
$y=\frac{24}{2}$
$y=12$

**m** $7a-4(a-2)=0$
$7a-4a+8=0$
$3a+8=0$
$3a=-8$
$a=\frac{-8}{3}$
$a=-2\frac{2}{3}$

**n** $15a-7(a-2)-5(a-3)=21$
$15a-7a+14-5a+15=21$
$3a+29=21$
$3a=21-29$
$3a=-8$
$a=\frac{-8}{3}$
$a=-2\frac{2}{3}$

**o** $7a-1(5a-2)=26$
$7a-5a+2=26$
$2a+2=26$
$2a=26-2$
$2a=24$
$a=\frac{24}{2}$
$a=12$

**p** $3a-1(7-2a)=42$
$3a-7+2a=42$
$5a-7=42$
$5a=42+7$
$=49$
$a=\frac{49}{5}$
$a=9\frac{4}{5}$

**q** $11-7(2-a)=18$
$11-14+7a=18$
$-3+7a=18$
$7a=18+3$
$=21$
$a=\frac{21}{7}$
$a=3$

**r** $3(3x+1)-2(x-3)=5(x+7)$
$9x+3-2x+6=5x+35$
$7x+9=5x+35$
$7x=5x+35-9$
$7x=5x+26$
$7x-5x=26$
$2x=26$
$x=\frac{26}{2}$
$x=13$

**s** $3(2x+5)-4(x-3)=3(5x+1)-2$
$6x+15-4x+12=15x+3-2$
$2x+27=15x+1$
$2x=15x+1-27$
$2x=15x-26$
$2x-15x=-26$
$-13x=-26$
$x=\frac{-26}{-13}$
$=2$

**t** $11(a-2)-4(2-a)-2(1-4a)$
$=19(a-3)$
$11a-22-8+4a-2+8a$
$=19a-57$
$23a-32=19a-57$
$23a=19a-57+32$
$=19a-25$
$23a-19a=-25$
$4a=-25$
$a=\frac{-25}{4}$
$a=-6\frac{1}{4}$

**9** **a** $\frac{(a-2)}{\cancel{5}}\times\cancel{5}=3\times 5$
$a-2=15$
$a=15+2$
$=17$

**b** $\cancel{3}\times\frac{(y+2)}{\cancel{3}}=4\times 3$
$y+2=12$
$y=12-2$
$=10$

**c** $\cancel{4}\times\frac{(3-2a)}{\cancel{4}}=5\times 4$
$3-2a=20$
$-2a=20-3$
$=17$
$a=\frac{17}{-2}$
$=-8\frac{1}{2}$

**d** $\cancel{3} \times \frac{(2a+5)}{\cancel{3}} = 7 \times 3$
$2a + 5 = 21$
$2a = 21 - 5$
$= 16$
$a = \frac{16}{2}$
$= 8$

**e** $\cancel{3} \times \frac{5(a-2)}{\cancel{3}} = 2 \times 3$
$5(a-2) = 6$
$5a - 10 = 6$
$5a = 6 + 10$
$= 16$
$a = \frac{16}{5}$
$a = 3\frac{1}{5}$

**f** $\cancel{5} \times \frac{4}{\cancel{5}}(x+3) = 6 \times 5$
$4(x+3) = 30$
$4x + 12 = 30$
$4x = 30 - 12$
$= 18$
$x = \frac{18}{4}$
$= 4\frac{1}{2}$

**10** **a** $\cancel{15}^5 \times \frac{y}{\cancel{3}} + \cancel{15}^3 \times \frac{y}{\cancel{5}} = 16 \times 15$
$5y + 3y = 240$
$8y = 240$
$y = \frac{240}{8}$
$= 30$

**b** $\cancel{15}^5 \times \frac{a}{\cancel{3}} - \cancel{15}^3 \times \frac{a}{\cancel{5}} = 12 \times 15$
$5a - 3a = 180$
$2a = 180$
$a = \frac{180}{2}$
$= 90$

**c** $\cancel{4} \times \frac{3y}{\cancel{4}} - \cancel{4}^2 \times \frac{y}{\cancel{2}} = 5 \times 4$
$3y - 2y = 20$
$y = 20$

**d** $2t + \cancel{2} \times \frac{t}{\cancel{2}} = 6 \times 2$
$2t + t = 12$
$3t = 12$
$t = \frac{12}{3}$
$t = 4$

**e** $4 \times a - \cancel{4} \times \frac{a}{\cancel{4}} = 9 \times 4$
$4a - a = 36$
$3a = 36$
$a = \frac{36}{3}$
$= 12$

**f** $\cancel{15}^3 \times \frac{a}{\cancel{5}} + \cancel{15}^5 \times \frac{(a-2)}{\cancel{3}} = 4 \times 15$
$3a + 5(a-2) = 60$
$3a + 5a - 10 = 60$
$8a - 10 = 60$
$8a = 60 + 10$
$= 70$
$a = \frac{70}{8}$
$a = 8\frac{3}{4}$

**g** $\cancel{12}^3 \times \frac{(m+2)}{\cancel{4}} - \cancel{12}^4 \times \frac{m}{\cancel{3}} = 2 \times 12$
$3(m+2) - 4m = 24$
$3m + 6 - 4m = 24$
$-m + 6 = 24$
$-m = 24 - 6$
$-m = 18$
$m = -18$

**h** $3 \times t + \cancel{3} \times \frac{(t-3)}{\cancel{3}} = 5 \times 3$
$3t + t - 3 = 15$
$4t - 3 = 15$
$4t = 15 + 3$
$4t = 18$
$t = \frac{18}{4}$
$t = 4\frac{1}{2}$

## WORKED SOLUTIONS to Chapter 6 **Practise, Practise**

**i**
$$\begin{aligned}
3 \times m - \cancel{3} \times \frac{(m-5)}{\cancel{3}} &= 5 \times 3 \\
3m - 1(m-5) &= 15 \\
3m - m + 5 &= 15 \\
2m + 5 &= 15 \\
2m &= 15 - 5 \\
&= 10 \\
m &= \tfrac{10}{2} \\
&= 5
\end{aligned}$$

**j**
$$\begin{aligned}
\cancel{12}^{4}\frac{(a+2)}{\cancel{3}} + \cancel{12}^{3}\frac{(a+3)}{\cancel{4}} &= 8 \times 12 \\
4(a+2) + 3(a+3) &= 96 \\
4a + 8 + 3a + 9 &= 96 \\
7a + 17 &= 96 \\
7a &= 96 - 17 \\
&= 79 \\
a &= \tfrac{79}{7} \\
a &= 11\tfrac{2}{7}
\end{aligned}$$

**k**
$$\begin{aligned}
\cancel{6}^{2} \times \frac{(3y-2)}{\cancel{3}} - \cancel{6}^{3} \times \frac{(y-3)}{\cancel{2}} &= 4 \times 6 \\
2(3y-2) - 3(y-3) &= 24 \\
6y - 4 - 3y + 9 &= 24 \\
3y + 5 &= 24 \\
3y &= 24 - 5 \\
&= 19 \\
y &= \tfrac{19}{3} \\
y &= 6\tfrac{1}{3}
\end{aligned}$$

**l**
$$\begin{aligned}
\cancel{20}^{4} \times \frac{2}{\cancel{5}}(x+4) - \cancel{20}^{5} \times \frac{3}{\cancel{4}}(x-2) & \\
&= 3 \times 20 \\
8(x+4) - 15(x-2) &= 60 \\
8x + 32 - 15x + 30 &= 60 \\
-7x + 62 &= 60 \\
-7x &= 60 - 62 \\
&= -2 \\
x &= \tfrac{-2}{-7} \\
x &= \tfrac{2}{7}
\end{aligned}$$

**11** **a**
$$\begin{aligned}
35 \times 7 - \cancel{35}^{7} \times \frac{3x}{\cancel{5}} &= 35 \times 3 - \cancel{35}^{5} \times \frac{2x}{\cancel{7}} \\
245 - 21x &= 105 - 10x \\
-21x &= 105 - 245 - 10x \\
&= -140 - 10x \\
-21x + 10x &= -140 \\
-11x &= -140 \\
x &= \tfrac{-140}{-11} \\
&= 12\tfrac{8}{11}
\end{aligned}$$

**b**
$$\begin{aligned}
6(x-3) - \cancel{6}^{3}\frac{(x-2)}{\cancel{2}_1} - \cancel{6}^{2}\frac{(x-4)}{\cancel{3}_1} &= 1 \times 6 \\
6(x-3) - 3(x-2) - 2(x-4) &= 6 \\
6x - 18 - 3x + 6 - 2x + 8 &= 6 \\
x - 4 &= 6 \\
x &= 6 + 4 \\
x &= 10
\end{aligned}$$

**c**
$$\begin{aligned}
&\cancel{20}^{5} \times \frac{x}{\cancel{4}_1} - \cancel{20}^{4}\frac{(x-2)}{\cancel{5}_1} \\
&= 20 \times 15 + \cancel{20}^{10}\frac{(7-x)}{\cancel{2}_1}
\end{aligned}$$
$$\begin{aligned}
5x - 4(x-2) &= 300 + 10(7-x) \\
5x - 4x + 8 &= 300 + 70 - 10x \\
x + 8 &= 370 - 10x \\
x &= 370 - 8 - 10x \\
&= 362 - 10x \\
x + 10x &= 362 \\
11x &= 362 \\
x &= \tfrac{362}{11} \\
x &= 32\tfrac{10}{11}
\end{aligned}$$

**d**
$$\begin{aligned}
&\cancel{60}^{20} \times \frac{1}{\cancel{3}_1}(x-3) - \cancel{60}^{12} \times \frac{1}{\cancel{5}_1}(x-5) \\
&= \cancel{60}^{15} \times \frac{3}{\cancel{4}_1}(3x-4) + 5 \times 60
\end{aligned}$$
$$\begin{aligned}
20(x-3) - 12(x-5) & \\
&= 45(3x-4) + 300 \\
20x - 60 - 12x + 60 & \\
&= 135x - 180 + 300 \\
8x &= 135x + 120 \\
8x - 135x &= 120 \\
-127x &= 120 \\
x &= \tfrac{120}{-127} \\
x &= -\tfrac{120}{127}
\end{aligned}$$

**e** $\cancel{30}^{6}\,\frac{(3x+2)}{\cancel{5}_1} - 30 \times 3(x-2) =$

$8 \times 30 - \cancel{30}^{10} \times \frac{2(3x+1)}{\cancel{3}_1} - \cancel{30}^{5} \times 5\frac{(4x-2)}{\cancel{6}_1}$

$6(3x+2) - 90(x-2)$
$= 240 - 20(3x+1) - 25(4x-2)$

$18x + 12 - 90x + 180$
$= 240 - 60x - 20 - 100x + 50$

$-72x + 192 = 270 - 160x$

$-72x = 270 - 192 - 160x$

$-72x = 78 - 160x$

$-72x + 160x = 78$

$88x = 78$

$x = \frac{78}{88}$

$x = \frac{39}{44}$

**12** **a** $a + b$ **b** $ab$

**c** $y + 2$ **d** $y + 1$

**e** $p + (p + 1) = 2p + 1$ **f** $\frac{x}{1000}$

**g** $1000x$ **h** $2b - a$

**13** **a** Let number be $x$.

$x + 19 = 37$

$x = 37 - 19$

$= 18$

The number is 18.

**b** Let number by $y$.

$6y = 84$

$y = \frac{84}{6}$

$= 14$

The number is 14.

**c** Let number be $a$.

$2a + 7 = 35$

$2a = 35 - 7$

$= 28$

$a = \frac{28}{2}$

The number is 14.

**d** Let number be $n$.

$5n - 11 = 64$

$5n = 64 + 11$

$= 75$

$n = \frac{75}{5}$

$= 15$

The number is 15.

**e** Let number be $m$.

$76 - 3m = 52$

$-3m = 52 - 76$

$= -24$

$m = \frac{-24}{-3}$

$= 8$

The number is 8.

**14** **a** Let number be $n$.

$6(n + 3) = 2n + 30$

$6n + 18 = 2n + 30$

$6n = 2n + 30 - 18$

$= 2n + 12$

$6n - 2n = 12$

$4n = 12$

$n = \frac{12}{4}$

$= 3$

The number is 3.

**b** Let smallest number be $x$.
Then numbers are $x$, $x + 1$, $x + 2$

$x + (x + 1) + (x + 2) = 42$

$3x + 3 = 42$

$3x = 42 - 3$

$= 39$

$x = \frac{39}{3}$

$= 13$

The numbers are 13, 14, 15.

**c** Let smaller number $n$.
Numbers are $n$, $n + 2$

$n + (n + 2) = 64$

$2n + 2 = 64$

$2n = 64 - 2$

$= 62$

$n = \frac{62}{2}$

$= 31$

The numbers are 31, 33.

**d** Let number of bears that Tony has be $n$.
Then Mai has $3n$ bears

$n + 3n = 24$

$4n = 24$

$n = \frac{24}{4}$

$= 6$

Tony has 6 and Mai has 18 bears.

**e** Let amount Duey receives be $\$x$ million.

$\therefore$ Huey receives $\$3x$ million

$\therefore$ Louie $\$(2x - 1)$ million

$$\begin{aligned} x + 3x + (2x - 1) &= 107 \\ 6x - 1 &= 107 \\ 6x &= 107 + 1 \\ &= 108 \\ x &= \tfrac{108}{6} \\ &= 18 \end{aligned}$$

Then Duey receives \$18 million, Huey \$(3 × 18) = \$54 million and Louie \$(2 × 18 − 1) = \$35 million.

**15** **a** $P = 4y + y + 4y + y = 10y$

$$\begin{aligned} 10y &= 105 \\ y &= \tfrac{105}{10} \\ &= 10.5 \end{aligned}$$

**b**

$$\begin{aligned} P &= (5a-3)+2a+(5a-3)+2a \\ &= 14a - 6 \\ 14a - 6 &= 78 \\ 14a &= 78 + 6 \\ &= 84 \\ a &= \tfrac{84}{14} \\ &= 6 \end{aligned}$$

**c**

$$\begin{aligned} A &= 4(2x + 3) \\ &= 8x + 12 \\ 8x + 12 &= 84 \\ 8x &= 84 - 12 \\ &= 72 \\ x &= \tfrac{72}{8} \\ &= 9 \end{aligned}$$

**d**

$$\begin{aligned} A &= \tfrac{1}{2} \times 6 \times (5x - 3) \\ &= 3(5x - 3) \\ &= 15x - 9 \\ 15x - 9 &= 51 \\ 15x &= 51 + 9 \\ &= 60 \\ x &= \tfrac{60}{15} \\ &= 4 \end{aligned}$$

**e** $5x - 4 = 2x + 11$ (opposite sides equal)

$$\begin{aligned} 5x &= 2x + 11 + 4 \\ &= 2x + 15 \\ 5x - 2x &= 15 \\ 3x &= 15 \\ x &= \tfrac{15}{3} \\ &= 5 \end{aligned}$$

**f**

$$\begin{aligned} (4a - 7) + 2a &= 41 \\ 6a - 7 &= 41 \\ 6a &= 41 + 7 \\ &= 48 \\ a &= \tfrac{48}{6} \\ &= 8 \end{aligned}$$

**16** **a** $3x + 13 = 5x - 3$

(vertically opposite angles)

$$\begin{aligned} 3x &= 5x - 3 - 13 \\ &= 5x - 16 \\ 3x - 5x &= -16 \\ -2x &= -16 \\ x &= \tfrac{-16}{-2} \\ &= 8 \end{aligned}$$

**b** $(a + 39) + a = 180$

($\angle$ sum of straight line)

$$\begin{aligned} 2a + 39 &= 180 \\ 2a &= 180 - 39 \\ &= 141 \\ a &= \tfrac{141}{2} \\ &= 70.5 \end{aligned}$$

**c** $(2x - 13) = (5x - 28)$

(alternate $\angle$s parallel lines)

$$\begin{aligned} 2x &= 5x - 28 + 13 \\ &= 5x - 15 \\ 2x - 5x &= -15 \\ -3x &= -15 \\ x &= \tfrac{-15}{-3} \\ x &= 5 \end{aligned}$$

**d** $(3a - 17) + (2a + 27) = 180$

(co-interior $\angle$s, parallel lines)

$$\begin{aligned} 5a + 10 &= 180 \\ 5a &= 180 - 10 \\ &= 170 \\ a &= \tfrac{170}{5} \\ a &= 34 \end{aligned}$$

**17**

$$\begin{aligned} n + 30 &= 4n \\ n &= 4n - 30 \\ n - 4n &= -30 \\ -3n &= -30 \\ n &= \tfrac{-30}{-3} \\ &= 10 \end{aligned}$$

## WORKED SOLUTIONS to Chapter 6 **Practise, Practise**

**18** Let David's age be $d$.
$\therefore$ Allyn is $(d + 31)$ years old.
Then $d + (d + 31) = 49$

$$\begin{aligned} 2d + 31 &= 49 \\ 2d &= 49 - 31 \\ &= 18 \\ d &= \tfrac{18}{2} \\ &= 9 \end{aligned}$$

Allyn $= d + 31 = 9 + 31 = 40$.
Allyn is 40 years old.

**19** Let number of \$5 be $n$.
$\therefore$ Number of \$1 is $9n$ and number of \$2 is $8n$
Value of coins is:

$$\begin{aligned} \$(5 \times n + 1 \times 9n + 2 \times 8n) &= \$(5n + 9n + 16n) \\ &= \$30n \end{aligned}$$

Then

$$\begin{aligned} 30n &= 90 \\ n &= \tfrac{90}{30} \\ &= 3 \end{aligned}$$

Number of \$2 coins is $8n$
i.e. $8 \times 3 = 24$
There are 24 \$2 coins.

**20** Let number of ducks be $d$.
$\therefore$ Number of sheep is $3d$

$$\begin{aligned} \text{Number of legs} &= (2 \times d) + (4 \times 3d) \\ &= 14d \\ 14d &= 56 \\ d &= \tfrac{56}{14} \\ &= 4 \end{aligned}$$

Number of sheep $= 3d$
The number of sheep is 12.

**21** Let distance walked $= d$ km
$\therefore$ Distance cycled $= 4d$ km
$\therefore$ Distance run $= \frac{1}{2}d$ km
$\therefore$ Distance swum $= \frac{1}{4}d$ km

$$\begin{aligned} \text{Distance travelled} &= d + 4d + \tfrac{1}{2}d + \tfrac{1}{4}d \\ &= 5\tfrac{3}{4}d \\ &= \tfrac{23}{4}d \\ \therefore \quad \tfrac{23}{4}d &= 46 \\ \therefore \quad 23d &= 46 \times 4 \\ &= 184 \\ \therefore \quad d &= \tfrac{184}{23} \\ &= 8 \end{aligned}$$

Distance swum $= \frac{1}{4}d$
The distance swum is 2 km.

**22** Let number of lions be $x$.
$\therefore$ Number of monkeys $= x + 17$
Number of lizards $= x + 30$

$$\begin{aligned} \text{Total number} &= x + (x + 17) + (x + 30) \\ &= 3x + 47 \\ 3x + 47 &= 161 \\ 3x &= 161 - 47 \\ &= 114 \\ x &= \tfrac{114}{3} \\ &= 38 \end{aligned}$$

$\therefore$ Number of lions $= 38$
Number of monkeys $= 38 + 17 = 55$
Number of lizards $= 38 + 30 = 68$

**23** Let the trip be $3d$ km.
$\therefore$ Distance for each part is $d$ km.

$$\left[\text{Time taken} = \frac{\text{Distance}}{\text{Speed}}\right]$$

Time 1st part $= \dfrac{d}{4}$

Time 2nd part $= \dfrac{d}{8}$

Time 3rd part $= \dfrac{d}{6}$

Time taken $= \dfrac{d}{4} + \dfrac{d}{8} + \dfrac{d}{6} = 9\dfrac{3}{4}$

$$^{6}\cancel{24} \times \frac{d}{\cancel{4}_1} + {}^{3}\cancel{24} \times \frac{d}{\cancel{8}_1} + {}^{4}\cancel{24} \times \frac{d}{\cancel{6}_1} = 24 \times 9\frac{3}{4}$$

$$\begin{aligned} 6d + 3d + 4d &= 234 \\ 13d &= 234 \\ \frac{13d}{13} &= \frac{234}{13} \\ d &= 18 \end{aligned}$$

$$\begin{aligned} \text{Complete trip} &= 3d \text{ km} \\ &= 3 \times 18 \text{ km} \end{aligned}$$

The trip is 54 km.

**24** **a** 
$$\begin{aligned} v &= 70 + 8 \times 6 \\ &= 70 + 48 \\ &= 118 \end{aligned}$$

**b** 
$$\begin{aligned} y &= 3 \times 5 + (-4) \\ &= 15 + (-4) \\ &= 11 \end{aligned}$$

**c** $E = \frac{1}{2} \times \frac{1}{2} \times 8 \times 8$
$= \frac{1}{4} \times 64$
$= 16$

**d** $D = \frac{72}{16}$
$= 4.5$

**e** $l = \pi rs$
$= 3.14 \times 10 \times 5$
$= 157$

**f** $C = 2 \times \frac{22}{\cancel{7}_1} \times \cancel{28}^4$
$= 176$

**g** $I = 500 \times 0.09 \times 3$
$= 135$

**h** $S = \frac{31}{2}(17 + 35)$
$= \frac{31}{\cancel{2}_1} \times \cancel{52}^{26}$
$= 806$

**i** $T = -8 + (14 - 1) \times 2$
$= -8 + 26$
$= 18$

**j** $S = \frac{16}{2}(10 + [16 - 1]4)$
$= 8(10 + 15 \times 4)$
$= 8 \times 70$
$= 560$

**k** $C = \frac{5}{9}(50 - 32)$
$= \frac{5}{\cancel{9}_1} \times \cancel{18}^2$
$= 10$

**l** $F = 0.25 \times 8$
$= 2$

**m** $S = \frac{12}{1 - 0.5}$
$= \frac{12}{0.5}$
$= 24$

**n** $F = 32 + \frac{9}{\cancel{5}_1} \times \cancel{10}^2$
$= 32 + 18$
$= 50$

**o** $V = 3.14 \times 4^2 \times 8$
$= 401.92$

**25** **a** $m = \frac{15}{15-9}$
$= \frac{15}{6}$
$= 2\frac{1}{2}$

**b** $y = \frac{\frac{1}{4} \times 12}{12-6}$
$= \frac{3}{6}$
$= \frac{1}{2}$

**c** $A = lb$
$36 = 9b$
$b = \frac{36}{9}$
$= 4$

**d** $V = lbh$
$90 = 3 \times b \times 5$
$90 = 15b$
$b = \frac{90}{15}$
$= 6$

**e** $I = Prn$
$90 = 450 \times 0.1 \times n$
$90 = 45n$
$n = \frac{90}{45}$
$n = 2$

**f** $F = \dfrac{L}{L + S}$
$4 = \dfrac{7}{7 + S}$
$4(7 + S) = 7$
$28 + 4S = 7$
$4S = 7 - 28$
$= -21$
$S = \frac{-21}{4}$
$S = -5\frac{1}{4}$

**g** $T = a + (n - 1)d$
$71 = 17 + (7 - 1)d$
$71 = 17 + 6d$
$71 - 17 = 6d$
$6d = 54$
$d = \frac{54}{6}$
$d = 9$

**h** $A = \frac{1}{2}xy$
$81 = \frac{1}{\cancel{2}_1} \times x \times \cancel{18}^9$
$81 = 9x$
$x = \frac{81}{9}$
$x = 9$

# WORKED SOLUTIONS to Chapter 7 **Practise, Practise**

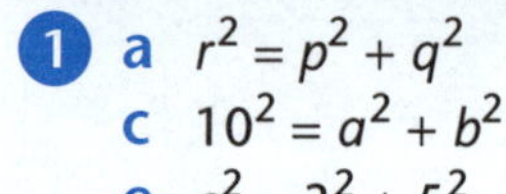

## Chapter 7—Pythagoras' Theorem

pp. 119–128

**1** a $r^2 = p^2 + q^2$  b $n^2 = m^2 + p^2$
c $10^2 = a^2 + b^2$  d $PR^2 = PQ^2 + QR^2$
e $c^2 = 3^2 + 5^2$  f $d^2 = 4^2 + 7^2$

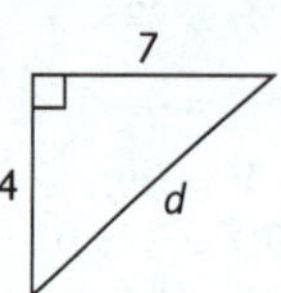


**2** a $d^2 = c^2 + e^2$  b $y^2 = x^2 + 5^2$
c $PQ^2 = PR^2 + RQ^2$  d $z^2 + 12^2 = 15^2$
$\therefore z^2 = 15^2 - 12^2$
e $w^2 + x^2 = y^2$  f $s^2 + 7^2 = t^2$

**3** a 49  b 121  c 289  d 841
e 961  f 12  g 30  h 16
i 25  j 130

**4** a
$$\begin{aligned} a^2 &= 6^2 + 8^2 \\ &= 36 + 64 \\ &= 100 \\ \therefore a &= \sqrt{100} \\ &= 10 \end{aligned}$$

b
$$\begin{aligned} b^2 &= 2^2 + 1.5^2 \\ &= 4 + 2.25 \\ &= 6.25 \\ \therefore b &= \sqrt{6.25} \\ &= 2.5 \end{aligned}$$

c
$$\begin{aligned} c^2 &= 8^2 + 15^2 \\ &= 64 + 225 \\ &= 289 \\ \therefore c &= \sqrt{289} \\ &= 17 \end{aligned}$$

d
$$\begin{aligned} d^2 &= 9^2 + 12^2 \\ &= 81 + 144 \\ &= 225 \\ \therefore d &= \sqrt{225} \\ &= 15 \end{aligned}$$

e
$$\begin{aligned} e^2 &= 15^2 + 20^2 \\ &= 225 + 400 \\ &= 625 \\ \therefore e &= \sqrt{625} \\ &= 25 \end{aligned}$$

f
$$\begin{aligned} f^2 &= 7^2 + 24^2 \\ &= 49 + 576 \\ &= 625 \\ \therefore f &= \sqrt{625} \\ &= 25 \end{aligned}$$

g
$$\begin{aligned} g^2 &= 9^2 + 40^2 \\ &= 81 + 1600 \\ &= 1681 \\ \therefore g &= \sqrt{1681} \\ &= 41 \end{aligned}$$

h
$$\begin{aligned} h^2 &= 10^2 + 24^2 \\ &= 100 + 576 \\ &= 676 \\ \therefore h &= \sqrt{676} \\ &= 26 \end{aligned}$$

i
$$\begin{aligned} i^2 &= 60^2 + 11^2 \\ &= 3600 + 121 \\ &= 3721 \\ \therefore i &= \sqrt{3721} \\ &= 61 \end{aligned}$$

j
$$\begin{aligned} j^2 &= 18^2 + 24^2 \\ &= 324 + 576 \\ &= 900 \\ \therefore j &= \sqrt{900} \\ &= 30 \end{aligned}$$

**5** a
$$\begin{aligned} x^2 + 3^2 &= 5^2 \\ \therefore x^2 &= 5^2 - 3^2 \\ &= 25 - 9 \\ &= 16 \\ \therefore x &= \sqrt{16} \\ &= 4 \end{aligned}$$

b
$$\begin{aligned} y^2 + 12^2 &= 13^2 \\ \therefore y^2 &= 13^2 - 12^2 \\ &= 169 - 144 \\ &= 25 \\ \therefore y &= \sqrt{25} \\ &= 5 \end{aligned}$$

c
$$\begin{aligned} t^2 + 24^2 &= 25^2 \\ \therefore t^2 &= 25^2 - 24^2 \\ &= 625 - 576 \\ &= 49 \\ \therefore t &= \sqrt{49} \\ &= 7 \end{aligned}$$

d
$$\begin{aligned} n^2 + 8^2 &= 17^2 \\ \therefore n^2 &= 17^2 - 8^2 \\ &= 289 - 64 \\ &= 225 \\ \therefore n &= \sqrt{225} \\ &= 15 \end{aligned}$$

e
$$\begin{aligned} m^2 + 40^2 &= 41^2 \\ \therefore m^2 &= 41^2 - 40^2 \\ &= 1681 - 1600 \\ &= 81 \\ \therefore m &= \sqrt{81} \\ &= 9 \end{aligned}$$

f
$$\begin{aligned} z^2 + 0.3^2 &= 0.5^2 \\ \therefore z^2 &= 0.5^2 - 0.3^2 \\ &= 0.25 - 0.09 \\ &= 0.16 \\ \therefore z &= \sqrt{0.16} \\ &= 0.4 \end{aligned}$$

g
$$\begin{aligned} p^2 + 0.8^2 &= 1^2 \\ \therefore p^2 &= 1^2 - 0.8^2 \\ &= 1 - 0.64 \\ &= 0.36 \\ \therefore p &= \sqrt{0.36} \\ &= 0.6 \end{aligned}$$

h Let AC = $x$ cm
$$\begin{aligned} \therefore x^2 + 16^2 &= 34^2 \\ \therefore x^2 &= 34^2 - 16^2 \\ &= 1156 - 256 \\ &= 900 \\ \therefore x &= \sqrt{900} \\ &= 30 \end{aligned}$$
Then AC = 30 cm.

**6** **a** $x^2 = 16^2 + 12^2$
$= 256 + 144$
$= 400$
$\therefore x = \sqrt{400}$
$= 20$

**b** $y^2 + 10^2 = 26^2$
$\therefore y^2 = 26^2 - 10^2$
$= 676 - 100$
$= 576$
$\therefore y = \sqrt{576}$
$= 24$

**c** $a^2 + 15^2 = 17^2$
$\therefore a^2 = 17^2 - 15^2$
$= 289 - 225$
$= 64$
$\therefore a = \sqrt{64}$
$= 8$

**d** $b^2 = 21^2 + 28^2$
$= 441 + 784$
$= 1225$
$\therefore b = \sqrt{1225}$
$= 35$

**e** $c^2 + 1.2^2 = 1.5^2$
$\therefore c^2 = 1.5^2 - 1.2^2$
$= 2.25 - 1.44$
$= 0.81$
$\therefore c = \sqrt{0.81}$
$= 0.9$

**f** $d^2 + 1.6^2 = 2^2$
$\therefore d^2 = 2^2 - 1.6^2$
$= 4 - 2.56$
$= 1.44$
$\therefore d = \sqrt{1.44}$
$= 1.2$

**g** $n^2 + 24^2 = 40^2$
$\therefore n^2 = 40^2 - 24^2$
$= 1600 - 576$
$= 1024$
$\therefore n = \sqrt{1024}$
$= 32$

**h** $q^2 + 14^2 = 50^2$
$\therefore q^2 = 50^2 - 14^2$
$= 2500 - 196$
$= 2304$
$\therefore q = \sqrt{2304}$
$= 48$

**i**



Let diagonal = $x$ cm
$\therefore x^2 = 1.5^2 + 2^2$
$= 2.25 + 4$
$= 6.25$
$\therefore x = \sqrt{6.25}$
$= 2.5$
The diagonal is 2.5 cm.

**j**



As OP = 25
then PR = 50
(diameter = 2 × radius)
Let PQ = $x$ cm
$\therefore x^2 + 30^2 = 50^2$
$x^2 = 50^2 - 30^2$
$= 2500 - 900$
$= 1600$
$\therefore x = \sqrt{1600}$
$= 40$
PQ is 40 cm.

**7** **a** To prove that:
$10^2 = 6^2 + 8^2$
LHS = 100
RHS = 36 + 64 = 100
$\therefore 10^2 = 6^2 + 8^2$
△ is right-angled.

**b** To prove that:
$14^2 = 12^2 + 9^2$
LHS = $14^2 = 196$
RHS = $12^2 + 9^2$
$= 144 + 81$
$= 225$
$\therefore 14^2 \neq 12^2 + 9^2$
△ is not right-angled.

**c** To prove that:

$$18^2 = 15^2 + 8^2$$
$$\text{LHS} = 18^2 = 324$$
$$\text{RHS} = 15^2 + 8^2 = 225 + 64 = 289$$
$$\therefore 18^2 \neq 15^2 + 8^2$$

△ is not right-angled.

**d** To prove that:

$$25^2 = 24^2 + 7^2$$
$$\text{LHS} = 625$$
$$\text{RHS} = 24^2 + 7^2 = 576 + 49 = 625$$
$$\therefore 25^2 = 24^2 + 7^2$$

△ is right-angled.

**e** To prove that:

$$6.1^2 = 6^2 + 1.1^2$$
$$\text{LHS} = 6.1^2 = 37.21$$
$$\text{RHS} = 6^2 + 1.1^2 = 36 + 1.21 = 37.21$$
$$\therefore 6.1^2 = 6^2 + 1.1^2$$

△ is right-angled.

**f** To prove that:

$$41^2 = 40^2 + 9^2$$
$$\text{LHS} = 41^2 = 1681$$
$$\text{RHS} = 40^2 + 9^2 = 1600 + 81 = 1681$$
$$\therefore 41^2 = 40^2 + 9^2$$

△ is right-angled.

**g** To prove that:

$$34^2 = 28^2 + 21^2$$
$$\text{LHS} = 34^2 = 1156$$
$$\text{RHS} = 28^2 + 21^2 = 784 + 441 = 1225$$
$$\therefore 34^2 \neq 28^2 + 21^2$$

△ is not right-angled.

**h** To prove that:

$$85^2 = 84^2 + 13^2$$
$$\text{LHS} = 85^2 = 7225$$
$$\text{RHS} = 84^2 + 13^2 = 7056 + 169 = 7225$$
$$\therefore 85^2 = 84^2 + 13^2$$

△ is right-angled.

**8** **a** $a^2 = 7^2 + 24^2 = 49 + 576 = 625$

$a = \sqrt{625} = 25$

Triad is {7, 24, 25}

**b** $y^2 = 9^2 + 12^2 = 81 + 144 = 225$

$\therefore\ y = \sqrt{225} = 15$

Triad is {9, 12, 15}

**c** $18^2 + n^2 = 30^2$

$\therefore\ n^2 = 30^2 - 18^2 = 900 - 324 = 576$

$n = \sqrt{576} = 24$

Triad is {18, 24, 30}

**d** $9^2 + t^2 = 41^2$

$\therefore\ t^2 = 41^2 - 9^2 = 1681 - 81 = 1600$

$\therefore\ t = \sqrt{1600} = 40$

Triad is {9, 40, 41}

**e** $13^2 + n^2 = 85^2$

$n^2 = 85^2 - 13^2 = 7225 - 169 = 7056$

$\therefore\ n = \sqrt{7056} = 84$

Triad is {13, 84, 85}

**f** $21^2 + 72^2 = p^2$

$\therefore\ p^2 = 441 + 5184 = 5625$

$\therefore\ p = \sqrt{5625} = 75$

Triad is {21, 72, 75}

**9** **a** $n^2 = 4^2 + 5^2$
$= 16 + 25$
$= 41$
$\therefore n = \sqrt{41}$
$= 6.4$ (to 1 decimal place)

**b** $w^2 = 2^2 + 7^2$
$= 4 + 49$
$= 53$
$\therefore w = \sqrt{53}$
$= 7.3$ (to 1 decimal place)

**c** $x^2 = 5^2 + 7^2$
$= 25 + 49$
$= 74$
$\therefore x = \sqrt{74}$
$= 8.6$ (to 1 decimal place)

**d** $m^2 = 8^2 + 3^2$
$= 64 + 9$
$= 73$
$\therefore m = \sqrt{73}$
$= 8.5$ (to 1 decimal place)

**e** $p^2 = 5^2 + 11^2$
$= 25 + 121$
$= 146$
$\therefore p = \sqrt{146}$
$= 12.1$ (to 1 decimal place)

**f** $q^2 = 1.5^2 + 4^2$
$= 2.25 + 16$
$= 18.25$
$\therefore q = \sqrt{18.25}$
$= 4.3$ (to 1 decimal place)

**g**

C
x
1
A
B
1

Let AC = $x$ cm
$\therefore x^2 = 1^2 + 1^2$
$= 2$
$\therefore x = \sqrt{2}$
$= 1.4$ (to 1 decimal place)
AC = 1.4 cm

**h**

S
3
y
P
Q
5

Let QS = $y$ cm
$y^2 = 3^2 + 5^2$
$= 9 + 25$
$= 34$
$\therefore y = \sqrt{34}$
$= 5.8$ (to 1 decimal place)
$\therefore$ QS = 5.8 cm

**10** **a** $a^2 + 4^2 = 7^2$
$\therefore a^2 = 7^2 - 4^2$
$= 49 - 16$
$= 33$
$\therefore a = \sqrt{33}$
$= 5.7$ (to 1 decimal place)

**b** $b^2 + 8^2 = 9^2$
$\therefore b^2 = 9^2 - 8^2$
$= 81 - 64$
$= 17$
$\therefore b = \sqrt{17}$
$= 4.1$ (to 1 decimal place)

**c** $c^2 + 3^2 = 4^2$
$\therefore c^2 = 4^2 - 3^2$
$= 16 - 9$
$= 7$
$\therefore c = \sqrt{7}$
$= 2.6$ (to 1 decimal place)

**d** $d^2 + 7^2 = 11^2$
$\therefore d^2 = 11^2 - 7^2$
$= 121 - 49$
$= 72$
$\therefore d = \sqrt{72}$
$= 8.5$ (to 1 decimal place)

**e** $e^2 + 9^2 = 12^2$
$\therefore e^2 = 12^2 - 9^2$
$= 144 - 81$
$= 63$
$\therefore e = \sqrt{63}$
$= 7.9$ (to 1 decimal place)

**f** $f^2 + 10^2 = 15^2$
$\therefore f^2 = 15^2 - 10^2$
$= 225 - 100$
$= 125$
$\therefore f = \sqrt{125}$
$= 11.2$ (to 1 decimal place)

**g** $g^2 + 9^2 = 17^2$
$\therefore g^2 = 17^2 - 9^2$
$= 289 - 81$
$= 208$
$\therefore g = \sqrt{208}$
$= 14.4$ (to 1 decimal place)

**h** $$\begin{aligned} h^2 + 4^2 &= 14^2 \\ \therefore \quad h^2 &= 14^2 - 4^2 \\ &= 196 - 16 \\ &= 180 \\ \therefore \quad h &= \sqrt{180} \\ &= 13.4 \quad \text{(to 1 decimal place)} \end{aligned}$$

**11** **a** $$\begin{aligned} p^2 &= 8^2 + 9.5^2 \\ &= 64 + 90.25 \\ &= 154.25 \\ \therefore p &= \sqrt{154.25} \\ &= 12.4 \quad \text{(to 3 significant figures)} \end{aligned}$$

**b** $$\begin{aligned} q^2 + 7^2 &= 10^2 \\ \therefore \quad q^2 &= 10^2 - 7^2 \\ &= 100 - 49 \\ &= 51 \\ \therefore \quad q &= \sqrt{51} \\ &= 7.14 \quad \text{(to 3 significant figures)} \end{aligned}$$

**c** $$\begin{aligned} p^2 + 2^2 &= 2.4^2 \\ \therefore \quad p^2 &= 2.4^2 - 2^2 \\ &= 5.76 - 4 \\ &= 1.76 \\ \therefore \quad p &= \sqrt{1.76} \\ &= 1.33 \quad \text{(to 3 significant figures)} \end{aligned}$$

**d** $$\begin{aligned} q^2 &= 7^2 + 1^2 \\ &= 49 + 1 \\ &= 50 \\ \therefore q &= \sqrt{50} \\ &= 7.07 \quad \text{(to 3 significant figures)} \end{aligned}$$

**12** **a** $$\begin{aligned} n^2 &= 5^2 + 5^2 \\ &= 25 + 25 \\ &= 50 \\ \therefore n &= \sqrt{50} \end{aligned}$$

**b** $$\begin{aligned} m^2 + 18^2 &= 25^2 \\ m^2 &= 25^2 - 18^2 \\ &= 625 - 324 \\ &= 301 \\ \therefore \quad m &= \sqrt{301} \end{aligned}$$

**c** $$\begin{aligned} r^2 + 8^2 &= 14^2 \\ \therefore \quad r^2 &= 14^2 - 8^2 \\ &= 196 - 64 \\ &= 132 \\ \therefore \quad r &= \sqrt{132} \end{aligned}$$

**d** $$\begin{aligned} s^2 + 8^2 &= 19^2 \\ \therefore \quad s^2 &= 19^2 - 8^2 \\ &= 361 - 64 \\ &= 297 \\ \therefore \quad s &= \sqrt{297} \end{aligned}$$

**13**

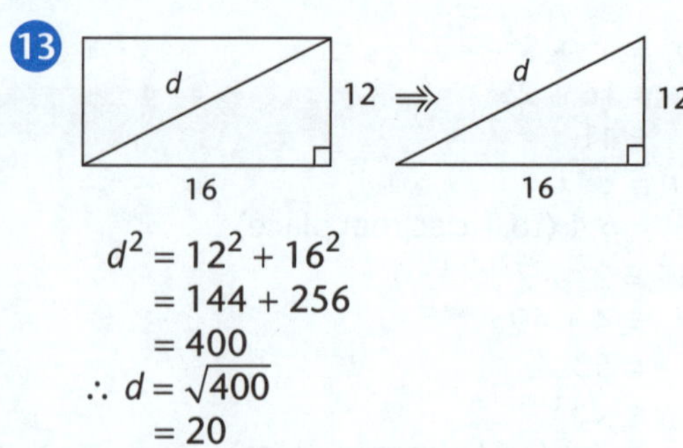


$$\begin{aligned} d^2 &= 12^2 + 16^2 \\ &= 144 + 256 \\ &= 400 \\ \therefore d &= \sqrt{400} \\ &= 20 \end{aligned}$$

The length of diagonal is 20 cm.

**14**

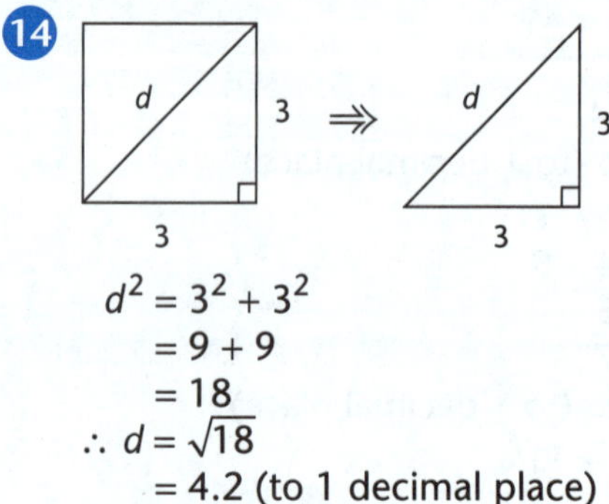


$$\begin{aligned} d^2 &= 3^2 + 3^2 \\ &= 9 + 9 \\ &= 18 \\ \therefore d &= \sqrt{18} \\ &= 4.2 \text{ (to 1 decimal place)} \end{aligned}$$

The diagonal is 4.2 cm.

**15** $$\begin{aligned} h^2 + 21^2 &= 35^2 \\ \therefore \quad h^2 &= 35^2 - 21^2 \\ &= 1225 - 441 \\ &= 784 \\ \therefore \quad h &= \sqrt{784} \\ &= 28 \end{aligned}$$

The brace reaches 28 cm up wall.

**16** $y^2 + 7.5^2 = 8^2$

$$\begin{aligned} \therefore \quad y^2 &= 8^2 - 7.5^2 \\ &= 64 - 56.25 \\ &= 7.75 \\ \therefore \quad y &= \sqrt{7.75} \\ &= 2.8 \text{ (to 1 decimal place)} \end{aligned}$$

The foot of the ladder is 2.8 m from the wall.

# WORKED SOLUTIONS to Chapter 7 **Practise, Practise**

**17** $h^2 + 21^2 = 40^2$

$$\begin{aligned} h^2 &= 40^2 - 21^2 \\ &= 1600 - 441 \\ &= 1159 \\ \therefore \quad h &= \sqrt{11.59} \\ &= 34.044\,089 \end{aligned}$$

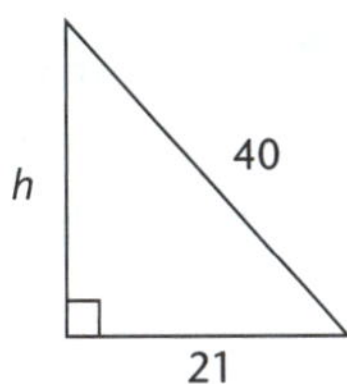


$$\begin{aligned} \text{Distance from top} &= 45 - 34.044\,089 \\ &= 10.955\,911 \\ &= 10.96 \text{ (to 4 significant figures)} \end{aligned}$$

Wires are attached 10.96 m from the top.

**18**

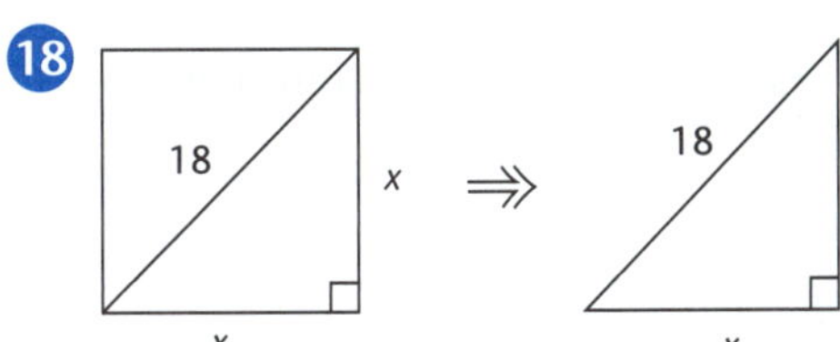


Let length of square be $x$ cm.

$$\begin{aligned} \therefore \quad x^2 + x^2 &= 18^2 \\ 2x^2 &= 324 \\ \therefore \quad x^2 &= 162 \\ x &= \sqrt{162} \\ &= 12.7 \text{ (to 1 decimal place)} \end{aligned}$$

The length of the side is 12.7 cm.

**19**

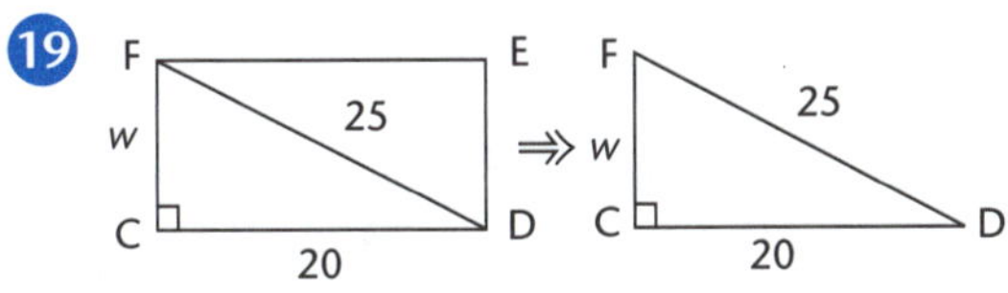


$$\begin{aligned} w^2 + 20^2 &= 25^2 \\ \therefore \quad w^2 &= 25^2 - 20^2 \\ &= 625 - 400 \\ &= 225 \\ \therefore \quad w &= \sqrt{225} \\ &= 15 \end{aligned}$$

The width is 15 cm.

**20**

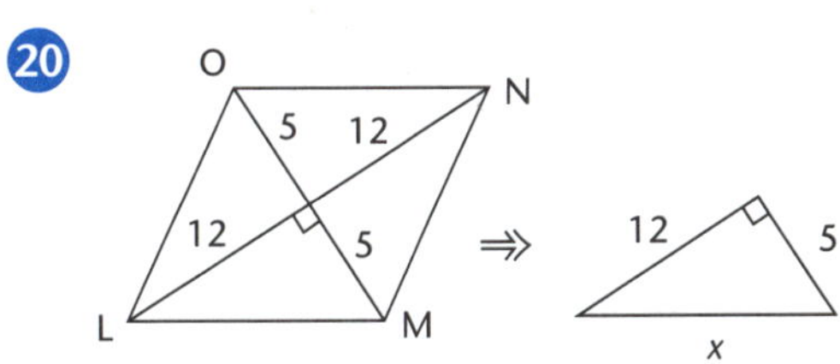


Let length of side be $x$ cm.

$$\begin{aligned} x^2 &= 5^2 + 12^2 \\ &= 169 \\ \therefore \quad x &= \sqrt{169} \\ &= 13 \end{aligned}$$

The length of each side is 13 cm.

**21**

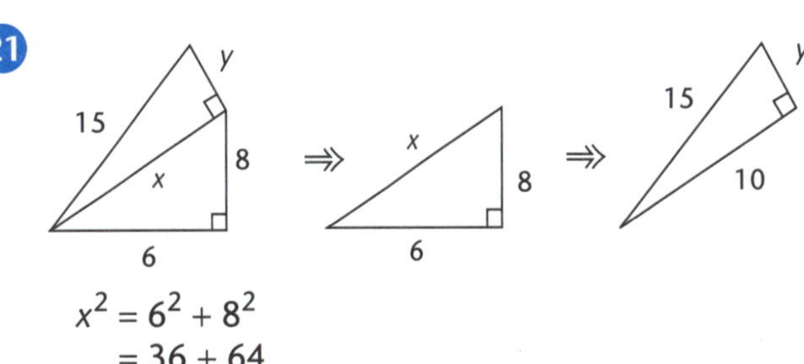


$$\begin{aligned} x^2 &= 6^2 + 8^2 \\ &= 36 + 64 \\ &= 100 \\ \therefore \quad x &= \sqrt{100} \\ &= 10 \end{aligned}$$

From second $\triangle$,

$$\begin{aligned} y^2 + 10^2 &= 15^2 \\ \therefore \quad y^2 &= 15^2 - 10^2 \\ &= 225 - 100 \\ &= 125 \\ \therefore \quad y &= \sqrt{125} \\ &= 11.2 \text{ (to 3 significant figures)} \end{aligned}$$

**22** Let third side by $y$ m.

$$\begin{aligned} y^2 + 7^2 &= 25^2 \\ \therefore \quad y^2 &= 25^2 - 7^2 \\ &= 625 - 49 \\ &= 576 \\ \therefore \quad y &= \sqrt{576} \\ &= 24 \end{aligned}$$

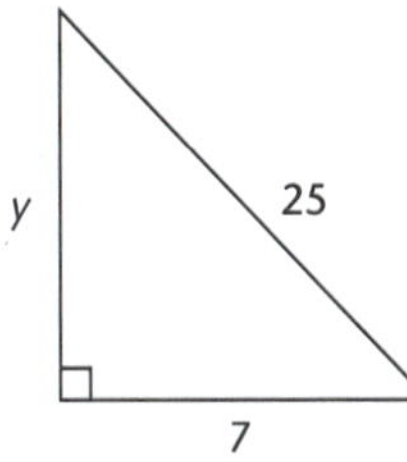


The third side is 24 m.

**23**

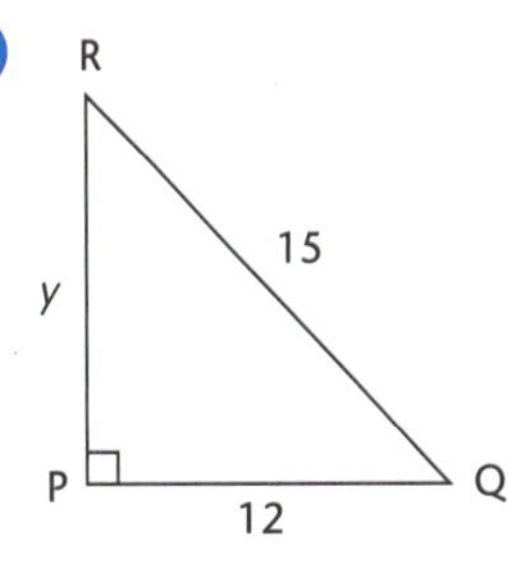

# WORKED SOLUTIONS to Chapter 7 **Practise, Practise**

Let PR be $y$ m.

$\therefore\ y^2 + 12^2 = 15^2$

$\therefore\ y^2 = 15^2 - 12^2$

$= 225 - 144$

$= 81$

$\therefore\ y = \sqrt{81}$

$= 9$

$\therefore$ PR = 9 m

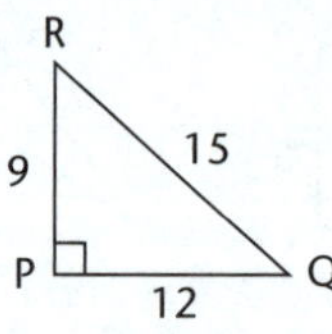


Perimeter = 9 + 12 + 15 = 36

The perimeter is 36 m.

**24**

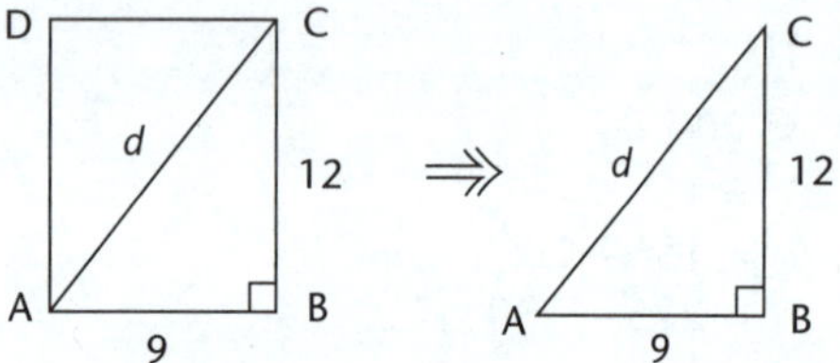


Consider rectangle ABCD:

Let AC = $d$ cm

$d^2 = 12^2 + 9^2$

$= 144 + 81$

$= 225$

$\therefore\ d = \sqrt{225}$

$= 15$

AC = 15 cm

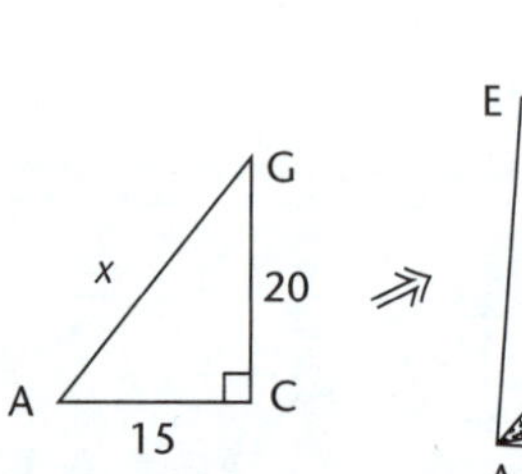


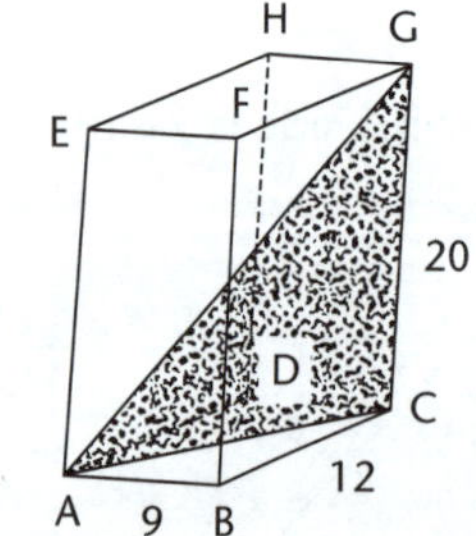


Now consider $\triangle$ACG:

Let AG = $x$ cm

$\therefore\ x^2 = 20^2 + 15^2$

$= 400 + 225$

$= 625$

$\therefore\ x = \sqrt{625}$

$= 25$

The shortest distance AG is 25 cm.

**25** Harry the Ant must crawl across the walls. To find the shortest route, fold out the net of the prism (part of it that you need):

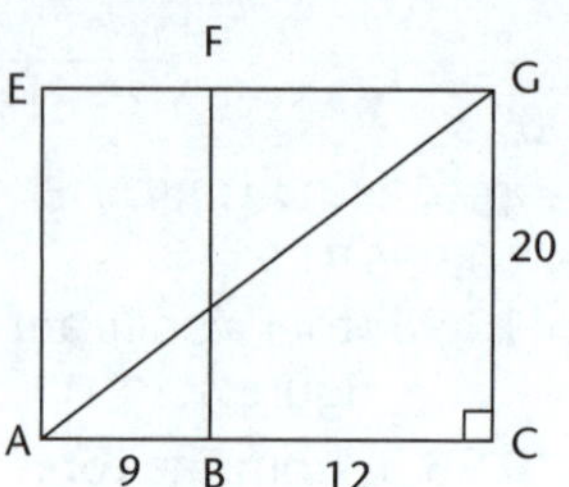


The shortest distance is the straight line between A and G.

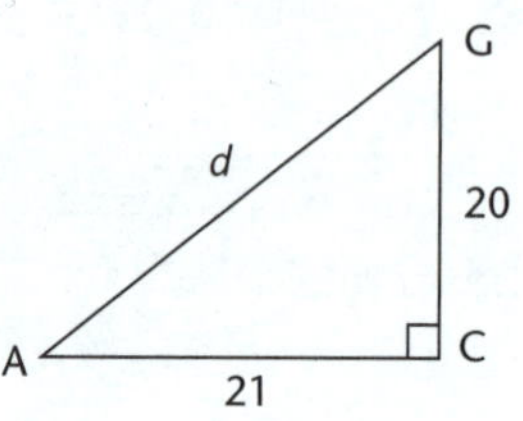


Let AG = $d$ cm

$\therefore\ d^2 = 21^2 + 20^2$

$= 441 + 400$

$= 441 + 400$

$= 841$

$\therefore\ d = \sqrt{841}$

$= 29$

The shortest distance for Harry is 29 cm. Harry's extra distance = 29 − 25 = 4. Harry travels an extra 4 cm.

**26**

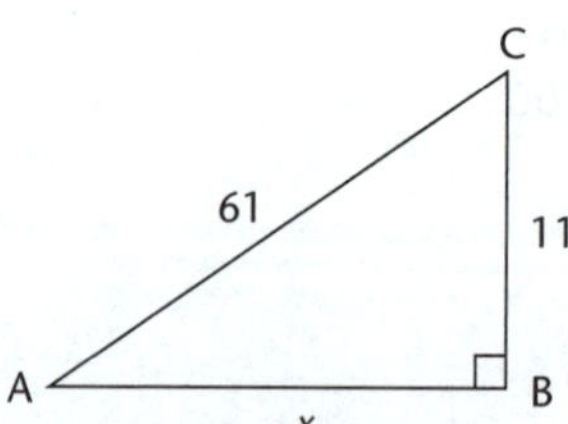


Let length AB be $x$ m.

$\therefore\ x^2 + 11^2 = 61^2$

$x^2 = 61^2 - 11^2$

$= 3721 - 121$

$= 3600$

$\therefore\ x = \sqrt{3600}$

$= 60$

AB is 60 m

$$\text{Gradient} = \frac{BC}{AB} = \frac{11}{60}$$

The gradient of the railway line is $\frac{11}{60}$.

27

PO = OQ
= OR (equal radii)
∴ PO = OQ
= 7.5 cm

Then PQ = 15 cm

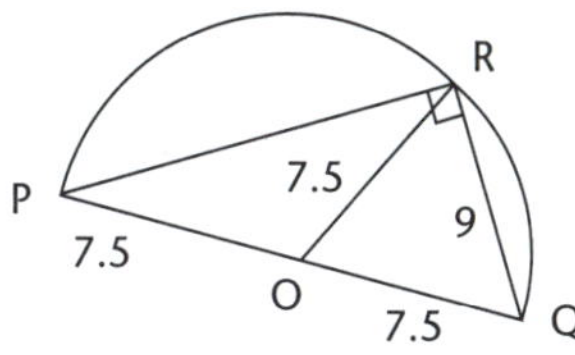


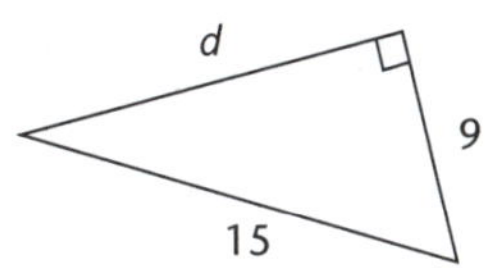


Let PR = $d$ cm.

$$\therefore d^2 + 9^2 = 15^2$$
$$d^2 = 15^2 - 9^2 = 225 - 81 = 144$$
$$\therefore d = \sqrt{144} = 12$$

∴ The length of PR is 12 cm.

28

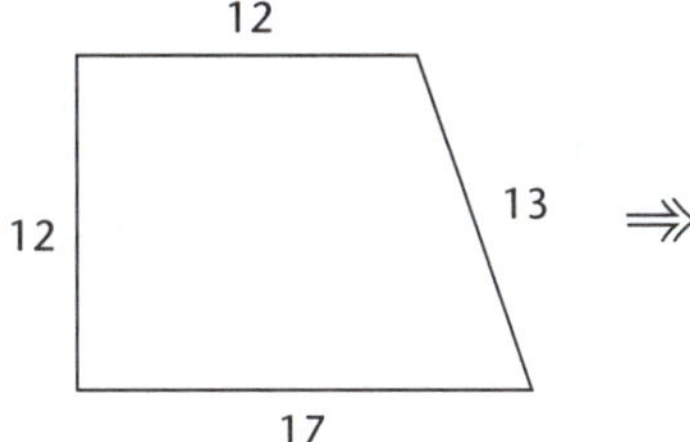


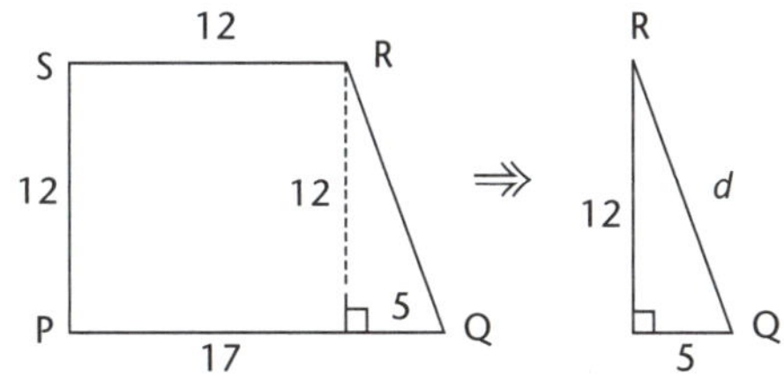


Let RQ = $d$ cm.

$$\therefore d^2 = 12^2 + 5^2 = 144 + 25 = 169$$
$$\therefore d = \sqrt{169} = 13$$

Perimeter = 12 + 12 + 13 + 17
= 54

The perimeter is 54 cm.

29

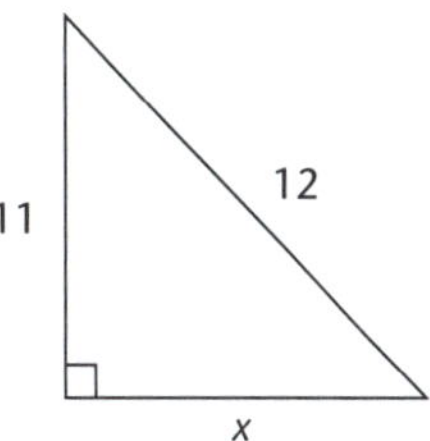


$$x^2 + 11^2 = 12^2$$
$$\therefore x^2 = 12^2 - 11^2 = 144 - 121 = 23$$
$$\therefore x = \sqrt{23} = 4.8 \text{ (to 1 decimal place)}$$

The foot of the ladder should be placed 4.8 m from the wall.

30

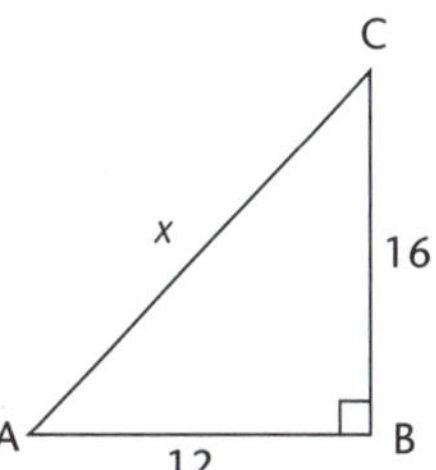


Let AC = $x$ cm.

$$\therefore x^2 = 12^2 + 16^2 = 144 + 256 = 400$$
$$\therefore x = \sqrt{400} = 20$$

Then diameter of circle, AC, is 20 cm.

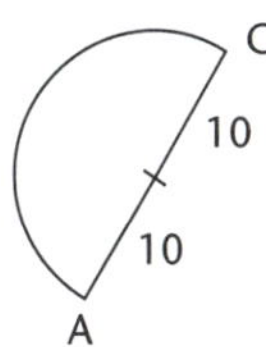

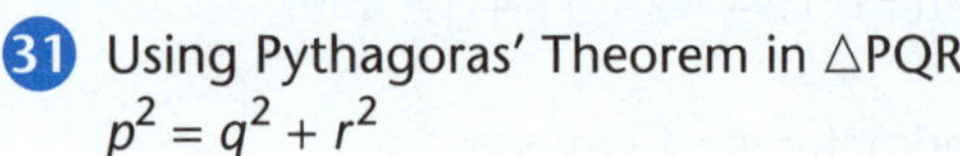


$$\text{Area} = \tfrac{1}{2}\pi r^2$$
$$= \tfrac{1}{2} \times \pi \times 10^2$$
$$= 157.079\,63$$
$$= 157 \text{ (to the nearest whole number)}$$

The area of the semicircle is 157 cm$^2$.

**31** Using Pythagoras' Theorem in $\triangle$PQR

$p^2 = q^2 + r^2$

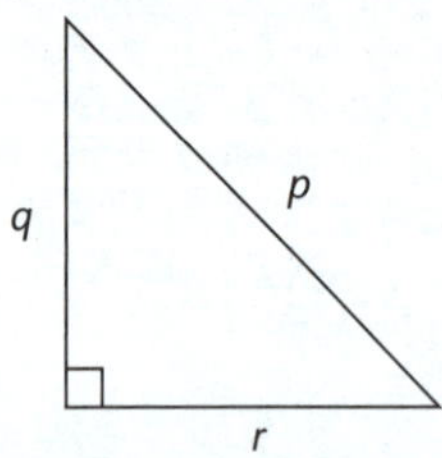


Now areas of semicircles:

$$A_1 = \tfrac{1}{2}\pi(\tfrac{1}{2}q)^2$$
$$= \tfrac{1}{8}\pi q^2$$

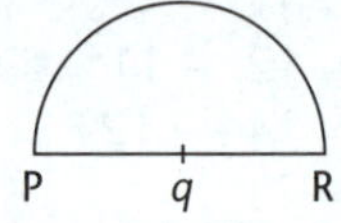


Radius = $\tfrac{1}{2}q$

$$A_2 = \tfrac{1}{2}\pi(\tfrac{1}{2}p)^2$$
$$= \tfrac{1}{8}\pi p^2$$

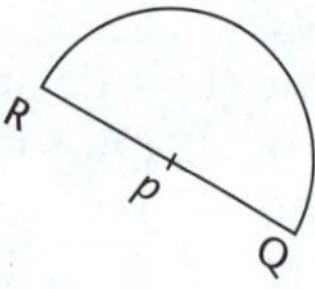


Radius = $\tfrac{1}{2}p$

$$A_3 = \tfrac{1}{2}\pi(\tfrac{1}{2}r)^2$$
$$= \tfrac{1}{8}\pi r^2$$

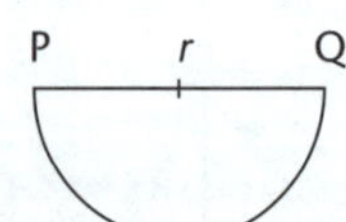


Radius = $\tfrac{1}{2}r$

$$\text{Total area} = \tfrac{1}{8}\pi p^2 + \tfrac{1}{8}\pi q^2 + \tfrac{1}{8}\pi r^2$$
$$A = \tfrac{1}{8}\pi(p^2 + q^2 + r^2)$$
$$= \tfrac{1}{8}\pi(p^2 + p^2) \qquad p^2 = q^2 + r^2$$
$$= \tfrac{1}{8}\pi(2p^2)$$
$$= \tfrac{1}{4}\pi p^2$$

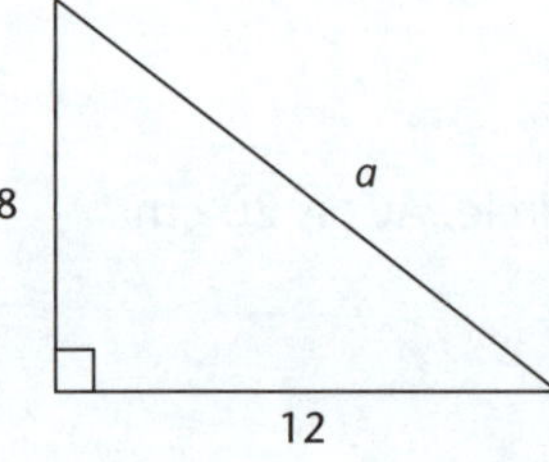


Let slant height be $a$ metres.

$$\therefore\ a^2 = 8^2 + 12^2$$
$$= 64 + 144$$
$$= 208$$
$$\therefore\ a = \sqrt{208}$$
$$= 14.422\,205$$
$$= 14.4 \text{ (to 1 decimal place)}$$

Slant height of roof is 14.4 metres.

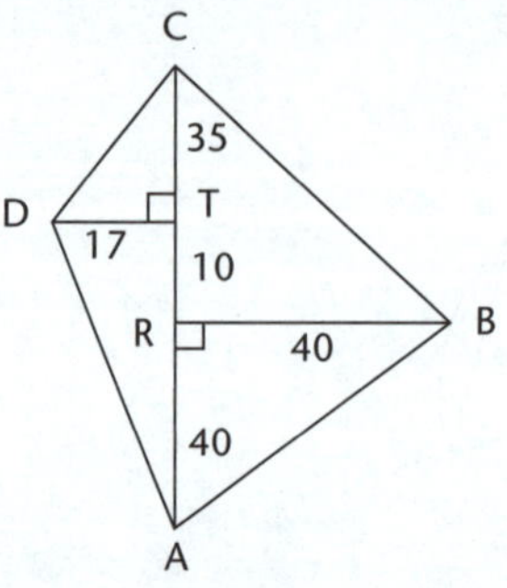


Consider $\triangle$ABR

$$AB^2 = 40^2 + 40^2$$
$$= 1600 + 1600$$
$$= 3200$$
$$\therefore\ AB = \sqrt{3200}$$

$\triangle$RBC

$$BC^2 = 45^2 + 40^2$$
$$= 2025 + 1600$$
$$= 3625$$
$$\therefore\ BC = \sqrt{3625}$$

$\triangle$DTC

$$DC^2 = 17^2 + 35^2$$
$$= 289 + 1225$$
$$= 1514$$
$$\therefore\ DC = \sqrt{1514}$$

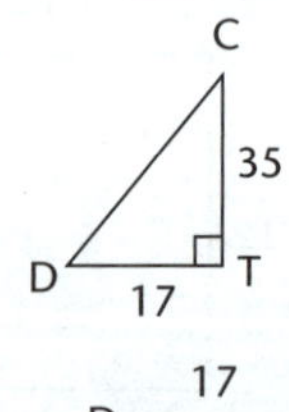


$\triangle$ATD

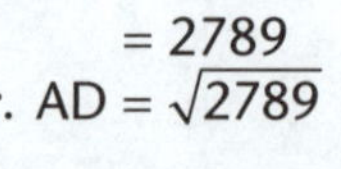

$$AD^2 = 50^2 + 17^2$$
$$= 2500 + 289$$
$$= 2789$$
$$\therefore\ AD = \sqrt{2789}$$

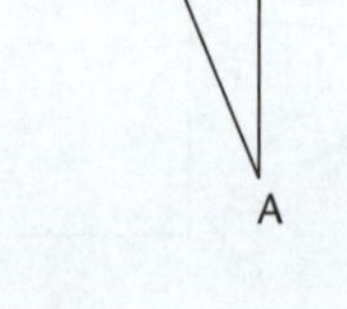


$$\text{Perimeter} = AB + BC + DC + AD$$
$$= \sqrt{3200} + \sqrt{3625} + \sqrt{1514} + \sqrt{2789}$$
$$= 208.4976\ldots$$

Perimeter is 208 metres.

# WORKED SOLUTIONS to Chapter 8 **Practise, Practise**

## Chapter 8—Measurement

**1** a Area $= \frac{1}{2}bh$

$= \frac{1}{2} \times 16.2 \times 7$

$= 56.7$

$\therefore$ The area is 56.7 $mm^2$.

b Convert 1 km → 1000 m

$\therefore$ Area $= lb$

$= 1000 \times 324$

$= 324\,000$

$\therefore$ The area is 324 000 $m^2$.

Or convert 324 m → 0.324 km

$\therefore$ Area $= lb$

$= 1 \times 0.324$

$= 0.324$

$\therefore$ The area is 0.324 $km^2$.

c Area $= \frac{1}{2}bh$

$= \frac{1}{2} \times 9 \times 17$

$= 76.5$ $cm^2$

$\therefore$ The area is 76.5 $cm^2$.

d

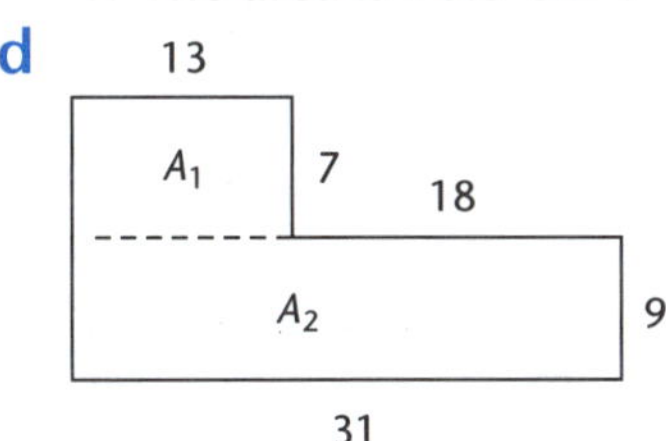


Area $= A_1 + A_2$

$= 13 \times 7 + 31 \times 9$

$= 91 + 279$

$= 370$

$\therefore$ The area is 370 $cm^2$.

e

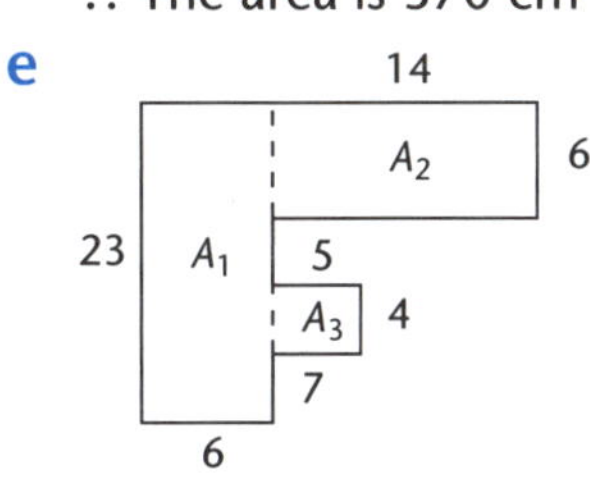


Area $= A_1 + A_2 + A_3$

$= 23 \times 6 + 14 \times 6 + 7 \times 4$

$= 138 + 84 + 20$

$= 250$

$\therefore$ The area is 250 $cm^2$.

f

12
$A_1$ 5 7
$A_2$ 5
$A_3$ 9
12

Area $= A_1 + A_2 + A_3$

$= 12 \times 5 + 19 \times 5 + 12 \times 9$

$= 60 + 95 + 108$

$= 263$

$\therefore$ The area is 263 $cm^2$.

**2** a Area $= \frac{1}{2} \times 8 \times 6 - \frac{1}{2} \times 4 \times 3$

$= 24 - 6$

$= 18$

$\therefore$ The area is 18 $cm^2$.

b

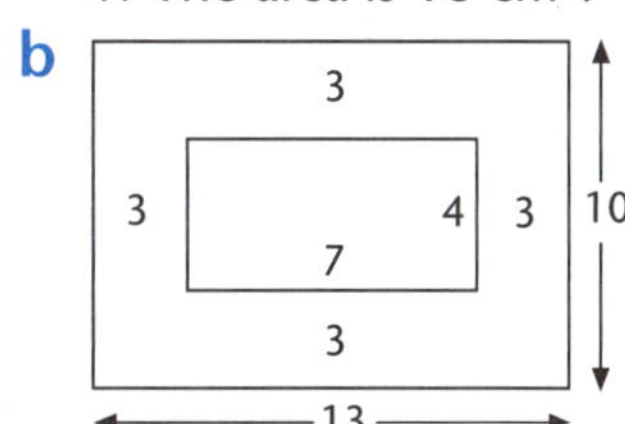


Area $= 13 \times 10 - 7 \times 4$

$= 130 - 28$

$= 102$

$\therefore$ The area is 102 $cm^2$.

c Area $= 28 \times 16 - (7 \times 7 + \frac{1}{2} \times 10 \times 8)$

$= 448 - (49 + 40)$

$= 448 - 89$

$= 359$

$\therefore$ The area is 359 $cm^2$.

**3** a

10
12

Area $= 12 \times 10$

$= 120$

$\therefore$ The area is 120 $cm^2$.

## WORKED SOLUTIONS to Chapter 8 **Practise, Practise**

**b**

15, 18, 30

Area $= \frac{1}{2} \times 18(15 + 30)$
$= 9(45)$
$= 405$
$\therefore$ The area is 405 $cm^2$.

**c**

Area $= \frac{1}{2} \times 14 \times 17$
$= 119$
$\therefore$ The area is 119 $cm^2$.

**d**

21, 16, 35

Area $= \frac{1}{2} \times 16(21 + 35)$
$= 8(56)$
$= 448$
$\therefore$ The area is 448 $cm^2$.

**e**

3, 12, 8

Area $= \frac{1}{2} \times 12(3 + 8)$
$= 6(11)$
$= 66$
$\therefore$ The area is 66 $cm^2$.

**4** **a**

705, 120, 425

Area $= \frac{1}{2} \times 120(425 + 705)$
$= 60(1130)$
$= 67\,800$
$\therefore$ area is 67 800 $m^2$
$\therefore$ area (ha) $= 67\,800 \div 10\,000$
$= 6.78$
$\therefore$ The area is 6.78 ha.

**b** Convert 1 km to 1000 m.

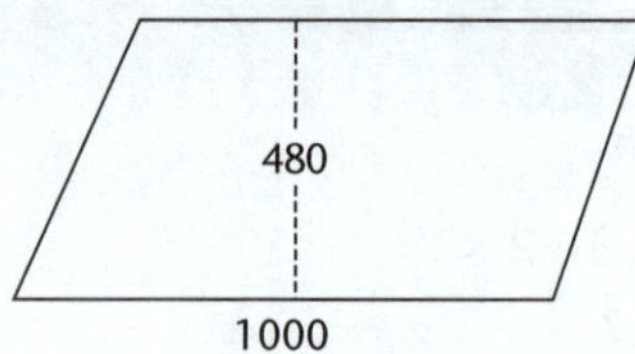


Area $= 1000 \times 480$
$= 480\,000$
$\therefore$ area is 480 000 $m^2$
$\therefore$ area (ha) $= 480\,000 \div 10\,000$
$= 48$
$\therefore$ The area is 48 ha.

**5** **a** 30 000 $m^2$ = 3 ha
**b** 4.72 ha = 47 200 $m^2$
**c** 65 400 $cm^2$ = 6.54 $m^2$
**d** 11.57 $m^2$ = 115 700 $cm^2$

**6** **a** Centre **b** Radius
**c** Sector **d** Segment
**e** Diameter **f** Quadrant
**g** Semicircle **h** Chord
**i** Circumference **j** Arc

**7** **a** Radius $= 18 \div 2 = 9$ cm
**b** Radius $= 30 \div 2 = 15$ m
**c** Radius $= 15 \div 2 = 7.5$ mm
**d** Radius $= 4.2 \div 2 = 2.1$ m
**e** Radius $= 0.1 \div 2 = 0.05$ cm

**8** **a** Diameter $= 2 \times 7 = 14$ cm
**b** Diameter $= 2 \times 15 = 30$ cm
**c** Diameter $= 2 \times 3\frac{1}{2} = 7$ m
**d** Diameter $= 2 \times 4.5 = 9$ mm
**e** Diameter $= 2 \times 2.05 = 4.1$ m

**9**

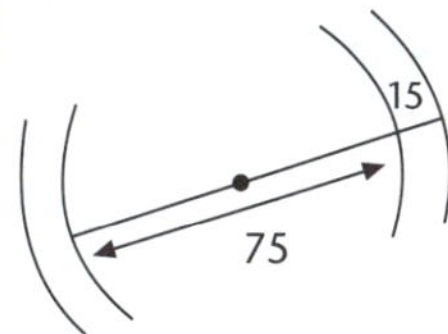


Inside radius = 75 ÷ 2 m
= 37.5 m

Outside radius = 37.5 + 15 m
= 52.5 m

**10** **a** Fraction = $\frac{90}{360}$ or $\frac{1}{4}$

**b** Fraction = $\frac{270}{360}$ or $\frac{3}{4}$

**c** Fraction = $\frac{30}{360}$ or $\frac{1}{12}$

**d** Fraction = $\frac{260}{360}$ or $\frac{2}{3}$

**e** Angle = 360° – 60°
= 300°

$\therefore$ Fraction = $\frac{300}{360} = \frac{5}{6}$

**11** **a** $C = \pi d$
$= 3.14 \times 10$ $\quad \begin{bmatrix} r = 5 \\ d = 10 \end{bmatrix}$
$= 31.4$

The circumference is 31.4 cm.

**b** $C = \pi d$
$= \frac{22}{\cancel{7}_1} \times \frac{\cancel{56}^8}{1}$ $\quad \begin{bmatrix} r = 28 \\ d = 56 \end{bmatrix}$
$= 176$

The circumference is 176 cm.

**c** $C = \pi d$
$= 3.142 \times 10$
$= 31.42$

The circumference is 31.42 cm.

**d** $C = \pi d$
$= 3\frac{1}{7} \times 35$
$= \frac{22}{\cancel{7}_1} \times \frac{\cancel{35}^5}{1}$
$= 110$

The circumference is 110 m.

**e** $C = \pi d$
$= 3.142 \times 4$ $\quad \begin{bmatrix} r = 2 \\ d = 4 \end{bmatrix}$
$= 12.568$

The circumference is 12.568 m.

**f** $C = \pi d$
$= \pi \times 12$
$= 12\pi$

The circumference is $12\pi$ cm.

**g** $C = \pi d$
$= \pi \times 16$
$= 16\pi$

The circumference is $16\pi$ cm.

**12** **a** $C = \pi d$
$= \frac{22}{\cancel{7}_1} \times \frac{\cancel{28}^4}{1}$ $\quad \begin{bmatrix} r = 14 \\ d = 28 \end{bmatrix}$
$= 88$ cm

**b** $C = \pi d$
$= \frac{22}{\cancel{7}_1} \times \frac{\cancel{140}^{20}}{1}$ $\quad \begin{bmatrix} d = 140 \end{bmatrix}$
$= 440$ mm

**c** $C = \pi d$
$= \frac{22}{\cancel{7}_1} \times \cancel{84}^{12}$ $\quad \begin{bmatrix} r = 21 \\ d = 42 \end{bmatrix}$
$= 264$ cm

**d** $C = \pi d$
$= \frac{22}{\cancel{7}_1} \times \frac{\cancel{42}^6}{1}$
$= 132$ cm

**e** $C = \pi d$
$= \frac{\cancel{22}^{11}}{\cancel{7}_1} \times \frac{\cancel{7}^1}{\cancel{2}_1}$ $\quad \begin{bmatrix} r = 1\frac{3}{4} \\ d = 1\frac{3}{4} \times 2 \\ = \frac{7}{4} \times 2 \\ = \frac{7}{2} \end{bmatrix}$
$= 11$ m

**f** $C = \pi d$
$= \frac{22}{\cancel{7}_1} \times \cancel{3.5}^{0.5}$
$= 11$ m

**13** **a** $C = \pi d$
$= 3.14 \times 200$ $\quad \begin{bmatrix} r = 100 \\ d = 200 \end{bmatrix}$
$= 628$ mm

**b** $C = \pi d$
$= 3.14 \times 4$
$= 12.56$ cm

**c** $C = \pi d$
$= 3.14 \times 40$ $\quad \begin{bmatrix} r = 20 \\ d = 40 \end{bmatrix}$
$= 125.6$ cm

**d** $C = \pi d$
$= 3.14 \times 9$
$= 28.26$ m

**e** $C = \pi d$
$= 3.14 \times 30$ $\quad \begin{bmatrix} r = 15 \\ d = 30 \end{bmatrix}$
$= 94.2$ m

**f** $C = \pi d$
$= 3.14 \times 30$
$= 94.2$ cm

# WORKED SOLUTIONS to Chapter 8 **Practise, Practise**

**g** $C = \pi d$
$= 3.14 \times 3$
$= 9.42$ m
$\begin{bmatrix} r = 1.5 \\ d = 3 \end{bmatrix}$

**14** **a** $C = \pi d$
$= \pi \times 2$
$= 6.28$ m
$\begin{bmatrix} r = 1 \\ d = 2 \end{bmatrix}$

**b** $C = \pi d$
$= \pi \times 1$
$= 3.14$ cm
$\begin{bmatrix} r = 0.5 \\ d = 1 \end{bmatrix}$

**c** $C = \pi d$
$= \pi \times 40$
$= 125.66$ cm

**d** $C = \pi d$
$= \pi \times 100$
$= 314.16$ m

**e** $C = \pi d$
$= \pi \times 300$
$= 942.48$ m
$\begin{bmatrix} r = 150 \\ d = 300 \end{bmatrix}$

**f** $C = \pi d$
$= \pi \times 200$
$= 628.32$ mm

**g** $C = \pi d$
$= \pi \times 5$
$= 15.71$ m
$\begin{bmatrix} r = 25 \\ d = 5 \end{bmatrix}$

**h** $C = \pi d$
$= \pi \times 90$
$= 282.74$ m
$\begin{bmatrix} r = 45 \\ d = 90 \end{bmatrix}$

**15** $C = \pi d$
$= \pi \times 24$
$= 75.398$
$= 75.4$ (1 decimal place)
Circumference is 75.4 cm.

**16** $C = \pi d$
$= \pi \times 140$
$= 140\pi$
$\begin{bmatrix} r = 70 \\ d = 140 \end{bmatrix}$
Circumference is $140\pi$ m.

**17** **a** $P = \frac{1}{2}C + d$
$= \frac{1}{2}\pi d + d$
$\begin{bmatrix} r = 8 \\ d = 16 \end{bmatrix}$
$= \frac{1}{2} \times \pi \times 16 + 16$
$= 41.132741$
$= 41.1$ cm (to 1 decimal place)

**b** $P = \frac{1}{2}C + 2 \times \text{width} + \text{length}$
$= \frac{1}{2}\pi d + 2 \times 5 + 10$ $[d = 10]$
$= \frac{1}{2} \times \pi \times 10 + 20$
$= 35.707963$
$= 35.7$ cm (to 1 decimal place)

**c** $P = \frac{1}{4}C + r + r$
$= \frac{1}{4}\pi d + 5 + 5$
$\begin{bmatrix} r = 5 \\ d = 10 \end{bmatrix}$
$= \frac{1}{4} \times \pi \times 10 + 10$
$= 17.853982$
$= 17.9$ cm (to 1 decimal place)

**d** $P = \frac{1}{4}C + r + 2 \times 4 + 8$
$= \frac{1}{4}\pi d + 8 + 8 + 8$
$\begin{bmatrix} r = 8 \\ d = 16 \end{bmatrix}$
$= \frac{1}{4}\pi + 16 + 24$
$= 4\pi + 24$
$= 36.566371$
$= 36.6$ cm (to 1 decimal place)

**e** $P = \frac{1}{6}C + 2r$
$= \frac{1}{6}\pi d + 2 \times 10$
$\begin{bmatrix} r = 10 \\ d = 20 \end{bmatrix}$
$= \frac{1}{6} \times \pi \times 20 + 20$
$= 30.471976$
$= 30.5$ cm (to 1 decimal place)

**f** $P = \frac{3}{4}C + 2r$
$= \frac{3}{4}\pi d + 2 \times 5$
$\begin{bmatrix} r = 5 \\ d = 10 \end{bmatrix}$
$= \frac{3}{4} \times \pi \times 10 + 10$
$= 33.561945$
$= 33.6$ cm (to 1 decimal place)

**18** **a** $d = \frac{C}{\pi}$
$\begin{bmatrix} C = \pi d \\ d = \frac{C}{\pi} \end{bmatrix}$
$= \frac{40}{\pi}$
$= 12.732395$
$= 12.7$ m (to 1 decimal place)

**b** $d = \frac{C}{\pi}$
$= \frac{250}{\pi}$
$= 79.577\,472$
$= 79.6$ cm (to 1 decimal place)

**c** $d = \frac{C}{\pi}$
$= \frac{4.5}{\pi}$
$= 1.432\,394\,5$
$= 1.4$ cm (to 1 decimal place)

**d** $d = \frac{C}{\pi}$
$= \frac{1000}{\pi}$
$= 318.309\,89$
$= 318.3$ m (to 1 decimal place)

**e** $d = \frac{C}{\pi}$
$= \frac{6400}{\pi}$
$= 2037.1833$
$= 2037.2$ km (to 1 decimal place)

**19** **a** $d = \frac{C}{\pi}$
$= \frac{50}{\pi}$
$= 15.915\,494$
$r = 15.915\,494 \div 2$
$= 7.957\,747\,2$
$= 8.0$ cm (to 1 decimal place)

Alternatively,
$r = \frac{1}{2}d$
$= \frac{1}{2} \times \frac{C}{\pi}$
$= \frac{1}{2} \times \frac{50}{\pi}$
$= 7.957\,747\,2$
$= 8.0$ cm (to 1 decimal place)

**b** $r = \frac{1}{2}d$
$= \frac{1}{2} \times \frac{C}{\pi}$
$= \frac{1}{2} \times \frac{2.5}{\pi}$
$= 0.397\,887\,3$
$= 0.4$ m (to 1 decimal place)

**c** $r = \frac{1}{2}d$
$= \frac{1}{2} \times \frac{C}{\pi}$
$= \frac{1}{2} \times \frac{0.4}{\pi}$
$= 0.063\,66$
$= 0.1$ cm (to 1 decimal place)

**d** $r = \frac{1}{2}d$
$= \frac{1}{2} \times \frac{C}{\pi}$
$= \frac{1}{2} \times \frac{100}{\pi}$
$= 15.915\,494$
$= 15.9$ m (to 1 decimal place)

**e** $r = \frac{1}{2}d$
$= \frac{1}{2} \times \frac{C}{\pi}$
$= \frac{1}{2} \times \frac{314}{\pi}$
$= 49.974\,652$
$= 50.0$ m (to the 1decimal place)

**20** $r = \frac{1}{2}d$
$= \frac{1}{2} \times \frac{C}{\pi}$
$= \frac{1}{2} \times \frac{200}{\pi}$
$= 31.830\,989$
$= 32$ (to the nearest metre)

The radius must be 32 metres.

**21** $P = 2 \times \frac{1}{2}C + 60 + 60$
$= C + 120$
$= \pi d + 120$
$= \pi \times 30 + 120$
$= 214.247\,78$
$= 214$ (to the nearest metre)

Distance around is 214 m.

Cost = 214.247 78 × $17.50
= $3749 (to nearest dollar)

**22** **a** $A = \pi r^2$ $[r = 14]$
$= \frac{22}{7_1} \times 14^2 \times 14$
$= 616$
The area is 616 $m^2$.

**b** $A = \pi r^2$
$= \frac{22}{7_1} \times 21^3 \times 21$
$= 1386$
The area is 1386 $cm^2$.

**c** $A = \pi r^2$ $\left[3\frac{1}{2} = \frac{7}{2}\right]$
$= \frac{22^{11}}{7_1} \times \frac{7^1}{2_1} \times \frac{7}{2}$
$= \frac{77}{2}$
$= 38\frac{1}{2}$ $cm^2$

**d** $A = \pi r^2$ $\left[\begin{matrix} d = 14 \\ r = 7 \end{matrix}\right]$
$= \frac{22}{7_1} \times 7^1 \times 7$
$= 154$ $m^2$

**e** $A = \pi r^2$
$= \frac{22}{7} \times \frac{7}{2} \times \frac{7}{2}$
$= 38\frac{1}{2}$ $m^2$

**f** $A = \pi r^2$
$= \frac{22}{7} \times 1 \times 1$
$= 3\frac{1}{7}$
The area is $3\frac{1}{7}$ $cm^2$.

**23** **a** $A = \pi r^2$
$= \pi \times 10 \times 10$
$= 314.2$
The area is 314.2 $m^2$.

**b** $A = \pi r^2$
$= \pi \times 5 \times 5$
$= 78.5$ $cm^2$

**c** $A = \pi r^2$
$= \pi \times 100 \times 100$
$= 31\,415.9$ $cm^2$

**d** $A = \pi r^2$
$= \pi \times 10 \times 10$
$= 314.2$ $cm^2$

**e** $A = \pi r^2$ $\left[\begin{matrix} d = 20 \\ r = 10 \end{matrix}\right]$
$= \pi \times 4 \times 4$
$= 50.265$
$= 50.3$ (to 1 decimal place)
The area is 50.3 $cm^2$.

**f** $A = \pi r^2$ $\left[\begin{matrix} d = 12 \\ r = 6 \end{matrix}\right]$
$= \pi \times 6 \times 6$
$= 113.097$
$= 113.1$ (to 1 decimal place)
The area is 113.1 $m^2$.

**g** $A = \pi r^2$ $\left[\begin{matrix} d = 50 \\ r = 25 \end{matrix}\right]$
$= \pi \times 25 \times 25$
$= 1963.485$
$= 1963.5$ $cm^2$ (to 1 decimal place)

**h** $A = \pi r^2$
$= \pi \times 2.5 \times 2.5$
$= 19.634\,95$
$= 19.6$ (to 1 decimal place)
The area is 19.6 $cm^2$.

**i** $A = \pi r^2$
$= \pi \times 18 \times 18$
$= 1017.876\,02$
$= 1017.9$ (to 1 decimal place)
The area is 1017.9 $cm^2$.

**j** $A = \pi r^2$
$= \pi \times 22.5 \times 22.5$
$= 1590.431\,281$
$= 1590.4$ (to 1 decimal place)
The area is 1590.4 $m^2$.

**24** **a** $r = \sqrt{\frac{A}{\pi}}$
$= \sqrt{\frac{35}{\pi}}$
$= 3.337\,791$
$= 3.34$ (to 2 decimal places)
The radius is 3.34 cm.

**b** $r = \sqrt{\frac{A}{\pi}}$
$= \sqrt{\frac{100}{\pi}}$
$= 5.641\,896$
$= 5.64$ (to 2 decimal places)
The radius is 5.64 cm.

**c** $r = \sqrt{\frac{A}{\pi}}$
$= \sqrt{\frac{155.6}{\pi}}$
$= 7.037\,686$
$= 7.04$ (to 2 decimal places)
The radius is 7.04 mm.

**d** $r = \sqrt{\frac{A}{\pi}}$
$= \sqrt{\frac{59.4}{\pi}}$
$= 4.348\,288$
$= 4.35$ (to 2 decimal places)
The radius is 4.35 cm.

**25** **a** $r = \sqrt{\frac{A}{\pi}}$
$= \sqrt{\frac{100}{\pi}}$
$= 5.641\,896$
$\therefore d = 11.283\,792$
$= 11.28$ (to 2 decimal places)
The diameter is 11.28 mm.

**b** $r = \sqrt{\frac{A}{\pi}}$
$= \sqrt{\frac{40}{\pi}}$
$= 3.568\,248$
$\therefore d = 7.136\,496$
$= 7.14$ (to 2 decimal places)
The diameter is 7.14 cm.

**26** **a** $A = \frac{1}{6}\pi r^2$
$= \frac{1}{6} \times \pi \times 5 \times 5$ $\left[\frac{60}{360} = \frac{1}{6}\right]$
$= 13.089\,969$
$= 13.1$ (to 1 decimal place)
The area is 13.1 cm$^2$.

**b** $A = \frac{3}{4}\pi r^2$
$= \frac{3}{4} \times \pi \times 10 \times 10$ $\left[\frac{270}{360} = \frac{3}{4}\right]$
$= 235.619\,45$
$= 235.6$ (to 1 decimal place)
The area is 235.6 cm$^2$.

**c** $A = \frac{1}{2}\pi r^2$
$= \frac{1}{2} \times \pi \times 5 \times 5$ $\begin{bmatrix} d = 10 \\ r = 5 \end{bmatrix}$
$= 39.269\,91$
$= 39.3$ (to 1 decimal place)
The area is 39.3 cm$^2$.

**d** $A = \frac{1}{4}\pi r^2$
$= \frac{1}{4} \times \pi \times 4 \times 4$
$= 12.566\,371$
$= 12.6$ (to 1 decimal place)
The area is 12.6 m$^2$.

**e** Angle $= 360° - 60°$
$= 300°$
Fraction $= \frac{300}{360} = \frac{5}{6}$
$A = \frac{5}{6}\pi r^2$
$= \frac{5}{6} \times \pi \times 6 \times 6$
$= 94.247\,78$
$= 94.2$ (to 1 decimal place)
The area is 94.2 cm$^2$.

**f** Area $= A$ semicircle $+ A$ rectangle
$= \frac{1}{2}\pi r^2 + lb$
$= \frac{1}{2} \times \pi \times 10^2 + 15 \times 20$
$= 457.079\,63$
$= 457.1$ (to 1 decimal place)
The area is 457.1 cm$^2$.

**g** Area $= 2 \times$ (area semicircle)
$= 2 \times \frac{1}{2}\pi r^2$
$= \pi r^2$
$= \pi \times 5 \times 5$
$= 78.539\,816$
$= 78.5$
The area is 78.5 cm$^2$.

**h** Area $= 2 \times$ (area semicircle)
$= 2 \times \frac{1}{2}\pi r^2$
$= \pi r^2$
$= \pi \times 4 \times 4$
$= 50.265\,482$
$= 50.3$ (to 1 decimal place)
The area is 50.3 cm$^2$.

**i** Area = $A$ semicircle + $A\,\triangle$

$= \frac{1}{2}\pi r^2 + \frac{1}{2}bh$

$= \frac{1}{2} \times \pi \times 3 \times 3 + \frac{1}{2} \times 6 \times 3$ [Base = 2 × 3 = 6]

$= \frac{1}{2} \times \pi \times 9 \times 9$

$= 23.137167$

$= 23.1$ (to 1 decimal place)

The area is 23.1 cm$^2$.

**j** Area = $A$ (large semicircle) + $A$ (small semicircle)

$A_{\text{Large}} = \frac{1}{2} \times \pi \times 8 \times 8$ [$d = 8$]

$= 100.53096$

$A_{\text{Small}} = \frac{1}{2}\pi r^2$ [$d = 8$, $r = 4$]

$= \frac{1}{2} \times \pi \times 4 \times 4$

$= 25.132741$

Total area $= 125.6637$

$= 125.7$ (to 1 decimal place)

The area is 125.7 m$^2$.

**k** Area = $A_{\text{square}} - A_{\text{semicircle}}$

$= lb - \frac{1}{2}\pi r^2$ [$d = 10$, $r = 6$]

$= 12 \times 12 - \frac{1}{2} \times \pi \times 6 \times 6$

$= 144 - \frac{1}{2} \times \pi \times 36$

$= 87.451332$

$= 87.5$ (to 1 decimal place)

The area is 87.5 cm$^2$.

**l** Area = $A_{\text{square}} + 4 \times$ (area of semicircle)

$= lb + 4 \times \frac{1}{2}\pi r^2$ [$4 \times \frac{1}{2} = 2$]

$= 12 \times 12 + 2\pi r^2$ [$d = 12$, $r = 6$]

$= 370.19467$

$= 370.2$ (to 1 decimal place)

The area is 370.2 cm$^2$.

**m** Area = $A_{\text{large circle}} - A_{\text{small circle}}$

$= \pi R^2 - \pi r^2$ [$R$ = radius large circle, $r$ = radius small circle]

$= \pi \times 6^2 - \pi \times 5^2$

$= 34.557519$

$= 34.6$ (to 1 decimal place)

The area is 34.6 cm$^2$.

**n** Consider the triangle.

Call the hypotenuse $r$.

$\therefore r^2 = 6^2 + 8^2$ (by Pythagoras' Theorem)

$= 36 + 64$

$= 100$

$\therefore r = \sqrt{100}$

$= 10$

There are three semicircles.

$A_1 = \frac{1}{2}\pi r^2$

$= \frac{1}{2}\pi \times 3^2$

$= 14.137167$

$A_2 = \frac{1}{2}\pi r^2$

$= \frac{1}{2}\pi \times 4^2$

$= 25.132741$

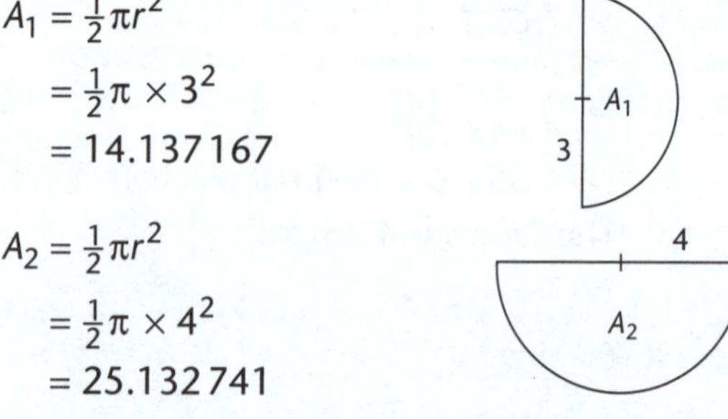


Now $A_3$ is semicircle on the hypotenuse.

$d = 10$

$\therefore r = 5$

$A_3 = \frac{1}{2}\pi r^2$

$= \frac{1}{2}\pi \times 5^2$

$= 39.269908$

Also $A\triangle = \frac{1}{2}bh$

$= \frac{1}{2} \times 8 \times 6$

$= 24$

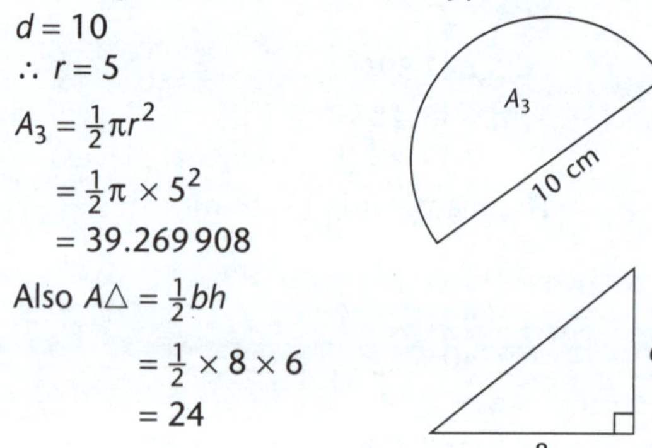


Total area $= A_1 + A_2 + A_3 + A\triangle$

$= 102.53982$

$= 102.5$ (to 1 decimal place)

The total area is 102.5 cm$^2$.

**27 a** Shaded $A = A_{\text{square}} - A_{\text{circle}}$ [$d = 16$, $\therefore r = 8$]

$= lb - \pi r^2$

$= 16 \times 16 - \pi \times 8 \times 8$

$= 54.93807$

$= 54.9$ (to 1 decimal place)

The shaded area is 54.9 cm$^2$.

**b** Shaded $A = A_{\text{large circle}} - A_{\text{small circle}}$

$= \pi R^2 - \pi r^2$ [$R = \frac{1}{2} \times 20 = 10$, $r = 6$]

$= \pi \times 10^2 - \pi \times 6^2$

$= 201.06193$

$= 201.1$

The shaded area is 201.1 cm$^2$.

**c**

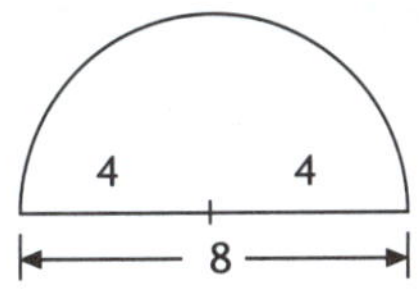


16

8

Shaded $A = A_{rectangle} - 4 \times (A_{semicircle})$

$= lb - 4 \times \frac{1}{2}\pi r^2$

$= 16 \times 8 - 2\pi \times 4^2$

$= 128 - 2 \times \pi \times 4^2$

$= 27.469035$

$= 27.5$ (to 1 decimal place)

The shaded area is 27.5 cm$^2$.

**d**

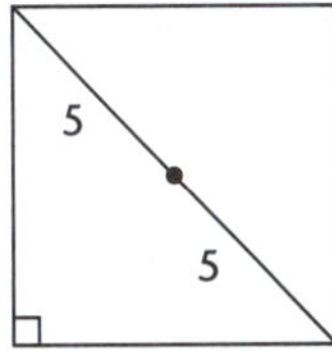


$\Rightarrow$

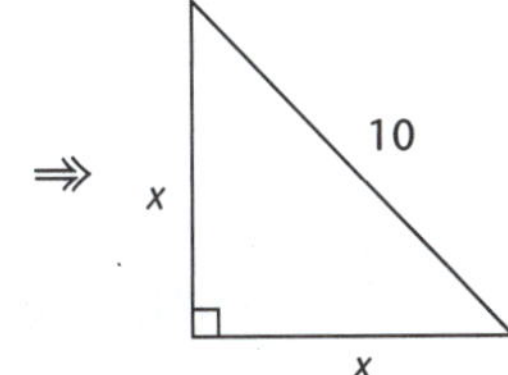


Let length of side of square = $x$ cm.

$\therefore x^2 + x^2 = 10^2$ (Pythagoras)

i.e. $2x^2 = 100$

$x^2 = 50$ *

Area of square $= L \times B$

$= x \times x$

$= x^2$

$= 50$ (from *)

Another solution:

The area of the triangle shown

$= \frac{1}{2}bh$

$= \frac{1}{2} \times 10 \times 5$

$= 25$

But the area of the square

$= 2 \times$ area of triangle

$= 2 \times 25$

$= 50$

Shaded $A = A_{circle} - A_{square}$

$= \pi r^2 - 50$

$= \pi \times 5^2 - 50$

$= 28.539816$

$= 28.5$ (to 1 decimal place)

The shaded area is 28.5 cm$^2$.

**28** Area $= A_{rectangle} + 2 \times A_{semicircle}$

$= l \times b + 2 \times \frac{1}{2}\pi r^2$ $\quad \begin{bmatrix} d = 20 \\ r = 10 \end{bmatrix}$

$= 60 \times 20 + \pi r^2$

$= 1200 + \pi \times 10^2$

$= 1514.1593$

$= 1514$ (to nearest metre$^2$)

The area is 1514 m$^2$.

**29** $A = \pi r^2$ but we do not have $r$. We have $C$.

Now $d = \dfrac{C}{\pi}$

$\therefore r = \dfrac{1}{2} \times \dfrac{C}{\pi}$

$= \dfrac{1}{2} \times \dfrac{110}{\pi}$

$= 17.507$

$A = \pi r^2$

$= \pi \times 17.507^2$

$= 962.887$

$= 963$ (to nearest cm$^2$)

The area of each disc is 963 cm$^2$.

**30** **a** Shaded $A = A_{square} - A_{circle}$ $\quad \begin{bmatrix} d = 5 \\ \therefore r = 2.5 \end{bmatrix}$

$= lb - \pi r^2$

$= 5 \times 5 - \pi \times 2.5^2$

$= 5.3650459$ cm$^2$

**b** Number of caps $= 100 \div 5$

$= 20$

20 caps fit across the width.

**c** Number of caps along length

$= 20$ m $\div 5$ cm

$= 2000 \div 5$

$= 400$

Total number of caps

= Number across $\times$ Number in length

$= 20 \times 400$

$= 8000$

The total number of caps is 8000.

**d** Wastage with 1 cap

$= 5.3650459$ cm$^2$ (from [a])

Total wastage $= 5.3650459 \times 8000$

$= 42920.367$

$= 42920$ (to nearest whole)

The wastage from roll is 42920 cm$^2$.

## WORKED SOLUTIONS to Chapter 8 **Practise, Practise**

**31** **a** Volume $= Ah$
$= 47 \times 6$
$= 282$
$\therefore$ The volume is 282 $m^3$.

**b** Volume $= Ah$
$= 312 \times 12$
$= 3744$
$\therefore$ The volume is 3744 $cm^3$.

**c** Volume $= 15 \times 15 \times 15$
$= 3375$
$\therefore$ The volume is 3375 $cm^3$.

**d** Convert 2 m to 200 cm
$\therefore$ Volume $= Ah$
$= 200 \times 60 \times 15$
$= 180\,000$
$\therefore$ The volume is 180 000 $cm^3$
(or $2 \times 0.6 \times 0.15 = 0.18$
$\therefore$ 0.18 $m^3$)

**e** Area of base $= \frac{1}{2} \times 25 \times 16$
$= 200$
$\therefore$ volume $= Ah$
$= 200 \times 8$
$= 1600$
$\therefore$ The volume is 1600 $cm^3$.

**f** Area of base $= 14 \times 6 + \frac{1}{2} \times 14 \times 10$
$= 154$
$\therefore$ Volume $= Ah$
$= 154 \times 8$
$= 1232$
$\therefore$ The volume is 1232 $cm^3$.

**g**

6
$A_2$ 4
7 7
8 $A_1$
20

Area of base $= A_1 + A_2$
$= 20 \times 8 + 6 \times 4$
$= 160 + 24 = 184$
Volume $= Ah$
$= 184 \times 5$
$= 920$
$\therefore$ The volume is 920 $cm^3$.

**h** Area of base $= 12 \times 10 + \frac{1}{2} \times 15 \times 12$
$= 210$
Volume $= Ah$
$= 210 \times 8$
$= 1680$
$\therefore$ The volume is 1680 $cm^3$.

**i**

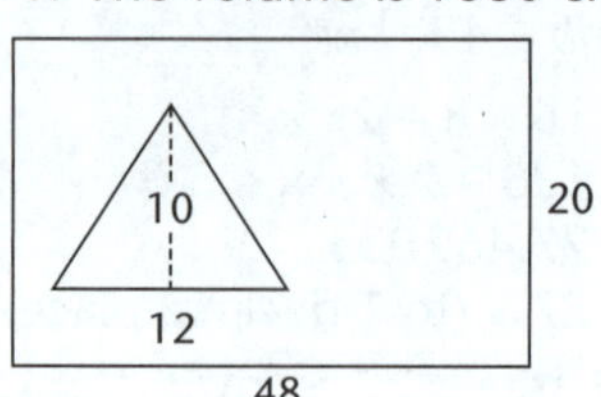


Area of base $= 48 \times 20 - \frac{1}{2} \times 12 \times 10$
$= 960 - 60$
$= 900$
$\therefore$ Volume $= Ah$
$= 900 \times 10$
$= 9000$
$\therefore$ The volume is 9000 $cm^3$.

**j** Area of base $= 20 \times 6$
$= 120$
Volume $= Ah$
$= 120 \times 7$
$= 840$
$\therefore$ The volume is 840 $cm^3$.

**32**

| Area of base | Height | Volume of prism |
|---|---|---|
| 36 $m^2$ | 11 m | 396 $m^3$ |
| 76 $mm^2$ | 10 mm | 760 $mm^3$ |
| 40 $mm^2$ | 24 mm | 960 $mm^3$ |

**33** Area of rectangular prism $= 12 \times 7 = 84$
$\therefore$ area of base $= 84$
Now, volume $= Ah$
$420 = 84h$
$\therefore \quad 84h = 420$
$\therefore \quad \frac{84h}{84} = \frac{420}{84}$
$h = 5$
$\therefore$ The height is 5 cm.

**34** Volume $= s^3$

$\therefore \quad s^3 = 512$

Taking cube root of both sides:

$\therefore s = \sqrt[3]{512}$

$= 8$

$\therefore$ The side length is 8 cm.

**35** **a** $V = \pi r^2 h$ $\quad \begin{bmatrix} d = 4 \\ r = 2 \end{bmatrix}$

$= \pi \times 2 \times 2 \times 8$

$= 100.531$

$= 100.5$ (to 1 decimal place)

The volume is 100.5 $m^3$.

**b** $V = \pi r^2 h$ $\quad \begin{bmatrix} d = 8 \\ \therefore r = 4 \end{bmatrix}$

$= \pi \times 4^2 \times 10$

$= 502.65$

$= 502.7$ (to 1 decimal place)

The volume is 502.7 $cm^3$.

**c** $V = \pi r^2 h$ $\quad \begin{bmatrix} d = 25 \\ \therefore r = 12.5 \end{bmatrix}$

$= \pi \times 12.5^2 \times 30$

$= 14\,726.216$

$= 14\,726.2$ (to 1 decimal place)

Volume is 14 726.2 $cm^3$.

**d** $V = \pi r^2 h$

$= \pi \times 2.4^2 \times 3$

$= 54.287$

$= 54.3$ (to 1 decimal place)

Volume is 54.3 $cm^3$.

**e** $V = \pi r^2 h$ $\quad \begin{bmatrix} d = 80 \text{ cm} \\ = 0.8 \text{ m} \\ \therefore r = 0.4 \text{ m} \end{bmatrix}$

$= \pi \times 0.4^2 \times 1.2$

$= 0.603\,186$

$= 0.6$ (to 1 decimal place)

Volume is 0.6 $m^3$.

**36** **a** $V = \pi r^2 h$ $\quad \begin{bmatrix} 3\frac{1}{7} = \frac{22}{7} \end{bmatrix}$

$= \frac{22}{\cancel{7}_1} \times \cancel{7}^1 \times 7 \times 20$

$= 3080 \text{ cm}^3$

**b** $V = \pi r^2 h$ $\quad \begin{bmatrix} d = 10 \\ \therefore r = 5 \end{bmatrix}$

$= \frac{22}{\cancel{7}_1} \times 5 \times 5 \times \cancel{14}^2$

$= 1100 \text{ m}^3$

**c** $V = \pi r^2 h$

$= \frac{22}{\cancel{7}_1} \times 12 \times 12 \times \cancel{21}^3$

$= 9504 \text{ cm}^3$

**d** $V = \pi r^2 h$ $\quad \begin{bmatrix} d = 7 \\ \therefore r = \frac{7}{2} \end{bmatrix}$

$= \frac{22}{\cancel{7}_1} \times \frac{7}{\cancel{2}_1} \times \frac{7}{\cancel{2}_1} \times \cancel{20}^{10}$

$= 770 \text{ cm}^3$

**e** $V = \pi r^2 h$ $\quad \begin{bmatrix} d = 40 \\ \therefore r = 20 \end{bmatrix}$

$= \frac{22}{\cancel{7}_1} \times 20 \times 20 \times \cancel{49}^7$

$= 61\,600 \text{ mm}^3$

**f** $V = \pi r^2 h$

$= \frac{22}{\cancel{7}_1} \times \cancel{3.5}^{0.5} \times 3.5 \times 15$

$= 577.5 \text{ m}^3$

**37** Shape is $\frac{1}{2}$ of a cylinder.

$\therefore V = \frac{1}{2} \pi r^2 h$ $\quad \begin{bmatrix} d = 1.5 \\ \therefore r = 0.75 \end{bmatrix}$

$= \frac{1}{2} \times \pi \times 0.75^2 \times 5$

$= 4.417\,864\,7$

$= 4.4$ (to 1 decimal place)

Volume is 4.4 $m^3$ $\quad [1 \text{ kL} = 1 \text{ m}^3]$

Capacity = 4.4 kL

Amount required to refill tank = 35% of 4.4 kL

= 1.54 kL

**38** $V = \pi r^2 h$

$= \pi \times 6^2 \times 6$

$= 678.584\,01$

$= 678.6$ (to 1 decimal place)

Volume of water is 678.6 $m^3$.

**39** $V =$ Volume of rectangular prism + Volume of $\frac{1}{2}$ cylinder

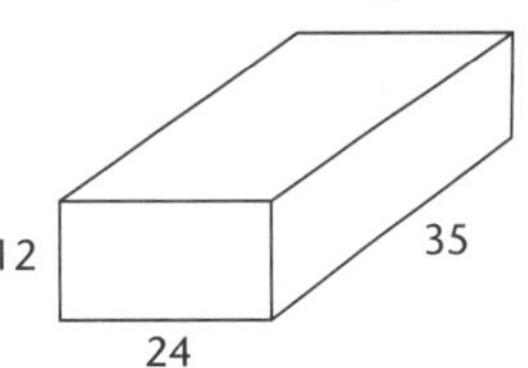


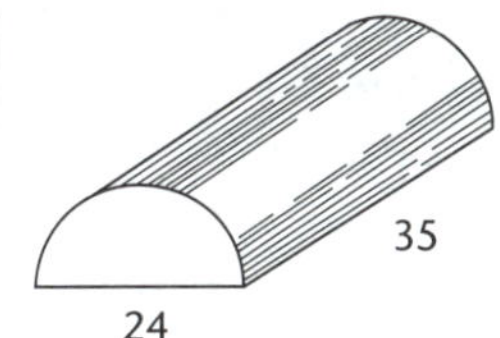


$V_{prism} = lbh$

$= 24 \times 12 \times 35$

$= 10\,080$

$V_{\text{half cyl.}} = \frac{1}{2} \times \pi r^2 h$ $\quad \begin{bmatrix} d = 24 \\ \therefore r = 12 \end{bmatrix}$

$= \frac{1}{2} \times \pi \times 12^2 \times 35$

$= 7916.8135$

Total volume $= 10\,080 + 7916.8135$

$= 17\,996.8135$

$= 17\,996.8$ (to 1 decimal place)

Volume is 17 996.8 $cm^3$.

# WORKED SOLUTIONS to Chapter 8 **Practise, Practise**

**40** $V_{cube} = 50 \times 50 \times 50$
$= 125\,000$
Volume of 4 cylinders $= 4 \times \pi r^2 h$
$= 4 \times \pi \times 10^2 \times 50$
$= 62\,831.853$

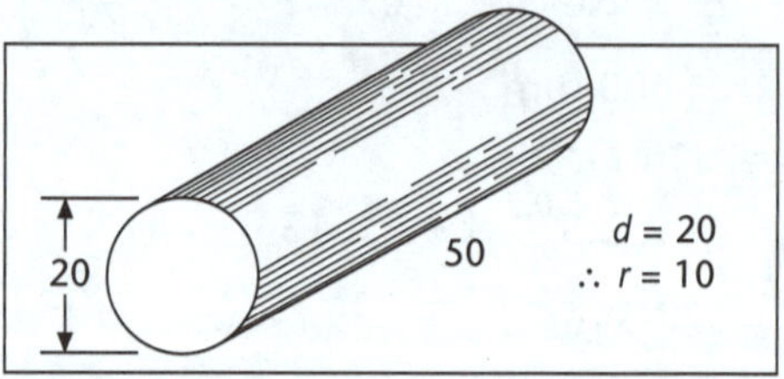


Volume remaining $= 125\,000 - 62\,831.853$
$= 62\,168.147$
$= 62\,168$ (to nearest cm$^3$)
The volume remaining is 62 168 cm$^3$.

**41** Capacity $= 1420 \div 1000$
$= 1.42$
∴ The capacity is 1.42 L.

**42**
**a** 3500 cm$^3$ = 3.5 L
**b** 28 cm$^3$ = 28 mL
**c** 760 mL = 760 cm$^3$
**d** 7850 mL = 7.85 L
**e** 58 L = 58 000 cm$^3$
**f** 1000 L = 1 000 000 cm$^3$

**43** Volume $= 100 \times 100 \times 100$
$= 1\,000\,000$
∴ The volume is 1 000 000 cm$^3$
Now, 1 m$^3$ = 1 000 000 cm$^3$
(because 1 m = 100 cm)
But 1 L = 1000 cm$^3$
∴ 1 m$^3$ = 1000 L
∴ 1 m$^3$ = 1 kL [Note: 1 kL = 1000 L]

**44**
**a** Area of base $= 20 \times 15$
$= 300$
∴ Volume $= Ah$
$= 300 \times 45$
$= 13\,500$
∴ The volume is 13 500 cm$^3$.
∴ Capacity $= 13\,500 \div 1000$
$= 13.5$
∴ The capacity is 13.5 L.

**b** Area of base $= \frac{1}{2}h(a + b)$
$= \frac{1}{2} \times 38(50 + 82)$
$= 19(132)$
$= 2508$
∴ Volume $= Ah$
$= 2508 \times 60$
$= 150\,480$
∴ The volume is 150 480 cm$^3$.
∴ Capacity $= 150\,480 \div 1000$
$= 150.48$
∴ The capacity is 150.48 L.

**c** Volume $= Ah$
$= 2 \times 2 \times 2$
$= 8$
∴ Volume is 8 m$^3$
∴ Capacity $= 8 \times 1000$
$= 8000$
∴ The capacity is 8000 L.

**45**
**a** $300 \div 60 = 5$ h
**b** $4 \times 24 = 96$ h
**c** $6\frac{1}{2} \times 60 = 390$ sec
**d** $\frac{3}{4} \times 60 = 45$ min
**e** $0.2 \times 60 = 12$ sec
**f** $60 \div 24 = 2\frac{1}{2}$ days
**g** $2000 \div 100 = 20$ centuries
**h** $60 \div 10 = 6$ decades

**46**
**a** 4:17 } 35 min
4:52
∴ 35 minutes

**b** 11:19 } 41 min
12:00 } 3 h
3:00 } 14 min
3:14

[3 h + 41 min + 14 min]

∴ 3 h 55 min

**c** 3:37 pm
} 23 min
4:00 pm
} 12 h
4:00 am
} 12 min
4:12 am

∴ 12 h 35 min

**47** **a** 5th Aug → 31st Aug: 27 days
September, October, November:
30, 31, 30 days
1st Dec → 24th Dec: 24 days
(The question says 'between', so we don't count Christmas Day.)
Now 27 + 30 + 31 + 30 + 24 = 142 days
∴ 142 days

**b** 2 Jan to 31 Jan = 30 days
Feb (leap year); March: 29, 31 days
1st April → 24th April: 24 days
Then 30 + 29 + 31 + 24 = 114 days
∴ 114 days

**48**

| | |
|---|---|
| 4:17 am | **0417** |
| **11:42** am | 1142 |
| 3:47 pm | **1547** |
| **9:48** pm | 2148 |

**49** 3:56

**50** **a** 8:45 + 0:15 = 9:00
∴ Asleep at 9 pm

**b** 9 pm + 9:20
∴ 6:20 am
∴ Wakes at 6:20 am.

**51** **a** 10:45 am
} 15 min
11:00 am
} 8 h
7:00 pm
} 15 min
7:15 pm

∴ 8 h 30 min

∴ Laura works for $8\frac{1}{2}$ hours.

**b** Pay = $12 × $8\frac{1}{2}$
Pay = $102
∴ Laura is paid $102.

**52** **a** Half an hour behind
∴ 3:40 – 0:30 = 3:10
∴ 3:10 pm in Adelaide

**b** Two hours ahead
∴ 4:52 + 2:00 = 6:52
∴ 6:52 pm in Sydney

**c** $1\frac{1}{2}$ h ahead
∴ 11:43 + 1:30 = 12:73, i.e. 1:13
∴ 1:13 am in Adelaide

**53** **a** 6:15 **b** 9:58

**54** **a** **i** 5:02
} 20 min
5:22

∴ Trip takes 20 minutes.

**ii** 4:48
} 12 min
5:00
} 36 min
5:36

∴ Trip takes 48 minutes.

**b** 5:36 + 0:05 = 5:41
∴ arrive at 5:41

**c** Grace St – City Centre
5:02 → 5:36 = 34 min
∴ 7:42 + 0:34 = 7:76, i.e. 8:16
∴ arrive at 8:16

**55** **a** 10 pm → midnight
∴ Auckland ahead by 2 hours

**b** 7 am → 2 pm
∴ New York behind 7 hours

**c** London – Los Angeles: Noon – 4 am
∴ LA is 8 hours behind
∴ 8 hours before 4 pm = 8 am

**d** Los Angeles – Sydney: 4 am – 10 pm
∴ Sydney is 18 h ahead
∴ 18 h ahead of 5:30 pm
∴ 5:30 pm + 18 h
∴ 5:30 am + 6 h
∴ 11:30 am
∴ Finishes 11:30 am Monday morning.

# WORKED SOLUTIONS to Chapter 9 **Practise, Practise**

## Chapter 9— Ratio and Rates

**1** **a** 25:15 = 5:3 (divide by 3)
**b** 3Ø:7Ø = 3:7
**c** 9:18:27 = 1:2:3 (divide by 9)
**d** 14:49 = 2:7 (divide by 7)
**e** 1ØØ:100 ØØØ = 1:1000
**f** 75:125 = 3:5 (divide by 25)

**2** **a** 6:2 = 3:1
∴ $x = 3$
**b** 15:45 = 1:3
∴ $x = 3$
**c** 12:8 = 3:2
∴ $x = 2$
**d** 7:5 = 49:35
∴ $x = 49$
**e** 13:3 = 39:9
∴ $x = 39$

**3** **a** 3.5:1.5 = 35:15
= 7:3
**b** 2:1.4 = 20:14
= 10:7
**c** 8.4:4.2 = 84:42
= 2:1
**d** 3.25:4.25 = 325:425
= 13:17
**e** $\frac{3}{4}:\frac{1}{2} = \frac{3}{4}:\frac{2}{4}$
= 3:2
**f** $\frac{2}{3}:\frac{5}{6} = \frac{4}{6}:\frac{5}{6}$
= 4:5
**g** $\frac{3}{5}:1\frac{1}{2} = \frac{3}{5}:\frac{3}{2}$
$= \frac{6}{10}:\frac{15}{10}$
= 6:15
= 2:5
**h** $2\frac{1}{4}:2\frac{1}{2} = \frac{9}{4}:\frac{5}{2}$
$= \frac{9}{4}:\frac{10}{4}$
= 9:10
**i** $\frac{4}{5}$:2.2 = 0.8:2.2
= 8:22
= 4:11

**4** **a** \$1.50:40c = 150c:40c
= 15Ø:4Ø
= 15:4
**b** \$2.40:\$1.60 = 240c:160c
= 24Ø:16Ø
= 3:2
**c** 8 hours:2 days = 8 h:48 h
= 8:48
= 1:6
**d** 1 mm:1 cm:1 m
= 1 mm:10 mm:1000 mm
= 1:10:1000
**e** 3 weeks:6 days = 21 days:6 days
= 21:6
= 7:2
**f** $1\frac{1}{2}$ min:45 sec = 90 sec:45 sec
= 90:45
= 2:1
**g** 1 mL:1 kL = 1 mL:1 000 000 mL
= 1:1 000 000
**h** 3 decades:2 centuries
= 3 decades:20 decades
= 3:20

**5** **a** $\frac{\cancel{12}^{2}\cancel{a}^{1}b}{\cancel{6}_{1}\cancel{a}_{1}}:\frac{\cancel{6}^{1}\cancel{a}^{1}}{\cancel{6}_{1}\cancel{a}_{1}} = 2b:1$
**b** $\frac{\cancel{3}^{1}\cancel{x}^{1}\cancel{y}^{1}}{\cancel{3}_{1}\cancel{x}_{1}\cancel{y}_{1}}:\frac{\cancel{15}^{5}\cancel{x}^{1}\cancel{y}^{1}}{\cancel{3}_{1}\cancel{x}_{1}\cancel{y}_{1}} = 1:5$
**c** $2a^2:(2a)^2 = \frac{\cancel{2a}^{1}}{\cancel{2a}^{1}}:\frac{\cancel{4}^{2}\cancel{a}^{1}}{\cancel{2a}^{1}}$
= 1:2
**d** $\frac{3}{p}:3p = 3:3p^2$ (multiply by $p$)
$= \frac{\cancel{3}}{\cancel{3}}:\frac{\cancel{3}p^2}{\cancel{3}}$
$= 1:p^2$

**6** **a** Profit = \$900 − \$700
= \$200
**b** **i** CP = \$700, SP = \$900
∴ \$7ØØ:\$9ØØ
= 7:9
**ii** Profit = \$200, CP = \$700
∴ \$2ØØ:\$7ØØ
= 2:7

**7** **a** Female = 56 − 31
= 25
∴ There are 25 female teachers.

**b** **i** male:female = 31:25
**ii** total:female = 56:25

**8** 10:24 = 5:12

**9** **a** 6 black, 10 red
$\therefore$ 6:10 = 3:5
**b** 10 red, 10 non-red
$\therefore$ 10:10 = 1:1

**10** **a** **i** AB = 3 cm
**ii** AC = 8 cm
**b** **i** AB:AC = 3 cm:8 cm
= 3:8
**ii** CD:AB = 2 cm:3 cm
= 2:3
**iii** AC:BD = 8 cm:7 cm
= 8:7

**11** Bad apples = $\frac{1}{3}$, good apples = $\frac{2}{3}$
$\therefore$ bad:good = $\frac{1}{3}:\frac{2}{3}$
= 1:2

**12** **a** Sides: 2 cm : 6 cm = 2:6
= 1:3
**b** Areas: $(2 \times 2)\ \text{cm}^2 : (6 \times 6)\ \text{cm}^2$
$= 4\ \text{cm}^2 : 36\ \text{cm}^2$
= 4:36
= 1:9
**c** Volumes: $(2 \times 2 \times 2)\ \text{cm}^3 : (6 \times 6 \times 6)\ \text{cm}^3$
$= 8\ \text{cm}^3 : 216\ \text{cm}^3$
= 8:216
= 1:27

**13** Perimeter = 45 cm
$\therefore$ length of side = 45 ÷ 5
= 9 cm
$\therefore$ ratio = 45 cm:9 cm
= 45:9
= 5:1

**14** **a** Netball + soccer = 360° − (90 + 150)
= 360° − 240°
= 120°
$\therefore$ Netball and soccer are both 60° sectors.
**b** **i** 90°:150° = 90:150
= 9:15
= 3:5
**ii** 60°:60° = 60:60
= 1:1
**iii** 90°:360° = 90:360
= 9:36
= 1:4

**15** Adults:children = 2:5
$\therefore$ 5 parts = 40
1 part = 8
2 parts = 16
$\therefore$ There are 16 adults.

**16** **a** 3:7 = smaller:longer
$\therefore$ 7 parts = 42 cm
1 part = 6 cm
3 parts = 18 cm
$\therefore$ The smaller piece of timber is 18 cm.
**b** Original length = 42 cm + 18 cm
= 60 cm
$\therefore$ The original length of timber is 60 cm.

**17** James:Nick:Ben = 4:3:2
3 parts = \$630
$\therefore$ 1 part = \$210
$\therefore$ 4 parts = \$840
$\therefore$ James contributed \$840.

**18** Old width:new width = 2:5
5 parts = 25 cm
1 part = 5 cm
$\therefore$ 2 parts = 10 cm
The original width of the photo was 10 cm.

**19** **a** Students:teachers = 20:1 = 80:4
Teachers:assistants = 4:1
$\therefore$ Students:teachers:assistants = 80:4:1
**b** 80:4:1
4 parts = 48
$\therefore$ 1 part = 12 ← assistants
80 parts = 960 ← students
$\therefore$ 960 students and 12 school assistants.

# WORKED SOLUTIONS to Chapter 9 **Practise, Practise**

**20** Josh = \$2.00 − \$0.80
= \$1.20

∴ Ratio of contributions:

Jo:Josh = 80c:\$1.20
= 80c:120c
= 80:120
= 2:3

∴ Total parts = 2 + 3
= 5

∴ $\frac{3}{5} \times 200\,000 = 120\,000$

∴ Josh will win \$120 000.

**21** Total parts = 11 + 4
= 15

∴ 15 parts = \$75

∴ $\frac{11}{15} \times 75 = 55$

$\frac{4}{15} \times 75 = 20$

∴ \$55 and \$20 are the parts.

**22** Total parts = 2 + 4 + 3
= 9

∴ 9 parts = \$270

∴ $\frac{4}{9} \times 270 = 120$

∴ The largest share is \$120.

**23** **a** 5:7 ∴ F **b** 1:11 ∴ B
**c** 1:1 = 6:6 ∴ G **d** 3:1 = 9:3 ∴ J

**24** **a** Cost price:profit = 21:5
5 parts = \$45 000
∴ 1 part = \$9000
∴ 61 parts = \$549 000
∴ Cost price is \$549 000.

**b** Selling price = cost price + profit
= \$549 000 + \$45 000
= \$594 000
∴ Selling price is \$594 000.

**c** Cost:selling = \$549 000:\$594 000
= 549:594
= 61:66 (divide by 9)

[Note: easier way is 61:61 + 5 (profit) = 61:66.]

**25** Total parts = 4 + 11
= 15

Also, $2\frac{1}{2}$ hours = 150 minutes

∴ 15 parts = 150 mins

∴ $\frac{4}{15} \times 150 = 40$

∴ 40 minutes is spent watching advertising.

**26** **a** Voted for:voted against = 3:7

**b** $\frac{3}{10} \times 40\,000 = 12\,000$
∴ 12 000 people voted for Kentish.

**27** Pay rate = \$99.60 ÷ 8
= \$12.45

Her pay rate is \$12.45/h.

**28** **a** Cost = 39.6 × \$0.75
= \$29.70
∴ Cost will be \$29.70.

**b** Consumption = 39.6 ÷ 30
= 1.32
∴ Household uses 1.32 kL/day.

**c** Daily cost = 1.32 × \$0.75
= \$0.99
∴ Water costs 99c/day.

**29** **a** Cost = 239.25 × \$0.72
= \$411.51
∴ Petrol for the trip costs \$411.51.

**b** Distance = 91 386 − 89 472
= 1914
∴ James travels 1914 km.

**c** Rate of consumption = 1914 ÷ 239.25
= 8
∴ Rate of petrol consumption is 8 km/L.

**d** Cost/km = \$411.51 ÷ 1914
= \$0.215
∴ The cost of travelling is 21.5c per km.

**30** Rates = 190 000 × \$0.009 275
= \$1762.25

The rates charged will be \$1762.25.

# WORKED SOLUTIONS to Chapter 9 **Practise, Practise**

**31** Bricks in 6 h = 480

Bricks in 1 h = 480 ÷ 6
= 80

Bricks in 10 h = 80 × 10
= 800

∴ He would lay 800 bricks in 10 hours.

**32** **a** Hours for 3 men on 20 m = 4

Hours for 1 man on 20 m = 4 × 3
= 12

∴ The time to build the fence will be 12 hours.

**b** Hours for 1 man on 20 m = 12

Hours for 1 man on 10 m = 6

Hours for 4 men on 10 m = 6 ÷ 4
$= 1\frac{1}{2}$ hrs

∴ The time to build the fence will be 90 minutes.

**33**

| | | Cost c/25 g |
|---|---|---|
| 250 g | $1.80 | 180 ÷ 10 = 18 ∴ 18c/25 g |
| 375 g | $2.55 | 255 ÷ 15 = 17 ∴ 17c/25 g |

∴ The better buy is 375 g honey.

**34**

| | | Cost c/500 g |
|---|---|---|
| 1 kg | $2.31 | 231 ÷ 2 = 115.5 ∴ 115.5c/500 g |
| 1.5 kg | $3.60 | 360 ÷ 3 = 120 ∴ 120c/500 g |
| 2 kg | $4.80 | 480 ÷ 4 = 120 ∴ 120c/500 g |

∴ The best buy is 1 kg washing powder.

**35** **a** 20 m/sec = 1200 m/min (multiply by 60)
= 1.2 km/min
= 72 km/h (multiply by 60)

∴ 20 m/sec = 72 km/h

**b** 16 L/h = 384 L/day (multiply by 24)
= 0.384 kL/day

∴ 16 L/h = 0.384 kL/day

**c** 3.4 cm/min = 204 cm/h (multiply by 60)
= 2.04 m/h
= 48.96 m/day (multiply by 24)
= 342.72 m/week (multiply by 7)

∴ 3.4 cm/min = 342.72 m/week

**36** **a** 84 km/h = 84 000 m/h
= 1400 m/min (divide by 60)
= $23.\dot{3}$ m/sec (divide by 60)
= $23\frac{1}{3}$ m/sec

∴ 84 km/h = $23\frac{1}{3}$ m/sec

**b** 10.08 kL/week = 10 080 L/week
= 1440 L/day (divide by 7)
= 60 L/h (divide by 24)
= 1 L/min (divide by 60)

∴ 10.08 kL/week = 1 L/min

**37** Speed = 100 km/h
= 100 km/60 min
= 2 km/1.2 min (divide by 50)

∴ Jordie will turn off in 1.2 minutes

i.e. Jordie will turn off in 1 min 12 sec.

[Calc: 1.2 [INV] [° ′ ″] or 2nd function [DMS]]

**38** Daily amount of water = 250 L × 2
= 500 L

∴ The daily amount is 500 L, i.e. 0.5 kL.

Now, daily cost = 0.5 × $0.75
= $0.375

Yearly cost = $0.375 × 365
= $136.875

∴ The yearly cost of Alice's showers is $136.88.

**39**

| Distance | Speed | Time |
|---|---|---|
| 300 km | 60 km/h | 5 h |
| 156 km | 52 km/h | 3 h |
| 760 km | 95 km/h | 8 h |
| 540 km | 216 km/h | $2\frac{1}{2}$ h |
| 20 km | 60 km/h | 20 min |
| 120 km | 240 km/h | 30 min |

**40** Time = $\frac{\text{Distance}}{\text{Speed}}$
$= \frac{320}{80}$
= 4 ∴ 4 hours

∴ 2:00 pm plus 4 hours

∴ 6:00 pm in Swan Hill

Francisco arrives in Swan Hill at 6 pm.

## WORKED SOLUTIONS to Chapter 9 **Practise, Practise**

**41** Time = 20 minutes
$= \frac{1}{3}$ hour
∴ Speed = distance ÷ time
$= 25 \div \frac{1}{3}$
= 75
∴ Her speed is 75 km/h.

**42** Time $= \frac{540}{80}$
= 6.75
∴ Time is $6\frac{3}{4}$ hours
i.e. time is 6 h 45 mins
∴ 1:30 pm plus 6 h 45 min = 8:15 pm
∴ Jenny arrives at 8:15 pm.

**43** Car A: Time $= \frac{975}{75}$
= 13
∴ Car A takes 13 hours and arrives at 1 pm
Car B: Time $= \frac{975}{90}$
$= 10.8\dot{3}$
$= 10\frac{5}{6}$
∴ Car B takes 10 h 50 mins and arrives at 12:50 pm.
∴ Car B arrives first and A arrives 10 minutes after B.

**44** **a** 4 cm:1 m = 4 cm:100 cm
= 4:100
= 1:25

**b** 1 cm:3 km
= 1 cm:300 000 cm
= 1:300 000

**c** 5 mm:1 km
= 5 mm:1 000 000 mm
= 5:1 000 000
= 1:200 000

**d** 3 cm:2 m
= 3 cm:200 cm
= 3:200

**45** **a** Length = 42 × 100 mm
= 4200 mm
∴ Length is 4200 mm, i.e. 4.2 m.

**b** Width = 31 × 100 mm
= 3100 mm
∴ Width is 3100 mm, i.e. 3.1 m.

**c** Length = 6.2 × 100 cm
= 620 cm
∴ Length is 620 cm, i.e. 6.2 m.

**46** **a** **i** Distance = 3 × 25 km
= 75
∴ The actual distance is 75 km.

**ii** Distance = 9.6 × 25
= 240
∴ The actual distance is 240 km.

**b** **i** Distance = 75 ÷ 25
= 3
∴ The distance on the map is 3 cm.

**ii** Distance = 125 ÷ 25
= 5
∴ The distance on the map is 5 cm.

**47** **a** 5 cm:35 m
= 5 cm:3500 cm
= 5:3500
= 1:700
The scale as a ratio is 1:700.

**b** Width = 3.5 × 700
= 2450
∴ The width of the hall is 2450 cm, i.e. 24.5 metres.

**c** Dance floor:
Length = 2.2 cm
Actual length = 2.2 × 700
= 1540
The length of the dance floor is 1540 cm, i.e. 15.4 m.
Width = 1.6 cm
Actual width = 1.6 × 700
= 1120
∴ The width of the dance floor is 1120 cm, i.e. 11.2 m.
The dimensions are 15.4 metres by 11.2 metres.

**48** **a** 1:500 = 1 cm:500 cm
= 1 cm:5 m
∴ Scale is 1 cm = 5 m

**b** Width = 3.5 cm
$\therefore$ width = $3.5 \times 5$ m
= 17.5 m

**c** 25 m by 12.5 m means 5 cm by 2.5 cm on the scale drawing—after measurement, we cannot find a position on the plan for a pool this size.

## Chapter 10—Geometry and Congruence

pp. 211–221

**1** **a** Complementary angles add up to 90°.
**b** Supplementary angles add up to 180°.
**c** Angles on a straight line add up to 180°.
**d** Vertically opposite angles are always equal.
**e** Angles at a point always add up to 360°.
**f** The angle sum of a triangle is 180°.
**g** The angle sum of a quadrilateral is 360°.
**h** The base angles of an isosceles triangle are equal.
**i** Any angle of an equilateral triangle is equal to 60°.
**j** For a pair of parallel lines cut by a transversal:
- **i** alternate angles are equal
- **ii** corresponding angles are equal
- **iii** co-interior angles are supplementary.

**2** **a** $x = 50$ (angles on a straight line)
**b** $x = 47$ (complementary angles)
**c** $x = 138$ (co-interior angles in parallel lines)
**d** $y = 80$ ($\angle$AOC is vertically opposite to $\angle$BOD)
$x = 100$ ($\angle$COB is supplementary to $\angle$BOD)
$z = 100$ ($\angle$DOA is vertically opposite to $\angle$COB)
**e** $x = 120$ (angles at a point)
**f** $x = 76$ (angles in a triangle)
**g** $\angle$BCA = $\angle$ABC = 61° (base angles of isosceles $\triangle$ABC)
$\therefore x + 61 + 61 = 180$ (angles sum of $\triangle$ABC)
$\therefore x + 122 = 180$
$\therefore x = 58$
**h** $y = 60$ (angle in an equilateral triangle)
**i** $\angle$RPQ = $\angle$QRP = $x$° (base angles of isosceles $\triangle$PQR)
$\therefore x + x + 70 = 180$ (angle sum of $\triangle$PQR)
$\therefore 2x + 70 = 180$
$\therefore 2x = 110$
i.e. $x = 55$
**j** $x = 63 + 44$ (exterior angle of $\triangle$ABC)
$x = 107$
**k** $x = 105$ (angle sum of quadrilateral ABCD)
**l** $\angle$BCQ = 72° (vertically opposite to $\angle$DCP)
$x + 72 = 180$ ($\angle$BCQ and $\angle$NBC are
$\therefore x = 108$ co-interior and MN || PQ)
**m** $\angle$FGA = 67° (vertically opposite to $\angle$HGB)
$\therefore x + 67 = 180$ ($\angle$CFG is co-interior to
$x = 113$ $\angle$FGA and AB || CD)
**n**

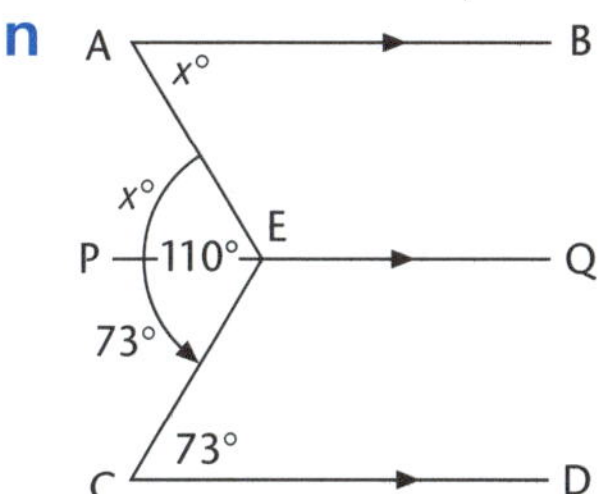


Construct a line PQ through E which is parallel to AB.
$\therefore \angle$PEA = $x$°
(alternate to $\angle$BAE and AB || PQ)
Also $\angle$CEP = 73°
(alternate to $\angle$ECD and PQ || CD)
$\therefore x + 73 = 110$
i.e. $x = 37$
**o** $x + 70 = 125$ ($\angle$PRS is exterior angle of
$\therefore x = 55$ $\triangle$PQR)
**p** $\angle$BCD = 71° (angle sum of quadrilateral ABCD)
$\therefore x = 109$ (supplementary to $\angle$BCD)

**3** **a** **i** A trapezium is a quadrilateral with one pair of opposite sides parallel.
**ii** Not necessarily.
**b** **i** Opposite sides are equal.
**ii** Opposite angles are equal.
**iii** Not unless it's a rhombus.

# WORKED SOLUTIONS to Chapter 10 **Practise, Practise**

**c** **i** All sides are equal.
**ii** 2
**iii** Yes
**iv** Yes, a special type of parallelogram.

**d** **i** 4
**ii** Equal
**iii** 90°
**iv** They are equal.

**e** **i** 2
**ii** Opposite sides are equal.
**iii** 90°
**iv** They are equal.

**f** **i** 1
**ii** 2 pairs of adjacent sides equal.

**g** A, C, D, E, F

**h** B, C, E, F

**i** **i** F  **ii** T
**iii** T  **iv** T
**v** F  **vi** T
**vii** T  **viii** T

**j** **i** Parallelogram

**4** **a** $\angle ABC = x°$ (alternate to $\angle DCB$ and AB || CD)
$\therefore x + 80 + 140 = 360$ (angles about point B)
$\therefore x = 140$

**b** $x = 110$ ($\angle DCB$ is co-interior to $\angle CBA$ and AB || DC)
$\angle ECD = 360° - (110 + 148)°$ (angles about point C)
$= 102°$
$\angle CDE = y°$ (= $\angle DEC$, base angles of isosceles $\triangle CDE$)
$\therefore y + y + 102 = 180$ (angle sum of $\triangle CDE$)
$\therefore 2y = 78$
$y = 39$

**c** $\angle BCA = 70°$ (supplementary to $\angle ACD$)
$\angle ABC = \angle BCA = 70°$ (base angles of isosceles $\triangle ABC$)
$x + 70 + 70 = 180$ (angle sum of $\triangle ABC$)
$\therefore x + 140 = 180$
i.e. $x = 40$

**d** $\angle RSQ = 48°$ (corresponding to $\angle CRB$ and AB || PQ)
$\angle QSD = 132°$ (supplementary to $\angle RSQ$)

**5** **a** $\angle GFD = a°$ (corresponding to $\angle HGB$ and AB || CD)
$\therefore$ EFC $= a°$ (vertically opposite to $\angle GFD$)

**b** $\angle ABC = 70°$ (vertically opposite to $\angle EBD$)
$\angle BCA = 70°$ (supplementary to $\angle ACF$)
$\angle ABC = \angle BCA$, therefore $\angle ABC$ is isosceles.
Therefore, AB = AC.

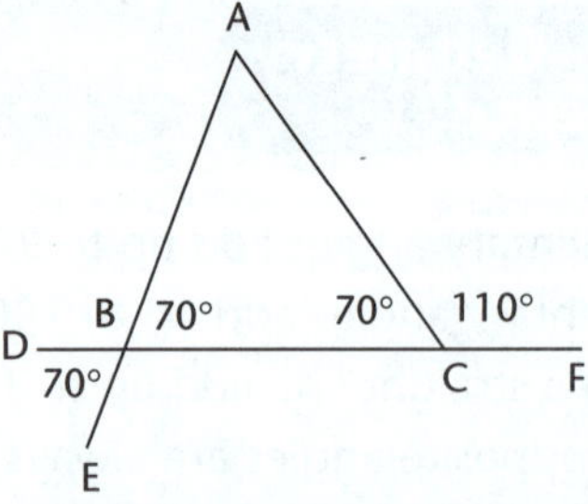


**c** $\angle BAE = a°$ (alternate to $\angle ABC$ and EF || BG)
$\angle FAC = (180 - x)°$ (co-interior to $\angle ACG$ and EF || BG)
$\angle BAE + \angle CAB + \angle FAC = 180°$ (angles in a line)
$\therefore a + b + 180 - x = 180$
$\therefore a + b - x = 0$
$\therefore x = a + b$

**d** $a + b + c + a + b + c = 180$ (angles about a line)
$2a + 2b + 2c = 180$
$2(a + b + c) = 180$
$\therefore a + b + c = 90$
But $\angle GBC = (a + b + c)°$
$= 90°$

**e** $a + b + a + b = 180$ (angles in a straight line)
$\therefore 2a + 2b = 180$
$2(a + b) = 180$
$\therefore a + b = 90$
$\angle DBF = (a + b)° = 90°$
Therefore, DB ⊥ FB.

[Note: ⊥ means 'is perpendicular to' (at 90°)]

**f** $\angle ABC = (180 - x)°$ (co-interior to $\angle DAB$ and AD || BC)
$\angle BCD + \angle ABC = 180°$ (co-interior angles and AB || DC)
$\angle BCD + (180 - x)° = 180°$
$\angle BCD - x° = 0°$
$\angle BCD = x°$

**6** 1 – 13, 2 – 21, 3 – 12, 4 – 11, 5 – 16, 6 – 17, 7 – 20, 8 – 15, 9 – 18, 10 – 19

**7** **c** Yes  **d** Yes

**8** c Yes d Yes e No f Yes

**9** a Yes b Yes c Yes

**10** a AD b QP c ∠QMN
d 25 cm e 34 $cm^2$

**11** a 7 cm b 154 $cm^2$

**12** a MN or NM
b AD or BC (depending on if you took Figure 2 as a reflection or rotation of Figure 1)
c ∠MQP or ∠NPQ
d ∠BAD or ∠ABC

**13** a BC b 10 cm
c ∠GHF d 90°
e 24 cm f 24 $cm^2$

**14** a QP b SR

**15** a AB = NM, AC = NP, BC = MP
b ∠A = ∠N, ∠B = ∠M, ∠C = ∠P

**16** d Yes
e AB = FH, AC = FG, BC = HG, ∠A = ∠F, ∠B = ∠H, ∠C = ∠G

**17** a 5 cm b 20 cm
c 25 $cm^2$ d No

**18** a SAS b AAS c SSS
d RHS e AAS

**19** a In △s PSR and PQR
PS = PQ (S) (given)
SR = QR (S) (given)
PR is common (S)
∴ △PSR ≡ △PQR (SSS)

[Note the setting out of the proof.
Note: ≡ means 'is congruent to'.]

b
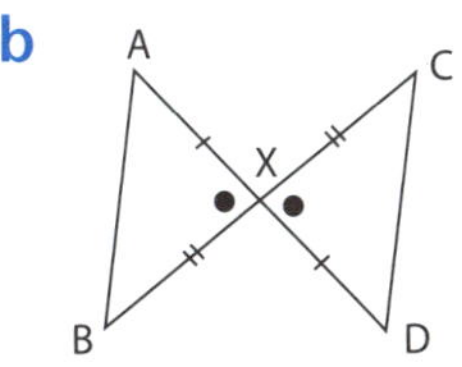


In △s ABX and DCX
AX = DX (S) (given)
∠BXA = ∠CXD (A) (vertically opposite)
BX = CX (S) (given)
∴ △ABX ≡ △DCX (SAS)

c
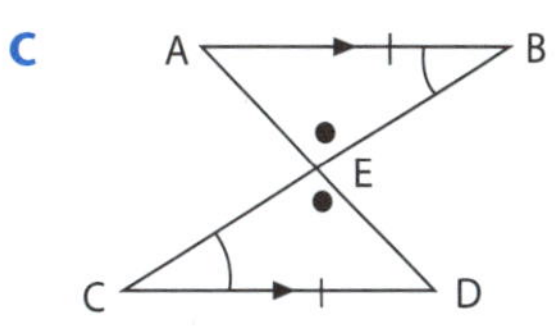


In △s BAE and CDE
∠AEB = ∠DEC (A)
(vertically opposite angles)
∠EBA = ∠DCE (A)
(alternate ∠s and AB || CD)
AB = DC (S) (given)
∴ △BAE ≡ △CDE (AAS)

d
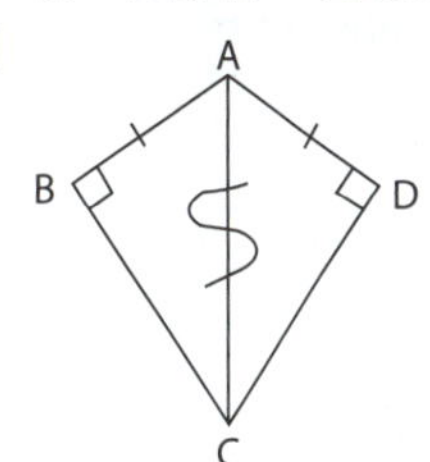


In △s ABC and ADC
∠ABC = ∠ADC (R) (= 90°, given)
AC is common (H)
AB = AD (S) (given)
∴ △ABC ≡ △ADC (RHS)

## Chapter 11—Statistics

pp. 239–244

**1**

| Number of ts | Tally | Frequency |
|---|---|---|
| 2 | \|\|\|\| | 4 |
| 3 | \|\|\|\| | 4 |
| 4 | ~~\|\|\|\|~~ | 5 |
| 5 | ~~\|\|\|\|~~ | 5 |
| 6 | \| | 1 |
| 7 | \| | 1 |
| | Σ | 20 |

a i 5 ii 18
iii 2

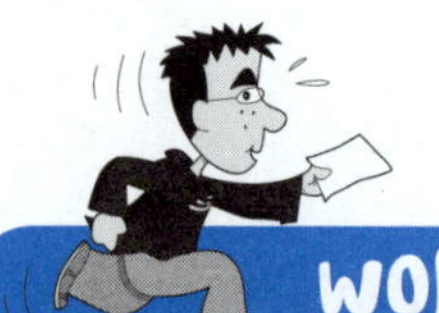

# WORKED SOLUTIONS to Chapter 11 **Practise, Practise**

**b** **i** $\frac{5}{20} = \frac{1}{4}$  **ii** $\frac{18}{20} = \frac{9}{10}$

**iii** $\frac{2}{20} = \frac{1}{10}$

**c** Range $= 7 - 2$
$= 5$

Mode = 4 and 5 (both have frequency of 5).

**d**

Frequency
5
4
3
2
1
2 3 4 5 6 7
Number of ts

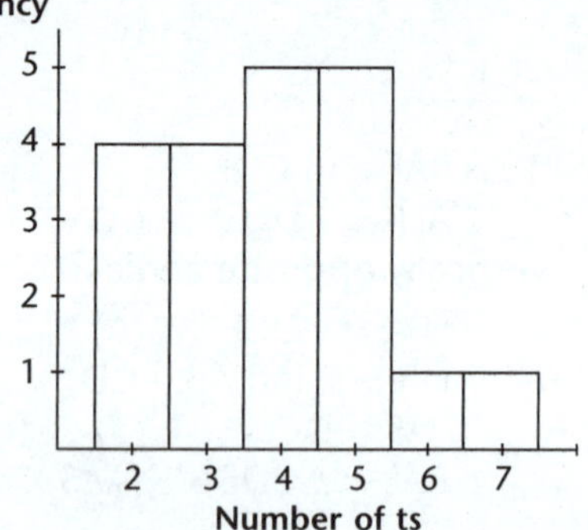

**2**

| Class | Class centre | Tally | Frequency |
|---|---|---|---|
| 141–145 | 143 | I | 1 |
| 146–150 | 148 | 卌 III | 8 |
| 151–155 | 153 | III | 3 |
| 156–160 | 158 | 卌 I | 6 |
| 161–165 | 163 | 卌 II | 7 |
| 166–170 | 168 | III | 3 |
| 171–175 | 173 | II | 2 |

**3**

| Stem | Leaf |
|---|---|
| 5 | 899 |
| 6 | 33578999 |
| 7 | 111234566689 |
| 8 | 0111238 |

**4**

| Number of lumps ($x$) | Tally | ($f$) Frequency |
|---|---|---|
| 0 | I | 1 |
| 1 | I | 1 |
| 2 | III | 3 |
| 3 | III | 3 |
| 4 | 卌 | 5 |
| 5 | IIII | 4 |
| 6 | 卌 | 5 |
| 7 | III | 3 |
| | $\Sigma$ | 25 |

**a** **i** 4  **ii** 17
**iii** 13  **iv** 8

**b** **i** $\frac{4}{25}$  **ii** $\frac{17}{25}$  **iii** $\frac{13}{25}$  **iv** $\frac{8}{25}$

**c**

Frequency
5
4
3
2
1
0
0 1 2 3 4 5 6 7 8
Number of lumps

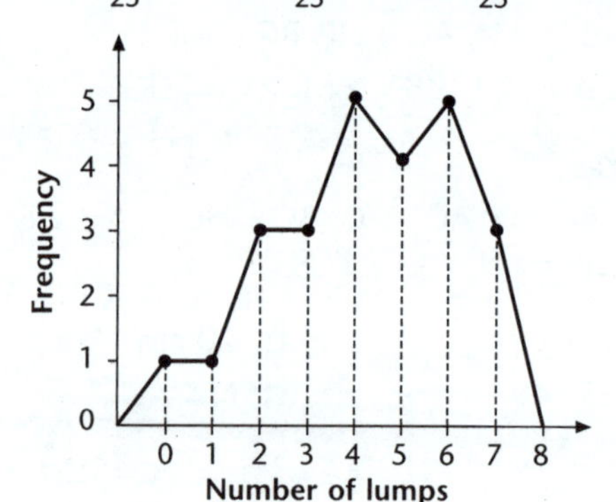

**5**

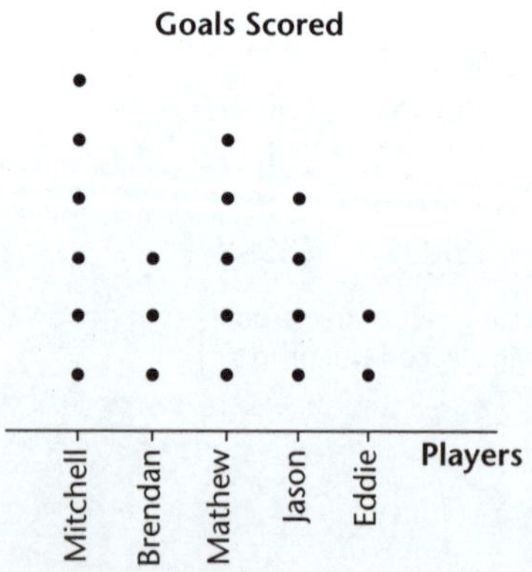

Goals Scored
Mitchell
Brendan
Mathew
Jason
Eddie
Players

**6**

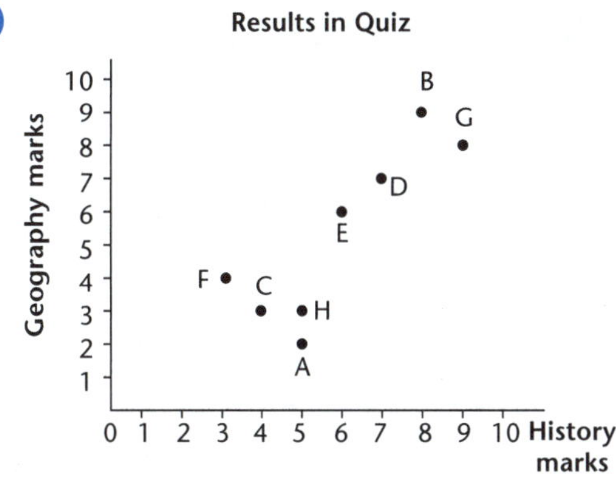


**7** Range = 8 − (−1)
= 8 + 1
∴ The range is 9.

**8** **a** Mean = $\frac{39}{7}$
= 5.6 (1 decimal place)
Mode = 6
3 scores 3 scores
For median: 3, 4, 6, (6), 6, 7, 7
∴ Median = 6
Range = 4

**b** Mean = $\frac{61}{8}$ = 7.625
Mode = 5
3 scores 3 scores
For median: 5, 5, 6, (7, 8), 9, 10, 11
Median is between 7 and 8.
∴ Median = $\frac{7+8}{2}$ = 7.5
Range = 6

**c** Mean = $\frac{36}{9}$ = 4
Mode = 1 and 4 and 7
For median: 1, 1, 1, 4, (4), 4, 7, 7, 7
∴ Median = 4
Range = 6

**d** Mean = $\frac{47}{12}$
= 3.9 (to 1 decimal place)
Mode = 7
5 scores 5 scores
For median: 0, 0, 1, 2, 2, (3, 4), 5, 7, 7, 7, 9
∴ Median = $\frac{3+4}{2}$
= 3.5 (median between 3 and 4)
Range = 9

**e** Mean = $\frac{105}{15}$
= 7
Mode = 7
For median: 3, 4, 4, 4, 5, 6, 7, (7), 7, 7, 9, 9, 11, 11, 11
∴ Median = 7
Range = 8

**9**

| Shoe size ($x$) | Number of children ($f$) | $x \times f$ |
|---|---|---|
| 3 | 8 | 24 |
| $3\frac{1}{2}$ | 8 | 28 |
| 4 | 6 | 24 |
| $4\frac{1}{2}$ | 12 | 54 |
| 5 | 10 | 50 |
| $5\frac{1}{2}$ | 8 | 44 |
| 6 | 6 | 36 |
| $6\frac{1}{2}$ | 4 | 26 |
| 7 | 2 | 14 |
| Σ | 64 | 300 |

**a** 64 children were surveyed.

**b** **i** 44 **ii** 20

**c** **i** $\frac{10}{64} = \frac{5}{32}$ **ii** $\frac{44}{64} = \frac{11}{16}$
**iii** $\frac{20}{64} = \frac{5}{16}$

**d** Range = 7 − 3
= 4
Mode = $4\frac{1}{2}$

**e** Mean = $\frac{300}{64}$
= 4.6875

**f** Median = $4\frac{1}{2}$ (look for 32nd and 33rd scores)

# WORKED SOLUTIONS to Chapter 11 **Practise, Practise**

**g**

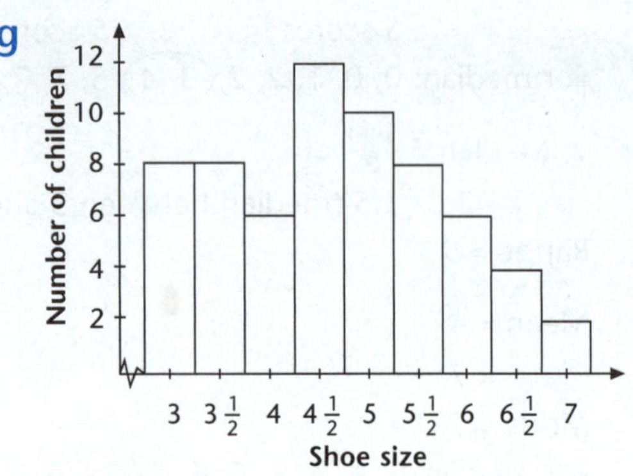


**h** $4\frac{1}{2}$—as this is the most common size worn. Mode.

**10** **a**

| Score ($x$) | Frequency ($f$) | $x \times f$ |
|---|---|---|
| 1 | 3 | 3 |
| 2 | 6 | 12 |
| 3 | 8 | 24 |
| 4 | 10 | 40 |
| 5 | 12 | 60 |
| 6 | 0 | 0 |
| 7 | 4 | 28 |
| 8 | 2 | 16 |
| 9 | 6 | 54 |
| 10 | 1 | 10 |
| | 52 | 247 |

**b** Range = 10 – 1
= 9

Mode = 5

**c** **i** 12 **ii** 39
**iii** 27 **iv** 52 – 39 = 13

**d** 52

**e** **i** $\frac{12}{52}$ or $\frac{3}{13}$ **ii** $\frac{39}{52}$ or $\frac{3}{4}$
**iii** $\frac{27}{52}$ **iv** $\frac{13}{52}$ or $\frac{1}{4}$

**f** Mean = $\frac{247}{52}$
= 4.75

**g** Median = 4 (26th and 27th scores)

**11** **a** **i** Mean = $\frac{8+6+5+9+5+3}{6}$
$= \frac{36}{6}$
$= 6$
Mean number of goals = 6

**ii** Range = 9 – 3 = 6
Mode = 5

**iii** For median: 3, 5, (5, 6), 8, 9
∴ Median = $\frac{5+6}{2}$ = 5.5

**b** Mean = $\frac{8+6+5+9+5+3+13}{6}$
$= \frac{49}{7}$
$= 7$
For median: 3, 5, 5, (6), 8, 9, 13
∴ Median is 6.
The mean increases by 1, while the median is 0.5 larger.

**c** Gai has scored (8 × 6.5) goals = 52 goals.
After 7 seasons Gai had scored 49 goals.
∴ Gai scored 3 goals in her 8th season.

**12** Zoran's total marks = 4 × 65 (after 4 results)
= 260
To average 70 Zoran must score (6 × 70) marks, i.e. 420 marks.
Zoran must average $\frac{160}{2}$ in his other two papers, i.e. average 80.
Zoran must average 80 in each of his final papers.

**13** **a** **i**

| Number of children (Score) $x$ | Number of families (Frequency) $f$ | $x \times f$ |
|---|---|---|
| 1 | 2 | 2 |
| 2 | 5 | 10 |
| 3 | 4 | 12 |
| 4 | 4 | 16 |
| 5 | 3 | 15 |
| 6 | 1 | 6 |
| 7 | 1 | 7 |
| Σ | 20 | 68 |

**ii** 20 **iii** 68

**b** Range = 7 − 1 = 6 Mode = 2

**c** **i** 4 **ii** 11

**iii** 5

**d** **i** $\frac{4}{20}$ or $\frac{1}{5}$ **ii** $\frac{11}{20}$

**iii** $\frac{5}{20}$ or $\frac{1}{4}$

**e** Mean = $\frac{68}{20}$ = 3.4

**f** Median = 3 (10th and 11th scores)

**14** **a** The outlier is 7.

**b** Range = 40 − 7 = 33

∴ Range is 33.

**c** The mode is 36.

**d**

| Stem | Leaf |
|---|---|
| 0 | 7 |
| 1 | 889 |
| 2 | 46789 |
| 3 | 44666889 |
| 4 | 0 |

The median is the middle of 29 and 34.

$$\therefore \frac{29+34}{2} = \frac{63}{2} = 31.5$$

∴ Median is 31.5.

**e** Mean = $\frac{527}{18}$

= 29.28 (2 dec. places)

**15** Mean = $\frac{8+3+7+4+3}{5} = \frac{25}{5} = 5$

**a** New numbers 14, 9, 13, 10, 9

$$\text{Mean} = \frac{14+9+13+10+9}{5} = \frac{55}{5} = 11$$

Mean has increased by 6.

**b** New numbers 48, 18, 42, 24, 18

$$\text{Mean} = \frac{48+18+42+24+18}{5} = \frac{150}{5} = 30$$

Mean has been multiplied also by 6.

**16**

| | Number ($x$) | Frequency ($f$) | $fx$ | Fraction | Decimal |
|---|---|---|---|---|---|
| **a** | 1 | 19 | 19 | $\frac{19}{100}$ | 0.19 |
| **b** | 2 | 21 | 42 | $\frac{21}{100}$ | 0.21 |
| | 3 | 20 | 60 | $\frac{20}{100}$ | 0.20 |
| | 4 | 19 | 76 | $\frac{19}{100}$ | 0.19 |
| | 5 | 21 | 105 | $\frac{21}{100}$ | 0.21 |
| | | 100 | 302 | | |

The decimals are almost all the same value. It shows there is an equal chance of the wheel stopping on all numbers.

**c** Mean = $\frac{302}{100}$

= 3.02

**d** Mode = 2 and 5

Median is between the 50th and 51st scores, i.e. median = 3.

**17**

Average (5 innings) = 15.8

∴ Total (5 innings) = 15.8 × 5

= 79

Average (next 10 innings) = 19.1

∴ Total = 19.1

= 191

Total over 15 innings = 191 + 79

= 270

Average = $\frac{270}{15}$

= 18

New total (after score of 34) = 270 + 34

= 304

New average (over 16 innings) = $\frac{304}{16}$ = 19

Season average = 19

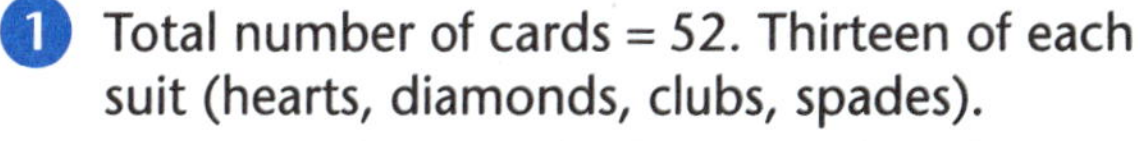

## Chapter 12—Probability

pp. 257–261

**1** Total number of cards = 52. Thirteen of each suit (hearts, diamonds, clubs, spades).

For example, consider hearts and we have:

2♥, 3♥, 4♥, 5♥, 6♥, 7♥, 8♥, 9♥, 10♥, J♥, Q♥, K♥, A♥

There are four of each number.

For example, consider Aces: A♥, A♦, A♠, A♣

## WORKED SOLUTIONS to Chapter 12 **Practise, Practise**

**a** $P(\text{Ace}) = \frac{4}{52} = \frac{1}{13}$

**b** $P(\text{King}) = \frac{4}{52} = \frac{1}{13}$

**c** $P(\heartsuit) = \frac{13}{52} = \frac{1}{4}$

**d** $P(\spadesuit) = \frac{13}{52} = \frac{1}{4}$

**e** $P(\text{Black}) = \frac{26}{52} = \frac{1}{2}$

**f** $P(\text{J, Q, K}) = \frac{12}{52} = \frac{3}{13}$

**g** $P(\text{Not picture card})$

$= 1 - \frac{3}{13}$ [total probability must add up to 1]

$= \frac{10}{13}$

**h** $P(2, 3, 4, 5, 6) = \frac{20}{52} = \frac{5}{13}$

**i** $P(6\heartsuit) = \frac{1}{52}$

**j** $P(\text{Red or } 6) = \frac{28}{52} = \frac{7}{13}$

[26 red cards + 2 other 6s: 6 clubs, 6 spades.
6 diamonds and 6 hearts have already been counted.]

**k** $P(\text{Red } 6) = \frac{2}{52} = \frac{1}{26}$

**2** $P(\text{Not rain}) = 1 - \frac{7}{10} = \frac{3}{10}$

**3** No. of jelly beans = 5 + 6 + 4 = 15

**a** $P(\text{Red}) = \frac{5}{15} = \frac{1}{3}$

**b** $P(\text{Not red}) = 1 - \frac{1}{3}$

$= \frac{2}{3}$

**c** $P(\text{Black}) = \frac{6}{15} = \frac{2}{5}$

**d** $P(\text{Red or white}) = P(\text{Not black})$

$= 1 - \frac{2}{5}$

$= \frac{3}{5}$

**e** $P(\text{Green}) = 0$ [i.e. impossible as there are no green jelly babies]

**f** $P(\text{White}) = \frac{4}{15}$

**g** $P(\text{Not white}) = 1 - \frac{4}{15}$

$= \frac{11}{15}$

**4** **a** $P(\text{Female}) = 1 - 0.48$

$= 0.52$

**b** Number of male babies $= 0.48 \times 1400$

$= 672$

672 male babies would have been expected.

**5**

| + | 1 | 2 | 3 | 4 | 5 | 6 |
|---|---|---|---|---|---|---|
| 1 | 2 | 3 | 4 | 5 | 6 | 7 |
| 2 | 3 | 4 | 5 | 6 | 7 | 8 |
| 3 | 4 | 5 | 6 | 7 | 8 | 9 |
| 4 | 5 | 6 | 7 | 8 | 9 | 10 |
| 5 | 6 | 7 | 8 | 9 | 10 | 11 |
| 6 | 7 | 8 | 9 | 10 | 11 | 12 |

(36 possible combinations)

**a** $P(8) = \frac{5}{36}$

**b** $P(5) = \frac{4}{36} = \frac{1}{9}$

**c** $P(\text{Odd}) = \frac{18}{36} = \frac{1}{2}$

**d** $P(>9) = P(10, 11, 12)$

$= \frac{6}{36} = \frac{1}{6}$

**e** $P(\text{9 or greater}) = P(9, 10, 11, 12)$

$= \frac{10}{36} = \frac{5}{18}$

**f** $P(\text{Not } 9) = 1 - P(9)$

$= 1 - \frac{4}{36}$

$= \frac{32}{36} = \frac{8}{9}$

**g** $P(\text{Odd or less than } 5) = P(2, 3, 4, 5, 7, 9, 11)$

$= \frac{22}{36}$

$= \frac{11}{18}$

**h** $P(\text{Odd and less than } 5) = P(3)$

$= \frac{2}{36} = \frac{1}{18}$

**6**

| + | 1 | 2 | 3 | 4 |
|---|---|---|---|---|
| **1** | 2 | 3 | 4 | 5 |
| **2** | 3 | 4 | 5 | 6 |
| **3** | 4 | 5 | 6 | 7 |
| **4** | 5 | 6 | 7 | 8 |

(16 combinations)

**a** $P(5) = \frac{4}{16} = \frac{1}{4}$

**b** $P(7) = \frac{2}{16} = \frac{1}{8}$

**c** $P(1) = 0$

**d** $P(>5) = \frac{6}{16} = \frac{3}{8}$

# WORKED SOLUTIONS to Chapter 12 **Practise, Practise**

**e** $P(\text{at least }5) = P(5, 6, 7, 8)$
$= \frac{10}{16} = \frac{5}{8}$

**7** **a** $P(9) = \frac{1}{10}$ [only one numbered nine]

**b** $P(\text{Even}) = P(2, 4, 6, 8, 10)$
$= \frac{5}{10} = \frac{1}{2}$

**c** $P(<6) = P(1, 2, 3, 4, 5)$
$= \frac{5}{10} = \frac{1}{2}$

**d** $P(6 \text{ or more}) = 1 - P(<6)$
$= 1 - \frac{1}{2}$
$= \frac{1}{2}$

**8** **a**

| Destinations | | |
|---|---|---|
| **State/Territory** | **Freq.** | **Rel. freq.** |
| SA | 18 | 0.18 |
| Victoria | 20 | 0.2 |
| NSW | 22 | 0.22 |
| Qld | 18 | 0.18 |
| Tasmania | 12 | 0.12 |
| WA | 6 | 0.06 |
| Northern Territory | 4 | 0.04 |
| Total | 100 | 1 |

**b** **i** $P(\text{NSW}) = 0.22$

**ii** $P(\text{WA}) = 0.06$

**iii** $P(\text{Vic or Tas}) = 0.2 + 0.12$
$= 0.32$

**iv** $P(\text{Not Qld}) = 1 - 0.18$
$= 0.82$

**v** $P(\text{Not NT or NSW}) = 1 - (0.04 + 0.22)$
$= 1 - 0.26$
$= 0.74$

**9**

| Score | Freq. | Rel. freq. |
|---|---|---|
| 7 | 5 | 0.1 |
| 8 | 15 | 0.3 |
| 9 | 20 | 0.4 |
| 10 | 10 | 0.2 |
| Total | 50 | 1 |

**10** **a**

| Seventeen-Year-Old Drivers | | | |
|---|---|---|---|
| | **Licensed** | **Not licensed** | **Total** |
| Female | 8 | 14 | 22 |
| Male | 12 | 16 | 28 |
| Total | 20 | 30 | 50 |

**b** 50 people

**c** **i** $\frac{8}{50} = \frac{4}{25}$ **ii** $\frac{16}{50} = \frac{8}{25}$

**d** $\frac{14}{22} = \frac{7}{11}$

**11** **a**

| Completed Assignment | | | |
|---|---|---|---|
| | **English completed** | **English not completed** | **Total** |
| Geography completed | 19 | 4 | 23 |
| Geography not completed | 2 | 3 | 5 |
| Total | 21 | 7 | 28 |

**b** **i** $\frac{4}{28} = \frac{1}{7}$ **ii** $\frac{3}{28}$

**12** **a** **i** 7 **ii** 10 **iii** 3

**b** $10 + 7 + 5 + 3 = 25$
$\therefore$ 25 people

**c** **i** $\frac{5}{25} = \frac{1}{5}$ **ii** $\frac{7}{25}$

**d**

| | **Likes Sport** | **Does not like Sport** | **Total** |
|---|---|---|---|
| **Likes Lifestyle** | 7 | 5 | 12 |
| **Does not like L/S** | 10 | 3 | 13 |
| **Total** | 17 | 8 | 25 |

**13** **a**



Firstly, $50 - 16 = 34$
Also, $25 + 18 - 34 = 9$
$\therefore$ 9 students like both.

**b** **i** $P(\text{Coffee and tea}) = \frac{9}{50}$

**ii** $P(\text{Tea, not coffee}) = \frac{9}{50}$

**14** **a**

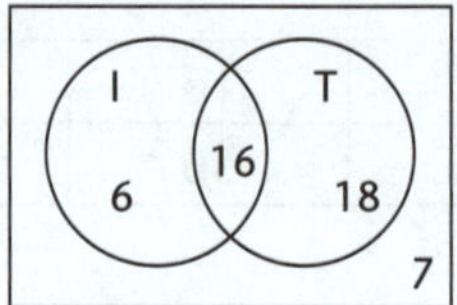


I: Indian
T: Thai

**b** $\text{Total} = 6 + 16 + 18 + 7$
$= 47$

$\therefore$ 47 interviewed

**c** $P(\text{Indian, not Thai}) = \frac{6}{47}$

**15** **a**

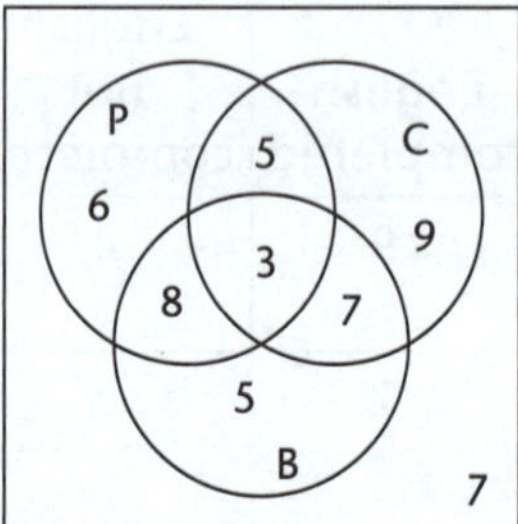


**b** **i** $P(\text{Chemistry}) = \frac{24}{50}$
$= \frac{12}{25}$

**ii** $P(\text{Biology and Physics}) = \frac{11}{50}$

**iii** $P(\text{Not Biology}) = 1 - \frac{23}{50}$
$= \frac{27}{50}$

# WORKED SOLUTIONS to Chapter 2 **Tests**

## Chapter 2—Number

pp. 22–24

### Level 1 Test

**1** a 16: 1, 2, 4, 8, 16 ✓ (1 mark)
b 35: 1, 5, 7, 35 ✓ (1 mark)
c 50: 1, 2, 5, 10, 25, 50 ✓ (1 mark)

**2** a 12: 1, (2), 3, 4, 6, 12 ✓
26: 1, (2), 13, 26 ✓
∴ Highest Common Factor of 2 (2 marks)
b 20: 1, 2, (4), 5, 10, 20 ✓
32: 1, 2, (4), 8, 16, 32 ✓
∴ Highest Common Factor of 4 (2 marks)
c 18: 1, 2, 3, (6), 9, 18 ✓
30: 1, 2, 3, 5, (6), 10, 15, 30 ✓
∴ Highest Common Factor of 6 (2 marks)

**3** a 3: 3, 6, 9, 12, 15 ✓ (1 mark)
b 5: 5, 10, 15, 20, 25 ✓ (1 mark)
c 7: 7, 14, 21, 28, 35 ✓ (1 mark)

**4** a 4: 4, 8, (12), 16, … ✓
6: 6, (12), 18, … ✓
∴ Lowest Common Multiple of 12 (2 marks)
b 3: 3, 6, 9, 12, (15), 18, … ✓
5: 5, 10, (15), 20, … ✓
∴ Lowest Common Multiple of 15 (2 marks)
c 6: 6, 12, 18, (24), 30, … ✓
8: 8, 16, (24), 32, … ✓
∴ Lowest Common Multiple of 24 (2 marks)

**5** a 4, 6, 8, 9, 10 ✓ (1 mark)
b 2, 3, 5, 7 ✓ (1 mark)

**6** a $2 \times 2 \times 2 \times 2 = 2^4$ ✓ (1 mark)
b $6 \times 6 \times 7 \times 7 \times 7 = 6^2 \times 7^3$ ✓ (1 mark)

**7** a $4 - (-2) = 4 + 2$
$= 6$ ✓ (1 mark)
b $-6 + (-2) = -6 - 2$
$= -8$ ✓ (1 mark)
c $3 - 7 = -4$ ✓ (1 mark)

**8** a $-3 \times 6 = -18$ ✓ (1 mark)
b $2 \times -4 = -8$ ✓ (1 mark)
c $-5 \times -3 = 15$ ✓ (1 mark)

**9** a $-12 \div 3 = -4$ ✓ (1 mark)
b $18 \div -6 = -3$ ✓ (1 mark)
c $-10 \div -2 = 5$ ✓ (1 mark)

**10** a $\frac{4}{5}$ $\quad 5\overline{)4.0}$ = 0.8
∴ 0.8 ✓ (1 mark)
b $\frac{1}{3}$ $\quad 3\overline{)1.00\ldots}$ = 0.33…
∴ $0.\dot{3}$ ✓ (1 mark)

**11** a $0.03 = \frac{3}{100}$ ✓ (1 mark)
b $0.8 = \frac{8}{10}$
$= \frac{4}{5}$ ✓ (1 mark)

**(Total 35 marks)**

### Level 2 Test

**1** a 20: 1, 2, (4), 5, 10, 20
24: 1, 2, 3, (4), 6, 8, 12, 24
∴ Highest Common Factor is 4 ✓ (1 mark)
b 25: 1, (5), 25
45: 1, 3, (5), 9, 15, 45
∴ Highest Common Factor is 5 ✓ (1 mark)
c 60: 1, 2, 3, 4, 5, 6, 10, 12, 15, (20), 30, 60
80: 1, 2, 4, 8, 10, (20), 40, 80
∴ Highest Common Factor is 20 ✓ (1 mark)

**2** a 4: 4, 8, 12, 16, (20), 24, …
10: 10, (20), 30, …
∴ Lowest Common Multiple is 20 ✓
(1 mark)
b 15: 15, 30, 45, 60, (75), 90, …
25: 25, 50, (75), 100, …
∴ Lowest Common Multiple is 75 ✓
(1 mark)
c 12: 12, 24, (36), 48, 60, 72, …
18: 18, (36), 54, 72, 90, …
∴ Lowest Common Multiple is 36 ✓
(1 mark)

# WORKED SOLUTIONS to Chapter 2 **Tests**

**3** a

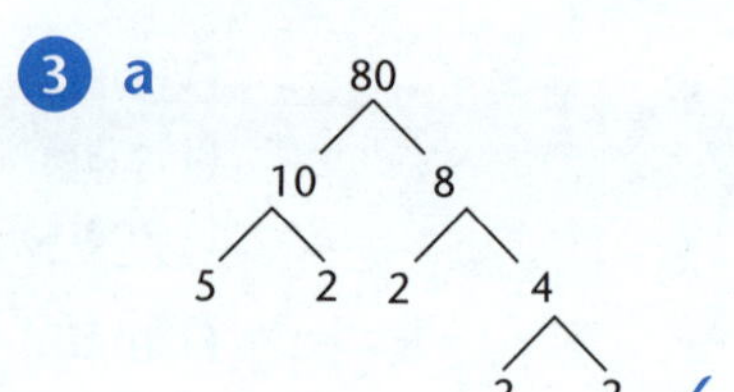


$\therefore 80 = 2 \times 2 \times 2 \times 2 \times 5$ ✓ (2 marks)

b

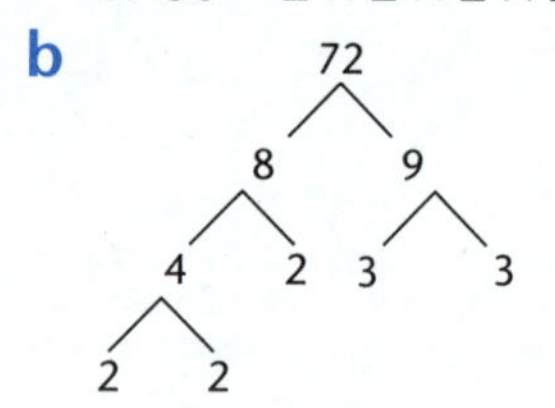


$\therefore 72 = 2 \times 2 \times 2 \times 3 \times 3$ ✓ (2 marks)

c

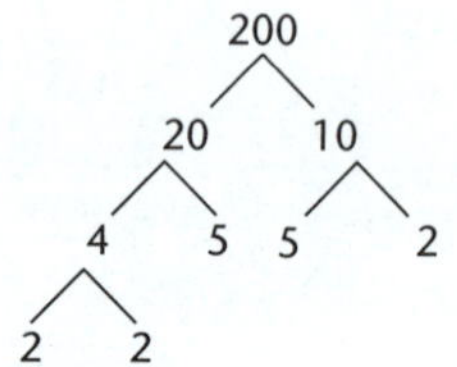


$\therefore 200 = 2 \times 2 \times 2 \times 5 \times 5$ ✓ (2 marks)

**4** a $3^4 \times 3^2 = 3^6$ ✓ (1 mark)
b $2^7 \div 2^5 = 2^2$ ✓ (1 mark)
c $(5^2)^3 = 5^6$ ✓ (1 mark)

**5** a $4^0 = 1$ ✓ (1 mark)
b $(6 - 2)^0 = 4^0$
$= 1$ ✓ (1 mark)
c $3^0 \times 5^0 = 1 \times 1$
$= 1$ ✓ (1 mark)

**6** a $-2 \times 4 = -8$ ✓ (1 mark)
b $\frac{10}{-5} = -2$ ✓ (1 mark)
c $4 - (-2) = 4 + 2 = 6$ ✓ (1 mark)

**7** a $12 - 4 \times 7 = 12 - 28$ ✓
$= -16$ ✓ (2 marks)
b $16 \div (-4) - 3 \times (-5) = -4 + 15$ ✓
$= 11$ ✓ (2 marks)
c $\frac{6-12}{3-2\times3} = \frac{-6}{3-6}$ ✓
$= \frac{-6}{-3}$
$= 2$ ✓ (2 marks)
d $(-4 - 2)(3 - 5) = -6 \times -2$ ✓
$= 12$ ✓ (2 marks)

**8** a $\frac{1}{7}$ $\quad 7\overline{)1.00000000\ldots}$ quotient $0.14285714\ldots$
$\therefore 0.\dot{1}4285\dot{7}$ ✓ (1 mark)
b $\frac{3}{11}$ $\quad 11\overline{)3.0000\ldots}$ quotient $0.2727\ldots$
$\therefore 0.\dot{2}\dot{7}$ ✓ (1 mark)
c $\frac{3}{20}$ $\quad 20\overline{)3.00}$ quotient $0.15$
$\therefore 0.15$ ✓ (1 mark)

**9** a $0.03 = \frac{3}{100}$ ✓ (1 mark)
b $0.033 = \frac{33}{1000}$ ✓ (1 mark)
c $0.3003 = \frac{3003}{10000}$ ✓ (1 mark)

**(Total 35 marks)**

## Level 3 Test

**1** a 60: 1, 2, 3, 4, 5, 6, 10, 12, 15, (20), 30, 60
100: 1, 2, 4, 5, 10, (20), 25, 50, 100
∴ Highest Common Factor of 20 ✓ (1 mark)
b 75: 1, 3, (5), 15, 25, 75
40: 1, 2, 4, (5), 8, 10, 20, 40
∴ Highest Common Factor of 5 ✓ (1 mark)

**2** 60: 1, 2, 3, 4, 5, 6, 10, 12, 15, (20), 30, 60
80: 1, 2, 4, 5, 8, 10, 16, (20), 40, 80
∴ Highest Common Factor of 20 ✓
∴ 20 stationery packs can be made. (1 mark)

**3** a 30: 30, 60, 90, (120), 150, …
40: 40, 80, (120), 160, …
∴ Lowest Common Multiple of 120 ✓ (1 mark)
b 16: 16, 32, (48), 64, …
24: 24, (48), 72, …
∴ Lowest Common Multiple of 48 ✓ (1 mark)

**4** 4: 4, 8, 12, …, (60), … ✓
5: 5, 10, …, (60), …
6: 6, 12, …, (60), …
∴ Lowest Common Multiple of 60
∴ 60 oranges ✓ (2 marks)

# WORKED SOLUTIONS to Chapters 2 and 3 **Tests**

**5** **a** Primes = 2, 3, 5, 7

Percentage $= \frac{4}{10} \times 100\%$

$= 40\%$ ✓ (1 mark)

**b** Composites = 4, 6, 8, 9, 10

Percentage $= \frac{5}{10} \times 100\%$

$= 50\%$ ✓ (1 mark)

**6** **a** $4^3 \times 4^2 \div 4^1 = 4^4$ ✓ (1 mark)

**b** $\frac{10^4}{10^3 \times 10} = \frac{10^4}{10^4}$

$= 1$ ✓ (1 mark)

**7** **a** $4^0 - 3^0 = 1 - 1$

$= 0$ ✓ (1 mark)

**b** $(4 - 3)^0 = 1^0$

$= 1$ ✓ (1 mark)

**c** $\frac{15^0 - 10^0}{5^0} = \frac{1-1}{1}$ ✓ (1 mark)

$= 0$

**8** **a** $3(-2 - 5) = 3 \times -7$ ✓

$= -21$ ✓ (2 marks)

**b** $(-16) \div (-4) + 3 \times (-2) = 4 - 6$ ✓

$= -2$ ✓ (2 marks)

**c** $\frac{15 - 4 \times 6}{3 - 6} = \frac{15 - 24}{-3}$ ✓

$= \frac{-9}{-3}$

$= 3$ ✓ (2 marks)

**d** $25 - (4 \times 3 - 6 \times 5)$

$= 25 - (12 - 30)$ ✓

$= 25 - (-18)$

$= 25 + 18$

$= 43$ ✓ (2 marks)

**9** **a** $(-1)^{2013} = -1$ ✓ (1 mark)

**b** $(-1)^{2014} = 1$ ✓ (1 mark)

**10** **a**

320
16 20
4 4 4 5
2 2 2 2 2 2 ✓

$\therefore 320 = 2 \times 2 \times 2 \times 2 \times 2 \times 2 \times 5$

$= 2^6 \times 5$ ✓ (2 marks)

**b**

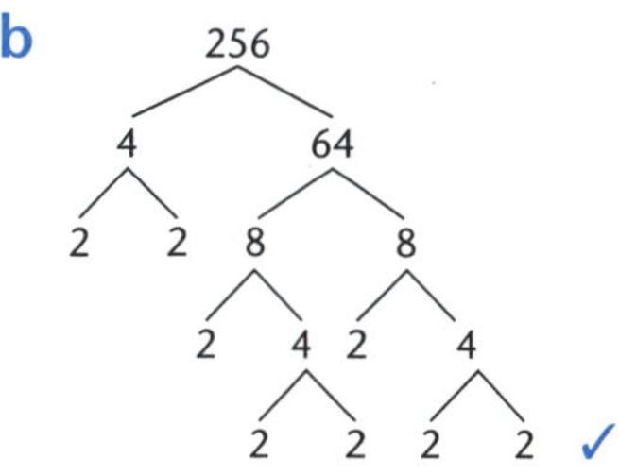


$\therefore 256 = 2 \times 2 \times 2 \times 2 \times 2 \times 2 \times 2 \times 2$

$= 2^8$ ✓ (2 marks)

**c**

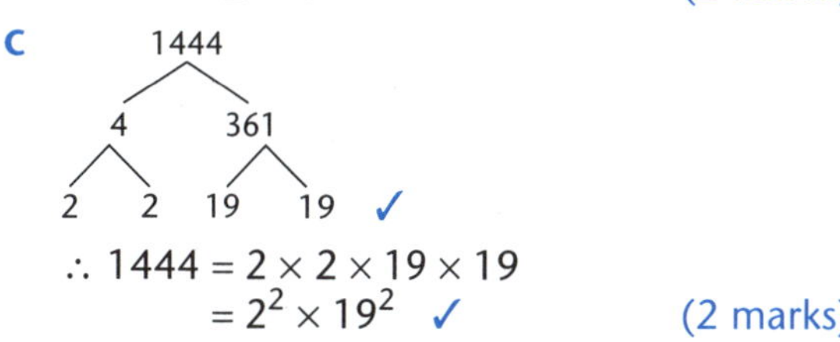


$\therefore 1444 = 2 \times 2 \times 19 \times 19$

$= 2^2 \times 19^2$ ✓ (2 marks)

**11** **a** $2a - 3b + 4c$

$= 2(-3) - 3(-4) + 4(-5)$ ✓

$= -6 + 12 - 20$

$= -14$ ✓ (2 marks)

**b** $\frac{bc + 1}{a} = \frac{(-4)(-5) + 1}{(-3)}$ ✓

$= \frac{20 + 1}{-3}$

$= \frac{21}{-3}$

$= -7$ ✓ (2 marks)

**12** $\frac{1}{99}$ $\qquad 99\overline{)1.00000}$ quotient $0.01010\ldots$

$\therefore 0.\dot{0}\dot{1}$ ✓ (1 mark)

**(Total 35 marks)**

## Chapter 3—Percentages and Applications

pp. 39–41

### Level 1 Test

**1** **a** $\frac{3}{5} \times 100\% = 60\%$ ✓ (1 mark)

**b** $1\frac{1}{2} \times 100\% = 150\%$ ✓ (1 mark)

**c** $0.45 \times 100\% = 45\%$ ✓ (1 mark)

**d** $0.8 \times 100\% = 80\%$ ✓ (1 mark)

**e** $0.06 \times 100\% = 6\%$ ✓ (1 mark)

**f** $0.355 \times 100\% = 35.5\%$ ✓ (1 mark)

**2** **a** $\frac{27}{100}$ ✓ (1 mark)

**b** $\frac{8}{100} = \frac{2}{25}$ ✓ (1 mark)

## WORKED SOLUTIONS to Chapter 3 **Tests**

c $\frac{40}{100} = \frac{2}{5}$ ✓ (1 mark)

**3** a 0.04 ✓ (1 mark)
b 0.075 ✓ (1 mark)
c 2.15 ✓ (1 mark)

**4** a $\frac{12}{48} \times 100\% = 25\%$ ✓ (1 mark)

b $\frac{320}{400} \times 100\% = 80\%$ ✓ (1 mark)

**5** a $0.15 \times 20 = 3$ ∴ \$3 ✓ (1 mark)
b $0.08 \times 600 = 48$ ∴ \$48 ✓ (1 mark)

**6** a GST = $0.1 \times 88 = 8.8$
∴ GST of \$8.80 ✓ (1 mark)
b Selling price = $88 + 8.8 = 96.8$
∴ Price is \$96.80. ✓ (1 mark)

**7** a Profit = \$3000 − \$2400
= \$600 ✓ (1 mark)

b Profit % = $\frac{600}{2400} \times 100\%$
= 25% ✓ (1 mark)

**8** 60% of capacity = 3600
10% of capacity = $3600 \div 6 = 600$ ✓
100% of capacity = $600 \times 10 = 6000$
∴ Tank holds 6000 L when full. ✓ (2 marks)

**9** Discount = $0.15 \times \$95$
= \$14.25
New price = \$95 − \$14.25
= \$80.75 ✓ (1 mark)

**10** New value = 120% of 6300
= $1.2 \times 6300$
= 7560
∴ New value of \$7560. ✓✓
or
Increase = 20% of 6300 = $0.2 \times 6300$
= 1260 ✓
∴ new value = $6300 + 1260 = 7560$
∴ New value of \$7560. ✓ (2 marks)

**(Total 25 marks)**

### Level 2 Test

**1** a 41%, $\frac{2}{5}$ = 40%, 0.42 = 42%

∴ $\frac{2}{5}$, 41%, 0.42 ✓ (1 mark)

b 37%, $\frac{3}{8}$ = 37.5%, 0.4 = 40%

∴ 37%, $\frac{3}{8}$, 0.4 ✓ (1 mark)

**2** a $0.04 \times \$290 = \$11.60$ ✓ (1 mark)
b $0.16 \times \$360 = \$57.60$ ✓ (1 mark)
c $0.085 \times \$7000 = \$595$ ✓ (1 mark)
d $0.0525 \times 4000 = 210$
∴ 210 mm ✓ (1 mark)

**3** a $\frac{30}{300} \times 100\% = 10\%$ ✓ (1 mark)

b $\frac{5}{2000} \times 100\% = 0.25\%$ ✓ (1 mark)

c $\frac{6}{120} \times 100\% = 5\%$ ✓ (1 mark)

d $\frac{35}{2000} \times 100\% = 1.75\%$ ✓ (1 mark)

**4** a 110% of \$25
= $1.1 \times \$25$ ✓
= \$27.50 ✓ (2 marks)
b 108% of \$650
= $1.08 \times \$650$ ✓
= \$702 ✓ (2 marks)

**5** a 92% of \$650
= $0.92 \times \$650$ ✓
= \$598 ✓ (2 marks)

b $90\frac{1}{2}\%$ of \$4000
= $0.905 \times \$4000$ ✓
= \$3620 ✓ (2 marks)

**6** a Discount = $0.15 \times \$45$ ✓
= \$6.75
∴ A discount of \$6.75. ✓ (2 marks)
b Discount = $0.2 \times \$2480$ ✓
= \$496
∴ A discount of \$496. ✓ (2 marks)

**7** a New price = 85% of 80
= $0.85 \times 80$
= 68
∴ New price of \$68. ✓✓
or Discount = 15% of 80
= $0.15 \times 80$
= 12 ✓
∴ new price = $80 - 12 = 68$
∴ New price of \$68. ✓ (2 marks)

**b** New price = 85% of 18
= 0.85 × 18
= 15.3
∴ New price of $15.30. ✓✓
or Discount = 15% of 18
= 0.15 × 18
= 2.7 ✓
∴ new price = 18 − 2.7 = 15.3
∴ New price of $15.30. ✓ (2 marks)

**8** Profit = 1500 − 1200 = 300
∴ profit of $300 ✓
% cost price = $\frac{300}{1200}$ × 100% = 25%
∴ profit % of 25% ✓ (2 marks)

**9** Cost of each = $1950 ÷ 30
= $65
∴ Profit = $90 − $65
= $25 ✓
∴ Profit % = $\frac{25}{65}$ × 100%
= 38.46% (2 decimal places)
∴ Profit percentage of 38.46%. ✓ (2 marks)
**(Total 30 marks)**

## Level 3 Test

**1** **a** $\frac{2}{3}$ × 100% = $66.\dot{6}$% ✓ (1 mark)
**b** $\frac{7}{8}$ × 100% = 87.5% ✓ (1 mark)

**2** **a** 0.085 ✓ (1 mark)
**b** 0.034 ✓ (1 mark)
**c** 0.1275 ✓ (1 mark)

**3** **a** 0.125 × $70 = $8.75 ✓ (1 mark)
**b** 0.008 × $50 = $0.40 ✓ (1 mark)
**c** 1.15 × 300 = 345 ✓ (1 mark)

**4** 120 × 1.10 × 0.9 = 118.8 ✓✓ (2 marks)

**5** GST = 120 ÷ 11 = 10.91 (2 dec. pl.) ✓
∴ Old price = 120 − 10.91 = 109.09
∴ Cost is $109 (to nearest dollar). ✓ (2 marks)

**6** Total capacity = 2800
= 2800 ÷ 35 × 100
= 8000
∴ $\frac{3}{4}$ full = $\frac{3}{4}$ × 8000 = 6000 ✓
∴ 6000 L when $\frac{3}{4}$ full. ✓ (2 marks)

**7** 120% of cost price = $9600 ✓
∴ 100% of cost price
= $9600 ÷ 120 × 100
= $8000
∴ Jack bought the stamp for $8000. ✓✓
(3 marks)

**8** First discount = 0.15 × 280 = 42
∴ Price after 1st discount = 280 − 42
= 238 ✓
2nd discount = 0.05 × 238 = 11.9
∴ Price after 2nd discount = 238 − 11.9
= 226.1
∴ Final price of $226.10. ✓
or
Final price = 280 × 0.85 × 0.95
= 226.1
∴ Final price of $226.10. ✓✓ (2 marks)

**9** Loss = $180 − $50
= $130 ✓
∴ % loss = $\frac{130}{180}$ × 100%
= $72.\dot{2}$%
∴ The percentage loss of $72.\dot{2}$%. ✓✓
(3 marks)

**10** Price with 10% GST = 75
Existing GST = 75 ÷ 11
= 6.82
∴ Price with no GST = 75 − 6.82
= 68.18 ✓
New 15% GST = 0.15 × 68.18
= 10.23 ✓
New price = 68.18 + 10.23
= 78.41
∴ New price would be $78.41. ✓
or
New price = 75 ÷ 1.1 × 1.15 = 78.41
∴ New price would be $78.41. ✓✓✓ (3 marks)
**(Total 25 marks)**

# WORKED SOLUTIONS to Chapter 4 **Tests**

## Chapter 4—Algebra

### Level 1 Test

**1** a $4x$ ✓ b $2n$ ✓
c $12y$ ✓ d $6a$ ✓
e $ab$ ✓ f $a^2$ ✓
g $8mn$ ✓ h $6bc$ ✓
i $2a^2$ ✓ j $10a^2$ ✓
(10 marks)

**2** a $4a = 4 \times a$ ✓ (1 mark)
b $2ab = 2 \times a \times b$ ✓ (1 mark)
c $3a^2 = 3 \times a \times a$ ✓ (1 mark)

**3** a $12a \div 4 = \frac{\cancel{12}^3 a}{\cancel{4}_1}$
$= 3a$ ✓ (1 mark)
b $20t \div 10 = \frac{\cancel{20}^2 t}{\cancel{10}_1}$
$= 2t$ ✓ (1 mark)
c $4t \div 2 = \frac{\cancel{4}^2 t}{\cancel{2}_1}$
$= 2t$ ✓ (1 mark)
d $5t \div t = \frac{5\cancel{t}}{\cancel{t}}$
$= 5$ ✓ (1 mark)

**4** a $a + b = 4 + 2$
$= 6$ ✓ (1 mark)
b $a - b = 4 - 2$
$= 2$ ✓ (1 mark)
c $ab = 4 \times 2$
$= 8$ ✓ (1 mark)
d $3a = 3 \times 4$
$= 12$ ✓ (1 mark)
e $b^2 = 2 \times 2$
$= 4$ ✓ (1 mark)

**5** a $5x + 3x = 8x$ ✓ (1 mark)
b $7x - 2x = 5x$ ✓ (1 mark)
c $6y - 3y + 2y = 5y$ ✓ (1 mark)

**6** a $2(a + 3) = 2a + 6$ ✓ (1 mark)
b $5(x - 2) = 5x - 10$ ✓ (1 mark)
c $3(2a - 4) = 6a - 12$ ✓ (1 mark)

**7** a $3 \times b = 3b$ ✓ (1 mark)
b $a + b$ ✓ (1 mark)
c $x - 2$ ✓ (1 mark)

**8** $y = x + 2$

| $x$ | 0 | 1 | 2 | 3 | 4 | 5 |
|---|---|---|---|---|---|---|
| $y$ | 2 | 3 | 4 | 5 | 6 | 7 |

✓✓ (2 marks)

**9** a $2b + 6 = 2(b + 3)$ ✓ (1 mark)
b $3x + 6y + 12 = 3(x + 2y + 4)$ ✓ (1 mark)
c $12x + 8 = 4(3x + 2)$ ✓ (1 mark)

**10** a $6a + 3b$ ✓✓ (2 marks)
b $6x + 5y$ ✓✓ (2 marks)
c $4x + 4y$ ✓✓ (2 marks)

**11** a $a^2 - b = (4 \times 4) - 2$
$= 16 - 2$
$= 14$ ✓✓ (2 marks)
b $M = 3t - 1$
$= 3 \times 3 - 1$
$= 9 - 1$
$= 8$ ✓✓ (2 marks)

**12** a $2(x + 3) + 4x$
$= 2x + 6 + 4x$ ✓
$= 6x + 6$ ✓ (2 marks)
b $3(x + 1) + 2(x + 3)$
$= 3x + 3 + 2x + 6$ ✓
$= 5x + 9$ ✓ (2 marks)
**(Total 50 marks)**

### Level 2 Test

#### Part A

**1** $5x - 3 + 2x = 7x - 3$ ✓ (1 mark)

**2** $3ab^2 = 3 \times a \times b \times b$ ✓ (1 mark)

**3** $2m \times n = 2mn$ ✓ (1 mark)

**4** $4x - 2 = 4 \times 5 - 2$
$= 20 - 2$
$= 18$ ✓ (1 mark)

# WORKED SOLUTIONS to Chapter 4 **Tests**

**5** $2a^2 = 2 \times 4^2$
$= 2 \times 16$
$= 32$ ✓ (1 mark)

**6** $x - 7 = 10,\ x = 17$ ✓ (1 mark)

**7** $-2 \times a \times b \times 3 = -6ab$ ✓ (1 mark)

**8** $4^0 + 2^0 - 3^0$
$= 1 + 1 - 1$
$= 1$ ✓ (1 mark)

**9** $(xy)^0 = (4 \times 3)^0$
$= 12^0$
$= 1$ ✓ (1 mark)

**10** $3a - 12b = 3(a - 4b)$ ✓ (1 mark)

**11** $-3(x + 2) = -3x - 6$ ✓ (1 mark)

**12** $2x - 6x = -4x$ ✓ (1 mark)

**13** $F = ma$
$= 10 \times 3$
$= 30$ ✓ (1 mark)

**14** $(2a)^2 = (2 \times 5)^2$
$= 10^2$
$= 100$ ✓ (1 mark)

**15** $3a \times 5a = 15a^2$ ✓ (1 mark)

## Part B

**1** a $2b \times 3a = 6ba$ or $6ab$ ✓ (1 mark)
b $4 \times b \times 3b = 12b^2$ ✓ (1 mark)
c $-3 \times 2y = -6y$ ✓ (1 mark)

**2** a $5x - 2 + 3x = 8x - 2$ ✓ (1 mark)
b $6a + 2b + 3a + b = 9a + 3b$ ✓ (1 mark)
c $9x + 4y - x + 2y = 8x + 6y$ ✓ (1 mark)

**3** a $8b \div 4 = \frac{\cancel{8}^2 b}{\cancel{4}_1}$
$= 2b$ ✓ (1 mark)

b $6ab \div 2b = \frac{\cancel{6}^3 a \cancel{b}^1}{\cancel{2}_1 \cancel{b}_1}$
$= 3a$ ✓ (1 mark)

c $\frac{\cancel{8}^4 x}{\cancel{6}_3 y} = \frac{4x}{3y}$ ✓ (1 mark)

**4** a Per. $= x + 1 + x + 1 + x$
$= 3x + 2$ ✓ (1 mark)
b Per. $= 3x \times 2 + 4 \times 2$
$= 6x + 8$ ✓ (1 mark)
c $A = \frac{1}{2}(x + 4) \times 2$
$= x + 4$ ✓ (1 mark)

**5** a $\frac{a}{b} = \frac{14}{2} = 7$ ✓ (1 mark)
b $y^2 + x = (-2)^2 + 5$
$= 4 + 5$
$= 9$ ✓ (1 mark)
c $k = \frac{3b - a}{c}$
$= \frac{3 \times 7 - 5}{4}$ ✓
$= \frac{16}{4}$
$= 4$ ✓ (2 marks)

**6** a $\$(a - b)$ ✓ (1 mark)
b $P = x + 3 + 5 + x + 3 + 5$
$= 2x + 16$ ✓✓ (2 marks)

**7** a $6a - 4 = 2(3a - 2)$ ✓ (1 mark)
b $a^2 - 4a = a(a - 4)$ ✓ (1 mark)
c $12xy - 4x = 4x(3y - 1)$ ✓ (1 mark)

**8** a $x(x + 7) = x^2 + 7x$ ✓ (1 mark)
b $4a(3 - a) = 12a - 4a^2$ ✓ (1 mark)
c $-(2y - 7) = -2y + 7$ ✓ (1 mark)

**9** a $5(x + 2) - 3$
$= 5x + 10 - 3$ ✓
$= 5x + 7$ ✓ (2 marks)
b $5(x + 1) + 2(x - 1)$
$= 5x + 5 + 2x - 2$ ✓
$= 7x + 3$ ✓ (2 marks)
c $2x + 3(x + 5)$
$= 2x + 3x + 15$ ✓
$= 5x + 15$ ✓ (2 marks)

# WORKED SOLUTIONS to Chapter 4 **Tests**

**10** $y = 3x - 2$ ✓

| $x$ | $-1$ | $0$ | $1$ | $2$ |
|---|---|---|---|---|
| $y$ | $-5$ | $-2$ | $1$ | $4$ |

✓ (2 marks)

**11** $y = x + 2$ ✓✓ (2 marks)

**(Total 50 marks)**

## Level 3 Test

### Part A

**1** $-4a \times -2a = 8a^2$ ✓ (1 mark)

**2** $3a + 5$ ✓ (1 mark)

**3** $2d^2 - 3 = 2(-5)^2 - 3$
$= 2 \times 25 - 3$
$= 50 - 3$
$= 47$ ✓ (1 mark)

**4** $3x - 4y + x - y = 4x - 5y$ ✓ (1 mark)

**5** $y - x = 6 - (-3)$
$= 6 + 3$
$= 9$ ✓ (1 mark)

**6** $\frac{x^2 \times x^3}{x^4} = \frac{x^5}{x^4}$
$= x$ ✓ (1 mark)

**7** $16ab \div 4b = \frac{\cancel{16}^4 a \cancel{b}^1}{\cancel{4}_1 \cancel{b}_1}$
$= 4a$ ✓ (1 mark)

**8** $3x(2 - x) = 6x - 3x^2$ ✓ (1 mark)

**9** $5t^2 - 30t = 5t(t - 6)$ ✓ (1 mark)

**10** $a = \frac{F}{m}$
$= \frac{3.2}{0.2}$
$= 16$ ✓ (1 mark)

**11** $x + 1, x + 3$ ✓ (1 mark)

**12** $(5x)^2 = 5x \times 5x$
$= 25x^2$ ✓ (1 mark)

**13** $8x \div 6y^2 = \frac{\cancel{8}^4 x}{\cancel{6}_3 y^2}$
$= \frac{4x}{3y^2}$ ✓ (1 mark)

**14** $\frac{1}{a} = 1 \div 1\frac{1}{4}$
$= 1 \div \frac{5}{4}$
$= 1 \times \frac{4}{5}$
$= \frac{4}{5}$ ✓ (1 mark)

**15** $(2y)^3 = 2y \times 2y \times 2y \div 4y$
$= 8y^3 \div 4y$
$= 2y^2$ ✓ (1 mark)

### Part B

**1** **a** $2a \times (-a) - 3a^2 \times -2$
$= -2a^2 + 6a^2$ ✓
$= 4a^2$ ✓ (2 marks)

**b** $2 - (a - 1) = 2 - a + 1$ ✓
$= 3 - a$ ✓ (2 marks)

**c** $15xy^2 \div 3xy = \frac{\cancel{15}^5 \cancel{x}^1 \cancel{y}^1 xy}{\cancel{3}_1 \cancel{x}_1 \cancel{y}_1}$
$= 5y$ ✓✓ (2 marks)

**2** $2y^2 - 3y = 2(-2)^2 - 3(-2)$ ✓
$= 8 + 6$
$= 14$ ✓ (2 marks)

**3** **a** $-5x - 25 = -5(x + 5)$ ✓ (1 mark)
**b** $3ay^2 - 6a^2 = 3a(y^2 - 2a)$ ✓ (1 mark)
**c** $t^3 - 2t^2 = t^2(t - 2)$ ✓ (1 mark)

**4** **a** $(x^2y^3)^2 \div xy = x^4y^6 \div xy$ ✓
$= x^3y^5$ ✓ (2 marks)

**b** $\frac{x^2y^3}{(xy)^0} = \frac{x^2y^3}{1}$ ✓
$= x^2y^3$ ✓ (2 marks)

**c** $\frac{(x^2)^0y^3}{(x^2y^3)^0} = \frac{y^3}{1}$ ✓
$= y^3$ ✓ (2 marks)

**5** **a** $3(x - 2) + 4(x + 1)$
$= 3x - 6 + 4x + 4$ ✓
$= 7x - 2$ ✓ (2 marks)

# WORKED SOLUTIONS to Chapters 4 and 5 **Tests**

**b** $2x - (x - 2) + 4x$
$= 2x - x + 2 + 4x$ ✓
$= 5x + 2$ ✓ (2 marks)

**6**

| $a$ | –1 | 0 | 1 | 2 |
|---|---|---|---|---|
| $a^2 - 2$ | –1 | –2 | –1 | 2 |

✓✓ (2 marks)

**7** $C = \frac{5}{9}(F - 32)$
$= \frac{5}{9}(104 - 32)$ ✓
$= \frac{5}{9} \times 72$
$= 40$ ✓ (2 marks)

**8** **a** $3x + 7$ ✓ (1 mark)
**b** $(x - 2)$ years old ✓ (1 mark)

**9** **a** $(6y)^2 = 6y \times 6y$
$= 36y^2$ ✓ (1 mark)
**b** $\sqrt{81y^2} = 9y$ ✓ (1 mark)

**10** **a** $a^2 - 4a + a^2 - a$
$= 2a^2 - 5a$ ✓ (1 mark)
**b** $5xy - 4x + 2yx$
$= 7xy - 4x$ ✓ (1 mark)

**11** **a** $y = 3x + 1$ ✓✓ (2 marks)
**b** $y = 3 - 2x$ ✓✓ (2 marks)

**(Total 50 marks)**

## Chapter 5—Patterns and Linear Relationships

pp. 81–86

### Level 1 Test

**1** **a** A(2, 2), B(–2, 1), C(–1, –3), D(3, –1), E(1, 0), F(0, 3) ✓✓✓✓✓✓ (6 marks)

**b**

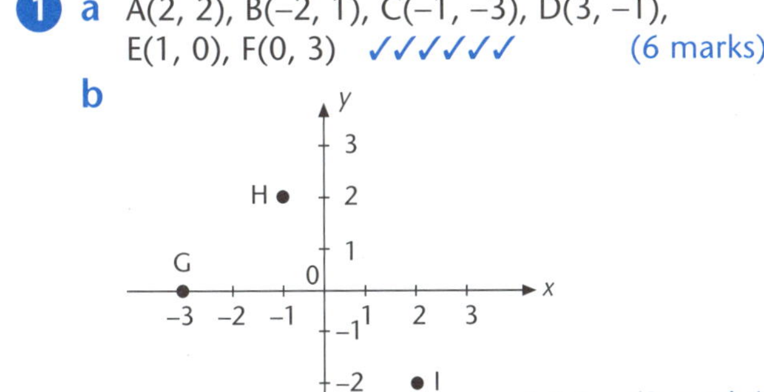


✓✓✓ (3 marks)

**2**

| $y = 11 + 3x$ | |
|---|---|
| $x$ | $y$ |
| 0 | 11 |
| 2 | 17 |
| 4 | 23 |

✓✓✓ (3 marks)

**3** Each $y$-value is three times the $x$-value, i.e. $y = 3x$ ✓ (1 mark)

**4** **a**

| $y = x + 3$ | |
|---|---|
| $x$ | $y$ |
| 0 | 3 |
| 1 | 4 |
| 2 | 5 |

✓✓✓

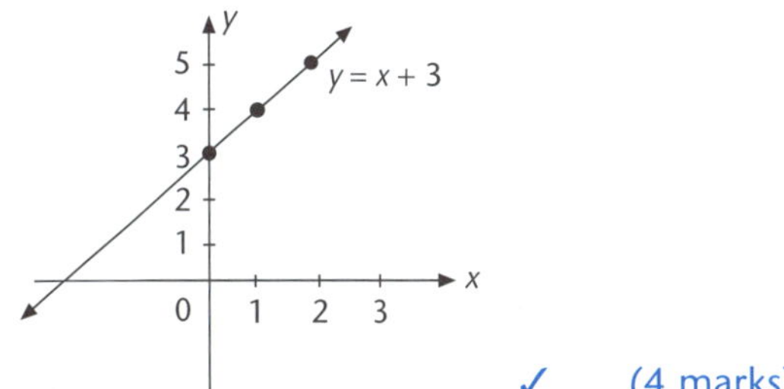


✓ (4 marks)

**b**

| $y = \frac{1}{2}x + 2$ | |
|---|---|
| $x$ | $y$ |
| 0 | 2 |
| 2 | 3 |
| 4 | 4 |

✓✓✓

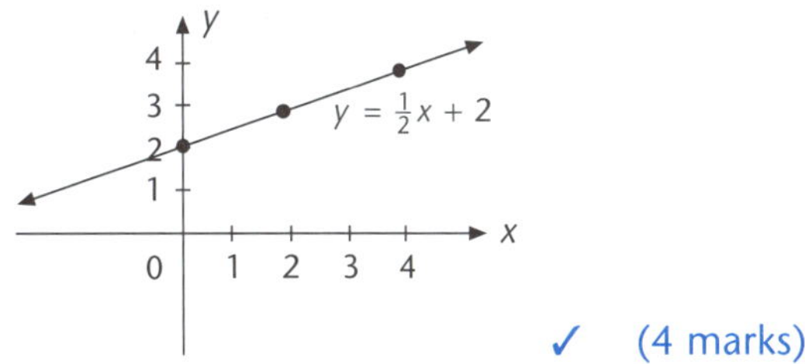


✓ (4 marks)

**5** $y = 10 - x$
**a** $6 = 10 - 4$ Yes ✓ (1 mark)
**b** $4 = 10 - 6$ Yes ✓ (1 mark)
**c** $-2 = 10 - 10$ No ✓ (1 mark)

# WORKED SOLUTIONS to Chapter 5 **Tests**

**d** $-1 = 10 - 11$ Yes ✓ (1 mark)

**e** $7 = 10 - (-3) = 10 + 3$ No ✓ (1 mark)

**6** Point of intersection $(-2, 5)$ ✓

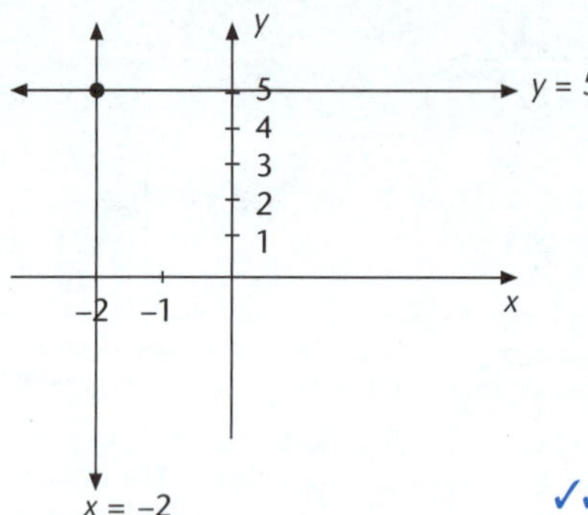


✓✓ (3 marks)

**7** $y = x + 2$

| $x$ | 0 | 1 | 2 |
|---|---|---|---|
| $y$ | 2 | 3 | 4 |

✓

$x - y = 3$

| $x$ | 0 | 1 | 2 |
|---|---|---|---|
| $y$ | −3 | −2 | −1 |

✓

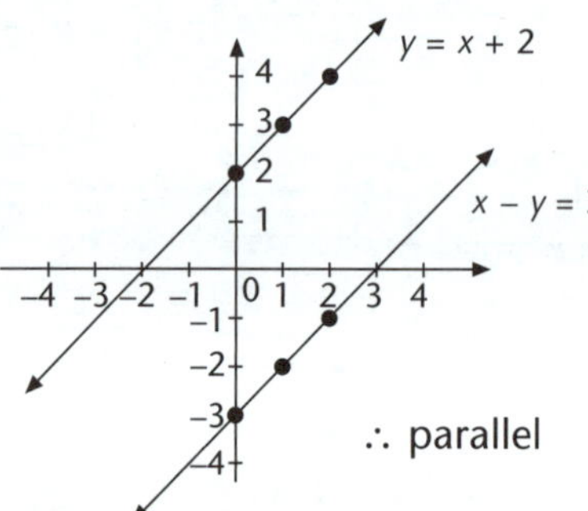


✓✓

∴ parallel ✓

(5 marks)

**8** **a**

| $n$ | 0 | 2 | 4 | 6 |
|---|---|---|---|---|
| $p$ | 0 | 6 | 12 | 18 |

✓✓ (2 marks)

**b**

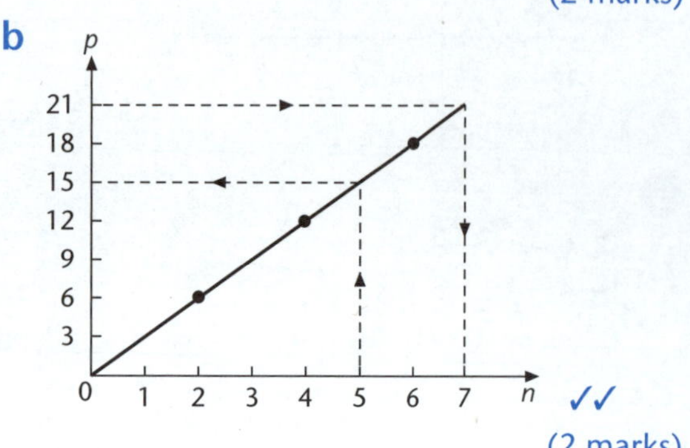


✓✓ (2 marks)

**c** Cost of 5 kg is \$15. ✓ (1 mark)

**d** She buys 7 kg of potatoes. ✓ (1 mark)

**(Total 40 marks)**

## Level 2 Test

**1**

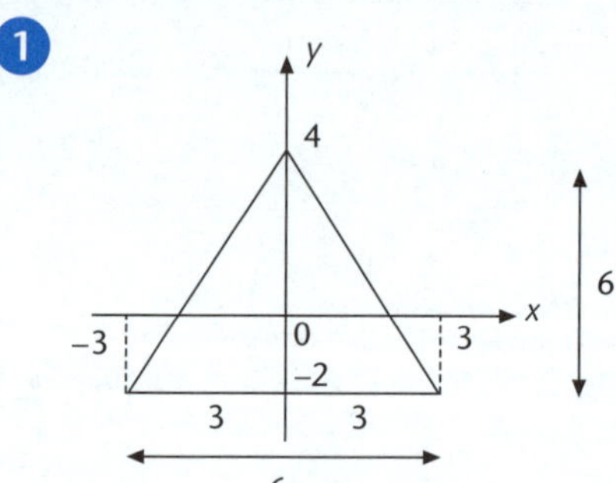


**a** A(−3, −2), B(3, −2), C(0, 4) ✓✓✓ (3 marks)

**b** Isosceles triangle ✓ (1 mark)

**c** $A = \frac{1}{2} \times \text{base} \times \text{height}$

$= \frac{1}{2} \times 6 \times 6$ ✓✓

$= 18 \text{ units}^2$ ✓ (3 marks)

**2**

| $y = 12 - 2x$ | | |
|---|---|---|
| $x$ | $y$ | |
| −1 | 14 | ✓ |
| 1 | 10 | ✓ |
| 3 | 6 | ✓ |
| 5 | 2 | ✓ |

(4 marks)

**3** **a**

| $x$ | −3 | −2 | −1 | 0 | 1 | 2 |
|---|---|---|---|---|---|---|
| $y$ | −1 | 0 | 1 | 2 | 3 | 4 |

✓✓✓✓✓✓ (6 marks)

**b** Each $y$ value is 2 more than each $x$ value.
$y = x + 2$ ✓ (1 mark)

**4** **a** $y = x - 1$

| $x$ | 0 | 1 | 2 |
|---|---|---|---|
| $y$ | −1 | 0 | 1 |

✓

$y = 2x - 3$

| $x$ | 0 | 1 | 2 |
|---|---|---|---|
| $y$ | −3 | −1 | 1 |

✓

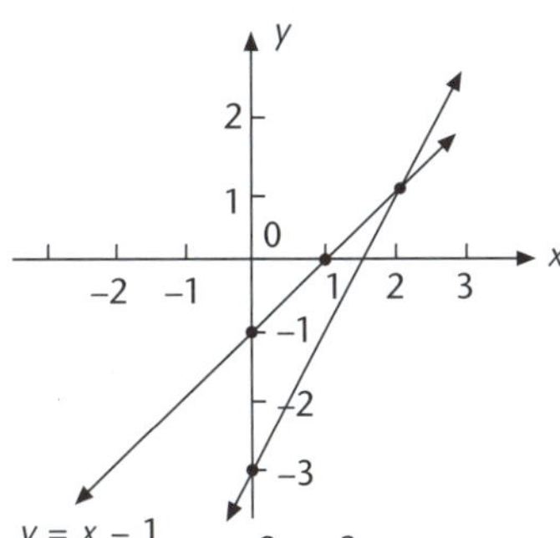


✓✓✓✓ (6 marks)

**b** Point of intersection is (2, 1). ✓✓ (2 marks)

**5** $y = 7 - 2x$

A(1, 5): $5 = 7 - 2$ Yes ✓

B(−1, 9): $9 = 7 - 2(-1)$

$= 7 + 2$ Yes ✓

$\therefore$ A and B lie on $y = 7 - 2x$

C(5, −3): $-3 = 7 - 10$ Yes ✓

$\therefore$ C lies on line $y = 7 - 2x$

$\therefore$ A, B and C are collinear ✓ (4 marks)

**6** **a** Gradient = 3 ✓ (1 mark)

**b** Gradient $= -\frac{1}{3}$ ✓ (1 mark)

[gradient is number in front of $x$]

Lines are perpendicular as gradients are negative reciprocals. ✓✓ (2 marks)

**7** **a**

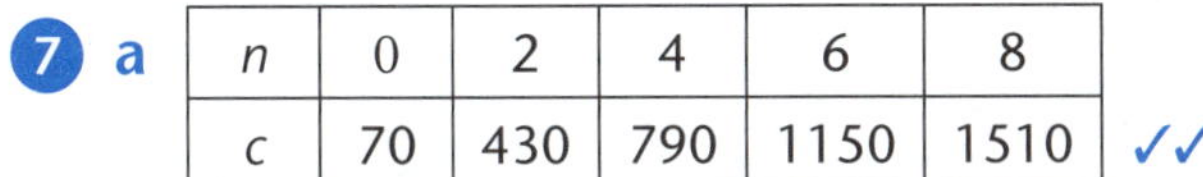

| $n$ | 0 | 2 | 4 | 6 | 8 |
|---|---|---|---|---|---|
| $c$ | 70 | 430 | 790 | 1150 | 1510 |

✓✓

**b**

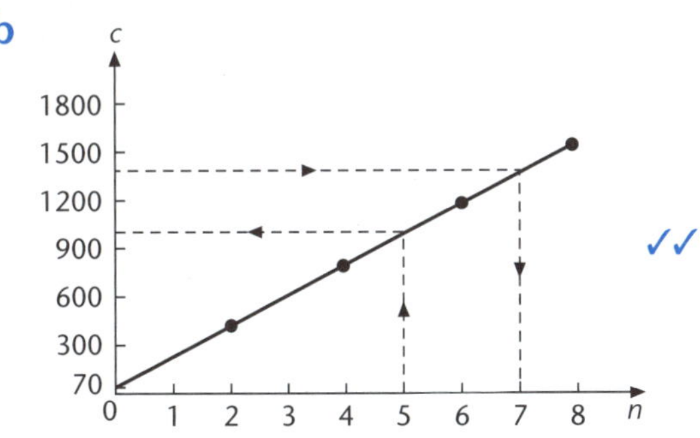


✓✓

**i** Cost for 5 h is \$970. ✓

**ii** For \$1330, the financial adviser works for 7 h. ✓ (6 marks)

**(Total 40 marks)**

## Level 3 Test

**1** **a, b**

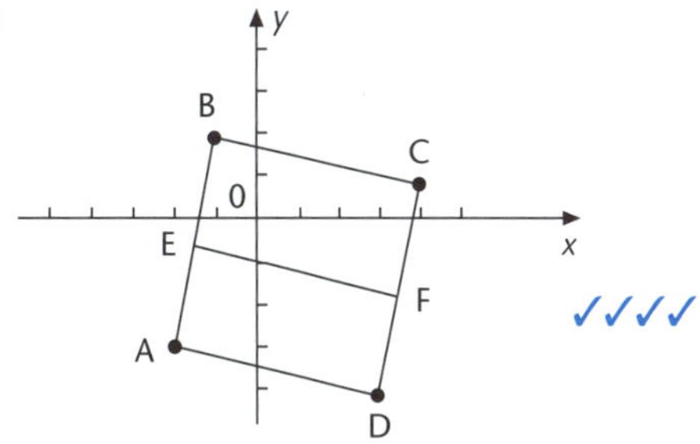


✓✓✓✓

D is (3, −4). ✓ (5 marks)

**c** Mid-points will be average of coordinates:

A(−2, −3) B(−1, 2)

$E = \left(\frac{-2 + -1}{2}, \frac{-3 + 2}{2}\right)$

Average of $x$ values; Average of $y$ values

$= \left(\frac{-3}{2}, \frac{-1}{2}\right) = \left(-1\frac{1}{2}, \frac{-1}{2}\right)$ ✓✓

C(4, 1) D(3, −4)

$F = \left(\frac{4 + 3}{2}, \frac{1 + -4}{2}\right)$

$= \left(\frac{7}{2}, \frac{-3}{2}\right) = \left(3\frac{1}{2}, -1\frac{1}{2}\right)$ ✓✓ (4 marks)

**d** AD || BC (opposite sides of parallelogram) ✓

AD || EF || BC (EF is same distance from both BC and AD—mid-points joined) ✓ (2 marks)

**2** **a**

| $2x + y = 7x$ | |
|---|---|
| $x$ | $y$ |
| −1 | 9 |
| 0 | 7 |
| 1 | 5 |

✓✓ (2 marks)

## WORKED SOLUTIONS to Chapter 5 **Tests**

**b**

| $y = \frac{4}{x+2}$ | |
|---|---|
| $x$ | $y$ |
| 0 | 2 ✓ |
| 2 | 1 ✓ |
| 4 | $\frac{2}{3}$ ✓ |

(3 marks)

**3** Each $y$ value is 3 + $x$ value squared, i.e. $y = x^2 + 3$ ✓✓ (2 marks)

**4** **a** $2x + y = 5$

| $x$ | 0 | 1 | 2 | 3 |
|---|---|---|---|---|
| $y$ | 5 | 3 | 1 | –1 |

✓

$2x - y = 7$

| $x$ | 0 | 1 | 2 | 3 |
|---|---|---|---|---|
| $y$ | –7 | –5 | –3 | –1 |

✓

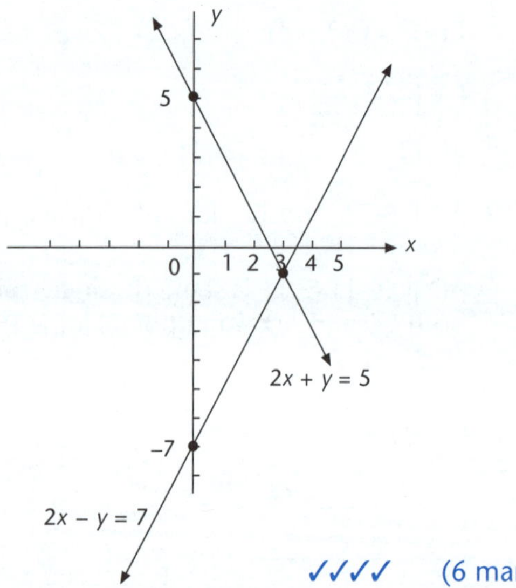


✓✓✓✓ (6 marks)

**b** Point of intersection is (3, –1). ✓✓ (2 marks)

**c** Does (3, –1) lie on line $y = 3x - 10$?
(3, –1): –1 = 9 – 10
Yes ✓
Then $y = 3x - 10$ passes through (3, –1)
∴ All three lines pass through (3, –1) ✓
∴ The lines are concurrent. ✓ (3 marks)

**5** $x - y = 5$
C(2, –3): 2 – (–3) = 5
2 + 3 = 5
Yes ✓
D(–2, –7): –2 – (–7) = 5
–2 + 7 = 5
Yes ✓
E(4, –1): 4 – (–1) = 5
4 + 1 = 5
Yes ✓
All points lie on line $x - y = 5$. (3 marks)

**6** **a** **i** Cycling

| $n$ | 0 | 1 | 2 | 3 | 4 | 5 |
|---|---|---|---|---|---|---|
| $d$ | 30 | 28 | 26 | 24 | 22 | 20 |

✓

**ii** Running

| $n$ | 0 | 1 | 2 | 3 | 4 | 5 |
|---|---|---|---|---|---|---|
| $d$ | 6 | 7 | 8 | 9 | 10 | 11 |

✓

**b**

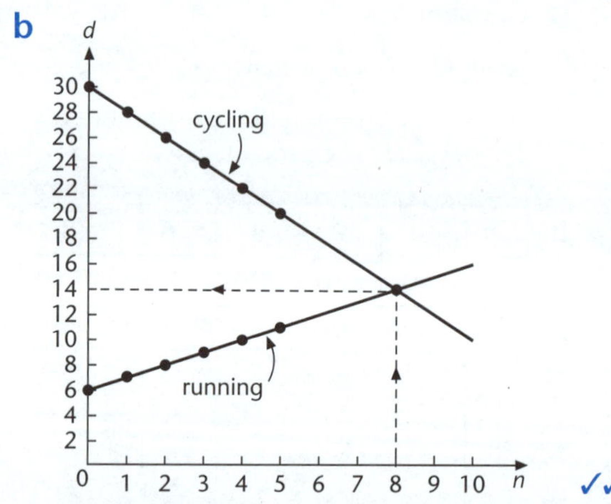


✓✓

**c** In 8 days' time Colin will run and cycle 14 km. ✓✓

**d** For cycling the rule is $d = 30 - 2n$.
If $d = 0$, then $2n = 30$. As $n = 15$,
in 15 days' time he will not cycle. ✓✓
(8 marks)

**(Total 40 marks)**

# WORKED SOLUTIONS to Chapter 6 **Tests**

## Chapter 6—Equations and Formulae

### Level 1 Test

**1** $4x - 1 = 19$

$\text{LHS} = 4 \times 5 - 1$
$= 19$
$= \text{RHS}$

$x = 5$ is a solution. ✓✓ (2 marks)

**2** **a** $x = \frac{35}{7} = 5$ ✓ (1 mark)

**b** $y = \frac{-12}{4} = -3$ ✓ (1 mark)

**c** $t = 19 - 11 = 8$ ✓ (1 mark)

**d** $m = 13 + 7 = 20$ ✓ (1 mark)

**3** **a** $3x - 5 = 16$
$3x = 16 + 5$
$3x = 21$ ✓
$x = \frac{21}{3}$
$x = 7$ ✓ (2 marks)

**b** $8y + 7 = 23$
$8y = 23 - 7$
$8y = 16$ ✓
$y = \frac{16}{8}$
$y = 2$ ✓ (2 marks)

**c** $4m - 8 = 12$
$4m = 12 + 8$
$4m = 20$ ✓
$m = \frac{20}{4}$
$m = 5$ ✓ (2 marks)

**d** $8a = 3a + 30$
$8a - 3a = 30$ ✓
$5a = 30$
$a = 6$ ✓ (2 marks)

**e** $\frac{x - 3}{4} = 5$
$x - 3 = 5 \times 4$ ✓
$x - 3 = 20$
$x = 20 + 3$
$x = 23$ ✓ (2 marks)

**4** **a** $6 \times \frac{y}{2} + 6 \times \frac{y}{3} = 10 \times 6$ ✓
$3y + 2y = 60$ ✓
$5y = 60$
$y = \frac{60}{5}$
$y = 12$ ✓ (3 marks)

**b** $\frac{2t - 5}{3} = 1$
$2t - 5 = 1 \times 3$ ✓
$2t - 5 = 3$
$2t = 3 + 5$
$2t = 8$ ✓
$t = \frac{8}{2}$
$t = 4$ ✓ (3 marks)

**5** **a** $m + n$ ✓✓ (2 marks)

**b** $100y$ ✓✓ (2 marks)

**c** $a + 1$ ✓✓ (2 marks)

**6** Let number $= N$
$\therefore 4N + 15 = 79$ ✓
$4N = 79 - 15$
$4N = 64$
$N = \frac{64}{4}$
$N = 16$
$\therefore$ The number is 16. ✓ (2 marks)

**7** $S = \frac{D}{T}$
$= \frac{840}{7}$ ✓
$= 120$ ✓ (2 marks)

**8** **a** $8y = -16$
$\frac{8y}{8} = \frac{-16}{8}$
$y = -2$ ✓ (1 mark)

**b** $-4y = -20$
$\frac{-4y}{-4} = \frac{-20}{-4}$
$y = 5$ ✓ (1 mark)

**c** $\frac{x}{-3} = -2$
$-3 \times \frac{x}{-3} = -3 \times -2$
$x = 6$ ✓ (1 mark)

**9** $2 - 3x = 13$

$-3x = 13 - 2$

$-3x = 11$ ✓

$\dfrac{-3x}{-3} = \dfrac{11}{-3}$

$x = -3\frac{2}{3}$ ✓ (2 marks)

**10** $14x + 13 = 29$

$4x = 29 - 13$

$4x = 16$ ✓

$\dfrac{4x}{4} = \frac{16}{4}$

$x = 4$ ✓

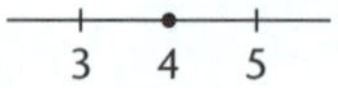


✓ (3 marks)

**(Total 40 marks)**

## Level 2 Test

**1** **a** $x = 3$, LHS $= 3 \times 3 - 12$

$= -3$

RHS $= 7 \times 3$

$= 21$

$x = 3$ is not a solution. ✓ (1 mark)

**b** $x = -3$, LHS $= 3 \times (-3) - 12$

$= -21$

RHS $= 7 \times (-3)$

$= -21$

$x = -3$ is a solution. ✓ (1 mark)

**2** **a** $x = \frac{12}{8}$

$= 1\frac{1}{2}$ ✓ (1 mark)

**b** $y = 5 - 7$

$= -2$ ✓ (1 mark)

**c** $m = -4 + 8$

$= 4$ ✓ (1 mark)

**d** $t = 4 \times 8$

$= 32$ ✓ (1 mark)

**3** **a** $4y + 6 = 20$

$4y = 20 - 6$

$4y = 14$

$y = \frac{14}{4}$

$y = 3\frac{1}{2}$ ✓ (1 mark)

**b** $7m - 3 = 39$

$7m = 39 + 3$

$7m = 42$

$m = \frac{42}{7}$

$m = 6$ ✓ (1 mark)

**c** $6m + 42 = 35$

$6m = 35 - 42$

$6m = -7$ ✓

$m = -\frac{7}{6}$

$m = -1\frac{1}{6}$ ✓ (2 marks)

**4** **a** $5y - 3 = 3y + 15$

$5y - 3y - 3 = 15$

$2y - 3 = 15$

$2y = 15 + 3$

$2y = 18$ ✓

$y = \frac{18}{2}$

$y = 9$ ✓ (2 marks)

**b** $7m + 14 = 3m + 30$

$7m - 3m + 14 = 30$

$4m + 14 = 30$

$4m = 30 - 14$

$4m = 16$ ✓

$m = \frac{16}{4}$

$m = 4$ ✓ (2 marks)

**c** $10 \times \dfrac{m}{2} - 10 \times \dfrac{m}{5} = 9 \times 10$ ✓

$5m - 2m = 90$

$3m = 90$ ✓

$m = \frac{90}{3}$

$m = 30$ ✓ (3 marks)

**d** $\dfrac{6(m + 2)}{2} + 6 \times \dfrac{m}{3} = 6 \times 5$ ✓

$3(m + 2) + 2m = 30$

$3m + 6 + 2m = 30$

$5m + 6 = 30$ ✓

$5m = 30 - 6$

$5m = 24$

$m = \frac{24}{5}$

$m = 4\frac{4}{5}$ ✓ (3 marks)

# WORKED SOLUTIONS to Chapter 6 **Tests**

**5** $P = 2(3x - 1) + 2x$
$= 6x - 2 + 2x$
$= 8x - 2$
$\therefore 8x - 2 = 22$ ✓
$8x = 22 + 2$
$8x = 24$
$x = \frac{24}{8}$
$x = 3$ ✓
∴ Width = 3 m ✓
Length = 3 × 3 − 1
= 8 m ✓ (4 marks)

**6** $v = u + at$
$= 18 + (-3) \times 5$ ✓
$= 18 - 15$ ✓
$= 3$ ✓ (3 marks)

**7** **a** $P = 2(l + b)$
$46 = 2(15 + b)$
$46 = 30 + 2b$
$2b + 30 = 46$ ✓
$2b = 46 - 30$
$2b = 16$
$b = 8$ ✓ (2 marks)

**b** $M = \frac{P}{x + y}$
$8 = \frac{P}{6 + 2}$
$8 = \frac{P}{8}$
$\frac{P}{8} = 8$ ✓
$8 \times \frac{P}{8} = 8 \times 8$
$P = 64$ ✓ (2 marks)

**c** $T = \frac{1}{2}p(a + b)$
$12 = \frac{1}{2} \times p(4 + 2)$ ✓
$12 = \frac{1}{2} \times p \times 6$
$12 = 3p$ ✓
$3p = 12$
$p = 4$ ✓ (3 marks)

**8** $x$ = Marcus' age
$x - 6$ = Rebecca's age ✓
$\therefore x + x - 6 = 28$
$2x - 6 = 28$ ✓
$2x = 28 + 6$
$2x = 34$
$x = 17$
∴ Marcus is 17, Rebecca is 11. ✓ (3 marks)

**9** Let the classes have $x$ and $(x + 8)$ students. ✓
$x + x + 8 = 50$
$2x + 8 = 50$ ✓
$2x = 50 - 8$
$2x = 42$
$x = 21$
∴ Classes have 21 and 29 students. ✓
(3 marks)

**(Total 40 marks)**

## Level 3 Test

**1** **a** LHS = $7 \times -2 + 9 = -5$
Solution ✓ (1 mark)

**b** LHS = $4 - 3(-2)$
$= 10$
Not a solution ✓ (1 mark)

**c** LHS = $\frac{9 \times -2}{2} - 4(-2)$
$= -9 + 8$
$= -1$
Solution ✓ (1 mark)

**2** **a** $x = -\frac{4}{5}$ ✓ (1 mark)

**b** $y = 7 - 11$
$= -4$ ✓ (1 mark)

**c** $t = -5 \times 3$
$= -15$ ✓ (1 mark)

**3** **a** $3m - 11 = 5$
$3m = 5 + 11$
$3m = 16$ ✓
$m = \frac{16}{3}$
$m = 5\frac{1}{3}$ ✓ (2 marks)

**b** $17 + 5y = 11$
$5y = 11 - 17$
$5y = -6$ ✓
$y = -\frac{6}{5}$
$y = -1\frac{1}{5}$ ✓ (2 marks)

## WORKED SOLUTIONS to Chapter 6 **Tests**

**c** $6a - 15 = 12$
$6a = 12 + 15$
$6a = 27$ ✓
$a = \frac{27}{6}$
$a = 4\frac{1}{2}$ ✓ (2 marks)

**d** $5 - 7m = 12 - 2m$
$5 - 7m + 2m = 12$
$5 - 5m = 12$
$-5m = 12 - 5$
$-5m = 7$ ✓
$m = \frac{7}{-5}$
$m = -1\frac{2}{5}$ ✓ (3 marks)

**e** $14t - 35 - 4t + 24 = 7$
$10t - 11 = 7$ ✓
$10t = 7 + 11$
$10t = 18$ ✓
$t = \frac{18}{10}$
$t = 1\frac{4}{5}$ ✓ (3 marks)

**4** **a** $2 \times 4y + 1(2y - 3) = 8 \times 2$ ✓
$8y + 2y - 3 = 16$
$10y - 3 = 16$ ✓
$10y = 16 + 3$
$10y = 19$
$y = \frac{19}{10}$
$y = 1\frac{9}{10}$ ✓ (3 marks)

**b** $\frac{10(x-3)}{2} - \frac{10(x-2)}{5} = 5 \times 10$
$5(x - 3) - 2(x - 2) = 50$ ✓
$5x - 15 - 2x + 4 = 50$ ✓
$3x - 11 = 50$
$3x = 50 + 11$
$3x = 61$ ✓
$x = \frac{61}{3}$
$x = 20\frac{1}{3}$ ✓ (4 marks)

**5** Let Brad's age be $x$ years.
Betty's age = $(2x + 9)$ years ✓
$x + 2x + 9 = 30$
$3x + 9 = 30$ ✓
$3x = 30 - 9$
$3x = 21$ ✓
$x = \frac{21}{3}$
$x = 7$ ✓
Betty' s age $= 2 \times 7 + 9$
$= 23$ years ✓ (5 marks)

**6** $v^2 = u^2 + 2as$
$13^2 = 11^2 + 2 \times a \times 3$ ✓
$169 = 121 + 6a$
$169 - 121 = 6a$
$48 = 6a$ ✓
$\frac{48}{6} = a$
$\therefore a = 8$ ✓ (3 marks)

**7** Let one angle = $x°$
$\therefore$ the other angles are $(x - 20)°$ and $(x - 10)°$ ✓
$\therefore x + x - 20 + x - 10 = 180$
$3x - 30 = 180$ ✓
$3x = 180 + 30$
$3x = 210$
$x = 70$
$\therefore$ Angles are 70°, 50°, 60°. ✓ (3 marks)

**8** Let the width = $x$
$\therefore$ length = $2x - 4$ ✓

As perimeter = 52
$\therefore$ sum of length and width = 26
i.e. $2x - 4 + x = 26$ ✓
$3x - 4 = 26$
$3x = 26 + 4$
$3x = 30$
$x = 10$ ✓
$\therefore$ Width is 10 cm, length is 16 cm. ✓
(4 marks)
**(Total 40 marks)**

# WORKED SOLUTIONS to Chapter 7 **Tests**

## Chapter 7—Pythagoras' Theorem

### Level 1 Test

**1** $l$ ✓ (1 mark)

**2** $l^2 = m^2 + n^2$ ✓ (1 mark)

**3** **a** 361 ✓ **b** 441 ✓ **c** 17 ✓ (3 marks)

**4** 7.07 ✓ (1 mark)

**5** **a** $y^2 = 0.8^2 + 1.5^2$ ✓
$= 2.89$
$y = \sqrt{2.89}$
$= 1.7$ ✓ (2 marks)

**b** $t^2 = 16^2 + 30^2$ ✓
$= 1156$
$t = \sqrt{1156}$
$= 34$ ✓ (2 marks)

**6** **a** $x^2 = 3^2 + 5^2$ ✓
$= 34$
$x = \sqrt{34}$
$= 5.8$ (to 1 decimal place) ✓ (2 marks)

**b** $m^2 = 7^2 + 3^2$ ✓
$= 58$
$m = \sqrt{58}$ ✓ (2 marks)

**c** $t^2 = 9^2 - 2^2$ ✓
$= 77$
$t = \sqrt{77}$
$= 8.77$ (to 2 decimal places) ✓ (2 marks)

**7** $BC^2 = 45^2 - 36^2$ ✓
$= 729$
$BC = \sqrt{729}$
$= 27$
$BC = 27$ cm ✓ (2 marks)

**8** **a** $15^2 = 12^2 + 9^2$ ✓
$225 = 225$
Yes, right-angled. ✓ (2 marks)

**b** $2.1^2 = 1.6^2 + 1.2^2$ ✓
$4.41 \neq 4$
No, not right-angled. ✓ (2 marks)

**9** $15^2 + 20^2 = 25^2$
$625 = 625$
Yes, Pythagorean triad. ✓ (1 mark)

**10**

3.5
3.1
$d$

$d^2 = 3.5^2 - 3.1^2$ ✓
$= 2.64$
$d = \sqrt{2.64}$
$= 1.6248\ldots$
$d = 1.6$ (to 1 decimal place)
The foot of ladder is 1.6 metres from base of wall. ✓ (2 marks)

**(Total 25 marks)**

### Level 2 Test

**1** $PR^2 = PQ^2 + QR^2$ ✓ (1 mark)

**2** **a** 1369 ✓ (1 mark)
**b** 41 ✓ (1 mark)

**3** 9.7 (to 1 decimal place) ✓ (1 mark)

**4** **a** $k^2 = 12^2 + 16^2$ ✓
$= 400$
$\therefore k = \sqrt{400}$
$= 20$ ✓ (2 marks)

**b** $h^2 = 2^2 + 2.1^2$ ✓
$= 8.41$
$h = \sqrt{8.41}$
$= 2.9$ ✓ (2 marks)

**5** **a** $m^2 = 4^2 + 7^2$ ✓
$= 65$
$m = \sqrt{65}$
$= 8.06$ (to 2 decimal places) ✓ (2 marks)

## WORKED SOLUTIONS to Chapter 7 **Tests**

**b** $t^2 = 8^2 + 5^2$ ✓
$= 89$
$t = \sqrt{89}$
$= 9.43$ (to 3 significant figures) ✓
(2 marks)

**c** $k^2 = 11^2 - 3^2$ ✓
$= 112$
$k = \sqrt{112}$
$= 10.6$ (to 1 decimal place) ✓
(2 marks)

**6** **a** $x^2 = 15^2 - 11^2$ ✓
$= 104$
$x = \sqrt{104}$ ✓ (2 marks)

**b** $y^2 = 8^2 - 3^2$ ✓
$= 55$
$y = \sqrt{55}$ ✓ (2 marks)

**7** **a** $16^2 = 8^2 + 14^2$
$256 \neq 260$
Not right-angled. ✓ (1 mark)

**b** $2.5^2 = 2^2 + 1.5^2$
$6.25 = 6.25$
Yes, right-angled. ✓ (1 mark)

**8** $x^2 = 21^2 + 28^2$ ✓
$= 1225$
$x = \sqrt{1225}$
$= 35$ ✓ (2 marks)

**9**

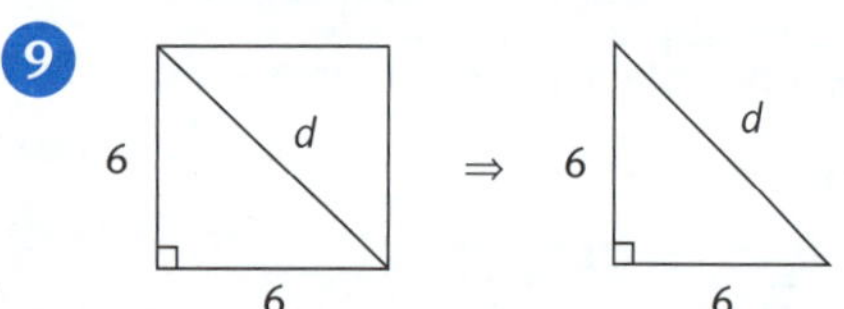


$d^2 = 6^2 + 6^2$ ✓
$= 72$
$d = \sqrt{72}$ ✓
$= 8.485$
$= 8.5$ (to 1 decimal place)
The diagonal is of length 8.5 cm. ✓
(3 marks)
**(Total 25 marks)**

### Level 3 Test

**1**

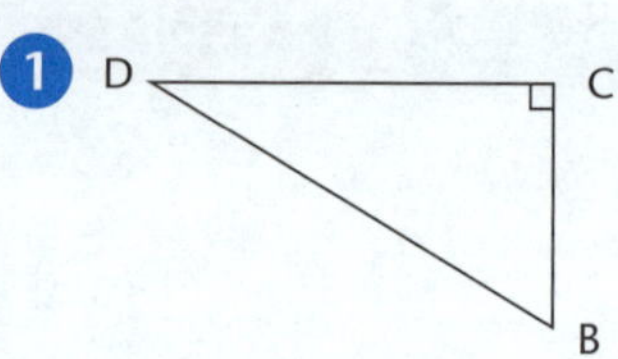


$BD^2 = BC^2 + DC^2$ ✓ (1 mark)

**2** **a** 808 ✓✓ (2 marks)
**b** 45 ✓✓ (2 marks)

**3** 6.082 76 = 6.08 (to 2 decimal places) ✓✓
(2 marks)

**4** **a** $m^2 = 18^2 + 24^2$ ✓
$= 900$
$m = \sqrt{900}$
$= 30$ ✓ (2 marks)

**b** $y^2 = 26^2 - 24^2$ ✓
$= 100$
$y = \sqrt{100}$
$= 10$ ✓ (2 marks)

**c** $x^2 = 2.8^2 + 2.1^2$ ✓
$= 12.25$
$x = \sqrt{12.25}$
$= 3.5$ ✓ (2 marks)

**d** $n^2 = 34^2 - 16^2$ ✓
$= 900$
$n = \sqrt{900}$
$= 30$ ✓ (2 marks)

**5** **a** $y^2 = 5^2 + 8^2$ ✓
$= 89$
$y = \sqrt{89}$
$= 9.4$ (to 1 decimal place) ✓ (2 marks)

**b** $t^2 = 15^2 - 14^2$ ✓
$= 29$
$t = \sqrt{29}$
$= 5.4$ (to 1 decimal place) ✓ (2 marks)

**c** $y^2 = 1.6^2 + 1.2^2$ ✓
$= 4$
$y = \sqrt{4}$
$= 2$ ✓
$x^2 = 2^2 + 5^2$ ✓
$x = \sqrt{29}$
$= 5.4$ (to 1 decimal place) ✓ (4 marks)

**6**

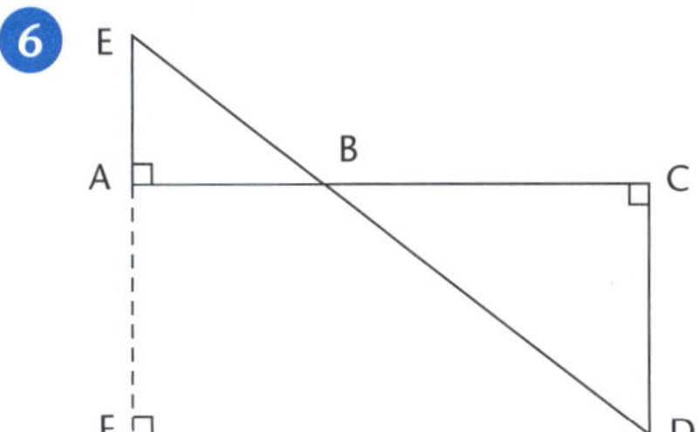


Complete rectangle ACDF, giving

DF = AC = 24 and EF = EA + AF (= CD)

$= 7 + 11$

$= 18$ ✓

Then, $DE^2 = EF^2 + DF^2$

$= 18^2 + 24^2$ ✓

$= 900$

$DE = \sqrt{900}$

$= 30$ cm ✓ (3 marks)

**7**

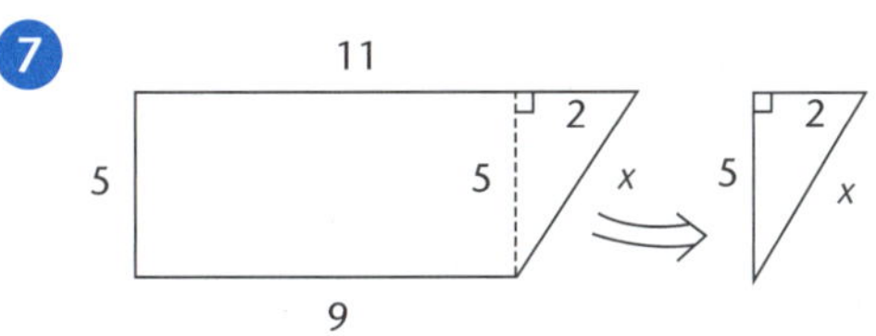


$x^2 = 2^2 + 5^2$

$= 29$ ✓

$\therefore x = \sqrt{29}$

$= 5.4$ (to 1 decimal place)

Perimeter $= 9 + 5 + 11 + 5.4$

$= 30.4$ cm ✓ (2 marks)

**8**

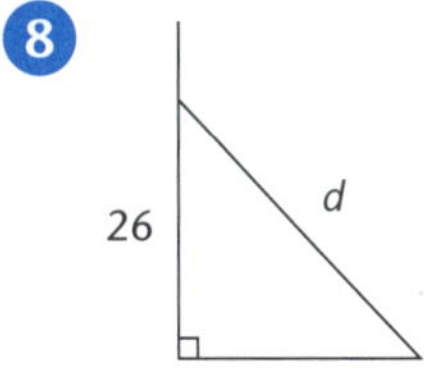


$d^2 = 26^2 + 15^2$

$= 901$

$d = \sqrt{901}$

$= 30.0166$ ✓

Four wires, so length

$= 4 \times 30.0166$

$= 120.066$

$= 120.1$ (to 1 decimal place)

Length = 120.1 metres ✓ (2 marks)

**(Total 30 marks)**

## Chapter 8—Measurement

pp. 169–174

### Level 1 Test

**1** **a**

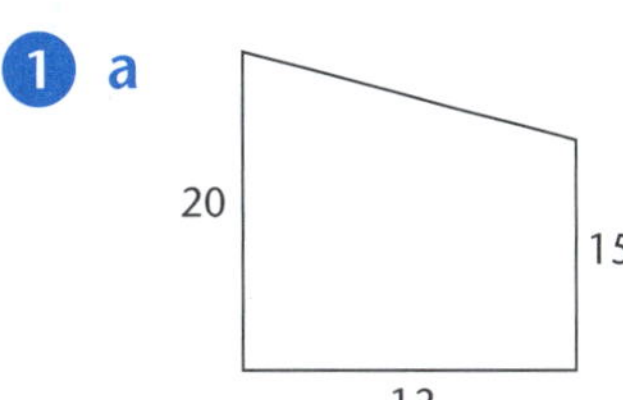


Area $= \frac{1}{2} \times 12(20 + 15)$ ✓

$= 6(35)$

$= 210$

$\therefore$ The area is 210 $cm^2$. ✓ (2 marks)

**b**

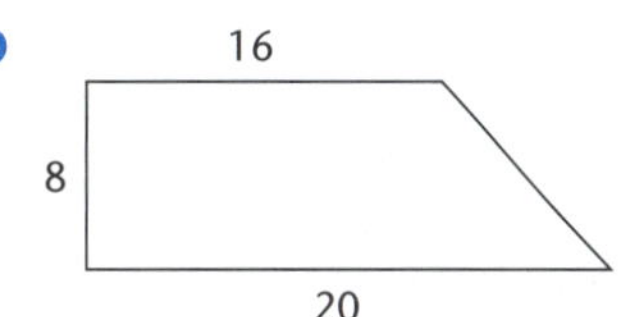


Area $= \frac{1}{2} \times 8(16 + 20)$ ✓

$= 4(36)$

$= 144$

$\therefore$ The area is 144 $cm^2$. ✓ (2 marks)

**2** **a** Volume $= 12 \times 7 \times 5 = 420$

$\therefore$ The volume is 420 $cm^3$. ✓✓ (2 marks)

**b** Volume $= \frac{1}{2} \times 12 \times 16 \times 14 = 1344$

$\therefore$ The volume is 1344 $cm^3$. ✓✓ (2 marks)

**3** Volume $= 12 \times 8 \times 10$

$= 960$ ✓

$\therefore$ The volume is 960 $cm^3$. ✓

Capacity $= 960 \div 1000$

$= 0.96$ L ✓ (3 marks)

**4** **a** $C = \pi d$

$= \pi \times 24$ ✓

$= 75.40$ (to 2 decimal places)

$\therefore$ The circumference is 75.40 cm. ✓

(2 marks)

**b** $C = \pi d$

$= \pi \times 16$ ✓

$= 50.27$ (to 2 decimal places)

$\therefore$ The circumference is 50.27 cm. ✓

(2 marks)

# WORKED SOLUTIONS to Chapter 8 **Tests**

**5** **a** $A = \pi r^2$
$= \pi \times 8^2$ ✓
$= 201.062$ (to 3 decimal places)
∴ The area is 201.062 $cm^2$. ✓ (2 marks)

**b** $A = \pi r^2$
$= \pi \times 5^2$ $(d = 10, r = 5)$ ✓
$= 78.540$ (to 3 decimal places)
∴ The area is 78.540 $cm^2$. ✓ (2 marks)

**6** $A = \pi r^2$
$= \pi \times 2^2$
$= 4\pi$ $cm^2$ ✓
$C = \pi d$
$= \pi \times 4$
$= 4\pi$ cm ✓ (2 marks)

**7** $V = \pi r^2 h$
$= \pi \times 12^2 \times 20$ ✓
$= 9047.79$ (to 2 decimal places)
∴ The volume is 9047.79 $cm^3$. ✓
(2 marks)

**8** **a** London is 8 h ahead of Los Angeles. ✓
**b** 3 am plus 8 h = 11 am
i.e. 11 am in Perth ✓
**c** 10 pm minus 15 h = 7 am
i.e. 7 am in New York ✓ (3 marks)

**9** **a** 0225 ✓
**b** 0.37 m ✓
**c** 2250 minus 0943 = 13 h 7 min ✓
**d** 0759 plus 30 min
i.e. 0829 ✓ (4 marks)
**(Total 30 marks)**

## Level 2 Test

**1** **a** Area $= \frac{1}{2} \times 6x \times (3x + 5)$
$= 3x(3x + 5)$
$= 9x^2 + 15x$
∴ The area is $(9x^2 + 15x)$ $cm^2$. ✓✓
(2 marks)

**b**



Area $= 2(3x + 4) + 3(x + 4)$
$= 6x + 8 + 3x + 12$
$= 9x + 20$
∴ The area is $(9x + 20)$ $cm^2$. ✓✓
(2 marks)

**2** **a**



Volume $= (16.4 \times 6.1 + 7.1 \times 3.9) \times 5$ ✓
$= 127.73 \times 5$
$= 638.65$
∴ The volume is 638.65 $cm^3$. ✓
(2 marks)

**b** Volume $= \frac{1}{2} \times 10.3 \times 7 \times 5$
$= 180.25$
∴ The volume is 180.25 $cm^3$. ✓✓
(2 marks)

**3** **a** Volume $= 18 \times 12 \times 9$
$= 1944$
∴ The volume is 1944 $cm^3$. ✓
∴ The capacity is 1944 mL,
i.e. 1.944 L. ✓ (2 marks)

**b** Volume $= 26 \times 16$
$= 416$
∴ The volume is 416 $m^3$. ✓
∴ The capacity is 416 kL. ✓ (2 marks)

# WORKED SOLUTIONS to Chapter 8 **Tests**

**4** **a**

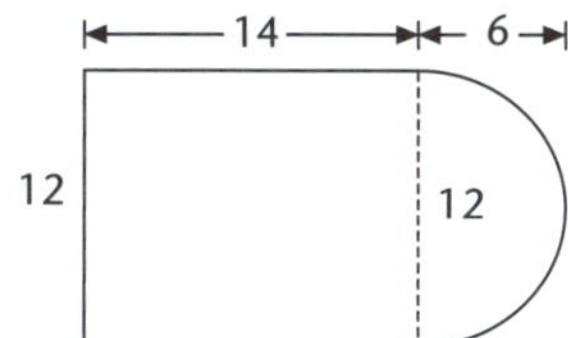


Per. $= 14 + 12 + 14 + \frac{1}{2} \times \pi \times 12$ ✓
$= 58.850$ (to 3 decimal places)
$\therefore$ The perimeter is 58.850 cm. ✓
(2 marks)

**b**

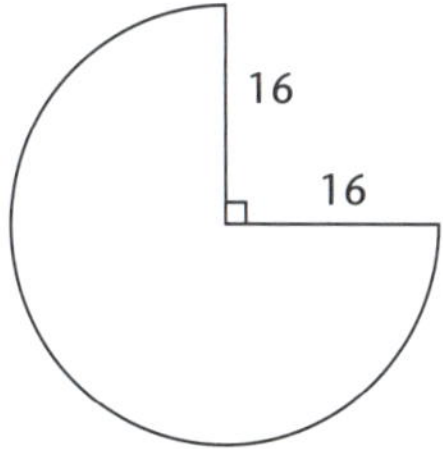


Per. $= 16 + 16 + \frac{3}{4} \times \pi \times 32$ ✓
$= 107.398$ (to 3 decimal places)
$\therefore$ The perimeter is 107.398 cm. ✓
(2 marks)

**c** Per. $= 2 + \frac{1}{2} \times \pi \times 2$ ✓
$= 5.142$ (to 3 decimal places)
$\therefore$ The perimeter is 5.142 m. ✓
(2 marks)

**d**

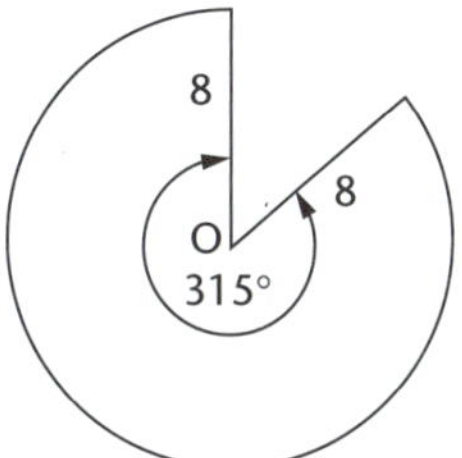


Per. $= 8 + 8 + \frac{315}{360} \times \pi \times 16$ ✓
$= 59.982$ (to 3 decimal places)
$\therefore$ The perimeter is 59.982 cm. ✓
(2 marks)

**5** $C = \pi d$
$\therefore d = \dfrac{C}{\pi}$
$= \dfrac{240}{\pi}$
$= 76.394\,372\,68$ ✓
$\therefore r = 38.2$ (to 1 decimal place)
$\therefore$ The radius is 38.2 cm. ✓ (2 marks)

**6** **a** $A = \pi r^2$
$= \frac{240}{360} \times \pi \times 6^2$ ✓
$= 75.40$ (to 2 decimal places)
$\therefore$ The area is 75.40 cm$^2$. ✓ (2 marks)

**b** Figure = rectangle + circle

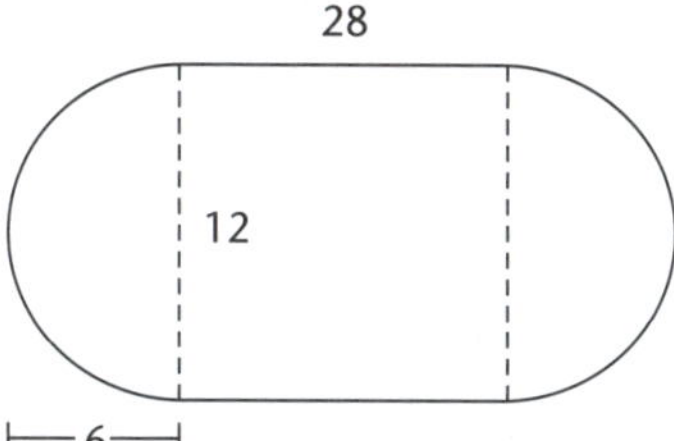


Area $= 28 \times 12 + \pi \times 6^2$ ✓
$= 449.10$ (to 2 decimal places)
$\therefore$ The area is 449.10 cm$^2$. ✓ (2 marks)

**7** $V = \pi r^2 h$
$= \pi \times 15^2 \times 36$ ✓
$= 25\,447$ (nearest whole cubic cm)
$\therefore$ The volume is 25 447 cm$^3$. ✓ (2 marks)

**8** $V = \pi r^2 h$
$= \pi \times 24^2 \times 60$ ✓
$= 108\,573$ (nearest whole)
$\therefore$ volume is 108 573 cm$^3$ ✓
As 1000 cm$^3$ = 1 L
$\therefore 108\,573 \div 1000 = 109$ (nearest whole L)
$\therefore$ The capacity is 109 L. ✓ (3 marks)

**9** **a** Time = 1214 minus 1131
= 29 + 14
= 43 ✓
i.e. 43 min

**b** Normal: 1223 minus 1131
= 23 + 29
= 52
Express: 1240 minus 1157
= 43
$\therefore$ The express is quicker by 9 min. ✓

**c** Arrived = 1146 plus 5 min
= 1151
$\therefore$ Time waiting = 1212 minus 1151
= 9 + 12
= 21
$\therefore$ Sandy waits 21 min. ✓

# WORKED SOLUTIONS to Chapter 8 **Tests**

**d** Scott cannot catch the express (as it does not stop at Nelson St).

He waits for the following bus.

$\therefore$ Arrives at Nelson St at 1305. ✓

(4 marks)

**(Total 35 marks)**

## Level 3 Test

**1 a**

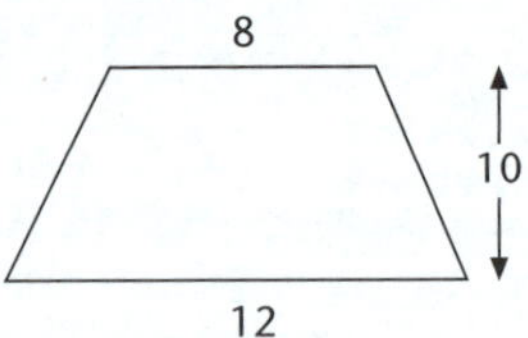


$$\text{Area} = \tfrac{1}{2} \times 10(8 + 12) \quad ✓$$
$$= 5(20)$$
$$= 100$$

$\therefore$ The area is 100 $cm^2$. ✓ (2 marks)

**b**

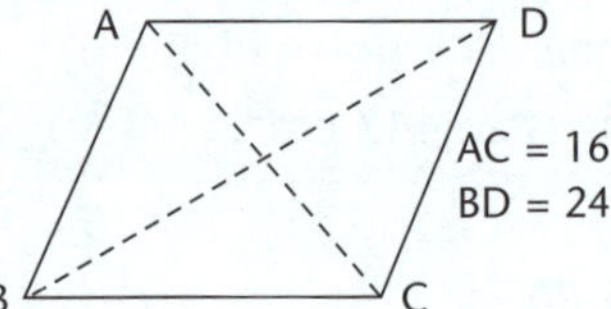


$$\text{Area} = \tfrac{1}{2} \times 16 \times 24 \quad ✓$$
$$= 192$$

$\therefore$ The area is 192 $cm^2$. ✓ (2 marks)

**2 a** $16 \times 12 - \tfrac{1}{2} \times 12 \times 9$ ✓

$= 192 - 54$

$= 138$

$\therefore$ The area is 138 $cm^2$. ✓ (2 marks)

**b** $64 \times 42 - \tfrac{1}{2} \times 10 \times 16$ ✓

$= 2608$

$\therefore$ The area is 2608 $cm^2$. ✓ (2 marks)

**3** Volume $= 8 \times 4 \times 1.8$

$= 57.6$ ✓

$\therefore$ volume is 57.6 $m^3$

$\therefore$ capacity is 57.6 kL ✓

$\therefore$ cost $= 57.6 \times \$1.25$

$= \$72$

$\therefore$ The cost of filling the pool is \$72. ✓

(3 marks)

**4**

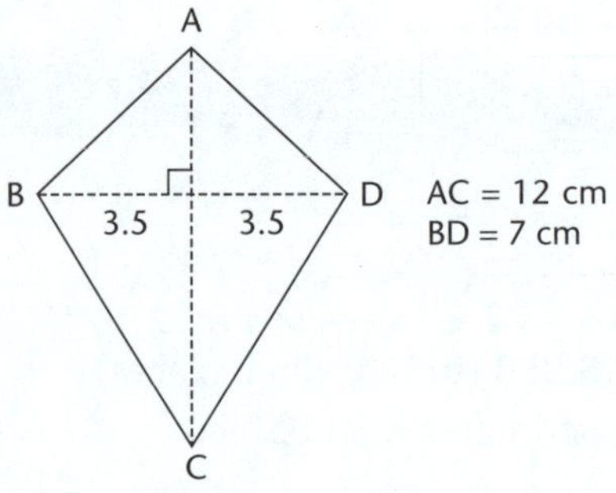


$$\text{Area} = \tfrac{1}{2} \times 12 \times 7 \quad ✓$$
$$= 42$$

$\therefore$ The area is 42 $cm^2$. ✓ (2 marks)

**5 a** $A = \pi r^2$

$\therefore r = \sqrt{\dfrac{A}{\pi}}$

$= \sqrt{\dfrac{36\pi}{\pi}}$

$= \sqrt{36} = 6$

$\therefore$ The radius is 6 cm. ✓ (1 mark)

**b** $A = \pi r^2$

$= \tfrac{120}{360} \times 36\pi$ ✓

$= 37.70$ (to 2 decimal places)

$\therefore$ The area is 37.70 $cm^2$. ✓ (2 marks)

**c** Length $= \tfrac{120}{360} \times \pi d$

$= \tfrac{120}{360} \times \pi \times 12$ ✓

$= 12.57$ (to 2 decimal places)

$\therefore$ The length of arc is 12.57 cm. ✓

(2 marks)

**6** $C = \pi d = \pi \times 72$

$= 226.1946$ cm

$$\text{Revolutions} = \frac{\text{Distance}}{\text{Circumference}}$$
$$= \frac{200\,000}{226.1946} \quad ✓$$
$$= 884.1944$$

[2 km = 200 000 cm]

That is, need 885 complete revolutions to travel past 2 km. ✓ (2 marks)

# WORKED SOLUTIONS to Chapter 8 **Tests**

**7** **a**

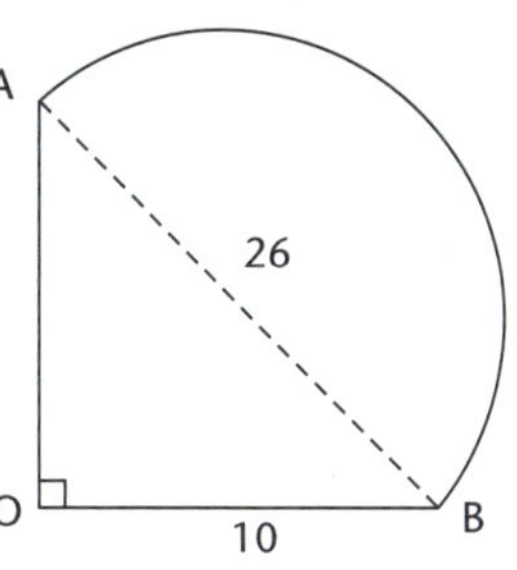


$AO^2 = 26^2 - 10^2$
$= 676 - 100$
$= 576$
$AO = \sqrt{576}$
$= 24$ ✓

$\therefore$ Perimeter $= 24 + 10 + \frac{1}{2} \times \pi \times 26$
$= 74.8407$ (to 4 decimal places)

$\therefore$ The perimeter is 74.8407 cm. ✓ (2 marks)

**b** Figure = semicircle + circle

Perimeter $= \frac{1}{2} \times \pi \times 20 + \pi \times 10$ ✓
$= 62.8319$ (to 4 decimal places)

$\therefore$ The perimeter is 62.8319 cm. ✓ (2 marks)

**8** **a** Area $= \frac{135}{360} \times \pi \times 10^2$ ✓
$= 117.81$ (to 2 decimal places)

$\therefore$ The area is 117.81 $cm^2$. ✓ (2 marks)

**b** Area $= \pi \times 8^2 - 2 \times \pi \times 4^2$ ✓
$= 100.53$ (to 2 decimal places)

$\therefore$ The area is 100.53 $cm^2$. ✓ (2 marks)

**c** Area $= \pi \times 4^2 - \pi \times 2^2$ ✓
$= 37.70$ (to 2 decimal places)

$\therefore$ The area is 37.70 $cm^2$. ✓ (2 marks)

**d** Diameter $= d$
$\therefore d^2 = 10^2 + 24^2$
$= 100 + 576$
$= 676$
$= \sqrt{676}$
$= 26$
$\therefore r = 13$ ✓

Area of shaded region
$= \frac{1}{2} \times \pi \times 13^2$
$= 265.46$ (to 2 decimal places)

$\therefore$ The area is 265.46 $cm^2$. ✓ (2 marks)

**e**

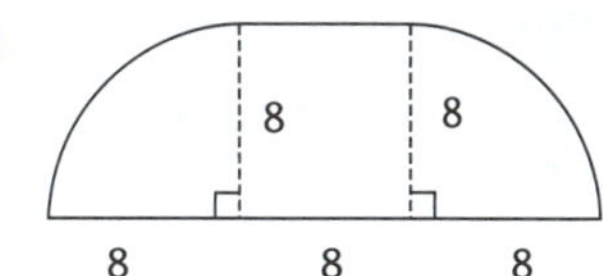


Area $= 8 \times 8 + \frac{1}{2} \times \pi \times 8^2$ ✓
$= 164.53$ (to 2 decimal places)

$\therefore$ The area is 164.53 $cm^2$. ✓ (2 marks)

**f**

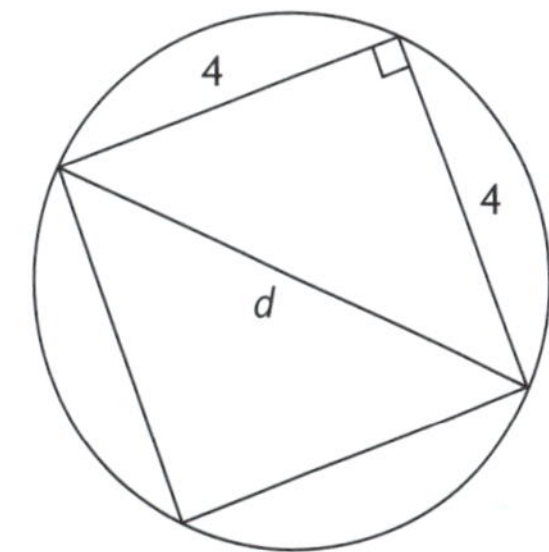


$d^2 = 4^2 + 4^2$
$= 16 + 16$
$= 32$
$d = \sqrt{32}$
$= 5.656854$
$\therefore r = 2.828427$ ✓

Area of shaded region
= Area of circle − Area of square
$= \pi \times 2.828427^2 - 4 \times 4$
$= 9.13$ (to 2 decimal places)

$\therefore$ The area is 9.13 $cm^2$. ✓ (2 marks)

**9** $V = \pi \times 0.6^2 \times 0.8$
$= 0.9047787$

i.e. volume is 0.904 7787 $m^3$ ✓

Now, 1 $m^3$ = 1000 L

$\therefore$ capacity $= 0.9047787 \times 1000$
$= 904.7787$
$= 904.8$ (to 1 decimal place)

$\therefore$ The capacity is 904.8 L. ✓ (2 marks)

**10** After 3 h, watch loses 6 min

$\therefore$ Time: 6:00 pm minus 6 min ✓
= 5:54 pm

$\therefore$ 5:54 on Kristie's watch. ✓ (2 marks)

**(Total 40 marks)**

# WORKED SOLUTIONS to Chapter 9 **Tests**

## Chapter 9—Ratio and Rates

pp. 195–197

### Test 1

**1** **a** 1:4 ✓ **b** 9:7 ✓ **c** 10:1 ✓ (3 marks)

**2** **a** AB: 1 unit, EF: 1 unit
∴ 1:1 ✓ (1 mark)

**b** CE: 2 units, AI : 8
∴ 2:8 = 1:4 ✓ (1 mark)

**c** BG: 5 units, AF: 5 units
∴ 5:5 = 1:1 ✓ (1 mark)

**3** **a** 16:14 = 8:7 ✓ (1 mark)

**b** 14:30 = 7:15 ✓ (1 mark)

**4** **a** 80:40 = 2:1 ✓ (1 mark)

**b** 120:80 = 3:2 ✓ (1 mark)

**5** 3 parts = 270 ✓
1 part = 90 ✓
∴ There is 90 mL of concentrate in the cordial. ✓ (3 marks)

**6** 5 + 3 = 8 ✓
∴ 8 parts ✓
∴ $\frac{5}{8} \times 464 = 290$ ✓
∴ There are 290 adults in the town. (3 marks)

**7** 2 + 3 = 5
∴ 5 parts ✓
∴ $\frac{3}{5} \times 30 = 18$
∴ The longer length is 18 m. ✓ (2 marks)

**8** Distance = 80 × 3 ✓
= 240
∴ The distance travelled was 240 km. ✓ (2 marks)

**9** Speed = 560 ÷ 8 = 70 ✓
∴ Her average speed was 70 km/h. ✓ (2 marks)

**10** **a** 1:40 000 ✓ (1 mark)

**b** 2 km = 2000 m
As 1 cm = 400 m
∴ 5 cm = 2000 m
∴ The distance on the directory is 5 cm. ✓ (1 mark)

**c** 6.4 × 400 = 2560
∴ 2560 m
∴ They are 2.56 km apart. ✓ (1 mark)

**(Total 25 marks)**

### Level 2 Test

**1** **a** 320:400 = 4:5 ✓ (1 mark)

**b** 15:12:6 = 5:4:2 ✓ (1 mark)

**c** 1 metre = 10 000 metres
i.e. 1:10 000 ✓ (1 mark)

**2** **a** 12:16 = 3:4
∴ $x = 4$ ✓ (1 mark)

**b** 5:15 = 10:30
∴ $x = 10$ ✓ (1 mark)

**3** Unsold = 72 − 24
= 48 ✓
∴ 72:48 = 3:2 ✓ (2 marks)

**4** 35 parts = 280 ✓
1 part = 8
∴ There are 8 teachers. ✓ (2 marks)

**5** 7 + 2 = 9
∴ 9 parts ✓
∴ $\frac{7}{9} \times 540 = 420$
$\frac{2}{9} \times 540 = 120$
∴ 420 g, 120 g ✓ (2 marks)

**6** 4 parts = 1.2 ✓
1 part = 0.3
∴ 2 parts = 0.6
i.e. shortest is 0.6 m ✓ (2 marks)

**7** 3 + 5 + 4 = 12
∴ 12 parts ✓
i.e. $\frac{5}{12} \times 480 = 200$
∴ The area sowed with oats is 200 ha. ✓ (2 marks)

# WORKED SOLUTIONS to Chapter 9 **Tests**

**8** Time = 2460 ÷ 1.2
= 2050 ✓
i.e. 2050 seconds
∴ It will take 34 min and 10 s for the tank to fill. ✓ (2 marks)

**9** Time = 1050 ÷ 90 ✓
$= 11\frac{2}{3}$
∴ It took 11 h 40 min. ✓ (2 marks)

**10** Time = 7:30 to 12:00
$= 4\frac{1}{2}$ h ✓
Speed = $360 \div 4\frac{1}{2}$
= 80
∴ 80 km/h ✓ (2 marks)

**11** Time = 15 ÷ 6 ✓
= 2.5
∴ The log would need to float for 2 hours and 30 minutes. ✓ (2 marks)

**12** Distance = 4.5 × 100 000
= 450 000 cm ✓
i.e. = 4500 m
∴ The towns are 4.5 km apart. ✓ (2 marks)
**(Total 25 marks)**

## Level 3 Test

**1** a $\frac{3}{4}:\frac{4}{4} = 3:4$ ✓ (1 mark)

b 5000 $m^2$: 20 000 $m^2$
= 1:4 ✓ (1 mark)

c $\frac{2}{3}:\frac{3}{4} = \frac{8}{12}:\frac{9}{12}$
= 8:9 ✓ (1 mark)

**2** a 48:64 = 3:4
∴ $x = 3$ ✓ (1 mark)

b 25:35 = 5:7
= 10:14
∴ $x = 14$ ✓ (1 mark)

**3** Angles are 60°, 40°, 80°
∴ 80°:40° = 2:1 ✓ (1 mark)

**4** Discount:original price
= 15:100
= 3:20 ✓ (1 mark)

**5** 2 + 3 + 5 = 10 ∴ 10 parts
∴ $\frac{2}{10} \times 360 = 72$
$\frac{3}{10} \times 360 = 108$
$\frac{5}{10} \times 360 = 180$
∴ \$72, \$108, \$180 ✓✓ (2 marks)

**6** Perimeter = 80 cm
∴ Sum of length + Breadth = 40 cm
∴ 5 parts = 40 ✓
∴ $\frac{2}{5} \times 40 = 16$
$\frac{3}{5} \times 40 = 24$
∴ The sides are 16 cm and 24 cm. ✓
(2 marks)

**7** 4 parts = 5.6 ✓
1 part = 1.4
3 parts = 4.2
∴ 4.2 kg copper would be needed. ✓
(2 marks)

**8** Time = 58 600 ÷ 240
= $244.1\dot{6}$ minutes
i.e. $4.069\dot{4}$ hours ✓
∴ It would take 4 hours and 4 minutes 10 s. ✓ (2 marks)

**9** Distance = $5.2 \times \frac{25}{60}$ ✓
= $2.1\dot{6}$ km
= 2167 m (to the nearest metre)
∴ Mark walks 2167 m. ✓ (2 marks)

**10** Time = 500 ÷ 60 ✓
= $8.\dot{3}$ h
∴ John's journey would take 8 hours 20 minutes. ✓ (2 marks)

**11** 3.6 km = 3600 m ✓
As 3600 ÷ 40 000 = 0.09, then the island would be 0.09 m or 9 cm on the map. ✓
(2 marks)

**12** a Train:bus = 3:2
= 15:10
Bus:car = 5:2
= 10.4
∴ Train:bus:car = 15:10:4 ✓✓ (2 marks)

**b** 4 parts = 48 ✓
1 part = 12
15 parts = 180
∴ 180 train travellers ✓ (2 marks)
**(Total 25 marks)**

## Chapter 10—Geometry and Congruence

### Level 1 Test

**1** **a** 90° ✓ **b** 180° ✓ **c** 180° ✓
**d** 360° ✓ **e** 360° ✓ (5 marks)

**2** **a** Square ✓ **b** Rhombus ✓
**c** Trapezium ✓ **d** Kite ✓
**e** Rectangle ✓ (5 marks)

**3** **a** $x + 140 = 180$
$x = 40$ ✓ (1 mark)

**b** $y + 35 = 90$
$y = 55$ ✓ (1 mark)

**c** $x + 90 + 120 = 360$
$x = 150$ ✓ (1 mark)

**d** $a = 113$ ✓ (1 mark)

**e** $b + 60 + 50 = 180$
$b = 70$ ✓ (1 mark)

**f** $k = 70$ ✓ (1 mark)

**g** $b = 60$ ✓ (1 mark)

**h** $z = 120$ ✓ (1 mark)

**i** $x = 72$ ✓ (1 mark)

**j** $c + 130 = 180$
$c = 50$ ✓ (1 mark)

**k** $x + 70 + 130 + 120 = 360$
$x = 40$ ✓ (1 mark)

**l**

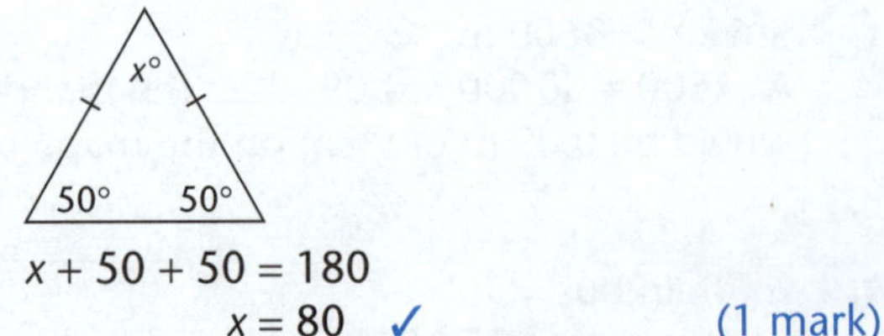


$x + 50 + 50 = 180$
$x = 80$ ✓ (1 mark)

**m**

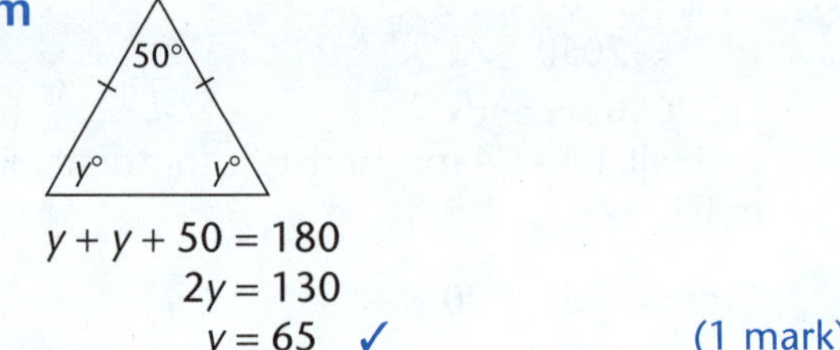


$y + y + 50 = 180$
$2y = 130$
$y = 65$ ✓ (1 mark)

**4** **a** MP ✓
**b** 5 cm ✓
**c** ∠PMN ✓ (3 marks)

**5** $x = 3$ ✓
$y = 83$ ✓
$z = 98$ ✓
$w = 6$ ✓ (4 marks)
**(Total 30 marks)**

### Level 2 Test

**1** **a** $a = 98$ ✓ $b = 82$ ✓ (2 marks)

**b** $p + 32 = 90$
$p = 58$ ✓ (1 mark)

**c** $x + 38 + 43 = 180$
$x + 81 = 180$
$x = 99$ ✓ (1 mark)

**d** $y + 94 + 90 + 76 = 360$
$y = 100$ ✓ (1 mark)

**e** $x + 116 = 180$
$x = 64$ ✓ (1 mark)

**f** $x + 100 + 80 + 70 = 360$
$x = 110$ ✓ (1 mark)

**2** **a** $x + 87 + 92 + 88 = 360$
$x = 93$ ✓ (1 mark)

**b**

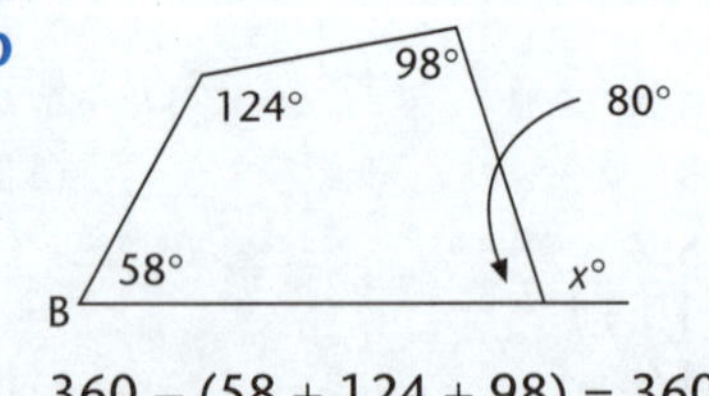


$360 - (58 + 124 + 98) = 360 - 280$
$= 80$ ✓
∴ $x + 80 = 180$
$x = 100$ ✓ (2 marks)

# WORKED SOLUTIONS to Chapter 10 **Tests**

**3** **a**

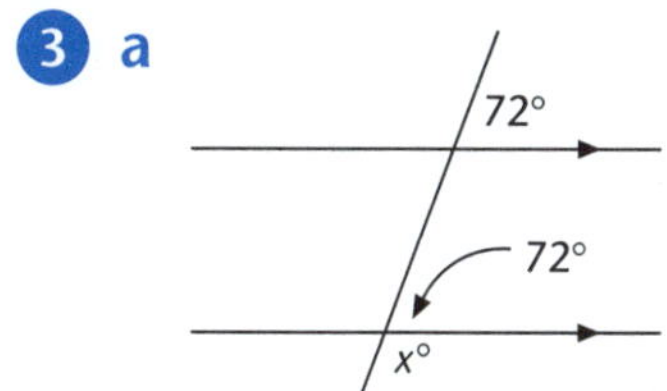


$x + 72 = 180$
$x = 108$ ✓ (1 mark)

**b** $x + 63 = 118$ ✓
$x = 55$ ✓
(exterior angle of triangle) (2 marks)

**4** $x + x + x = 180$
$x = 60$ ✓ (1 mark)

**5**

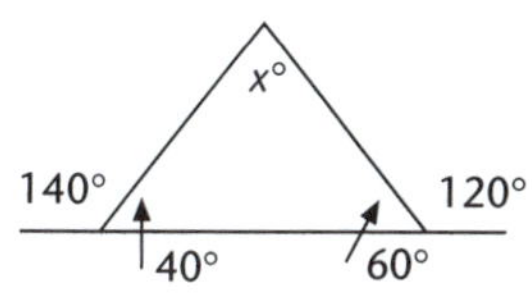


$x + 40 + 60 = 180$ ✓
$x = 80$ ✓ (2 marks)

**6** **a** True, a pair of co-interior angles are supplementary. ✓ (1 mark)

**b** False, a pair of alternate angles is *not* equal. ✓ (1 mark)

**c** True, a pair of corresponding angles is equal. ✓ (1 mark)

**7** **a** $x + 52 + 52 = 180$
$x = 76$ ✓ (1 mark)

**b** $x + x + 82 = 180$
$x = 49$ ✓ (1 mark)

**c** $x = 60$ ✓ (1 mark)

**8**

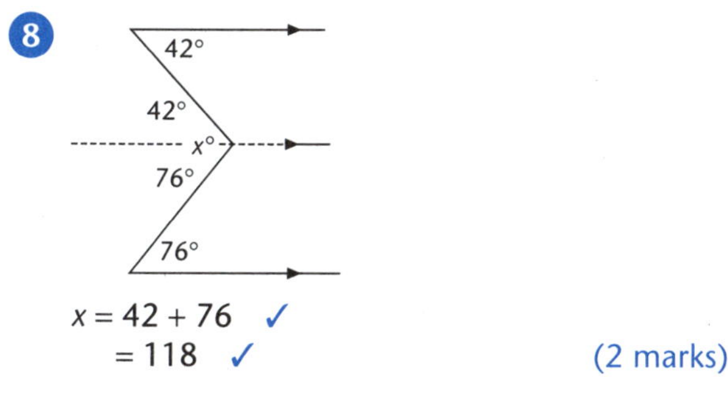


$x = 42 + 76$ ✓
$= 118$ ✓ (2 marks)

**9** $x = 60$ (angle sum of △ABC) ✓✓
$y = 120$ (co-interior to ∠ABD and AB || ED) ✓✓
(4 marks)

**10** **a** Congruent ✓ **b** True ✓
**c** NP ✓ **d** ∠NPM ✓
**e** All the same areas. ✓ (5 marks)

**11** **a** Yes, SAS test ✓
**b** Yes, RHS test ✓
**c** Yes, SSS test ✓
**d** Yes, AAS test (∠s are 50°, 70°, 60°) ✓
(4 marks)

**12**

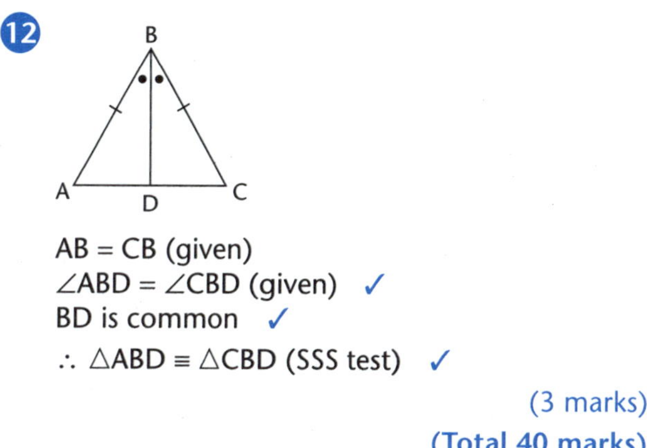


AB = CB (given)
∠ABD = ∠CBD (given) ✓
BD is common ✓
∴ △ABD ≡ △CBD (SSS test) ✓
(3 marks)

**(Total 40 marks)**

## Level 3 Test

**1** **a** T ✓ **b** F ✓ **c** T ✓
(3 marks)

**2** **a** $x + 46 + 69 = 180$
$x = 65$ ✓ (1 mark)

**b** $x + x + 88 = 360$
$x = 136$ ✓ (1 mark)

**c** $x + 46 = 82$
$x = 36$ ✓ (1 mark)

**d** $x = 83$ ✓ (1 mark)

**e**

58° 58° x°

$x + 58 = 180$
$x = 122$ ✓ (1 mark)

**f** $x + x + 90 = 180$
$x = 45$ ✓ (1 mark)

# WORKED SOLUTIONS to Chapter 10 **Tests**

**g**

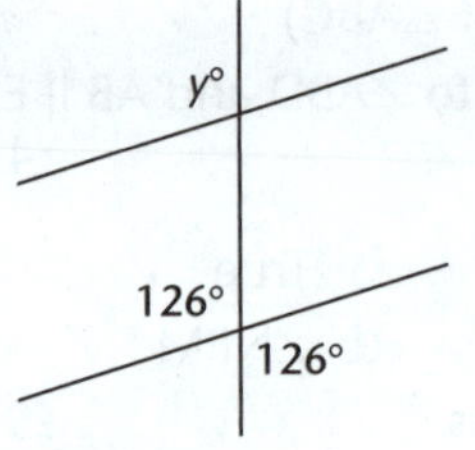


$y = 126$ ✓ (1 mark)

**h** $x + 72 = 113$ (exterior angle of triangle)
$x = 41$ ✓ (1 mark)

**i** $2x + x + 66 = 180$
$3x + 66 = 180$
$3x = 114$
$x = 38$ ✓ (1 mark)

**j**

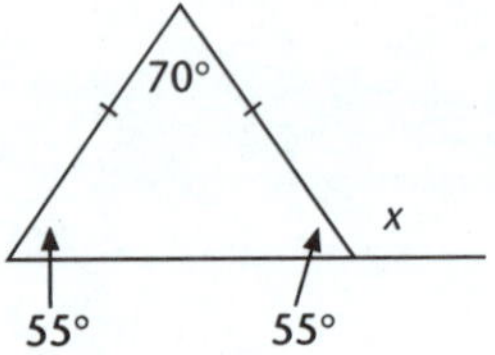


$x + 55 = 180$
$x = 125$ ✓ (1 mark)

**3** **a**

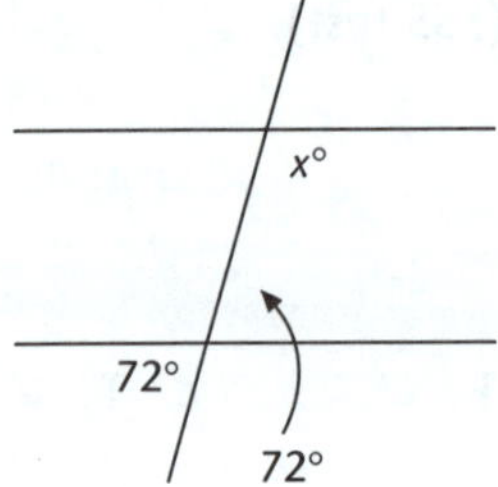


$x + 72 = 180$
$x = 108$ ✓ (1 mark)

**b**

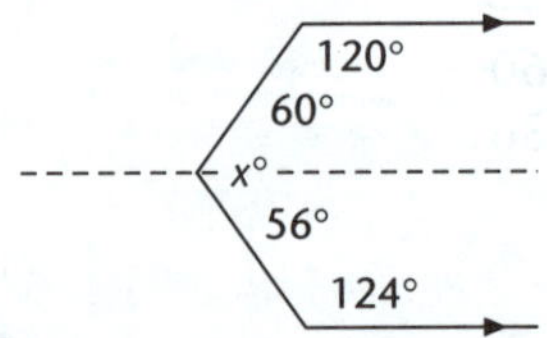


$x = 60 + 56$
$x = 116$ ✓ (1 mark)

**c**

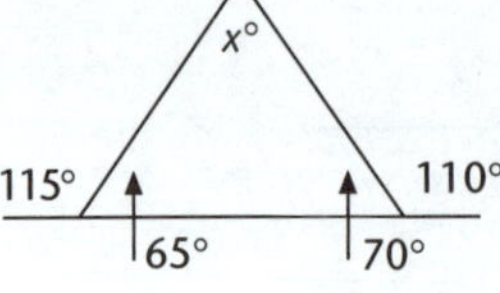


$x + 65 + 70 = 180$ ✓
$x = 45$ ✓ (2 marks)

**d**

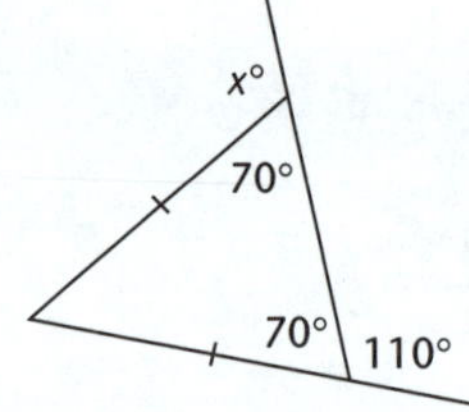


$x + 70 = 180$ ✓
$x = 110$ ✓ (2 marks)

**4**

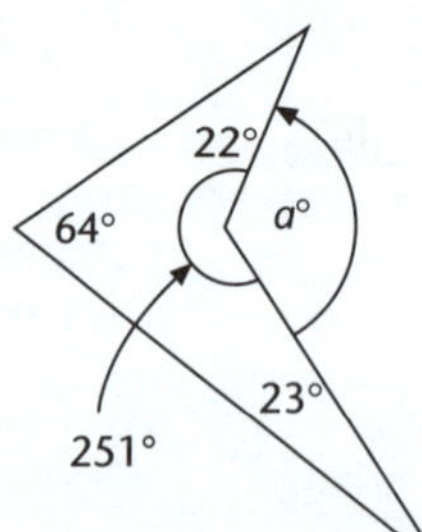


$a + 251 = 360$ ✓
$a = 109$ ✓ (2 marks)

**5** $x = 54$ ✓
$y = 72$ ✓
$z = 18$ ✓ (3 marks)

**6** **a** $a + 94 = 180$
$a = 86$ ✓ (1 mark)

**b** $b + 116 = 180$
$b = 64$ ✓ (1 mark)

**7** **a** ∠PYX = 46° (alternate to ∠YPQ and XZ // PQ) ✓✓ (2 marks)

**b** ∠ZYP = 134° (supplementary to ∠PXY) ✓✓ (2 marks)

**c** ∠PQY = 53° (angle sum of △PQY) ✓✓ (2 marks)

**8** **a** ∠HND = 110° (co-interior to ∠BMI and AB // CD) ✓
∠FQD = 70° (vertically opposite to ∠GQC) ✓ (2 marks)

**b** No, because a pair of co-interior angles HND and FQD are not supplementary. ✓✓ (2 marks)

**9** $b + c + d = 180$ (angle sum of △ABC)
$a + d = 180$ (supplementary angles)
$b + c + d = a + d$
$\therefore a = b + c$ ✓✓✓ (3 marks)

**10** a Yes, AAS test ✓
b No ✓
c No ✓ (3 marks)

**11** ∠BCM = ∠DCM (given)
CM is common ✓
∠BMC = ∠DMC (given) ✓
∴ △BCM ≡ △DCM (RHS test) ✓ (3 marks)
(Total 45 marks)

## Chapter 11—Statistics

pp. 246–250

### Level 1 Test

**1** a

| Score ($x$) | Tally | Frequency ($f$) |
|---|---|---|
| 11 | \|\| | 2 |
| 12 | \|\| | 2 |
| 13 | ~~\|\|\|\|~~ | 5 |
| 14 | \|\|\|\| | 4 |
| 15 | \|\|\|\| | 4 |
| 16 | ~~\|\|\|\|~~ \| | 6 |
| 17 | ~~\|\|\|\|~~ \| | 6 |
| 18 | ~~\|\|\|\|~~ \| | 6 |
| 19 | ~~\|\|\|\|~~ | 5 |

✓✓✓ (3 marks)

b

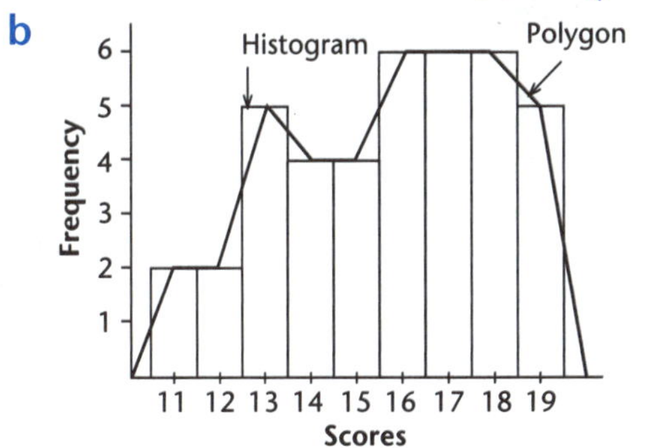


✓✓ (2 marks)

**2** a

| Score ($x$) | Frequency ($f$) | $x \times f$ |
|---|---|---|
| 22 | 5 | 110 |
| 23 | 7 | 161 |
| 24 | 10 | 240 |
| 25 | 8 | 200 |
| 26 | 5 | 130 |
| Total | 35 | 841 |

✓✓ (2 marks)

b Mean $= \dfrac{\Sigma fx}{\Sigma f}$

$= \frac{841}{35}$ ✓

$= 24.03$ (to 2 decimal places) ✓
(2 marks)

c Mode: 24 ✓
Median = 18th score
i.e. 24 ✓
Range = 4 ✓ (3 marks)

d 12 ✓ (1 mark)

e $5 + 10 + 5 = 20$ ✓ (1 mark)

**3** a Mean = 12.5 (by calculator) ✓✓
or $\bar{x} = (12 + 18 + 14 + 3 + 5 + 19 + 9 + 11 + 14 + 20) \div 10$

$= \frac{125}{10}$

$= 12.5$ ✓✓ (2 marks)

b Median 3, 5, 9, 11, 12, 14, 14, 18, 19, 20 ✓
i.e. middle of 12 and 14
∴ The median is 13. ✓ (2 marks)

c Mode = 14 ✓ (1 mark)

d Range is $20 - 3 = 17$ ✓ (1 mark)

**4** a The outlier is 45. ✓ (1 mark)

b $\bar{x} = 19.85$ (to 2 decimal places) ✓✓
(2 marks)

c

| Stem | Leaf |
|---|---|
| 0 | 69 |
| 1 | 244789 |
| 2 | 3489 |
| 3 | |
| 4 | 5 |

∴ The median is 18. ✓ (1 mark)

d The mode is 14. ✓ (1 mark)

e The range is $45 - 6 = 39$. ✓ (1 mark)

# WORKED SOLUTIONS to Chapter 11 **Tests**

**5** a The outlier is 2. ✓ (1 mark)

b The mode is 6. ✓ (1 mark)

c Range = 7 − 2 = 5

∴ The range is 5. ✓ (1 mark)

d

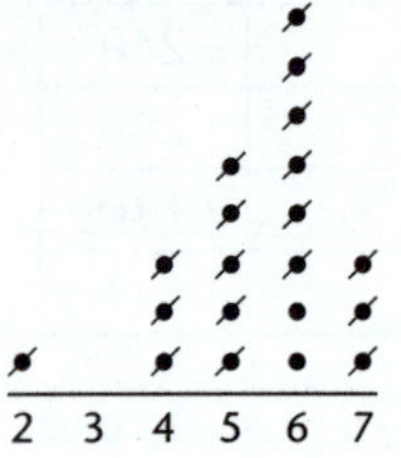


The middle of 6 and 6 is 6

∴ Median is 6. ✓ (1 mark)

**(Total 30 marks)**

## Level 2 Test

**1** a

| Class | Class centre ($x$) | Tally | Frequency ($f$) | $fx$ |
|---|---|---|---|---|
| 51–55 | 53 | \|\|\|\| | 4 | 212 |
| 56–60 | 58 | \|\|\| | 3 | 174 |
| 61–65 | 63 | ~~\|\|\|\|~~ \|\|\| | 8 | 504 |
| 66–70 | 68 | ~~\|\|\|\|~~ \| | 6 | 408 |
| 71–75 | 73 | ~~\|\|\|\|~~ \|\|\| | 8 | 584 |
| 76–80 | 78 | \| | 1 | 78 |
| | | Sum | 30 | 1960 |

✓✓✓ (3 marks)

b $\bar{x} = \dfrac{\Sigma fx}{\Sigma f} = \dfrac{1960}{30}$

$= 65.\dot{3}$ ✓ (1 mark)

c 61–65 and 71–75 ✓ (1 mark)

d

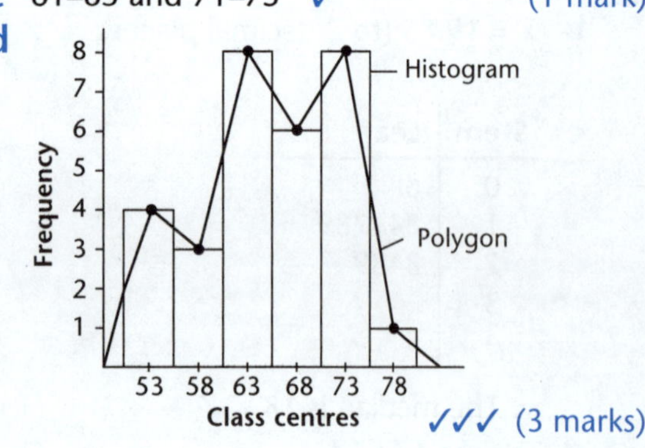


✓✓✓ (3 marks)

**2** a $\bar{x} = 7.\dot{2}\dot{7}$ ✓ (1 mark)

b 2, 3, 4, 5, 5, 6, 7, 8, 11, 13, 16

∴ The median is 6. ✓ (1 mark)

c The mode is 5. ✓ (1 mark)

d The range is 16 − 2 = 14. ✓ (1 mark)

**3** a

| People | Houses |
|---|---|
| 1 | 2 |
| 2 | 4 |
| 3 | 5 |
| 4 | 4 |
| 5 | 2 |
| 6 | 3 |

✓✓ (2 marks)

b Number of people

$= 1 \times 2 + 2 \times 4 + 3 \times 5 + 4 \times 4 + 5 \times 2 + 6 \times 3$ ✓

$= 2 + 8 + 15 + 16 + 10 + 18$

$= 69$ ✓ (2 marks)

c Number of houses $= 2 + 4 + 5 + 4 + 2 + 3$

$= 20$ ✓ (1 mark)

d Mean $= \frac{69}{20}$

$= 3.45$ ✓ (1 mark)

**4** a Sum of five scores $= 5 \times 8$

$= 40$ ✓ (1 mark)

b Sum of six scores $= 6 \times 10$

$= 60$ ✓

∴ New score $= 60 - 40$

$= 20$

∴ The new score is 20. ✓ (2 marks)

**5** a

| Stem | Leaf |
|---|---|
| 2 | 0269 |
| 3 | 24578 |
| 4 | 12234688 |
| 5 | 19 |

✓✓ (2 marks)

b i $\bar{x} = 38.6\dot{1}$ ✓✓ (2 marks)

ii Mode = 48 ✓ (1 mark)

iii Middle of 38 and 41

∴ The median is 39.5. ✓ (1 mark)

iv 59 − 20 = 39 ✓ (1 mark)

# WORKED SOLUTIONS to Chapter 11 **Tests**

**6** **a** Sample A: 10 students
Sample B: 10 students
$\therefore$ Total of 20 students. ✓ (1 mark)

**b** Mode of Sample A: 2
Mode of Sample B: 3
$\therefore$ The mode of Sample B is higher than the mode of Sample A. ✓ (1 mark)

**c** **i** Median of Sample A: 2 ✓
**ii** Median of Sample B: 3 ✓ (2 marks)

**d** Mean of Sample A $= \frac{2+12+3+4}{10}$
$= \frac{21}{10}$
$= 2.1$ ✓

Mean of Sample B $= \frac{1+6+15}{10}$
$= \frac{26}{10}$
$= 2.6$ ✓

$\therefore$ Sample B had higher mean by 0.5.
✓ (3 marks)
**(Total 35 marks)**

## Level 3 Test

**1** **a**

| Score ($x$) | Frequency ($f$) | $fx$ |
|---|---|---|
| 3 | 4 | 12 |
| 5 | 4 | 20 |
| 8 | 6 | 48 |
| 10 | 4 | 40 |
| 12 | 5 | 60 |
| Sum | 23 | 180 |

✓✓ (2 marks)

**b** Mean $= \frac{180}{23}$
$= 7.83$ (to 2 decimal places) ✓
Mode = 8 ✓
The median is the 12th score $\therefore$ 8 ✓
Range is $12 - 3 = 9$ ✓ (4 marks)

**c**

Frequency
Score

✓✓✓ (3 marks)

**2** **a**

| Stem | Leaf |
|---|---|
| 3 | 24789 |
| 4 | 00246 |
| 5 | 258 |
| 6 | 36 |
| 7 | □ |

The median = 43 ✓ (1 mark)

**b** $\therefore$ Range = 43
$\therefore$ $32 + 43 = 75$
$\therefore$ $\square = 5$ ✓ (1 mark)

**c**

| Class | Class centre | Frequency ($f$) | $fx$ |
|---|---|---|---|
| 31–35 | 33 | 2 | 66 |
| 36–40 | 38 | 5 | 190 |
| 41–45 | 43 | 2 | 86 |
| 46–50 | 48 | 1 | 48 |
| 51–55 | 53 | 2 | 106 |
| 56–60 | 58 | 1 | 58 |
| 61–65 | 63 | 1 | 63 |
| 66–70 | 68 | 1 | 68 |
| 71–75 | 73 | 1 | 73 |
| | Sum | 16 | 758 |

✓✓✓ (3 marks)

**d** Mean from stem-and-leaf is 47.5625

Mean from frequency table is $\frac{\Sigma fx}{\Sigma f} = \frac{758}{16}$
$= 47.375$

The mean using the frequency table is less than the mean using the stem-and-leaf plot. ✓✓ (2 marks)

**3**

| Score ($x$) | Frequency ($f$) | $fx$ |
|---|---|---|
| 2 | 4 | 8 |
| 4 | 6 | 24 |
| Sum | 10 | 32 |

$\therefore$ Mean $= \frac{32}{10}$
$= 3.2$ ✓✓ (2 marks)

**4** Sum of goals in 5 games $= 5 \times 22$
$= 110$
Sum of goals in 6 games $= 6 \times 19$
$= 114$
$\therefore$ There were 4 goals scored in the 6th game. ✓✓ (2 marks)

# WORKED SOLUTIONS to Chapters 11 and 12 **Tests**

**5** **a** The outlier is 143 ✓

**b** If we ignore 143 then

**i** mode unchanged at 162 ✓

**ii** median unchanged at 162 ✓

**iii** range drops from 28 to 15 ✓

**iv** old mean = 161.5 ✓
new mean = 163.6 (1 dec. pl.) ✓

∴ Mean increases by 2.1. (6 marks)

**6**

| Item ($x$) | Customers ($f$) | $fx$ |
|---|---|---|
| 1 | 1 | 1 |
| 2 | 3 | 6 |
| 3 | 6 | 18 |
| 4 | 6 | 24 |
| 5 | 5 | 25 |
| 6 | 3 | 18 |
| 7 | 2 | 14 |
| 8 | 4 | 32 |
| Sum | 30 | 138 |

**a** Median: middle of 15th, 16th customer,
= 4 ✓
Mode: 3 and 4 ✓
Range: 7 ✓ (3 marks)

**b** Mean $= \frac{138}{30}$
$= 4.6$ ✓✓ (2 marks)

**7** **a**

| Score | Tally | Freq. |
|---|---|---|
| 12 | \| | 1 |
| 13 | \| | 1 |
| 14 | | 0 |
| 15 | \|\|\| | 3 |
| 16 | \|\|\| | 3 |
| 17 | ~~\|\|\|\|~~ | 5 |
| 18 | \|\| | 2 |
| 19 | \|\|\| | 3 |
| 20 | \|\| | 2 |
| | Total | 20 |

✓✓ (2 marks)

**b** **i** As mode = 17
∴ % $= \frac{5}{20} \times 100\% = 25\%$ ✓

**ii** As median = 17
∴ 25% ✓

(2 marks)

**(Total 35 marks)**

## Chapter 12—Probability

pp. 263–268

### Level 1 Test

**1** **a** $P(\text{Red}) = \frac{4}{9}$ ✓

**b** $P(\text{Green}) = \frac{3}{9} = \frac{1}{3}$ ✓

**c** $P(\text{Yellow}) = 0$ ✓

**d** $P(\text{Red or green}) = \frac{7}{9}$ ✓

**e** $P(\text{Not green}) = 1 - \frac{3}{9}$
$= \frac{6}{9}$
$= \frac{2}{3}$ ✓

**f** $P(\text{Not red or green}) = P(\text{Blue}) = \frac{2}{9}$ ✓

(6 marks)

**2** **a** $P(\text{Red}) = \frac{26}{52} = \frac{1}{2}$ ✓

**b** $P(\text{Not red}) = 1 - \frac{1}{2}$
$= \frac{1}{2}$ ✓

**c** $P(\text{Green}) = 0$ ✓

**d** $P(9) = \frac{4}{52} = \frac{1}{13}$ ✓

**e** $P(\text{Red } 4) = \frac{2}{52} = \frac{1}{26}$ ✓

**f** $P(7 \text{ of clubs}) = \frac{1}{52}$ ✓

**g** $P(\text{Heart}) = \frac{13}{52} = \frac{1}{4}$ ✓

**h** $P(7 \text{ or } 8) = \frac{8}{52} = \frac{2}{13}$ ✓ (8 marks)

# WORKED SOLUTIONS to Chapter 12 **Tests**

**3**

| Score | Freq. | Rel. freq. |
|---|---|---|
| 4 | 6 | 0.3 |
| 5 | 7 | 0.35 |
| 6 | 2 | 0.1 |
| 7 | 5 | 0.25 |
| **Totals** | **20** | **1** |

✓✓✓ (3 marks)

**4** **a**

| | Males | Females | Total |
|---|---|---|---|
| Smokers | 5 | 8 | 13 |
| Non-smokers | 4 | 6 | 10 |
| Totals | 9 | 14 | 23 |

✓

**b** 9 ✓

**c** $\frac{6}{10} = \frac{3}{5}$ ✓

**d** **i** $\frac{8}{23}$ ✓

**ii** $\frac{4}{23}$ ✓ (5 marks)

**5** **i** $\frac{4}{21}$ ✓

**ii** $\frac{2}{21}$ ✓

**iii** $\frac{8}{21}$ ✓ (3 marks)

**(Total 25 marks)**

## Level 2 Test

**1** **a** $\frac{1}{10}$ ✓ **b** $\frac{5}{10} = \frac{1}{2}$ ✓ **c** $\frac{7}{10}$ ✓

(3 marks)

**d** $P(\text{Composite}) = P(4, 6, 8, 9, 10)$

$= \frac{1}{2}$ ✓ (1 mark)

**e** $P(\text{Factor of } 12) = P(1, 2, 3, 4, 6)$

$= \frac{5}{10}$

$= \frac{1}{2}$ ✓ (1 mark)

**f** $P(\text{Mult. of } 3) = P(3, 6, 9)$

$= \frac{3}{10}$ ✓ (1 mark)

**g** $P(\text{Not a } 7) = 1 - P(7)$

$= 1 - \frac{1}{10}$

$= \frac{9}{10}$ ✓ (1 mark)

**2**

| Score | Freq. | Rel. freq. |
|---|---|---|
| 1 | 9 | $\frac{9}{60}$ |
| 2 | 11 | $\frac{11}{60}$ |
| 3 | 7 | $\frac{7}{60}$ |
| 4 | 12 | $\frac{12}{60}$ |
| 5 | 8 | $\frac{8}{60}$ |
| 6 | 13 | $\frac{13}{60}$ |
| **Total** | **60** | **1** |

✓✓ (2 marks)

**3** **a** **i** As $15 - (6 + 4) = 5$, then

$P(\text{Green}) = \frac{5}{15} = \frac{1}{3}$ ✓

**ii** $P(\text{Not blue}) = 1 - \frac{6}{15}$ ✓

$= \frac{9}{15}$

$= \frac{3}{5}$

**iii** $P(\text{Red, blue or green}) = 1$ ✓

**b** The removed ball must have been red.

∴ 6 blue, 3 red, 5 green

**i** $P(\text{Green}) = \frac{5}{14}$ ✓

**ii** $P(\text{Blue or red}) = \frac{9}{14}$ ✓ (5 marks)

**4** **a**

| | Plays guitar | Not play guitar | Total |
|---|---|---|---|
| Plays piano | 6 | 5 | 11 |
| Does not play piano | 11 | 9 | 20 |
| Total | 17 | 14 | 31 |

✓

**b** 11 ✓

**c** **i** $P(\text{Guitar, not piano}) = \frac{11}{31}$ ✓

**ii** $P(\text{Guitar and piano}) = \frac{6}{31}$ ✓

**d** 14 students do not play guitar

∴ $P(\text{No piano}) = \frac{9}{14}$ ✓ (5 marks)

**5** **a**



As $12 + 14 - 20 = 6$,
then 6 girls play both. ✓ (1 mark)

## WORKED SOLUTIONS to Chapter 12 **Tests**

**b** **i** $P(\text{Both}) = \frac{6}{20} = \frac{3}{10}$ ✓

**ii** $P(\text{Netball, not hockey}) = \frac{8}{20} = \frac{2}{5}$ ✓

(2 marks)

**6** **a** 17 ✓

**b** **i** $P(\text{Both}) = \frac{5}{45} = \frac{1}{9}$ ✓

**ii** $P(\text{Train, not ferry}) = \frac{21}{45} = \frac{7}{15}$ ✓

(3 marks)

**(Total 25 marks)**

### Level 3 Test

**1** **a** $\frac{1}{2}$ ✓ **b** $\frac{1}{2}$ ✓

**c** $\frac{12}{13}$ ✓ **d** $\frac{8}{52} = \frac{2}{13}$ ✓

**e** $\frac{2}{52} = \frac{1}{26}$ ✓ **f** $\frac{50}{52} = \frac{25}{26}$ ✓

(6 marks)

**2** **a** $P(\text{Less than 10}) = \frac{9}{20}$ ✓

**b** $P(\text{Greater than 16}) = \frac{4}{20} = \frac{1}{5}$ ✓

**c** $P(\text{Prime}) = \frac{8}{20} = \frac{2}{5}$ ✓

**d** $P(\text{Composite}) = \frac{11}{20}$ ✓

**e** $P(\text{Div. by 3}) = \frac{6}{20} = \frac{3}{10}$ ✓

**f** $P(\text{Mult. of 4}) = \frac{5}{20} = \frac{1}{4}$ ✓

**g** $P(\text{Factor of 24}) = \frac{7}{20}$ ✓

**h** $P(\text{Even and mult. of 3}) = \frac{3}{20}$ ✓ (8 marks)

**3** **a**

| | Eat breakfast | Do not eat breakfast | Total |
|---|---|---|---|
| Male | 42 | 48 | 90 |
| Female | 46 | 64 | 110 |
| Total | 88 | 112 | 200 |

✓

**b** **i** $P(\text{Male, no breakfast}) = \frac{48}{200} = \frac{6}{25}$ ✓

**ii** $P(\text{Female, eats breakfast}) = \frac{46}{200} = \frac{23}{100}$ ✓

**c** **i** $P(\text{Female}) = \frac{110}{200} = \frac{11}{20}$ ✓

**ii** $P(\text{Eats breakfast}) = \frac{88}{200} = \frac{11}{25}$ ✓

(5 marks)

**4** **a**

| Support for pension increase | | | | |
|---|---|---|---|---|
| | For | Against | No op. | Total |
| 18–35 | 12 | 8 | 5 | 25 |
| 36–50 | 13 | 9 | 3 | 25 |
| 51–65 | 19 | 6 | 0 | 25 |
| Over 65 | 23 | 2 | 0 | 25 |
| Total | 67 | 25 | 8 | 100 |

✓

**b** **i** $P(51\text{–}65) = \frac{25}{100} = \frac{1}{4}$ ✓

**ii** $P(\text{Against increase}) = \frac{25}{100} = \frac{1}{4}$ ✓

**iii** $P(\text{18–35 and 'for'}) = \frac{12}{25}$ ✓

(4 marks)

**5** **a**

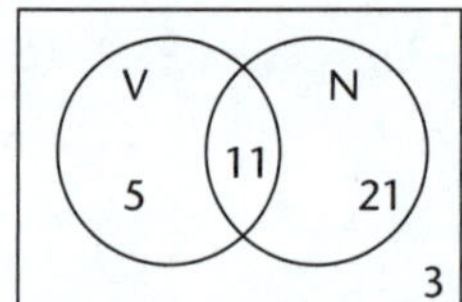


V: Volunteered
N: Netball

✓✓

As $40 - 3 = 37$, and $16 + 32 - 37 = 11$, then 11 women have volunteered and played netball.

**b** $P(\text{Volunteered, not played}) = \frac{5}{40} = \frac{1}{8}$ ✓

**c**

| | Volunteered | Not volunteered | Total |
|---|---|---|---|
| Netball | 11 | 21 | 32 |
| No netball | 5 | 3 | 8 |
| Total | 16 | 24 | 40 |

✓✓

**d** **i** $P(\text{Never played netball}) = \frac{8}{40} = \frac{1}{5}$

$\therefore$ Netball $= \frac{1}{5} \times 120 = 24$

$\therefore$ 24 women ✓

**ii** $P(\text{Played netball, no volunteer}) = \frac{21}{40}$

$\therefore$ Number $= \frac{21}{40} \times 120 = 63$

$\therefore$ 63 women ✓

(7 marks)

**(Total 30 marks)**

# WORKED SOLUTIONS to Sample Examination 1

## Sample Examination 1 pp. 270–275

### Part A (30 marks)

**1** $28:18 = 14:9$ ✓ (1 mark)

**2** $\text{Area} = \pi r^2$
$= \pi \times 4.8^2$
$= 72.382\,295$
$= 72.38$ (to 2 decimal places)
∴ Area is $72.38\text{ cm}^2$. ✓ (1 mark)

**3** Substitute (2, 4) in $y = 3x - 2$
∴ $y = 3x - 2$
$4 = 3(2) - 2$
$4 = 6 - 2$
$4 = 4$ Yes
∴ $y = 3x - 2$ passes through (2, 4) ✓ (1 mark)

**4** $4a - 2b - 3a + 4b = a + 2b$ ✓ (1 mark)

**5** Range = highest score – lowest score
$= 39° - 23°$
$= 16°$
∴ The range of temperatures was 16°. ✓ (1 mark)

**6**

130°
130°
50°

∴ $x = 50$ ✓ (1 mark)

**7** $0.12 \times 740 = 88.8$
∴ 12% of \$740 is \$88.80 ✓ (1 mark)

**8** $\frac{4.76}{3.91 - 1.007} = 1.639\,6831$
$= 1.64$ (to 2 decimal places)
∴ 1.64 ✓ (1 mark)

**9** Temperature $= 4 - 6 - 3$
$= -5$
∴ The temperature at 7 am in Orange was −5°. ✓ (1 mark)

**10** $1.4 - (0.2)^2 = 1.4 - 0.04 = 1.36$ ✓ (1 mark)

**11** 125% of \$320
Amount $= 1.25 \times 320$
$= 400$
∴ The new amount is \$400. ✓ (1 mark)

**12** Time = 2.30 to 6.00
∴ $3\frac{1}{2}$ hours

Distance = speed × time
$= 80 \times 3\frac{1}{2}$
$= 280$
∴ The car has travelled 280 km. ✓ (1 mark)

**13** Area of trapezium $= \frac{1}{2} \times 16(14 + 24)$
$= 8(38)$
$= 304$
∴ The area of the quadrilateral is $304\text{ cm}^2$. ✓ (1 mark)

**14** $a(3 - 4a) = 3a - 4a^2$ ✓ (1 mark)

**15** Rate $= \frac{1680}{20}$
$= 84$
∴ She types 84 words/minute. ✓ (1 mark)

**16** $P(1, 2, 3, 4, 6) = \frac{5}{6}$
∴ The probability is $\frac{5}{6}$. ✓ (1 mark)

**17** As $180 - (72 + 63) = 45$, then both △s have 63°, 10 cm and 45°
∴ ASA ✓ (1 mark)

**18**

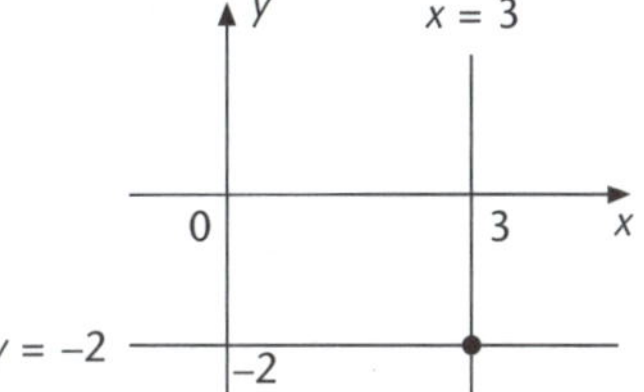


∴ (3, −2) ✓ (1 mark)

**19** $4x - 2 = 3x + 5$
$4x - 3x = 5 + 2$
∴ $x = 7$ ✓ (1 mark)

**20** $12x^6$ ✓ (1 mark)

**21** $24\% = \frac{24}{100}$
$= \frac{6}{25}$ ✓ (1 mark)

# WORKED SOLUTIONS to Sample Examination 1

**22** $y$ is 3 times $x$ plus 2

$\therefore y = 3x + 2$ ✓ (1 mark)

**23** Hectares $= 43\,670 \div 10\,000$
$= 4.367$

$\therefore$ 4.367 ha ✓ (1 mark)

**24** Profit $= 240 - 200 = 40$

$\therefore$ Profit is \$40

$\therefore$ Percentage $= \frac{40}{200} \times 100\% = 20\%$

$\therefore$ Profit is 20%. ✓ (1 mark)

**25** $12x - 18 = 6(2x - 3)$ ✓ (1 mark)

**26** AF = 5 units
DI = 5 units
$\therefore$ AF:DI = 5:5
i.e. AF:DI = 1:1 ✓ (1 mark)

**27** GST $= 440 \div 11 = 40$

$\therefore$ \$40 is the included GST ✓ (1 mark)

**28** $(2 \times 3)^0 = 6^0 = 1$ ✓ (1 mark)

**29** The outlier is 3. ✓ (1 mark)

**30** Debt $= \frac{2}{3}$ of $\frac{1}{2}$ of \$1200

$= \frac{2}{3} \times \frac{1}{2} \times 1200$

$= 400$

$\therefore$ Maria's credit card debt was \$400. ✓ (1 mark)

## Part B (40 marks)

**1** **a** Volume $= Ah$
$= 160 \times 60 \times 50$
$= 480\,000$

$\therefore$ The volume is $480\,000$ cm$^3$. ✓ (1 mark)

**b** Capacity $= 480\,000 \div 1000$
$= 480$

$\therefore$ The capacity is 480 L. ✓ (1 mark)

**2** $3(2x - 1) = 4(x + 1)$

$6x - 3 = 4x + 4$ ✓

$6x - 4x = 4 + 3$

$\frac{\cancel{2}^1 x}{\cancel{2}_1} = \frac{7}{2}$

$x = 3\frac{1}{2}$ ✓ (2 marks)

**3**

| $n$ | 0 | 1 | 2 | 3 |
|---|---|---|---|---|
| $c$ | 60 | 130 | 200 | 270 |

✓✓ (2 marks)

**4** The shaded area is a trapezium around a triangle.

$35 \times 40 + \frac{1}{2} \times 35 \times 5 - \frac{1}{2} \times 20 \times 14$ ✓
$= 1347.5$

$\therefore$ The shaded area is 1347.5 cm$^2$. ✓ (2 marks)

**5** $3(a - 4) - 2(2a - 5) = 3a - 12 - 4a + 10$ ✓
$= -a - 2$ ✓ (2 marks)

**6**

| $x$ | 0 | 1 | 2 |
|---|---|---|---|
| $y$ | 4 | 2 | 0 |

✓

$y = 4 - 2x$

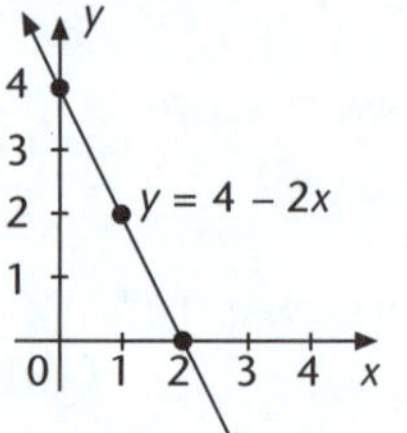


✓ (2 marks)

**7** $a^2 - bc = (-3)^2 - (-2)(-4)$
$= 9 - 8$ ✓
$= 1$ ✓ (2 marks)

**8** Volume $= \pi r^2 h$
$= \pi \times 5^2 \times 7$ ✓
$= 549.778\,71$
$= 549.78$

$\therefore$ The volume of the cylinder is 549.78 cm$^3$. ✓ (2 marks)

**9** $x^2 = 9^2 - 5^2$ ✓
$= 81 - 25$
$= 56$

$\therefore x = \sqrt{56}$ ✓ (2 marks)

**10** **a** \$40 000:\$50 000

$\therefore$ 4:5 ✓ (1 mark)

**b** Total parts = 9

$\therefore$ 9 parts = \$58 500
1 part = \$6500
5 parts = \$32 500

$\therefore$ Trevor's share is \$32 500. ✓ (1 mark)

# WORKED SOLUTIONS to Sample Examination 1

**11** **a**

| Score ($x$) | Frequency ($f$) | $fx$ |
|---|---|---|
| 4 | 3 | 12 |
| 5 | 2 | 10 |
| 6 | 4 | 24 |
| 7 | 3 | 21 |
| Total | 12 | 67 |

✓ (1 mark)

**b** $\text{Mean} = \frac{\text{Sum of } fx}{\text{Sum of } f}$

$= \frac{67}{12}$

$= 5\frac{7}{12}$ ✓ (1 mark)

**12** **a** 1:2 500 000

1 cm = 2 500 000 cm

∴ The scale on the map is 1 cm:25 km. ✓ (1 mark)

**b** Distance on map = 475 ÷ 25

= 19

∴ Nyngan and Sydney are 19 cm apart on the map. ✓ (1 mark)

**13** **a** $\text{Time} = \frac{\text{Distance}}{\text{Speed}}$

$= \frac{360}{80}$

= 4.5 hours

= 4 h 30 min ✓

∴ The time from Kununurra to Halls Creek was 4 hours, 30 minutes. (1 mark)

**b** 9:30 plus 4 h 30 min = 2:00

∴ The arrival time in Sydney would be 2:00 pm. ✓ (1 mark)

**14** **a** 15% of \$89 = 0.15 × 89

= 13.35

∴ The saving on jeans will be \$13.35. ✓ (1 mark)

**b** Price paid = 85% of original

∴ price = 0.85 × 349

= 296.65

∴ The microwave will cost \$296.65. ✓ (1 mark)

**15** % spent = 36 + 8 + 14 + 18

= 76

∴ % not spent = 100 − 76

= 24

∴ 24% of pay = \$336 ✓

1% of pay = \$336 ÷ 24

= \$14

36% of pay = \$14 × 36

= \$504

∴ Fiona spent \$504 on groceries for the fortnight. ✓ (2 marks)

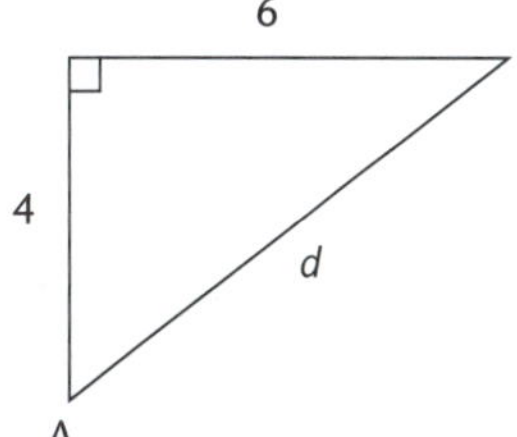


Let $d$ be the distance from A.

∴ $d^2 = 4^2 + 6^2$

$= 16 + 36$

$= 52$ ✓

∴ $d = \sqrt{52}$

= 7.211 026

= 7.211 (to 3 decimal places)

The bushwalker is 7.211 km from his starting point (i.e. 7 km 211 m). ✓ (2 marks)

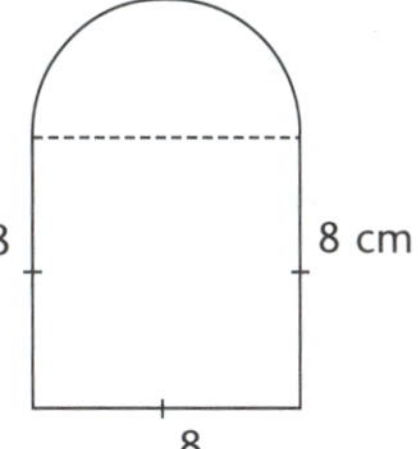


The shape is a rectangle and semicircle.

∴ Perimeter = 8 + 8 + 8 + $\frac{1}{2}$ circumference

$= 24 + \frac{1}{2} \times 2\pi r$

$= 24 + \frac{1}{\cancel{2}} \times \cancel{2} \times \pi \times 4$ ✓

= 24 + 12.566 371

= 36.566 371

= 37 (to the nearest cm)

∴ The perimeter is 37 cm. ✓ (2 marks)

**18** a

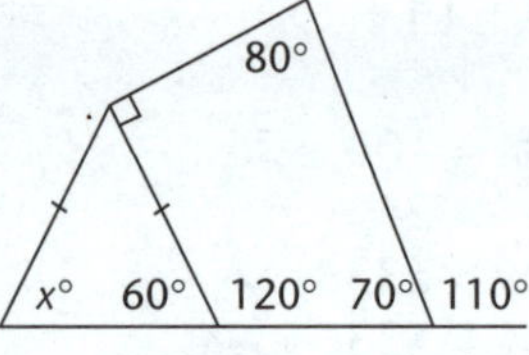


$\therefore x = 60$ ✓ (1 mark)

b

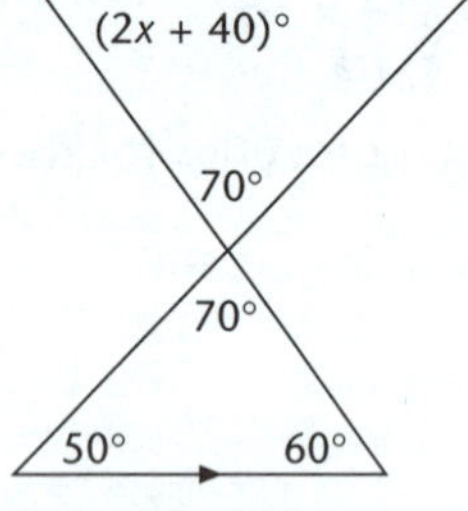


$2x + 40 = 60$
(Alternate ∠s equal with parallel lines.)
$2x = 60 - 40$
$2x = 20$
$x = 10$ ✓ (1 mark)

**19** Cost = \$240 + \$80
= \$320

∴ Profit = selling − cost
= \$400 − \$320
= \$80

a Profit:cost = \$80:\$320
= 1:4 ✓

The ratio of Ken's profit to cost price is 1:4. (1 mark)

b % profit = $\frac{1}{4} \times 100$ (from **a**)
= 25% ✓

Ken's percentage profit was 25%. (1 mark)

**20** a

| Student allowance (weekly) | | | | |
|---|---|---|---|---|
| | None | \$0–\$10 | More than \$10 | Total |
| **Males** | 3 | 7 | 2 | 12 |
| **Females** | 5 | 9 | 4 | 18 |
| **Total** | 8 | 16 | 6 | 30 |

✓ (1 mark)

b $P$(Female, more than \$10)

$= \frac{4}{30} = \frac{2}{15}$ ✓ (1 mark)

## Part C (15 marks)

**1** 4, 7, 2, 3, 8, 9, 2, 8

a The two modes are 2, 8 ✓ (1 mark)

b 2̸, 2̸, 3̸, 4, 7, 8̸, 8̸, 9̸

The median is the middle of 4 and 7

$= \frac{4+7}{2}$

$= 5\frac{1}{2}$ ✓ (1 mark)

c Mean $= \frac{4+7+2+3+8+9+2+8}{8}$

$= \frac{43}{8}$

$= 5\frac{3}{8}$ ✓ (1 mark)

**2** a

| Children in family | Frequency |
|---|---|
| 1 | 3 |
| 2 | 5 |
| 3 | 4 |
| 4 | 1 |
| 5 | 2 |

∴ Students surveyed = sum of frequencies
= 15

∴ 15 students were surveyed. ✓ (1 mark)

b The most common sized family was 2. ✓ (1 mark)

c From ($a$), $fx$ column:
3
10
12
4
10
Total = 39

∴ Average (mean) $= \frac{\text{Sum of } fx}{\text{Sum of } f}$

$= \frac{39}{15}$

$= 2\frac{3}{5}$

∴ The average number of children is $2\frac{3}{5}$. ✓ (1 mark)

**3** a % overweight $= \frac{144}{360} \times 100$
= 40% ✓

∴ 40% believed they were overweight. (1 mark)

**b** 35% of 360° = 126°

∴ $B$ is 35% ✓ (1 mark)

∴ 35% considered their weight 'normal'.

**c** Underweight = 100 – (40 + 35)
= 100 – 75
= 25

∴ Underweight is 25%

∴ Overweight:underweight = 40:25
= 8:5 ✓
(1 mark)

**4** **a** $AB^2 = 25^2 - 24^2$
$= 625 - 576$
$= 49$

∴ $AB = \sqrt{49}$
$= 7$

∴ $AB$ is 7 cm ✓ (1 mark)

**b** Volume = area of base × height

$672 = \frac{1}{2} \times 24 \times 7 \times \text{height}$

$672 = 84h$

$\therefore \frac{\cancel{84}h}{8\cancel{4}} = \frac{672}{84}$

$h = 8$

∴ $CD$ is 8 cm ✓ (1 mark)

**c** Total area = 2 × front + 3 other faces
$= 2 \times \frac{1}{2} \times 24 \times 7 + 25 \times 8$
$+ 24 \times 8 + 7 \times 8$
$= 616$

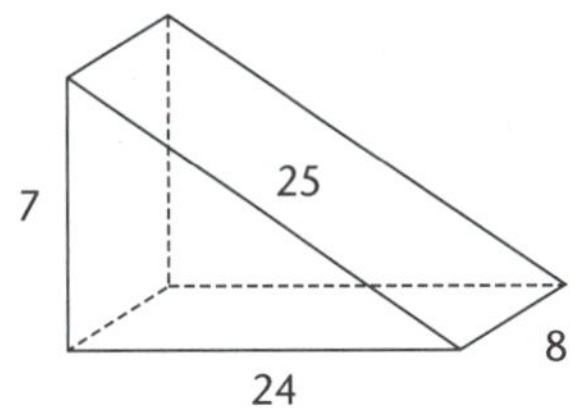


∴ The total area of the five faces of the prism would be is 616 m$^2$. ✓ (1 mark)

**5**

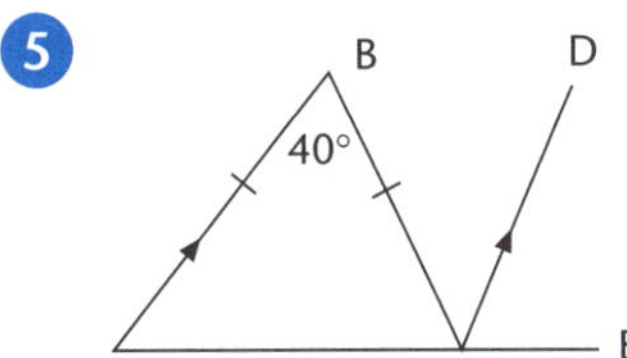


**a** ∠BCD = 40° ✓ (1 mark)

**b** ∠BAC = 70° ✓ (1 mark)

**c** ∠DCE = 70° ✓ (1 mark)

## Sample Examination 2 pp. 276–281

### Part A (30 marks)

**1** 5.19 ✓ (1 mark)

**2** $\frac{35}{100} = \frac{7}{20}$ ✓ (1 mark)

**3** $8a + 7 - 3a + 4 = 5a + 11$ ✓ (1 mark)

**4** $-3 + 4 - 2 = -1$ ✓ (1 mark)

**5** $3(2x - 4) = 6x - 12$ ✓ (1 mark)

**6** 2.4 metres = 240 cm ✓ (1 mark)

**7**

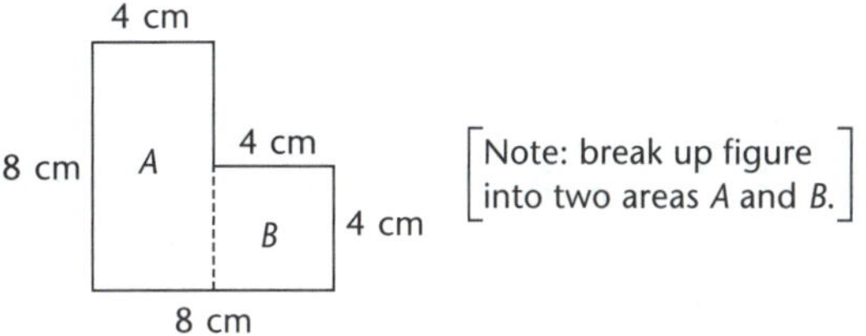


[Note: break up figure into two areas $A$ and $B$.]

Area $A = 8 \times 4 = 32$ cm$^2$

Area $B = 4 \times 4 = 16$ cm$^2$

Total area = (32 + 16) cm$^2$
= 48 cm$^2$ ✓ (1 mark)

**8** 12:15 = 4:5 ✓ (1 mark)

**9** $\frac{5}{8} = 0.625$ ✓ (1 mark)

**10** $(1\frac{1}{2})^2 = 1\frac{1}{2} \times 1\frac{1}{2}$
$= 2\frac{1}{4}$ ✓ (1 mark)

**11** $2\frac{2}{3} = \frac{8}{3}$ [Convert to improper fraction and invert.]

∴ Reciprocal of $2\frac{2}{3} = \frac{3}{8}$ ✓ (1 mark)

**12** $2a \times 3a = 6a^2$ ✓ [Note: $a \times a = a^2$] (1 mark)

**13** $3b^2 = 3 \times b \times b$
$= 3 \times 2 \times 2$
$= 12$ ✓ (1 mark)

**14** $\frac{12}{100} \times \$9.50 = \$1.14$,
or $0.12 \times \$9.50 = \$1.14$ ✓ (1 mark)

# WORKED SOLUTIONS to Sample Examination 2

**15** $x + 8 = 2$
$x = 2 - 8$
$x = -6$ ✓ (1 mark)

**16** I and III ✓ (1 mark)

**17** Note: convert all four numbers into decimal form then arrange in ascending order (from smallest to largest).

$13\% = 0.13$ (2)

$0.15 = 0.15$ (3)

$\frac{3}{25} = 0.12$ (1)

$\sqrt{0.9} = 0.948\,683\,3$ (4)

Therefore, in ascending order:

$\frac{3}{25}$, 13%, 0.15, $\sqrt{0.9}$ ✓ (1 mark)

**18** $\frac{-5+9}{-2} = \frac{4}{-2} = -2$ ✓ (1 mark)

**19** Average $= \frac{8+7+6+5+9+7}{6}$
$= \frac{42}{6}$
$= 7$ ✓ (1 mark)

**20** $\frac{17}{20} \times 100 = 85\%$ ✓ (1 mark)

**21**

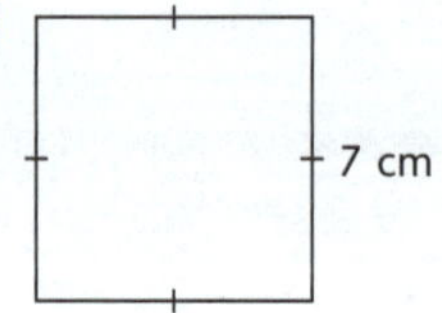


If the square has perimeter of 28 cm then the side is 7 cm long.

$\therefore$ Area $= 7 \times 7$
$= 49 \text{ cm}^2$ ✓ (1 mark)

**22** Suppose there were 8 blue balls
$\therefore$ 16 red balls, 4 green balls
$\therefore$ Total of 28 balls.

$P(\text{Red}) = \frac{16}{28} = \frac{4}{7}$ ✓ (1 mark)

**23** $15ab \div 5a = \frac{\cancel{15}^3 \cancel{a}^1 b}{\cancel{5}_1 \cancel{a}_1}$ [Cancel]
$= 3b$ ✓ (1 mark)

**24** Increase $3a$ by $14 = 3a + 14$ ✓ (1 mark)

**25** 2 hours 43 minutes ✓ (1 mark)

**26** $\frac{5x + 15y}{5} = \frac{\cancel{5}(x + 3y)}{\cancel{5}}$ [Factorise then cancel.]
$= x + 3y$ ✓ (1 mark)

**27** $a + 46 = 180$
$a = 134$ ✓ [Co-interior angles and parallel lines.] (1 mark)

**28** $2000 \div 250 = 8$ ✓ [2 L = 2000 mL] (1 mark)

**29** Number of students $= (1 + 3 + 5 + 4 + 2 + 1)$
$= 16$ ✓ (1 mark)

**30**

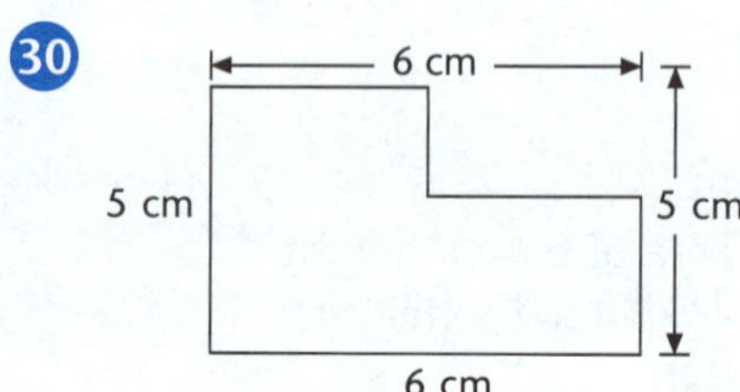


Perimeter $= (5 + 5 + 6 + 6)$
$= 22 \text{ cm}$ ✓ (1 mark)

## Part B (40 marks)

**1** 4:3 means there are 7 parts:
boys = 4 parts
girls = 3 parts
3 parts = 12 ✓
$\therefore$ 1 part = 4
boys = 4 parts
$= 4 \times 4$
$= 16$

$\therefore$ There are 16 boys. ✓ (2 marks)

**2** $\frac{2x - 1}{3} = 5$
$2x - 1 = 15$ ✓
$2x = 16$
$x = 8$ ✓ (2 marks)

**3** $3(2x - 5) + 2(x + 3) = 6x - 15 + 2x + 6$ ✓
$= 8x - 9$ ✓ (2 marks)

# WORKED SOLUTIONS to Sample Examination 2

**4** Discount = 15% of \$830

$= \frac{15}{100} \times \$830$

$= \$124.50$ ✓

Cost = \$830 − \$124.50

= \$705.50 ✓ (2 marks)

**5** $A = \pi r^2$

$= \pi \times 7^2$ ✓

$= \pi \times 7 \times 7$

$= 153.938\,04$

$= 154\text{ cm}^2$ (to the nearest $\text{cm}^2$) ✓ (2 marks)

**6**

| $x$ | 0 | 1 | 2 |
|---|---|---|---|
| $y$ | −1 | 1 | 3 |

✓

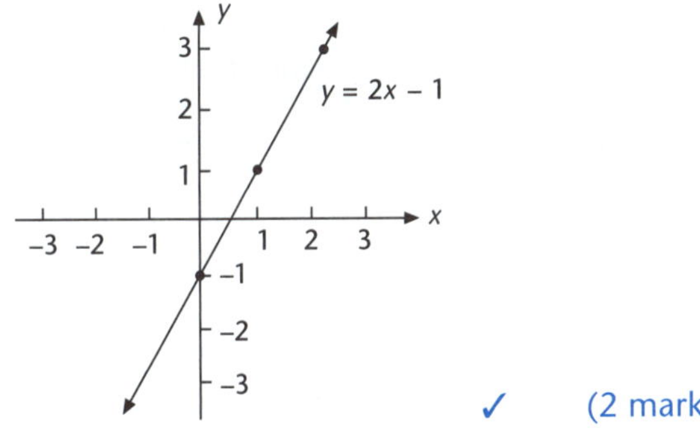


✓ (2 marks)

**7**

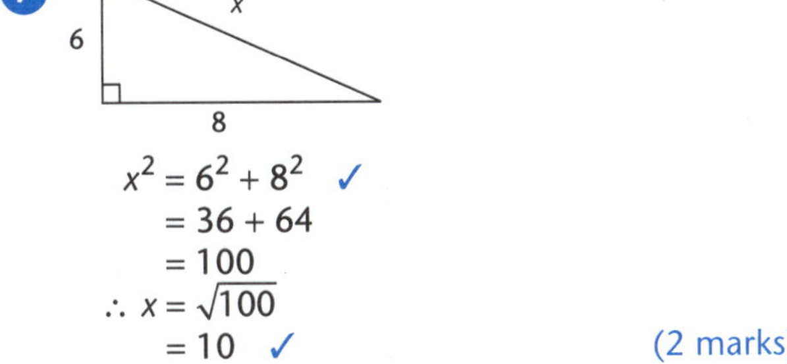


$x^2 = 6^2 + 8^2$ ✓

$= 36 + 64$

$= 100$

$\therefore x = \sqrt{100}$

$= 10$ ✓ (2 marks)

**8** $V = lbh$

$= (8 \times 5 \times 2)\text{ cm}^3$ ✓

$= 80\text{ cm}^3$ ✓ (2 marks)

**9**

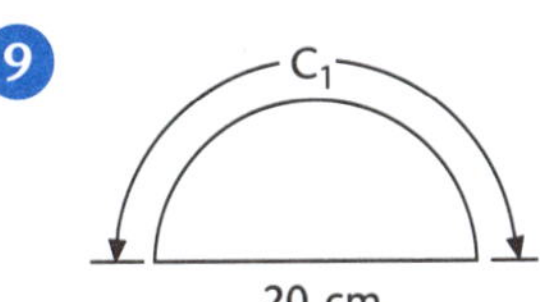


Let $C_1$ = curved part

$C_1 = \frac{1}{2}\pi d$

$= \frac{1}{2} \times \pi \times 20$

$= 31.415\,927$

$= 31$ (nearest cm) ✓

$\therefore$ Perimeter = (31 + 20) cm

= 51 cm ✓ (2 marks)

**10** Plane leaves:

- New York = 7:00 am
- Hawaii = 1:00 am ✓

Now, add 10 h 20 min to 1:00 am

$\therefore$ arrives 11:20 am ✓ (2 marks)

**11**

Total = 12 balls

**a** $P(\text{White}) = \frac{6}{12} = \frac{1}{2}$ ✓ (1 mark)

**b** $P(\text{Pink}) = \frac{0}{12} = 0$ ✓ [Note: there are no pink balls in the bag.] (1 mark)

**12** $F = \frac{mv^2}{r}$

$= \frac{12 \times 4^2}{8}$ ✓

$= \frac{12 \times 16}{8}$

$= 24$ ✓ (2 marks)

**13** Mode = 23, median = 23 ✓

$\therefore$ The difference is 0. ✓ (2 marks)

**14** **a** $x = 40$ (alternate to $\angle ACD$, parallel lines) ✓ (1 mark)

**b** $y = 75$ (angle sum of $\triangle ABC$) ✓ (1 mark)

**15** Eleni spends $\frac{1}{3} + \frac{2}{5} = \frac{11}{15}$ of what she earns.

$\therefore$ She saves $1 - \frac{11}{15} = \frac{4}{15}$

Savings $= \frac{4}{15} \times \$28\,365$ ✓

$= \$7564$

$\therefore$ Eleni saves \$7564 p.a. ✓ (2 marks)

# WORKED SOLUTIONS to Sample Examination 2

**16** Average speed $= \dfrac{\text{distance travelled}}{\text{time taken}}$

$= \frac{306}{4\frac{1}{2}}$ km/h ✓

$= 68$ km/h ✓ (2 marks)

**17** ~~1~~, ~~1~~, ~~1~~, ~~2~~, (3, 3), ~~4~~, ~~4~~, ~~4~~, ~~5~~ ✓

There are two 3s in the middle.

∴ Median = 3 ✓ (2 marks)

**18** $3x + 5 = 8x - 35$

$3x - 8x = -35 - 5$

$-5x = -40$ ✓

$x = \frac{-40}{-5}$

$x = 8$ ✓ (2 marks)

**19** Volume $= \pi r^2 h$

$= \pi \times 5^2 \times 14$ ✓

$= 1099.5574$

$= 1100\ \text{cm}^3$ (to the nearest $\text{cm}^3$) ✓

The volume of the cylinder is $1100\ \text{cm}^3$.

(2 marks)

**20**

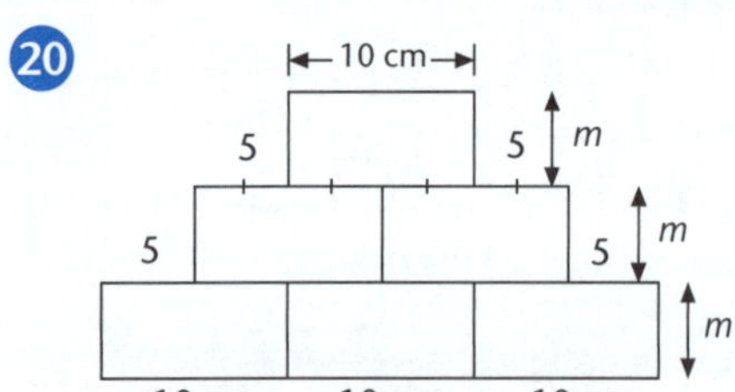


**a** Perimeter $= (4 \times 10) + (4 \times 5) + (6 \times m)$

$= (60 + 6m)$ cm ✓ (1 mark)

**b** Area of one brick $= (10 \times m) = 10m\ \text{cm}^2$

Total area $= (6 \times 10m)$

$= 60m\ \text{cm}^2$ ✓ (1 mark)

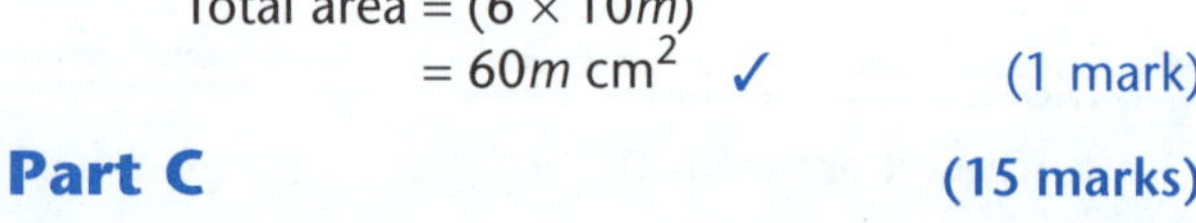

## Part C (15 marks)

**1**

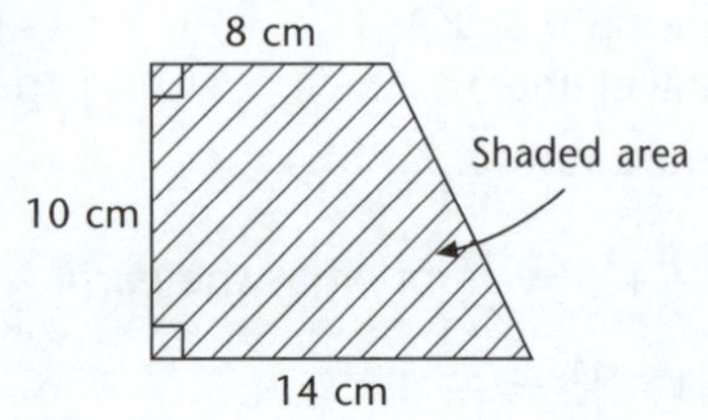


**a** Shaded area $= 10 \times 8 + \frac{1}{2} \times 6 \times 10$

$= 80 + 30$

$= 110$

∴ The shaded area of the prism is $110\ \text{cm}^2$. ✓ (1 mark)

**b** Volume = shaded area × length

$= (110 \times 3)$

$= 330$

∴ The volume of the prism is $330\ \text{cm}^3$. ✓

(1 mark)

**2 a**

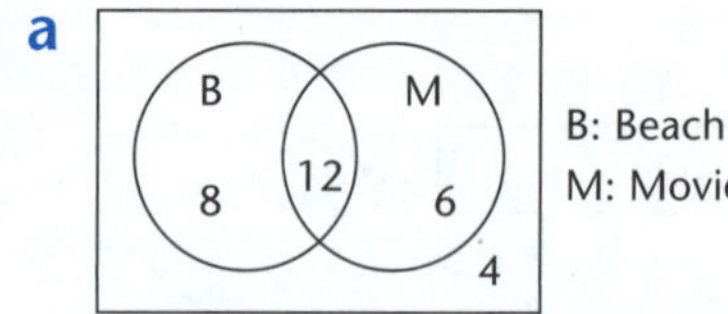


B: Beach
M: Movie

Firstly, $30 - 4 = 26$, and then $20 + 18 - 26 = 12$

∴ 12 students gone to beach and seen movie. ✓ (1 mark)

**b** 20 students went to beach

∴ $P(\text{Seen movie}) = \frac{12}{20} = \frac{3}{5}$ ✓ (1 mark)

**3 a**

| Number of passengers $x$ | Tally | Frequency $f$ | $f \times x$ |
|---|---|---|---|
| 0 | \|\|\|\| | 4 | 0 |
| 1 | \|\|\|\| | 4 | 4 |
| 2 | ~~\|\|\|\|~~ | 5 | 10 |
| 3 | ~~\|\|\|\|~~ | 5 | 15 |
| 4 | \|\| | 2 | 8 |
| 5 | ~~\|\|\|\|~~ \|\| | 7 | 35 |
| 6 | \|\|\| | 3 | 18 |
| | Totals (Σ) | 30 | 90 |

✓✓ (2 marks)

**b** Mean $= \dfrac{\Sigma fx}{\Sigma f}$ or $\dfrac{\text{Total } fx}{\text{Total } f}$

$= \frac{90}{30}$

$= 3$ ✓ (1 mark)

**c** **i** Mode = 5 ✓

**ii** Range = 6 ✓ (2 marks)

## WORKED SOLUTIONS to Sample Examination 2

**4** **a** Perimeter $= 2 \times 2x + 2 \times (x + 7)$
$= 4x + 2x + 14$
$= 6x + 14$ ✓ (1 mark)

**b** If perimeter = 44, then
$6x + 14 = 44$ ✓
$6x = 30$
$x = 5$ ✓ (2 marks)

**5**

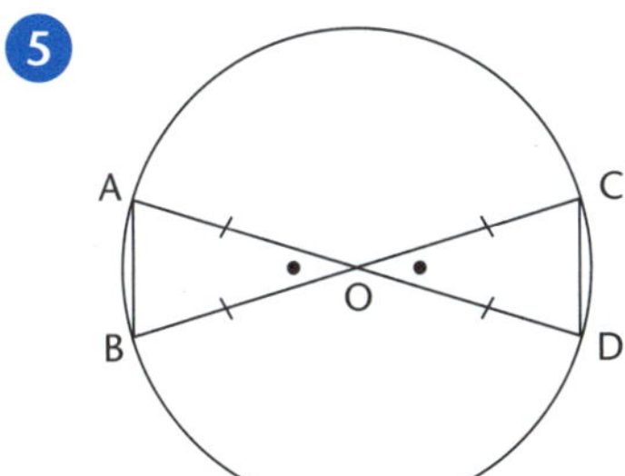


**i** AO = DO (equal radii)
BO = CO (equal radii) ✓
$\angle$AOB = $\angle$DOC (vert. opp. $\angle$s)
$\therefore$ $\triangle$ABO $\equiv$ $\triangle$DCO (SAS test) ✓

**ii** $\angle$ABO = 70°
$\angle$BAO = 70° (base $\angle$s, isosceles $\triangle$)
$\therefore$ $\angle$AOB = 40° ($\angle$ sum of $\triangle$)
$\therefore$ $\angle$COA = 140° (BC is diameter) ✓
(3 marks)

# INDEX

## E



## F



## G



## H



## I



## K



## L



## M

## S



## T



## U



## V



## X



## Y

# NOTES

# NOTES

# NOTES